Cultivated Meat

Carlos Ricardo Soccol • Carla Forte Maiolino Molento • Germano Glufke Reis • Susan Grace Karp
Editors

# Cultivated Meat

## Technologies, Commercialization and Challenges

Springer

*Editors*
Carlos Ricardo Soccol
Bioprocess Engineering and Biotechnology Department
Federal University of Paraná
Curitiba, Paraná, Brazil

Carla Forte Maiolino Molento
Animal Welfare Laboratory &
Cellular Animal Science Laboratory
Federal University of Paraná
Curitiba, Paraná, Brazil

Germano Glufke Reis
School of Business Administration
Federal University of Paraná
Curitiba, Paraná, Brazil

Susan Grace Karp
Bioprocess Engineering and Biotechnology Department
Federal University of Paraná
Curitiba, Paraná, Brazil

ISBN 978-3-031-55967-9 ISBN 978-3-031-55968-6 (eBook)
https://doi.org/10.1007/978-3-031-55968-6



The book cover image was created with BioRender.com

This Springer imprint is published by the registered company Springer Nature Switzerland AG
The registered company address is: Gewerbestrasse 11, 6330 Cham, Switzerland

# Contents

**Introduction to New Food Systems** .................................. 1
Carla Forte Maiolino Molento, Germano Glufke Reis, Susan Grace Karp, and Carlos Ricardo Soccol

**Introduction to Animal Cell Culture** ................................ 7
Carlos Ricardo Soccol, Vanete Thomaz Soccol, Gabriela Dos Santos Costa, Igor Negreiros Piazenski, and João Pedro Manica Candelario

**Cell Lines for Cultivated Meat Production** ........................ 29
Vivian Feddern, Ana Paula Almeida Bastos, Vanessa Gressler, Diana M. C. Marques, Frederico C. Ferreira, Carlos André Vitorino Rodrigues, Marcus Vinicius Telles Teixeira, and Camila Luna da Silva

**Culture Medium for Cultivated Meat** ............................... 55
Gustavo Roncoli Reigado, Izabela Daniel Sardinha Caldeira, Pedro Oliveira Marques, Mayara Santos, Felipe Santiago Chambergo Alcalde, and Viviane Abreu Nunes

**Microcarriers and Scaffolds in Cultivated Meat Production** ........... 83
Raíssa Canova, Luiza Meurer Brand, Juliana do Canto Olegário, and Bibiana Matte

**Bioreactors for Cultivated Meat Production** ........................ 107
Luciana Porto de Souza Vandenberghe, Ariane Fátima Murawski de Mello, Giuliana Biagini, Patricia Beatriz Gruening de Mattos, Igor Negreiros Piazenski, João Pedro Manica Candelario, and Carlos Ricardo Soccol

**Modeling and Simulation of Cell Cultures in Cultivated Meat Production** ........................................ 131
Luiz Alberto Junior Letti, Andressa Tedesco Andretta, Walter José Martinez Burgos, Fernando Enrique Rosas Vega, and Carlos Ricardo Soccol

**Downstream Processes for Cultivated Meat** . . . . . . . . . . . . . . . . . . . . . . . . . 145
Julio Cesar de Carvalho, Susan Grace Karp, Luis Daniel Goyzueta Mamani, Giuliana Biagini, Gabriela dos Santos Costa, Leonardo Wedderhoff Herrmann, and Carlos Ricardo Soccol

**Ingredients and Formulation of Cultivated Meat Products** . . . . . . . . . . . . 167
Paula Toshimi Matumoto-Pintro, Fernando Antônio Anjo, and Bianka Rocha Saraiva

**Packaging, Conservation, and Shelf Life of Cultivated Meat** . . . . . . . . . . . 183
Débora Gonçalves Bortolini, Luane de Oliveira Maior, Gustavo Henrique Couto, Marjorie Benegra, Maria Giovana Binder Pagnoncelli, and Luiz Gustavo Lacerda

**Quality and Risk Control in Cultivated Meat Production** . . . . . . . . . . . . . 209
Renata Ernlund Freitas de Macedo, Guilherme Agostinis Ferreira, Luiza Poniewas, Fabiane Barchiki, Carmen Lúcia Kuniyoshi Rebelatto, Débora Regina Daga, Leandro Batista Costa, and Edvaldo Antonio Ribeiro Rosa

**Scaling-Up of Cultivated Meat Production Process** . . . . . . . . . . . . . . . . . . 241
Maria Clara Manzoki, Ariane Fátima Murawski de Mello, Walter José Martinez-Burgos, Alexander da Silva Vale, Giuliana Biagini, Igor Negreiros Piazenski, Vanete Thomaz Soccol, and Carlos Ricardo Soccol

**Waste Management in Cultivated Meat Production** . . . . . . . . . . . . . . . . . . 265
Adenise Lorenci Woiciechowski, Thamarys Scapini, Maria Clara Manzoki, Walter José Martínez-Burgos, Denisse Tatiana Molina-Aulestia, Susan Grace Karp, and Carlos Ricardo Soccol

**Environmental Impacts of Cultivated Meat** . . . . . . . . . . . . . . . . . . . . . . . . . 277
Hanna L. Tuomisto and Toni Ryynänen

**Regulatory Aspects of Cultivated Meat** . . . . . . . . . . . . . . . . . . . . . . . . . . . 299
Alexandre Cabral, Isabela de Oliveira Pereira, and Amanda Leitolis

**Forecasting the Market Potential and Consumer Acceptance of Cultivated Meat** . . . . . . . . . . . . . . . . . . . . . . . . . . . . . . . . . . . . . . . . . . . . . 319
Marina Sucha Heidemann, Stéphanie Massaki, and Germano Glufke Reis

**Cultivated Meat and Bioeconomy in Tandem: Unlocking the Sustainability Implications of Transformative Food Supply Chains** . . . . . 337
Vinicius Picanço Rodrigues and Mariel Alem Fonseca

**Global and Regional Policies for Cultivated Meat** . . . . . . . . . . . . . . . . . . . 359
Maria Marques, Rodrigo Luiz Morais-da-Silva, Jennifer C. Biscarra-Bellio, Mariana Hase Ueta, and Carla Forte Maiolino Molento

**Patents and Innovations in Cultivated Meat Production** . . . . . . . . . . . . . . 385
Susan Grace Karp, Leonardo Wedderhoff Herrmann, Giuliana Biagini, Ana Paula Boligon, and Carlos Ricardo Soccol

**Future Feast: Mapping the Cultivated Meat Sector** .................. 407
Cyntia Vilasboas Calixto Casnici and David Schulzmann

**Challenges and Prospects for Cultivated Meat Production** ............ 425
Germano Glufke Reis, Susan Grace Karp, Carla Forte Maiolino Molento, and Carlos Ricardo Soccol

# Introduction to New Food Systems

**Carla Forte Maiolino Molento, Germano Glufke Reis, Susan Grace Karp, and Carlos Ricardo Soccol**

**Abstract** In this chapter, we introduce the guiding ideas for this book, as well as a list of its content. The motivations for new food systems are diverse and complex and are presented here with the indication of selected further reading materials for those interested in gaining a deeper understanding of the so-called "elephants in the room." Then, introductory concepts regarding alternative proteins as well as health and nutritional aspects are briefly commented, as an invitation for the reader to delve into each chapter that follows, for what we hope will be a useful, enjoyable, and exciting trip through the world of cultivated meat.

**Keywords** Alternative proteins · Animal ethics · Cultivated meat · Cell-based meat · Public health · Sustainability

Human food security remains an unsolved issue, although the agricultural Green Revolution and its chain of incremental innovation have been in place for more than 70 years. While it is true that human abilities to produce food have increased significantly, the modern means for producing protein-rich foods of animal origin generated what Phillips et al. (2024) have called "elephants in the room": obvious major problems or issues that people avoid discussing or even acknowledging. The

C. F. M. Molento (✉)
Animal Welfare Laboratory & Cellular Animal Science Laboratory, Federal University of Paraná, Curitiba, Paraná, Brazil
e-mail: carlamolento@ufpr.br

G. G. Reis
School of Management, UFPR Federal University of Parana, Curitiba, Brazil

SFoodLab – Sustainable Food Value Chains Laboratory, UFPR Federal University of Parana, Curitiba, Brazil

S. G. Karp · C. R. Soccol
Bioprocess Engineering and Biotechnology Department, Federal University of Paraná, Curitiba, Paraná, Brazil



C. R. Soccol et al. (eds.), *Cultivated Meat*,
https://doi.org/10.1007/978-3-031-55968-6_1

elephants in the room regarding animal production are (1) the absurd reality that food for more than 80 billion farm animals is produced in an effort to feed 8 billion humans, (2) forest destruction for pasture and crops to feed farm animals, (3) greenhouse gas emissions, (4) water use and waste management problems, (5) the unfair treatment of animals in intensive production systems, (6) the inefficiency of the current animal production systems as compared to new technologies, (7) public health risks of novel pandemics and antibiotic resistance, and (8) the predominant use of Earth's carbon atoms by human and, mostly, farm animal bodies (Phillips et al. 2024).

It is known that humans may follow a healthy diet without animal protein and that vegetarian and vegan foods are more efficient in the use of environment resources (Poore and Nemecek 2018) and free of animal cruelty. However, mostly due to cultural reasons, it is difficult to imagine a vegan humanity, and it may be easier instead to find new ways of producing meat. As put by Burgat (2017), alternative meats may gradually substitute conventional meat, with no change requirements of the predominant structure of a carnivore humanity. In fact, the search for innovative meat production systems has emerged as a new gold rush, attracting the attention of scientists, investors, conventional big meat industries, as well as the government in many countries. When the new systems become refined to the point of price and flavor parity with the conventional products, the new meat production systems will likely supply a significant proportion of the global meat market, for example, of around 60% according to Gerhardt et al. (2020). Collectively, the descriptions and forecasts related to alternative meats suggest that we are going through a great revolution in meat production systems (Biscarra-Bellio et al. 2023).

When simultaneously considered, the issues with conventional meat may be understood as problems within a one-welfare scenario. One welfare is a collaborative approach for integrating animal welfare, human well-being, and the environment, with an end point of improving global welfare and achieving gains in development (Pinillos 2018). Alternative meats may be considered, thus, contributions to one welfare goals. However, there remain challenges for best one-welfare outcomes, such as biotechnological bottlenecks and social issues that need attention if the benefits of the new food production chains are to be maximized and reach all stakeholders. Indeed, we have only started scrutinizing the potential of new food systems and alternative proteins to provide original one-welfare improvements and the required forecasting, planning, and actions for best outcomes. For instance, it has been postulated that cultivated meat systems may effectively unleash a virtuous cycle of Sustainable Development Goals (SDGs) and well-being within and across the North-South (Nobre 2022). Thus, research on how governance of new food production systems may further contributions to the achievement of multiple SDGs warrants further attention. As for animals, the benefits of uncoupling meat from animal raising and killing seem to go beyond farmed animals to reach wild animals (Heidemann et al. 2020). Besides, considering that normal meat may become stranger as cultivated meat becomes more normal (Van der Weele and Driessen 2019), a potential growing estrangement in relation to farm animal instrumentalization and killing brings completely new possibilities for animal protection.

The new meat production systems are inserted in the broader context of alternative proteins, which are protein-rich food items produced with the goal of replacing conventional animal products such as meat, eggs, and dairy products. More specifically, they are called alternative proteins because they originate from production methods that constitute an alternative to the raising and killing of animals.

## 1 Introduction to Alternative Proteins

The term "alternative proteins" includes food products that are developed to substitute conventional meat, eggs, and milk and all their processed forms. As such, alternative proteins are produced with a variety of methods, which can be summarized into three main categories: plant-based meat, milk, and egg analogues; fermentation products including animal proteins produced via precision fermentation; and cultivated products, which are real animal products resulting from original animal cells multiplied ex vivo, i.e., out of animal bodies. The category of cultivated products has only recently become possible, as the maintenance of live and functioning animal cells in human-made environments required more knowledge than the mastering of other cell types. In reality, multiplying cells to produce food is not a new process, and humans have benefitted from it for millennia, by producing wine, beer, cheese, and bread, for instance.

The products resulting from ex vivo animal cell multiplication and differentiation into muscle and adipose tissues are called cultivated or cell-based meat. Throughout its development, this type of meat has received different names, such as laboratory, synthetic, clean, and slaughter-free meat, which are all synonyms. For different reasons, the latter are considered obsolete, and a brief commented list of main nomenclature and related terminology is available in Reis et al. (2020). In addition, cultivated meat may be understood as an umbrella terminology, embracing all different types of meat, including beef, poultry, pork, fish, shrimp, and octopus. The term may also allude to an array of cultivated ingredients, such as adipose cells which may be used with cultivated muscle for real animal meat or blended with plant proteins, to originated hybrid meats made of plant-based and animal cell-based ingredients.

## 2 Health and Nutritional Value of Cultivated Meat

Different from plant-based meat alternatives, which are more advanced in marketing and discussions with the general public, cultivated meat is made of the very same animal proteins that are found in conventional meat. This is so because it is the original animal DNA that sets the protein amino acid profile in cultivated meat, in the exact same way as when the cells are in vivo, i.e., in the animal body. Thus, nutritionally thinking, the first question that frequently comes regarding alternative meats, about protein intake, is easily answered for cultivated meats. However, meat is not only protein, and other components are relevant for its nutritional value and for the new possibilities brought by cultivated meat.

There are great expectations regarding human health and the nutritional value of cultivated meat, and they originate at the high control level that can be exerted in its production. Cultivated meat does not depend on the anatomy of animals, which gives enormous room for creating shapes and composition of final products. Such new possibilities are so innovative that it is difficult to foresee what meat products will be in demand in the future.

Cultivated products can be tailored for a diversity of objectives. For instance, meat composition in terms of fat percentage becomes programmable. Not only the percentage can be set with high precision, but also whether final products will have marbled fat or an outer layer of fat. It is also possible to produce cultivated meat with plant-based fats, for consumers willing to decrease their cholesterol intake. The opposite is also possible, cultivated animal fat can be added to a plant-based protein matrix, so that flavor is improved in plant-based meat analogues. It is likely that many different combinations will emerge as both cultivated meat food engineers and consumer demand become more creative, given further contact with such new food production systems. Thus, an interplay where demand changes meat as changing meat reshapes demand is expected (Biscarra-Bellio et al. 2023).

Additional implications for human health are also covered in the following chapters, such as benefits in food security, food safety, and diminished risks of new pandemics and antibiotic resistance.

## 3 A Textbook for Cultivated Meat

A brief introduction of the current context surrounding conventional and alternative meats suffices to bring a perception of the relevance of improving knowledge and education regarding cultivated meat. Thus, the motivations for a complete volume on the topic seem evident. Having this starting point, what follows is a selected group of experts in each field, compiling the best of available knowledge to help clarifying the most relevant aspects of cultivated meat and related issues. What is needed in terms of technological development, and where is current science in this spectrum? Who are the stakeholders, what is on the horizon for each of them, and what decisions can maximize benefits for all? How can new food production systems be best organized, with effective and appropriate regulations, within an environment of adequate public policies and governance that allow for a sound entrepreneurial ecosystem? What are possible futures, geopolitics, and landscapes regarding meat production, commercialization, and consumption?

The technological aspects are presented in Chaps. 2–7. They start from the initial historical steps for animal cell cultivation to delve into detailed presentation of current knowledge of the process, including the description of cell lines, culture media, microcarriers and scaffolds, bioreactors, as well as modeling and simulation. In

Chap. 8–12, the processes closely related to final products are discussed, touching on issues of ingredients and formulation of final products; packaging, conservation, and shelf-life characteristics; quality and risk control and the intricacies of scaling cultivated meat production up. Then, we exit the production facilities to enter other realms, looking at waste production and management, life-cycle assessment, regulatory aspects, and market issues in cultivated meat chains, in Chaps. 13–16. The exploratory thinking in terms of global views follows with bioeconomy, public policies, patent and innovation, and a complete mapping of the cultivated meat sector, to be finalized with our perception of key drivers regarding the innovative industry of cultivated meat.

The overall goal of this book is to provide robust information, organized to support teaching and research activities in the field of cultivated meat. As the field develops with increasing speed, in an exponential movement that seems far from its maximum, this book provides the basic knowledge and scientific details, which register a picture of the current biotechnological reality and additional relevant issues in relation to the production, commercialization, and challenges of cultivated meat. We hope professionals, professors, students, and all those interested in cultivated meat will find that the information and discussions presented in this volume are useful and also that the reading is enjoyable and exciting.

**Acknowledgments** This book was edited in the context of the Novel Arrangement for Research and Innovation (NAPI) on Alternative Proteins, which is part of a broader research funding strategy adopted by the government of the State of Parana, Brazil, through its research support agency Fundação Araucária and State Secretariat for Science and Technology (SETI PR).

## References

Biscarra-Bellio JC, de Oliveira GB, Marques MCP, Molento CFM (2023) Demand changes meat as changing meat reshapes demand: the great meat revolution. Meat Sci 196:109040. https://doi.org/10.1016/j.meatsci.2022.109040

Burgat F (2017) L'humanité carnivore. Le Seuil, Paris, p 466

Gerhardt C, Suhlmann G, Ziemßen F, Donnan D, Warschun M, Kühnle HJ (2020) How will cultured meat and meat alternatives disrupt the agricultural and food industry? Ind Biotechnol 16:262–270. https://doi.org/10.1089/ind.2020.29227.cge

Heidemann MS, Molento CFM, Reis GG, Philips CJC (2020) Uncoupling meat from animal slaughter and its impacts on human-animal relationships. Front Psychol 11:1824. https://doi.org/10.3389/fpsyg.2020.01824

Nobre F (2022) Cultured meat and the sustainable development goals. Trends Food Sci Technol 124:140–153. https://doi.org/10.1016/j.tifs.2022.04.011

Phillips C, Molento C, Bogueva D, Marinova D (2024) Planetary health and animal and plant dietary proteins. In: Preedy VR (ed) Plant and animal proteins in health and disease prevention. Taylor & Francis, Milton Park, UK

Pinillos R (2018) One welfare - a framework to improve animal welfare and human well-being. CAB International, Wallingford, Oxfordshire, p 90

Poore J, Nemecek T (2018) Reducing food's environmental impacts through producers and consumers. Science 360:987–992. https://doi.org/10.1126/science.aaq0216

Reis GG, Heidemann MS, Borini FM, Molento CFM (2020) Livestock value chain in transition: cultivated (cell-based) meat and the need for breakthrough capabilities. Technol Soc 62:101286. https://doi.org/10.1016/j.techsoc.2020.101286

Van der Weele C, Driessen C (2019) How normal meat becomes stranger as cultured meat becomes more normal: ambivalence and ambiguity below the surface of behavior. Front Sustain Food Syst 3:69. https://doi.org/10.3389/fsufs.2019.00069

# Introduction to Animal Cell Culture

**Carlos Ricardo Soccol, Vanete Thomaz Soccol, Gabriela Dos Santos Costa, Igor Negreiros Piazenski, and João Pedro Manica Candelario**

**Abstract** Animal cell culture has marked a significant revolution in science, allowing different fields to benefit from its use. The aim of this chapter is to introduce animal cell culture, covering its main advances over the years, its main applications, such as the use of animal cell culture for vaccine production, and to address the main challenges facing the technology. Several challenges still need to be addressed, such as the development of serum-free media, the optimization of culture due to shear sensitivity in bioreactors, cost implications, among others. The field is constantly evolving, and the outlook is positive for the emergence of new applications and the achievement of new production scales.

**Keywords** Cell culture · Challenges · First cell lines · Media development

## 1 Introduction

The cultivation of animal cells is a subject that has accumulated advancements since the middle ages, when animal tissues became a subject of more in-depth study, continuing to evolve to this day. This progress has led to various applications and discoveries in the fields of biology, medicine, and the bioprocess industry, including a deeper understanding of cellular physiology and biochemistry, drug screening and development, disease modeling, in vitro fertilization (IVF) technology, stem cell and cancer research, monoclonal antibody production, regenerative medicine, therapeutic protein production, tissue engineering, and more (Bhatia et al. 2019; Jedrzejczak-Silicka 2017; Rodríguez-Hernández et al. 2014; Verma et al. 2020). Continuous efforts were being made to enhance bioprocessing techniques for the cultivation of animal cells. This included improvements in media formulations,

C. R. Soccol (✉) · V. T. Soccol · G. Dos Santos Costa · I. N. Piazenski · J. P. M. Candelario
Bioprocess Engineering and Biotechnology Department, Federal University of Paraná, Curitiba, Paraná, Brazil
e-mail: soccol@ufpr.br


C. R. Soccol et al. (eds.), *Cultivated Meat*,
https://doi.org/10.1007/978-3-031-55968-6_2

bioreactor designs, and process control strategies to optimize cell growth and productivity. The field of 3D cell culture and organoids saw significant progress. These technologies aim to mimic the natural cellular environment more closely, allowing for better representation of in vivo conditions. Equally efforts were made to develop more sustainable practices in cell culture, including the use of alternative and plant-based media components to reduce the environmental impact of cell culture processes. In this chapter, we revise the animal cell culture history, traditional applications of animal cell culture, challenges, and future prospects for animal cell culture.

## *1.1 Early Stages of Cell Culture Development*

The exploration of cells and tissues begin with the advent of microscopes in the late sixteenth century, enabling doctors and scientists to explore human cells and tissues with greater details (Jedrzejczak-Silicka 2017). The "father of microbiology," Antonie van Leeuwenhoek, was the first to conduct examinations on numerous tissue samples. From 1676 onward, he described insect specimens, microorganisms, and various animal tissues, conducting in-depth investigations of blood cells, sperm cells, skeletal muscle fibers, epithelial cells, teeth, and circulatory system structures (Jedrzejczak-Silicka 2017; Karamanou et al. 2010). After the description of cells, this stage of science has not stopped growing, and new fields of science are opening up, one of them being the development of artificial cell cultivation.

## *1.2 Development of the First Cell Culture Techniques*

Extensive observations and descriptions of cells and tissues have been made over the centuries. However, cell culture itself, i.e., the in vitro maintenance and propagation of cells, tissues, or organs under optimal conditions, only began in the mid-nineteenth or twentieth century (Bhatia et al. 2019; Verma et al. 2020). During this period, pioneering scientists showed that cells could remain viable outside the body. For instance, in 1885, Wilhelm Roux successfully maintained parts of chicken embryos, allowing them to develop in a heated saline solution for several days (Wessel 2011; Zurlo et al. 1994). These early experiments not only influenced but also propelled further discoveries related to the nutritional and physicochemical requirements of cells (Bhatia et al. 2019; Wessel 2011). Ross G. Harrison is renowned for overcoming fundamental challenges in cell culture and pioneering the first reproducible techniques of in vitro cell cultivation in the early nineteenth century. His methods became routine and fundamental components in experimental research (Bhatia et al. 2019; Wessel 2011; Rodríguez-Hernández et al. 2014). The experiments involved isolating pieces of living embryonic tissue. The cells were grown in hanging drops of coagulated lymph enclosed in glass slides (Landecker

2002). Harrison conducted extensive studies on maintaining nerve cells and monitoring fiber development. However, he faced difficulties due to inevitable bacterial contamination, leading Harrison to introduce the aseptic technique. This innovation allowed the obtainment of sterile preparations that could be maintained in vitro for over 5 weeks (Jedrzejczak-Silicka 2017).

In 1910, Montrose Burrows and Alexis Carrel, utilizing the methods proposed by Harrison, successfully established cell cultures of embryonic and adult tissues (such as connective, cartilage, bone, skin, kidneys, and others) from various species (e.g., dog, cat, chicken, guinea pig, rat) that could be maintained in vitro. They also began working with cancerous human cells (Jedrzejczak-Silicka 2017; Rodríguez-Hernández et al. 2014). The success was attributed to the use of chicken plasma, a more accessible and consistent medium, ensuring a reliable preparation process (Jedrzejczak-Silicka 2017). Seeking further improvement, Carrel evaluated other mediums, including diluted plasma with varying concentrations of salt solution and the use of serum. Consequently, cells could be maintained in vitro for several months (Rodríguez-Hernández et al. 2014), allowing the maintenance of cell lines.

## *1.3 Establishment and Development of the First Cell Lines*

The progress in animal cell and tissue culture faced challenges until the early 1950s due to the unavailability of a suitable cell line and the necessary conditions for their cultivation (Bhatia et al. 2019). The pursuit for cells with the ability to sustain viability and functionality in the long term was crucial for a deeper understanding of cell proliferation. Carrel's directed cell cultivation achieved noteworthy results, maintaining certain cells for extended periods with periodic exchange of media (Carrel 1912; Witkowski 1980). His success can be attributed to his knowledge, emphasis on rigorous asepsis, and a specifically equipped laboratory with well-trained staff (Wessel 2011; Jedrzejczak-Silicka 2017; Kelz 2022).

Several researchers dedicated their efforts to enhancing the properties of culture media, creating more suitable conditions, and deepening our understanding of cells' physiological needs. The addition of components to the culture media, such as glucose, amino acids, hormones, and vitamins, has enabled significant breakthroughs (Lewis and Lewis 1912; Lewis 1922; Baker and Carrel 1926; Vogelaar and Erlichman 1933). Various cell lines, both normal and cancerous, have been stabilized due to these advancements. Normal adult cells have finite growth, with many being stabilized through subculturing cells originating from disrupted adult and embryonic tissues (Jedrzejczak-Silicka 2017). In contrast, cell lines derived from cancerous tissues can exhibit indefinite proliferation, exemplified by the successful growth of cells Henrietta Lacks' (HeLa) cervical cancer, in 1951. These cell lines demonstrated robust and consistent growth, enabling numerous discoveries in cellular, molecular, and therapeutic fields (Bhatia et al. 2019; Jedrzejczak-Silicka 2017; Yao and Asayama 2017). These discoveries were made possible because of the opportunity to work with a consistent and uniform population of cells, rather than relying on samples from living animals. In summary, working with immortal cell lines provided researchers with a stable and reliable experimental platform (Yao and Asayama 2017). Another well-established cell line worth mentioning is the Chinese hamster ovary (CHO) cells, which were first reported in 1957 by Dr. Theodore T. Puck (Xu et al. 2017). These cells are widely utilized in the production of biopharmaceuticals due to their robust and high-density growth, coupled with their ability to synthesize a significant amount of protein (Omasa et al. 2010; Xu et al. 2017).

In certain scenarios, researchers have induced tumorous behavior in noncancerous cells through viral infections like the *Polyomavirus*, discovered in 1953 as a tumorigenic agent (Oliveira et al. 1999). This method led to the establishment of the baby hamster kidney (BHK) cells in 1964 (Stoker and Macpherson 1964), which have found valuable applications in the development of veterinary viral vaccines.

With the establishment of the first cell lines, the design of culture media was enhanced, along with cultivation techniques, enabling the exploration of new tissues and cells. This expansion significantly deepened our understanding of animal cell culture. These advancements laid the foundation for further progress in the field, fostering the connection between cell culture and technological developments in industry and human health. Figure 1 shows a timeline from microscopy to the establishment of cell culture in synthetic media.

## 1.4 Media Development and Supplements

Animal cells, from the early attempts at cultivation, have proven to be highly demanding regarding their food source, requiring various nutritional factors to proliferate and differentiate (Soccol et al. 2022). To achieve successful cultivation, the conditions of cell concentration and nutritional and hormonal balance must be spotless to ensure optimal growth and a satisfactory response (Pinnamaneni et al. 2019). To illustrate, in the present day, the formulation of a culture medium designed for animal cell cultivation must include a complex combination of elements (for details, see Table 1).

From the very beginning, media compositions relied on animal-derived biological fluids and tissue extracts (Lobo-Alfonso et al. 2008; Yao and Asayama 2017). However, due to various concerns related to animal welfare and the operational challenges in the extraction of those components, the quest for new alternatives has intensified, a pursuit that opens the possibility of large-scale production, independent of animal sources (Soccol et al. 2022). Initially, synthetic media consisted of a basal composition of salts and glucose, but serum components were still required for optimal cell growth and proliferation (Yao and Asayama 2017; Lobo-Alfonso et al. 2008). The development of basal media has played a crucial role in advancing cell culture techniques and facilitating various applications in biomedical research. Several key basal media formulations have been developed over the years, each with their own specific compositions and applications (Table 2).

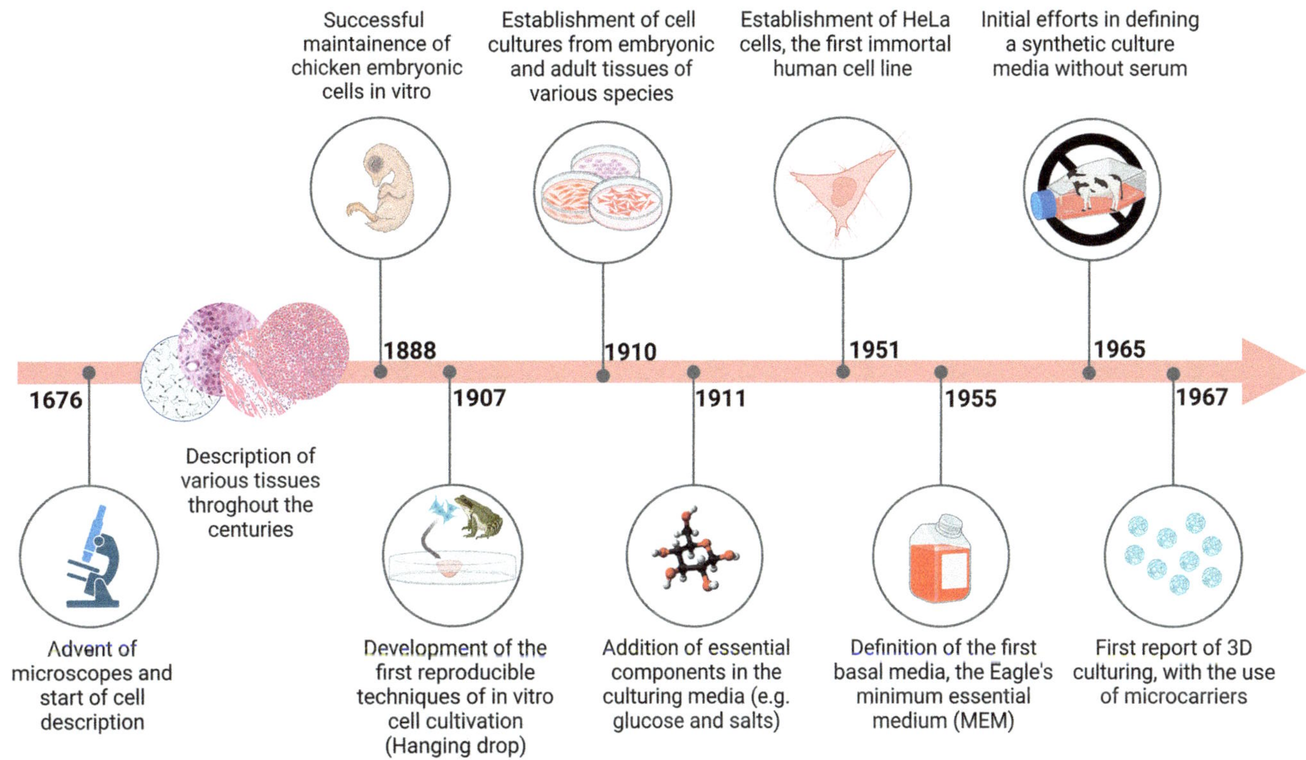


**Fig. 1** Beginning with the advent of microscopy and the initial observation of various tissues and cells, this timeline highlights the early development of animal cell culture and its key initial steps. It progresses through the early successful techniques and cultures, the establishment of diverse cultures including immortal cells, and advancements in formulating culture media by avoiding animal-derived components. The timeline culminates in the onset of improvements in cultivation efficiency, such as the use of microcarriers. All these developments enabled a rapid escalation of technology, leading to industrialization and continued progress (Created with BioRender.com)

**Table 1** Basic components of a typical medium for culturing animal cells

| Amino acids | Vitamins | Salts | Others | Proteins |
|---|---|---|---|---|
| • Arginine<br>• Cystine<br>• Glutamine<br>• Histidine<br>• Isoleucine<br>• Leucine<br>• Lysine<br>• Methionine<br>• Phenylalanine<br>• Threonine<br>• Tryptophan<br>• Tyrosine<br>• Valine | • Biotin<br>• Coline<br>• Folate<br>• Nicotinamide<br>• Pantothenate<br>• Pyridoxal<br>• Thiamine<br>• Riboflavin | •NaCl<br>•KCl<br>•$NaH_2PO_4$<br>•$NaHCO_3$<br>•$CaCl_2$<br>•$MgCl_2$ | • Glucose<br>• Penicillin<br>• Streptomycin<br>• Phenol red<br>• Serum | • Insulin<br>• Transferrin<br>• Specific growth factors |

Source: Adapted from Freshney (2010)

Harry Eagle created the minimum essential medium (MEM), which contained essential components necessary for cell survival and growth (Eagle 1955). Building upon this, Dulbecco's modified Eagle's medium (DMEM) and Ham's media were developed, each with their own specific compositions. Ham's media provided a serum-free option for cell culture, while DMEM optimized the basal medium with the inclusion of additional nutrients and growth factors (Ham 1965; Dulbecco and Freeman 1959). Furthermore, the fusion of Ham's media and DMEM resulted in a versatile medium known as DMEM/F-12, which combined the strengths of both formulations (Barnes and Sato 1979). These advancements in synthetic media enabled the cultivation of cells in a defined and controlled environment, reducing dependence on serum components. Besides that, ongoing improvements in novel media are continuously enhancing their effectiveness and reliability. The development of culture media is vital for maintaining cells, and the appropriate selection of media is a decisive factor in the success of cultivation. Each cell has specific requirements and purposes, necessitating the tailoring of media accordingly (Verma et al. 2020). In line with this, there is currently a strong emphasis on the development of nutrient media that are free from animal-derived components. The primary objective is to ensure robust cell development while maintaining an animal-free environment. This focus on animal-free media aims to facilitate the scalability of the cultivation process for large-scale applications in animal cell cultivation (Soccol et al. 2022).

## 1.5 Improvement of Cultivation Techniques

It is well established that animal cells are extremely sensitive, requiring precise operational parameters during cultivation for optimal growth, including factors such as pH, temperature, humidity, gas transfer, shear rates, and dissolved nutrients (Verma et al. 2020; Bhatia et al. 2019). Except for cells derived from the

**Table 2** Types of culture medium for animal cells

| Media | Description | References |
|---|---|---|
| Medium 199 | Produced by the sequential addition of amino acids, vitamins, and nucleic acid precursors. Developed for culturing chicken embryonic cells in protein-free conditions. Often used in organ culture. | Morgan et al. (1950) |
| Basal medium Eagle (BME) | Supplemented medium, but considered simple, for the proliferation of mouse L cells and human HeLa cells, containing 13 amino acids and 8 vitamins. | Eagle (1955) |
| NCTC109 | Medium with a complex composition, created to cultivate protein-free L cells. Amino acid composition based on horse serum ultrafiltrate and chicken embryonic tissue extract. | Evans et al. (1956) |
| CMRL1066 | Created from Morgan 199 medium, altered to grow mouse L cells in the absence of proteins. The modifications made included an increase in substances such as cysteine, glutathione, and ascorbic acid, elimination of fat-soluble vitamins, changes in the types of nucleic acid precursors, and addition of coenzymes. | Yao and Asayama (2017) |
| Minimum essential medium (MEM) | It is the modified BME medium, where the amino acid concentration is doubled. The MEM formulation does not contain essential amino acids, which can be added by users. | Eagle (1959) |
| Dulbecco's modified MEM (DMEM) | Medium modified to have four times the concentration of amino acids and proteins of BME. Developed to study the plaque formation capacity of the polyoma virus in mouse embryonic cells. It has a high supplementation capacity and changes in the concentration of components depending on the demand of the cultivated cell. | Dulbecco and Freeman (1959) |
| Ham's F-10 | Serum-free culture medium. As substitutes we find albumin and fetuin. Capable of proliferating CHO cells, but less significantly when compared to serum-containing media. | Ham (1963) |
| Ham's F-12 | Medium with linoleic acid and putrescine as albumin and fetuin substitutes allows colony formation by a single CHO cell in protein-free conditions. Cited as the world's first chemically defined medium. | Ham (1965) |
| RPMI 1640 | Modified medium based on 5A medium. Intended for the cultivation of peripheral blood lymphocytes, having high concentrations of phosphate and low concentrations of calcium and magnesium. Widely used in suspension cultures. | Moore et al. (1966) |
| α-MEM | Modified MEM medium for culturing mouse and hamster hybrid cells. Supplemented with nonessential amino acids. | Stanners et al. (1971) |
| Kaighn's modified Ham's F-12 (Ham's F-12 K) | Medium with higher concentrations of amino acids and other components, when compared to Ham F-12 medium, with the aim of improving the proliferation of primary cells. | Yao and Asayama (2017) |
| Iscove's modified DMEM (IMDM) | Medium supplemented with vitamins and amino acids that are not present in DMEM, containing no serum and using transferrin, bovine serum albumin, and soy lipids as substitutes. Suitable for high-density cultures and rapidly proliferating cell cultures. | Iscove and Melchers (1978) |

hematopoietic system, all normal tissue-derived cells are anchorage-dependent and necessitate a surface or cell culture support for proliferation; this property results in the formation of a monolayer at the bottom of the culture vessel (Merten 2015; Bhatia et al. 2019; Verma et al. 2020). This monolayer configuration ensures uniform nutrient and growth factor distribution from the medium during growth (Edmondson et al. 2014). However, this monolayer setup comes with limitations. Cultivation is confined to a specific area, restricting the final cell yield to the adhesion space within a particular flask (Merten 2015). Furthermore, monolayer cultures display low expansion and low maintenance of cell morphology (Li et al. 2015). These spatial and morphological constraints hamper the scalability of cell growth, leading to the exploration of alternative solutions to expand the adhesion surface in cell culture (Merten 2015). A small-scale solution for surface limitation involves using traditional culture systems like culture plates, roller bottles, and multitray systems, providing larger surface areas compared to standard culture flasks. However, there is limited scalability, and controlling cultivation conditions such as pH and oxygenation proves challenging (Merten 2015). When envisioning large-scale applications, the conceptualization of a "3D cell culture" was well received. This innovative approach, as reported by Van Wezel (1967), involved cells adhering to the surface of microcarriers—micrometric plastic beads that could be suspended in the culture. In this setup, even though the maximum growth rate of cells in monolayer was compared to cultures on microcarriers, higher cell concentrations could be achieved (Van Wezel 1967). This was due to the significant mechanical activity of the cells, allowing cell transfer and inducing proliferation across multiple microcarriers simultaneously (Wang and Ouyang 1999). Another approach was the incorporation of the cells into a matrix or scaffold, permitting the spheroid formation of cells (Edmondson et al. 2014). The technique involves using hydrogels or synthetic materials to create a three-dimensional space for the cells; this matrix can be in suspension or fixed, depending on the setup proposed (Edmondson et al. 2014). With the emergence of these technologies, it became possible to utilize different bioreactors to scale cells in suspension, such as stirred-tank bioreactors, hollow fiber bioreactors, fixed-bed and fluidized-bed reactors, bubble columns, airlift bioreactors, and the wave bioreactor (Menshutina et al. 2021). These advancements in culturing technology have facilitated the transition from traditional 2D cell culture to more physiologically relevant 3D culture systems. In particular, the use of 3D culture systems has found applications in diverse fields, including drug testing, as they provide an environment that more accurately mimics the in vivo conditions (Edmondson et al. 2014).

## 2 Applications of Animal Cell Culture

Cell culture is an invaluable cornerstone of modern scientific research, particularly in cellular and molecular biology. Within the controlled laboratory environments of cell culture, scientists can nurture and manipulate living cells, shedding light on complex biological phenomena. This method has played a pivotal role in advancing our comprehension of diseases, enabling the development of novel therapeutic strategies and pharmaceuticals. The versatility of cell culture extends beyond health and medicine, with applications in diverse fields such as agriculture and biotechnology. The development in the field over the years has opened doors to a range of practical applications.

Before the concept of cultured meat, many biotechnology products already fundamentally rely on the mass culture of animal cell lines (Verma et al. 2020). Animal cell culture finds application in various fields, ranging from basic to advanced research. It has provided a model system for various lines such as basic cell biology studies, drug testing, toxicity assessments, gene therapy, vaccine production, recombinant protein production, and more benefit from this technology and continue to advance, making animal cell cultivation indispensable for scientific and technological development (Fig. 2) (Verma et al. 2020; Rodríguez-Hernández et al. 2014). Several advantages are associated with the use of animal cell cultures that make it suitable for the applications mentioned. The main benefit is the consistency and reproducibility of results that can be achieved when employing a defined cell lineage. In cell cultures, it is possible to maintain a high degree of control over the physicochemical environment, including parameters such as pH, temperature, osmotic pressure, oxygen levels, and carbon dioxide tension (Arango et al. 2013). With this level of control and a defined medium, cell cultivation can be optimized. In the production of biopharmaceuticals, for example, this technology has become indispensable due to its capacity to facilitate the precise, scalable, and reproducible

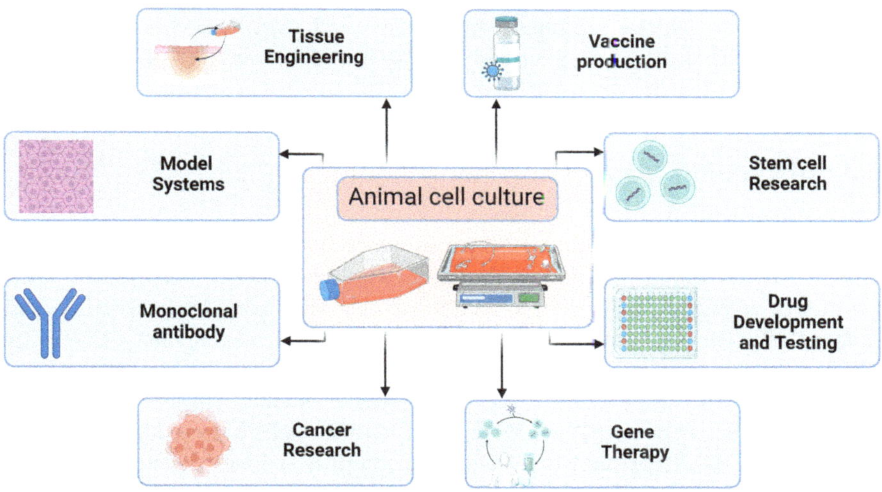


**Fig. 2** Traditional applications of animal cell cultivation, including tissue engineering, vaccine production, monoclonal antibody production, use of model systems, cancer and stem cell research, gene therapy, and drug testing and development are the main traditional applications of the technology (Created with BioRender.com)

large-scale generation of biologically active molecules (O'Flaherty et al. 2020). In the following topics, we are going to discuss some of the most important applications of animal cell culture.

## 2.1 Model Systems in Basic Research, Health, and Disease

In vitro animal cell culture systems are foundational tools in biological research, enabling scientists to cultivate and study animal cells in a controlled laboratory environment. This versatile technique is employed to investigate the principles of cellular biology, biochemistry, physiology, and dynamics between cells and pathogens such as bacteria and viruses, assess the impacts of drugs, examine the complex process of aging, explore the triggers behind aging, among other applications (Arango et al. 2013; Segeritz and Vallier 2017). Further, cell culture minimizes the confounding influences of environmental and biological variability that may be present in other experimental models (Arango et al. 2013). But these models might not fully or accurately anticipate the impacts on the whole organism (Nikolic et al. 2018). Additionally, the use of genetically defined and well-characterized cell lines simplifies data analysis, promoting accuracy and reproducibility (Arango et al. 2013). As an example, cancer cell lines are highly valued as in vitro model systems, being essential in cancer research and especially in drug discovery. Their importance primarily stems from their unique ability to provide a source of biological material for experimental investigations. Recent discoveries, aided by technologies such as genomics, transcriptomics, and proteomics, have underscored the invaluable contribution of these cell lines to medical research (Mirabelli et al. 2019). Liver and lung cells have also been extensively used as in vitro models (Nikolic et al. 2018).

## 2.2 Vaccine Production

Vaccinating the population is one of the most powerful interventions in the realm of global health. Through vaccination programs worldwide, a significant reduction in the number of deaths caused by infectious agents has been achieved (Rodrigues et al. 2015). The most recent case of mass vaccination was against COVID-19, which helped the world return to normality after the lockdowns that occurred in 2020/2021 (WHO 2023). Given the COVID-19 pandemic, the pace of vaccine development has been accelerated like never before. As of March 2023, over 5.55 billion individuals across the globe have been administered with at least one dose of a COVID-19 vaccine (Holder 2021). For the study and development of vaccines and solutions for COVID-19, cell culture has been of paramount importance, as it enables the production of the virus and subsequent isolation and study (Chen et al. 2023; Fang et al. 2022).

While this is achievable today, vaccine production has not always relied on studies involving cell culture. An example is rabies, one of the oldest known diseases in history with cases dating back 4000 years, for most of the human history, a bit by a rabid animal was invariably fatal (Koury and Warrington 2023). Vaccination used to rely on the basic "Pasteur Treatment," which involved a brain tissue vaccine with the addition of formaldehyde (Hoenig 1986). These vaccines offer limited protection and are frequently associated with undesirable side effects. The use of cell cultures has revolutionized vaccine production. Rabies virus, obtained from inoculated cell cultures, can be obtained in a pure and abundant form. Concentrated and purified viruses can be employed to create highly immunogenic and well-tolerated vaccines suitable for both pre-exposure immunization and post-exposure rabies treatment (Sureau 1987). Most of the modern flu vaccines now available are cell based, using animal cells (in vitro) for growing most viruses meant for vaccine manufacture. For example, vaccines for deadly diseases like polio, rabies, chicken pox, measles, and hepatitis B are produced using animal cell culture (CDC 2023). Besides the production of viruses, it is possible to use animal cell culture to produce recombinant antigens as vaccines.

Production of vaccines by recombinant proteins can use bacterial expression systems, that are the most used due to the ease of handling and to their capacity for high level expression. However, for antigens in which post-translational modifications (e.g., glycosylation) are necessary, the use of mammalian or insect cells should be considered (Nascimento and Leite 2012).

## *2.3 Monoclonal Antibody Production*

Another utility of animal cell culture is monoclonal antibody production. Monoclonal antibodies (mAbs) serve as vital agents used in biomedical research, disease diagnosis, and the treatment of conditions such as infections and cancer. These antibodies are generated from cell lines or clones derived from animals previously immunized with the substance of interest under study (Quinteros et al. 2017). The application of mAbs in the field of biomedical research is a crucial factor in the identification of proteins, carbohydrates, and nucleic acids. The use of these antibodies has been significant in uncovering the functions of numerous molecules that regulate cell replication and differentiation, thus contributing to an enhanced understanding of the molecular structure and function of various molecules (National Research Council (US) Committee 1999). In the process of antibody manufacturing, maintaining desired quality attributes, reducing development time, achieving cost efficiency, and ensuring production flexibility are critical considerations in a highly competitive market. In this context, numerous companies often focus on therapies targeting similar clinical indications and targets. The use of animal cells to produce these antibodies enables rapid progress in the field as it allows production of antibodies similar to those found in the natural source (Li et al. 2010a, b). The mAb production in mammalian cells is a well-known technique for synthesis of identical antibodies (Kaysfeld et al. 2023). The choice of an expression system is guided by its capacity to achieve elevated productivity while maintaining acceptable product quality characteristics (Li et al. 2010a, b).

## 2.4 Tissue Engineering

Animal cell culture is a cornerstone in the development and future of innovative technologies within tissue engineering and regenerative medicine. The reconstruction of damaged or degenerated tissues and organs, vital for the advancement of these disciplines, is directly reliant on animal cell culture. Beyond addressing urgent healthcare challenges, ranging from injury treatment to overcoming the critical organ transplant shortage, these technologies hold the promise of groundbreaking solutions (Berthiaume and Yarmush 2003). Tissue engineering consists of the construction of bioartificial tissues, in vitro and through in vivo manipulation of cellular growth and function. Cells serve as foundational building blocks in tissue formation and play a vital part in facilitating tissue healing and regeneration. The types of cells used in tissue engineering are sourced from various origins, including a patient's autologous cells, allogeneic cells from another human, and xenogeneic cells from a different species. In the field of tissue engineering, cells can be a component of the in vitro construct or recruited in vivo with the assistance of immobilized or soluble signals (Belleghem et al. 2020).

One example of tissue that has been used in cellular culture, a focus of extensive research, is cartilaginous tissue. Cartilage tissue engineering is emerging as a technique for regenerating damaged cartilaginous tissue due to diseases or traumas. Given that cartilage lacks natural regenerative capabilities, it is crucial to develop methods that provide the appropriate cells, biomaterials, and signaling factors to the site of the injury (Chung and Burdick 2008). Another example is skin production, aiming to reconstruct the structural and functional components of the skin, reduce scar formation, and enhance wound healing quality (Chen et al. 2009). An organ under exploration for engineering is the kidney, as the ability to regenerate a new kidney from a patient's own cells would offer significant relief for the countless individuals afflicted by kidney disease (Geuens et al. 2020).

## 2.5 Stem Cell Research

Cell culture has played a substantial impact in advancing stem cell research, leaving a lasting imprint on the field. Stem cell research is experiencing a remarkable progress, with the potential to create therapeutic treatments for addressing diseases and investigating illnesses from their early stages. Stem cells have the potential to drive advancements in cell therapy, tissue engineering, regenerative medicine, and find valuable applications in the pharmaceutical and biotechnological sectors (McKee and Chaudhry 2017). Stem cells are special cells with the ability to both self-renew and differentiate into several tissues. These cells can be found in various forms and are obtained from embryonic tissues or adult tissues. Their potential ranges from the ability to develop into various cell types (pluripotent) to a more limited capacity to transform into specific cell types (unipotent), depending on the specific type of stem cell (Ylostalo 2020). One of the most promising aspects of human embryonic stem cells (hESC) is their remarkable capacity to generate specialized cells capable of serving as substitutes for damaged tissue in patients suffering from a diverse array of degenerative conditions. This potential offers the hope of groundbreaking treatments and regenerative medicine approaches for a multitude of diseases, including neurodegenerative disorders, heart diseases, and spinal cord injuries, by providing a renewable source of replacement cells tailored to address individual patient needs (Vazin and Freed 2010).

Over the past three decades, cultured stem cells have found their way into clinical applications, and it is worth noting that the early studies on mammalian cell culture were essential in bridging the path to bringing stem cells into clinical practice. Additionally, cell culture technology has not only been instrumental in obtaining induced pluripotent stem cells (iPSCs) but has also been crucial in refining the processes of their derivation and cultivation. This technological synergy has further accelerated the therapeutic potential and applicability of iPSCs in personalized medicine, regenerative therapies, and disease modeling, opening new horizons for innovative medical interventions (Fuchs 2012).

## 2.6 *Drug Development and Testing*

The utilization of in vitro cell cultures has been instrumental in the progress of developing new drugs. Its significance in cell biology research lies in providing a foundation that ensures mimicking the in vivo microenvironment and its intricate interactions, particularly in 3D culture models. It is employed for the assessment of the toxicity of novel medications and to determine the appropriate and safe dosage levels for these new drugs (Cardoso et al. 2023).

As the pharmaceutical industry continues to experience steady growth in the development of new and effective drugs and their large-scale production, the demand for in vitro drug testing is rising (Zhang et al. 2022). Cell-based assays have a significant contribution to the pharmaceutical industry, given its heavy reliance on drug research and development. This need for fostering ongoing innovation is underscored by the significant investments of time and money, often spanning over 12 years from target identification to market approval (Jaroch et al. 2018). Furthermore, for ethical and cost-related reasons, the use of animals to assess the mode of action, metabolism, and toxicity of new drug candidates, both in research and industrial applications, has been avoided. The use of animal cell culture increasingly mitigates issues related to new drug testing, as it allows for testing drugs and their effects before animal testing is necessary, yielding results that enable the progression or cessation of research, thus preventing unnecessary resource expenditure.

In addition, large-scale cell cultures offer the opportunity to generate genetically engineered proteins, antibodies, hormones, and biopharmaceuticals, which can be harvested and harnessed for therapeutic applications (Segeritz and Vallier 2017). Biotherapeutic products derived from cell culture have now firmly established themselves as an indispensable component of the pharmaceutical industry (O'Flaherty et al. 2020). The approval of human tissue plasminogen activator as the first therapeutic protein, produced from recombinant mammalian cells in 1986, marked the beginning of mammalian cell culture becoming a method for biopharmaceutical production in the industry (Kim et al. 2012). Mammalian cells are a suitable choice for performing posttranslational modifications that closely resemble those in native human proteins, in contrast to microbial and yeast systems (Kuystermans and Al-Rubeai 2015). Out of the 159 approved monoclonal antibodies products created using recombinant techniques in cell-based systems, the majority (107, which accounts for 67%) are manufactured in mammalian cells (Walsh and Walsh 2022).

## 2.7 Gene Therapy

Gene therapy is developed with the aim of treating a wide range of conditions stemming from genetic abnormalities. The process begins by extracting cells from the patient who either lacks a functional gene or has a defective one. These problematic genes are then substituted with healthy, functional genes, and the modified cells are cultured and expanded under controlled laboratory conditions. Finally, these engineered cells are reintroduced into the patient's body. This innovative approach offers a promising avenue for potentially curing genetic diseases by rectifying the underlying genetic defects through targeted gene replacement (Shupe et al. 2022; Segeritz and Vallier 2017). Cell culture is important in gene therapy as it provides a controlled environment for the modification and expansion of therapeutic gene-carrying cells before reintroducing them into patients. This pivotal step ensures the safety, efficiency, and precision of gene therapy, offering a promising avenue for treating genetic diseases and disorders at their source.

Gene therapy offers a direct method for addressing genetic and various other medical conditions. Additionally, there are related techniques, such as gene editing, which come in various forms and approaches. The foundation for all these methods is a thorough comprehension of gene functionality and how alterations in genes can impact our well-being, with a primary focus on understanding these processes in animal cell culture (Scheller and Krebsbach 2009). On the other hand, gene therapies are more expensive to research and manufacture than well-established treatments relying on small-molecule drugs. However, they offer a distinct advantage in directly aiding patients with diseases that conventional medications often fail to address effectively. This cost disparity is offset by the potential to provide transformative solutions for individuals whose conditions have proven resistant to common pharmaceutical interventions (Nature 2023).

### 2.8 *Cancer Research*

Cell culture is extremely important in cancer research, providing a controlled and versatile environment for studying the complexities of cancer at the cellular and molecular levels. By culturing cancer cells in vitro, researchers gain insights into the mechanisms of tumorigenesis, disease progression, and the response to potential therapies. This controlled setting allows for the systematic investigation of cancer cell behavior, including proliferation, migration, invasion, and drug resistance. Moreover, cell culture models offer the flexibility to work with a wide range of cancer cell lines, enabling the study of diverse cancer types and their unique characteristics. Cell culture not only serves as a foundational tool in understanding the molecular underpinnings of cancer but also accelerates the development of novel therapeutic strategies, advancing our ability to combat this formidable disease (Mirabelli et al. 2019; Lovitt et al. 2014). The history of cancer research and the establishment of continuous cell lines are closely related. Different investigators have begun to understand the complex mechanisms that transform a normal cell into a cancer cell and subsequent tumor development due to the availability of these precious models. Human cancer cell lines remain essential in modern cancer research, serving as preclinical model systems to gain mechanistic and therapeutic insights. The online accessibility of information derived from these studies has created a valuable resource for the study of cancer cell lines, assisting researchers in selecting the most suitable in vitro model system for their specific research projects. This has advanced our understanding of cancer and its potential treatment strategies (Mirabelli et al. 2019).

Furthermore, the fundamental contrast between regular and cancer cells becomes apparent through the utilization of animal cell culture techniques, facilitating the cultivation of both cell types within a laboratory setting. By subjecting normal cells to various stimuli like radiation, chemicals, and viruses, researchers gain insights into the transformative mechanisms that lead to the development of cancer. This approach not only aids in understanding the etiology of cancer but also paves the way for studying potential interventions. Cell culture, for instance, serves as a valuable tool for identifying drugs that can selectively target and destroy cancer cells while sparing healthy ones, offering a promising avenue for more precise and effective cancer therapies (Han et al. 2014; Oliveri 2022; Fatima et al. 2022).

## 3 Challenges and Future Prospects for Animal Cell Culture

The cultivation of animal cells, when compared to the cultivation of microorganisms, requires greater care, since the characteristics of animal cells are considerably different from those of microorganisms. Animal cells grow slowly, the cellular structure is more fragile, and the growth medium requires greater complexity of components, and in many cases, the cells require supports that provide fixation for their proliferation and differentiation. Therefore, in a way, cultivating animal cells requires greater investment by the research laboratory or industry, and this investment is partly justified by the fact that normally the final product has a high added value.

Cells obtained from living tissues for cell culture need to be selected from reliable sources, being certified animals or trusted biobanks. This selection reduced the risks of introducing pathogenic and infectious prions. Prions are infectious proteins that cause neurodegenerative and lethal diseases. In mammals, prions reproduce by recruiting the normal cellular isoform of the prion protein (PrPC) and stimulating its conversion to the disease-causing isoform (PrPSc) (Colby and Prusiner 2011; Li et al. 2010a, b).

Cell culture contamination is a recurring concern and takes longer to realize when working with animal cells. In cases of bacterial and fungal contamination in cell cultures, contamination is quickly detected because the cell culture is more quickly destroyed. The types of cellular contaminants that may occur include cell cross-contamination, microbial (bacterial, fungal, mycobacterial, mycoplasma, and virus), and chemical contaminants (e.g., free radicals, endotoxins, heavy metals, and detergents) (Baust et al. 2017). The use of antibiotics in the cell cultivation stage is also extremely important to avoid contamination of the process by bacteria.

In cell cultivation, the culture medium, and supplements play a key role. Animal-derived serum, one of the most notable components of the medium, stands out for being complete (Subbiahanadar Chelladurai et al. 2021). It serves as a source of amino acids, hormones, growth factors, vitamins, proteins, inorganic salts, carbohydrates, and lipids for cellular metabolism and cell growth (Yao and Asayama 2017). Initially, it was believed that fetal bovine serum (FBS) only helped cell growth, but it was later identified that FBS has essential components for cell fixation, proliferation, and maintenance such as serum albumin and fetuin, in addition to containing low levels of γ-globulin (a cell growth inhibitor) compared to horse and calf serum (Gstraunthaler et al. 2013; Yao and Asayama 2017). In recent years, the FBS production method has been heavily criticized due to animal welfare concerns. FBS is harvested from bovine fetuses removed from pregnant cows during slaughter. The common method of harvesting the fetus is by cardiac puncture without any anesthesia. This practice of harvesting FBS is inhumane as it exposes the fetus to pain and/or discomfort. In addition to moral concerns, there are numerous scientific and technical problems regarding the use of FBS in cell culture. Therefore, the concern of the scientific community is to reduce the use of FBS and replace it with synthetic alternatives, serum-free media (Verma et al. 2020). Some substitutes for FBS have been reported in the literature, including human platelet lysate (HPL), bovine ocular fluid, earthworm heat-inactivated coelomic fluid, sericin protein, and chemically defined medium (Duarte et al. 2023).

Another main challenge in culturing animal cells, faced by the industry, is increasing the yield of the process. This is because low cell concentrations result in low yields, increasing the cost of the process and consequently the product. Being able to scale production processes that use animal cells in cultivation, without loss of cell viability and with minimal risk of contamination of the process, is the objective of industries. To scale up this technology, certain issues are inherent to the process due to the nature of the cells. The bioreactor design must address various needs of the culture, including effective gas transfer, lower shear rates, optimized carriers/scaffolds for improved adhesion and purification, cost-effective inputs, and their periodic exchanges, which is necessary to reduce the inhibitory effects caused by metabolites (Soccol et al. 2022).

It is extremely important that cell cultivation planning considers all the points outlined above. The development of alternatives to FBS is already a reality and should be encouraged within the scientific community to make the cultivation of animal cells cleaner. The risks of contamination of cultivation must be accounted for and prevented by applying good laboratory and/or manufacturing practices, in addition to the conscious use of antibiotics in the cultivation medium. It is also important that when developing a method for culturing animal cells, the technique can guarantee the maintenance of cells and tissues in the media, overcoming the discrepancy between in vivo cell kinetics and in vitro cell kinetics.

## 4 Conclusion and Future Perspectives

In conclusion, animal cell culture is a cornerstone in various fields of fundamental research and practical applications in health and disease. Its controlled environment not only enhances precision and reproducibility but also facilitates deeper insights into cell biology, physiology, and interactions with pathogens. This foundational technique proves indispensable in various contexts. It is crucial in vaccine production, exemplified by the rapid development of COVID-19 vaccines, and essential for generating antibodies used in biomedical research and therapies. Furthermore, in the fields of tissue engineering and regenerative medicine, animal cell culture stands at the forefront, offering innovative solutions for the repair of damaged tissues and organs. Simultaneously, advances in stem cell research, including the use of induced pluripotent stem cells (iPSCs), hold great promise for personalized medicine and the modeling of diseases. In drug development and testing, animal cell culture assesses drug toxicity and efficacy, reducing reliance on animal testing and supporting gene therapy for genetic disease treatment. In the realm of cancer research, it provides a controlled environment to study cancer complexities, improving our understanding and treatment of the disease. However, it is important to acknowledge that animal cell culture encounters its own set of challenges. These include the need for meticulous handling, the prevention of contamination, and the establishment of reliable cell lines. Moreover, the industry is actively engaged in efforts to increase yields and reduce costs, especially in large-scale production processes. Looking forward, the development of alternatives to FBS and the adoption of synthetic substitutes represent promising steps toward making animal cell culture more sustainable and ethical. As the field continues to evolve, it is imperative that we rise to these challenges. Doing so will allow us to fully harness the potential of animal cell culture, driving advancements in biomedical research and delivering innovative healthcare solutions.

## References


Arango M-T, Quintero-Ronderos P, Castiblanco J, Montoya-Ortíz G (2013) Cell culture and cell analysis. In: Autoimmunity: from bench to bedside. El Rosario University Press

Baker LE, Carrel A (1926) Action on fibroblasts of the protein fraction of embryonic tissue extract. J Exp Med 44:387–395. https://doi.org/10.1084/jem.44.3.387

Barnes D, Sato G (1979) Growth of a human mammary tumour cell line in a serum-free medium. Nature 281:388–389. https://doi.org/10.1038/281388a0

Baust JM, Buehring GC, Campbell L et al (2017) Best practices in cell culture: an overview. In Vitro Cell Dev Biol Anim 53:669–672. https://doi.org/10.1007/s11626-017-0177-7

Belleghem SMV, Mahadik B, Snodderly KL, Fisher JP (2020) Overview of tissue engineering concepts and applications. In: Biomaterials science. Elsevier, New York, pp 1289–1316

Berthiaume F, Yarmush ML (2003) Tissue engineering. In: Meyers RA (ed) Encyclopedia of physical science and technology, 3rd edn. Academic Press, New York, pp 817–842

Bhatia S, Naved T, Sardana S (2019) Introduction to animal tissue culture science. In: Introduction to pharmaceutical biotechnology, Volume 3: Animal tissue culture and biopharmaceuticals. IOP Publishing

Cardoso BD, Castanheira EMS, Lanceros-Méndez S, Cardoso VF (2023) Recent advances on cell culture platforms for in vitro drug screening and cell therapies: from conventional to microfluidic strategies. Adv Healthc Mater 12:2202936. https://doi.org/10.1002/adhm.202202936

Carrel A (1912) On the permanent life of tissues outside of the organism. J Exp Med 15:516–528. https://doi.org/10.1084/jem.15.5.516

CDC (2023) Cell-based flu vaccines. Centers for Disease Control and Prevention. https://www.cdc.gov/flu/prevent/cell-based.htm. Accessed 25 Oct 2023

Chen M, Przyborowski M, Berthiaume F (2009) Stem cells for skin tissue engineering and wound healing. Crit Rev Biomed Eng 37:399–421

Chen D-Y, Turcinovic J, Feng S et al (2023) Cell culture systems for isolation of SARS-CoV-2 clinical isolates and generation of recombinant virus. iScience 26:106634. https://doi.org/10.1016/j.isci.2023.106634

Chung C, Burdick JA (2008) Engineering cartilage tissue. Adv Drug Deliv Rev 60:243–262. https://doi.org/10.1016/j.addr.2007.08.027

Colby DW, Prusiner SB (2011) Prions. Cold Spring Harb Perspect Biol 3:a006833. https://doi.org/10.1101/cshperspect.a006833

Duarte AC, Costa EC, Filipe HAL et al (2023) Animal-derived products in science and current alternatives. Biomater Adv 151:213428. https://doi.org/10.1016/j.bioadv.2023.213428

Dulbecco R, Freeman G (1959) Plaque production by the polyoma virus. Virology 8:396–397. https://doi.org/10.1016/0042-6822(59)90043-1

Eagle H (1955) Nutrition needs of mammalian cells in tissue culture. Science 122:501–504. https://doi.org/10.1126/science.122.3168.501

Eagle H (1959) Amino acid metabolism in mammalian cell cultures. Science 130:432–437. https://doi.org/10.1126/science.130.3373.432

Edmondson R, Broglie JJ, Adcock AF, Yang L (2014) Three-dimensional cell culture systems and their applications in drug discovery and cell-based biosensors. Assay Drug Dev Technol 12:207–218. https://doi.org/10.1089/adt.2014.573

Evans VJ, Bryant JC, Fioramonti MC et al (1956) Studies of nutrient media for tissue cells in vitro. I. A protein-free chemically defined medium for cultivation of strain L cells. Cancer Res 16:77–86
Fang Z, Lyu J, Li J et al (2022) Application of bioreactor technology for cell culture-based viral vaccine production: present status and future prospects. Front Bioeng Biotechnol 10
Fatima M, Abourehab MAS, Aggarwal G et al (2022) Advancement of cell-penetrating peptides in combating triple-negative breast cancer. Drug Discov Today 27:103353. https://doi.org/10.1016/j.drudis.2022.103353
Freshney RI (2010) Culture of Animal Cells: A Manual of Basic Technique and Specialized Applications, 1st edn. Wiley
Fuchs E (2012) The impact of cell culture on stem cell research. Cell Stem Cell 10:640–641. https://doi.org/10.1016/j.stem.2012.03.010
Geuens T, van Blitterswijk CA, LaPointe VLS (2020) Overcoming kidney organoid challenges for regenerative medicine. NPJ Regen Med 5:8. https://doi.org/10.1038/s41536-020-0093-4
Gstraunthaler G, Lindl T, Van Der Valk J (2013) A plea to reduce or replace fetal bovine serum in cell culture media. Cytotechnology 65:791–793. https://doi.org/10.1007/s10616-013-9633-8
Ham RG (1963) Putrescine and related amines as growth factors for a mammalian cell line. Biochem Biophys Res Commun 14:34–38. https://doi.org/10.1016/0006-291X(63)90206-7
Ham RG (1965) Clonal growth of mammalian cells in a chemically defined, synthetic medium. Proc Natl Acad Sci U S A 53:288–293. https://doi.org/10.1073/pnas.53.2.288
Han Y, Li S, Cao X et al (2014) Different inhibitory effect and mechanism of hydroxyapatite nanoparticles on normal cells and cancer cells in vitro and in vivo. Sci Rep 4:7134. https://doi.org/10.1038/srep07134
Hoenig LJ (1986) Triumph and controversy. Pasteur's preventive treatment of rabies as reported in JAMA. Arch Neurol 43:397–399. https://doi.org/10.1001/archneur.1986.00520040075024
Holder J (2021) Tracking coronavirus vaccinations around the world. The New York Times. https://doi.org/10.1017/S0025727300040126
Iscove NN, Melchers F (1978) Complete replacement of serum by albumin, transferrin, and soybean lipid in cultures of lipopolysaccharide-reactive B lymphocytes. J Exp Med 147:923–933. https://doi.org/10.1084/jem.147.3.923
Jaroch K, Jaroch A, Bojko B (2018) Cell cultures in drug discovery and development: the need of reliable in vitro-in vivo extrapolation for pharmacodynamics and pharmacokinetics assessment. J Pharm Biomed Anal 147:297–312. https://doi.org/10.1016/j.jpba.2017.07.023
Jedrzejczak-Silicka M (2017) History of cell culture. In: Gowder SJT (ed) New insights into cell culture technology. IntechOpen
Karamanou M, Poulakou-Rebelakou E, Tzetis M, Androutsos G (2010) Anton van Leeuwenhoek (1632-1723): father of micromorphology and discoverer of spermatozoa. Rev Argent Microbiol 42:311–314. https://doi.org/10.1590/S0325-75412010000400013
Kaysfeld MW, Kumar D, Nielsen MK, Jørgensen JB (2023) Dynamic optimization for monoclonal antibody production. https://doi.org/10.48550/ARXIV.2302.09932
Kelz R (2022) Tissue culture and biological time: Alexis Carrel, Henri Bergson and the plasticity of living matter. BioSocieties 17:442–460. https://doi.org/10.1057/s41292-020-00224-2
Kim JY, Kim Y-G, Lee GM (2012) CHO cells in biotechnology for production of recombinant proteins: current state and further potential. Appl Microbiol Biotechnol 93:917–930. https://doi.org/10.1007/s00253-011-3758-5
Koury R, Warrington SJ (2023) Rabies. In: StatPearls. StatPearls Publishing, Treasure Island, FL
Kuystermans D, Al-Rubeai M (2015) Biopharmaceutical products from animal cell culture. In: Al-Rubeai M (ed) Animal cell culture. Springer, Cham, pp 717–757
Landecker H (2002) New times for biology: nerve cultures and the advent of cellular life in vitro. Stud Hist Philos Sci C 33:667–694. https://doi.org/10.1016/S1369-8486(02)00026-2
Lewis MR (1922) The importance of dextrose in the medium of tissue cultures. J Exp Med 35:317–322. https://doi.org/10.1084/jem.35.3.317

Lewis WH, Lewis MR (1912) The cultivation of chick tissues in media of known chemical constitution. Anat Rec 6:207–211. https://doi.org/10.1002/ar.1090060503

Li F, Vijayasankaran N, Shen A et al (2010a) Cell culture processes for monoclonal antibody production. MAbs 2:466–479. https://doi.org/10.4161/mabs.2.5.12720

Li J, Browning S, Mahal SP et al (2010b) Darwinian evolution of prions in cell culture. Science 327:869–872. https://doi.org/10.1126/science.1183218

Li B, Wang X, Wang Y et al (2015) Past, present, and future of microcarrier-based tissue engineering. J Orthop Transl 3:51–57. https://doi.org/10.1016/j.jot.2015.02.003

Lobo-Alfonso J, Price P, Jayme D (2008) Benefits and limitations of protein hydrolysates as components of serum-free media for animal cell culture applications. In: Pasupuleti VK, Demain AL (eds) Protein hydrolysates in biotechnology. Springer, Dordrecht, The Netherlands, pp 55–78

Lovitt CJ, Shelper TB, Avery VM (2014) Advanced cell culture techniques for cancer drug discovery. Biology (Basel) 3:345–367. https://doi.org/10.3390/biology3020345

McKee C, Chaudhry GR (2017) Advances and challenges in stem cell culture. Colloids Surf B Biointerfaces 159:62–77. https://doi.org/10.1016/j.colsurfb.2017.07.051

Menshutina N, Guseva E, Batyrgazieva D, Mitrofanov I (2021) Information system for selection of conditions and equipment for mammalian cell cultivation. Data 6:23. https://doi.org/10.3390/data6030023

Merten O-W (2015) Advances in cell culture: anchorage dependence. Philos Trans R Soc B 370:20140040. https://doi.org/10.1098/rstb.2014.0040

Mirabelli P, Coppola L, Salvatore M (2019) Cancer cell lines are useful model systems for medical research. Cancers (Basel) 11:1098. https://doi.org/10.3390/cancers11081098

Moore GE, Ito E, Ulrich K, Sandberg AA (1966) Culture of human leukemia cells. Cancer 19:713–723. https://doi.org/10.1002/1097-0142(196605)19:5<713::AID-CNCR2820190518>3.0.CO;2-Y

Morgan JF, Morton HJ, Parker RC (1950) Nutrition of animal cells in tissue culture. I. Initial studies on a synthetic medium. Exp Biol Med 73:1–8. https://doi.org/10.3181/00379727-73-17557

Nascimento IP, Leite LCC (2012) Recombinant vaccines and the development of new vaccine strategies. Braz J Med Biol Res 45:1102–1111. https://doi.org/10.1590/S0100-879X2012007500142

National Research Council (US) Committee (1999) Monoclonal antibody production. National Academies Press, Washington, DC

Nature (2023) The gene-therapy revolution risks stalling if we don't talk about drug pricing. Nature 616:629–630. https://doi.org/10.1038/d41586-023-01389-z

Nikolic M, Sustersic T, Filipovic N (2018) In vitro models and on-chip systems: biomaterial interaction studies with tissues generated using lung epithelial and liver metabolic cell lines. Front Bioeng Biotechnol 6:120. https://doi.org/10.3389/fbioe.2018.00120

O'Flaherty R, Bergin A, Flampouri E et al (2020) Mammalian cell culture for production of recombinant proteins: a review of the critical steps in their biomanufacturing. Biotechnol Adv 43:107552. https://doi.org/10.1016/j.biotechadv.2020.107552

Oliveira MLS, Brochado SM, Sogayar MC (1999) Mechanisms of cell transformation induced by polyomavirus. Braz J Med Biol Res 32:861–865. https://doi.org/10.1590/S0100-879X1999000700010

Oliveri V (2022) Selective targeting of cancer cells by copper ionophores: an overview. Front Mol Biosci 9

Omasa T, Onitsuka M, Kim W-D (2010) Cell engineering and cultivation of Chinese hamster ovary (CHO) cells. Curr Pharm Biotechnol 11:233–240. https://doi.org/10.2174/138920110791111960

Pinnamaneni R, Vemuri PK, Rao KRSS (2019) Animal cell culture technology: principles and applications. Scholar's Press

Quinteros DA, Bermúdez JM, Ravetti S et al (2017) Therapeutic use of monoclonal antibodies: general aspects and challenges for drug delivery. Nanostruct Drug Deliv 2017:807–833. https://doi.org/10.1016/B978-0-323-46143-6.00025-7

Rodrigues AF, Soares HR, Guerreiro MR et al (2015) Viral vaccines and their manufacturing cell substrates: new trends and designs in modern vaccinology. Biotechnol J 10:1329–1344. https://doi.org/10.1002/biot.201400387
Rodríguez-Hernández CO, Torres-García SE, Olvera-Sandoval C et al (2014) Cell culture: history, development and prospects. Int J Curr Res Acad Rev 2:188–200
Scheller EL, Krebsbach PH (2009) Gene therapy. J Dent Res 88:585–596. https://doi.org/10.1177/0022034509337480
Segeritz C-P, Vallier L (2017) Cell culture. In: Basic science methods for clinical researchers. Academic Press, London, pp 151–172. https://doi.org/10.1016/B978-0-12-803077-6.00009-6
Shupe J, Zhang A, Odenwelder DC, Dobrowsky T (2022) Gene therapy: challenges in cell culture scale-up. Curr Opin Biotechnol 75:102721. https://doi.org/10.1016/j.copbio.2022.102721
Soccol CR, Karp SG, Letti LAJ et al (2022) Is the development of low-cost media one of the greatest challenges to produce cultivated meat on an industrial scale? Biotechnol Res Innov 6:e2022005. https://doi.org/10.4322/biori.20226201
Stanners CP, Eliceiri GL, Green H (1971) Two types of ribosomes in Mouse–Hamster hybrid cells. Nat New Biol 230:52–54. https://doi.org/10.1038/newbio230052a0
Stoker M, Macpherson I (1964) Syrian hamster fibroblast cell line BHK21 and its derivatives. Nature 203:1355–1357. https://doi.org/10.1038/2031355a0
Subbiahanadar Chelladurai K, Selvan Christyraj JD, Rajagopalan K et al (2021) Alternative to FBS in animal cell culture - an overview and future perspective. Heliyon 7:e07686. https://doi.org/10.1016/j.heliyon.2021.e07686
Sureau P (1987) Rabies vaccine production in animal cell cultures. In: Vertebrate cell culture I. Springer, Berlin, pp 111–128
Van Wezel AL (1967) Growth of cell-strains and primary cells on micro-carriers in homogeneous culture. Nature 216:64–65. https://doi.org/10.1038/216064a0
Vazin T, Freed WJ (2010) Human embryonic stem cells: derivation, culture, and differentiation: a review. Restor Neurol Neurosci 28:589–603. https://doi.org/10.3233/RNN-2010-0543
Verma A, Verma M, Singh A (2020) Animal tissue culture principles and applications. Anim Biotechnol 2020:269–293. https://doi.org/10.1016/B978-0-12-811710-1.00012-4
Vogelaar JPM, Erlichman E (1933) A feeding solution for cultures of human fibroblasts. Am J Cancer 18:28–38. https://doi.org/10.1158/ajc.1933.28
Walsh G, Walsh E (2022) Biopharmaceutical benchmarks 2022. Nat Biotechnol 40:1722–1760. https://doi.org/10.1038/s41587-022-01582-x
Wang Y, Ouyang F (1999) Bead-to-bead transfer of Vero cells in microcarrier culture. Cytotechnology 31:221–224. https://doi.org/10.1023/A:1008079013375
Wessel GM (2011) The sub-culture of cell culture. Mol Reprod Dev 78(2):Fm i. https://doi.org/10.1002/mrd.21292
WHO (2023) Coronavirus (COVID-19) dashboard. https://covid19.who.int. Accessed 25 Oct 2023
Witkowski JA (1980) Dr. Carrel's immortal cells. Med Hist 24:129–142
Xu N, Ma C, Ou J et al (2017) Comparative proteomic analysis of three Chinese hamster ovary (CHO) host cells. Biochem Eng J 124:122–129. https://doi.org/10.1016/j.bej.2017.05.007
Yao T, Asayama Y (2017) Animal-cell culture media: history, characteristics, and current issues. Reprod Med Biol 16:99–117. https://doi.org/10.1002/rmb2.12024
Ylostalo JH (2020) 3D stem cell culture. Cell 9:2178. https://doi.org/10.3390/cells9102178
Zhang L, Ma X-J-N, Fei Y-Y et al (2022) Stem cell therapy in liver regeneration: focus on mesenchymal stem cells and induced pluripotent stem cells. Pharmacol Ther 232:108004. https://doi.org/10.1016/j.pharmthera.2021.108004
Zurlo J, Rudacille D, Goldberg AM (1994) Animals and alternatives in testing: history, science, and ethics. Mary Ann Liebert, New York. https://repository.library.georgetown.edu/handle/10822/869663. Accessed 27 Oct 2023

# Cell Lines for Cultivated Meat Production

**Vivian Feddern, Ana Paula Almeida Bastos, Vanessa Gressler, Diana M. C. Marques, Frederico C. Ferreira, Carlos André Vitorino Rodrigues, Marcus Vinicius Telles Teixeira, and Camila Luna da Silva**

**Abstract** Cellular agriculture is an emerging field that aims to provide a more sustainable alternative for the production of animal-derived products, without the need to sacrifice animals. Cultivated meat is one of the main products of cell agriculture and consists in the production of meat from animal cell cultures in a growth medium. A cell line with reproducible and consistent properties is essential for any bioprocess and is crucial for the production of cultivated meat. Developing cell lines from farm animal sources with the necessary proliferative capacity and differentiation potential is a critical technical component that must be optimized to enable scale-up and commercialization of cultivated meat. There are two major strategies to establish these cell lines: (1) using a sample of the tissue of interest (primary cell sources), with further isolation of the muscle-resident progenitor cells; and (2) by the use of pluripotent cell sources, like embryonic stem cells (ESCs) and induced pluripotent stem cells (iPSCs), which have the capability to differentiate toward muscle-resident progenitor cells. As muscle is composed of different tissue types (nerves, blood ves-

V. Feddern (✉)
Embrapa Suínos e Aves [Embrapa Swine and Poultry], Concórdia, Santa Catarina, Brazil

Embrapa Clima Temperado [Embrapa Temperate Agriculture], Pelotas, Rio Grande do Sul, Brazil
e-mail: vivian.feddern@embrapa.br

A. P. A. Bastos · V. Gressler
Embrapa Suínos e Aves [Embrapa Swine and Poultry], Concórdia, Santa Catarina, Brazil
e-mail: ana.bastos@embrapa.br; vanessa.gressler@embrapa.br

D. M. C. Marques · F. C. Ferreira · C. A. V. Rodrigues
Department of Bioengineering and Institute for Bioengineering and Biosciences, Instituto Superior Técnico, Universidade de Lisboa, Lisbon, Portugal

Associate Laboratory i4HB—Institute for Health and Bioeconomy, Instituto Superior Técnico, Universidade de Lisboa, Lisbon, Portugal
e-mail: diana.c.marques@tecnico.ulisboa.pt; frederico.ferreira@tecnico.ulisboa.pt; carlos.rodrigues@tecnico.ulisboa.pt

M. V. T. Teixeira · C. L. da Silva
Sustineri Piscis, Campus Tecnológico do Inmetro, Duque de Caxias, Rio de Janeiro, Brazil


C. R. Soccol et al. (eds.), *Cultivated Meat*,
https://doi.org/10.1007/978-3-031-55968-6_3

sels, adipose and connective tissue), the production of cultured meat requires the use of progenitor cells with the capability to differentiate into myocytes, adipocytes, fibroblasts, chondrocytes, and hematopoietic cells. This chapter will cover potential strategies for cell line development and their advantages and disadvantages, regarding the isolation protocol of cultivated fish, chicken and mammalian cells. In addition, sampling techniques, cell line characterization, conservation and quality control, as well as challenges and future prospects in the field will be addressed.

**Keywords** Cultivated fish · Cultivated chicken · Cultivated mammalian cells · Muscle cells · Satellite cells

## 1 Introduction

Cell lines from farm animal sources are critical components to enable scale-up and commercialization of cultivated meat. However, the lack of access to cell lines is a major barrier to cultivated meat research.

The main cellular components present in meat are skeletal myocytes, with adipocytes, fibroblasts, chondrocytes, and hematopoietic cell types (Reiss et al. 2021). The culture conditions to produce cultivated meat in an efficient way must be identified, so as to optimize proliferation and differentiation of muscle cells into myotubes and myofibers while maintaining a similar texture and taste as real meat (Kim et al. 2023b).

The creation of food using animal cells encompasses a diverse range of cell types that initiate the production process, resulting in the development of cellular products such as proteins, lipids, and tissues derived from complete animal cells sourced from many species including poultry, cattle, pigs, fish, shellfish, and game animals. The production procedures for each cell-based food product exhibit significant variability, contingent upon the exact cell line and species employed. Hence, the primary emphasis of this chapter is placed on the procedures that are prevalent for cell line selection, sourcing, isolation, and preparation processes in most animal cell-based food production chains.

## 2 Obtaining Tissue Samples for Cultivated Meat Production

In the context of cultured meat production, it is imperative that the initial cell types possess the capacity for self-renewal to attain sufficient quantities. Additionally, these cells must possess the capability to differentiate into the fully developed cell types that comprise meat (Bentzinger et al. 2012). In order to fulfill these specified criteria, stem cells emerge as the most viable contenders for utilization as an initial cell source. There are two primary categories of stem cells that possess the requisite capacity for proliferation and differentiation in the context of cultured meat production: multipotent stem cells and pluripotent stem cells (Roberts et al. 2015).

As an emerging technology, the infrastructure for the production of cultured meat is currently being constructed, and the production process still has room for innovation and cost-effectiveness enhancement. The procedure to produce cultivated meat consists basically of the following steps: (1) initial phase involving selection, isolation/acquisition of cells, laboratory-based cell proliferation with appropriate stimulation and nourishment, and subsequent freezing or storage of cells; (2) large-scale proliferation and expansion of desired cells, along with their storage; (3) induction of differentiation or development of muscle/adipose tissue to achieve maturation; (4) dissociation and collection of the desired cells; (5) production of the desired end product, which involves processing into final food items, either in an unstructured form (e.g., hamburgers, sausages, nuggets) or as "whole cut" steaks (Fig. 1). The completion of each step in the process may vary depending on factors such as the kind of commodity, the desired final product, and the utilization of biomaterials, potentially resulting in the inclusion of distinct substeps within each stage.

The initial step in producing cell-based meals involves meticulous selection of preferred cell sources, which can span a variety of species, including cattle, chicken, pork, fish, and shellfish. In addition, certain types of cells that are essential for the formation of the final product are also chosen, such as undifferentiated stem cells, muscle precursor cells, fibroblasts, or adipose tissue. Initially, the acquisition of cellular material can be performed by excision of tissue from living or post-slaughtered animals, allowing the acquisition of minute tissue samples that have the potential to

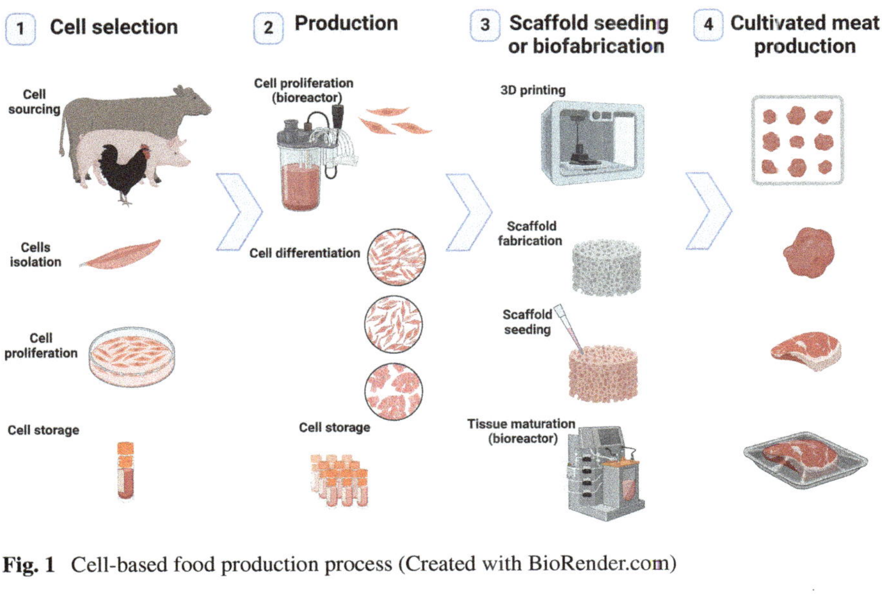


**Fig. 1** Cell-based food production process (Created with BioRender.com)

generate adult stem cells. Importantly, biopsies should only be performed on sound animals. Additionally, one can harvest pluripotent cells as ESCs or cells to reprogram into iPSCs. In this manner, the desired cell type is isolated or reprogrammed in the laboratory under controlled conditions, also known as in vitro cultivation.

Isolation of primary cells is the initial step toward efficient cultured meat production, where crucial parameters include the viability of the obtained cells and the cost associated with the method (Guan et al. 2022). In all methods, it has been observed that the origin, sex, and especially the age of the animal are critical for the successful acquisition of proliferative cells that can be scaled up for industrial applications (Choi et al. 2021). Tissues obtained from biopsies are either explanted (a technique in which a sample adheres to a plate, promoting cell migration to the culture surface) or further processed via mechanical and enzymatic processes that liberate the cells. However, some digestion of cell surface antigens may occur. As with all cell isolation techniques, there is still the possibility of cross-contamination. Therefore, the initial stage involves obtaining the tissue sample, followed by the isolation of the specific tissue of interest by dissection and/or disaggregation, in order to subsequently seed the cells in a culture medium. Following the isolation of tissue, the subsequent culture that will be generated is classified as a primary cell culture. Primary cell culture involves the extraction of cells from tissue fragments, which are then either attached to a suitable substrate or physically or enzymatically disrupted to generate a cell suspension. Subsequently, a portion of these cells will finally adhere to the substrate. It is imperative to aseptically extract the tissue and expeditiously transport it to the tissue culture facility. It is advised to refrain from doing dissections or biopsies on animals within the tissue culture laboratory due to the potential presence of microbial contamination in animals. Nevertheless, the manipulation of embryonic or fetal samples, such as chicken eggs, necessitates the utilization of precise, specific techniques.

Complementary approaches are required for further purification and extraction of the target cell from these initial extracts. Hence, numerous techniques have been devised for the purpose of cell purification, each with its own set of merits and drawbacks that need careful evaluation in relation to the specific intended use. Current methodologies for cell separation employ surface proteins, selective plating, differential adhesion, detachment of cells (Choi et al. 2021), genetic expression, and principles of physics to effectively isolate certain cells based on their phenotypic characteristics, like cell capture using magnetic beads with cell-specific antibodies, cell sorting (FACS) (Post et al. 2020; Rubio et al. 2019), and Percoll density gradient centrifugation. The effectiveness of cell purification is influenced by the availability of cellular reagents and the specific application, since these variables might limit the utilization of particular mechanisms involved in cell labeling and sorting processes. Therefore, it may be necessary to develop alternative precoating methods that can be mass-produced (Guan et al. 2022). Depending on the desired cell categories, specific cell isolation techniques can be utilized for adipose tissue-derived stem cells (Lu et al. 2014; Sampaio et al. 2015), mesenchymal stem cells (Vassiliev and Nottle 2013; Feyen et al. 2016), or fibroblasts (Park et al. 2022).

## *2.1 Cell Sources for Cultured Fish Production*

In fish, the embryonic skeletal development generally commences with myoblasts in the somatic mesoderm, resulting in the formation of two distinct populations of muscle cells: a superficial layer giving rise to slow-twitch muscle fibers and another contributing to fast-twitch muscle tissue (Sänger and Stoiber 2001). This process forms the most abundant tissue in fish, myotomal muscle, which makes up 50%–70% of body mass in most species. The myotomal muscle consists of these two types of fibers located in specific parts of the fish's body, unlike other vertebrates that have a mixture of both fiber types within each muscle. The superficial red muscle layers on both sides of the body consist of slow-twitch fibers, containing a high density of mitochondria and a network of blood capillaries used during low-speed swimming. The white muscle layers, constituting over 90% of the myotomal muscle, are composed of fast-contracting muscle fibers dependent on anaerobic metabolic pathways (Sänger and Stoiber 2001; Funkenstein et al. 2006).

The existence of muscle satellite cells (MuSC) was initially reported by Alexander Mauro in the 1960s (Mauro 1961). MuSCs, a type of adult stem cells of mesodermal origin, reside in muscle tissue and play a vital role in organism development while promoting muscle tissue regeneration and hypertrophy. These cells are positioned strategically at the periphery of myofibrils, between the basal lamina and the sarcolemma. In adult individuals, this cell type remains quiescent until the need for muscle regeneration arises after an injury. In this context, the cells exit from their quiescent state and proliferate, generating cells committed to differentiation into mononucleated proliferative myoblasts. These descendant myoblasts then multiply and fuse, forming new myofibrils and myotubes, which bundle together to form muscle fibers (Yin et al. 2013; Knežić et al. 2022).

In fish, the selection of the biopsy site in juvenile animals is critical and generally involves taking a sample from the lateral white epaxial muscle region of the animal after sample cleaning and asepsis (Powell et al. 1989; Koumans et al. 1990; Saad et al. 2023). Once a tissue sample is obtained under aseptic conditions, the process begins by cutting the tissue into 2–4 $mm^3$ pieces. From here, there are two options: obtaining cells through explants or enzymatic digestion of the sample. Explant methods are the simplest and can be applied by adhering the diced pieces to the culture surface on serum or collagen-coated film, followed by coverage with basal medium (Oestbye and Ytteborg 2019; Kong et al. 2021). However, the yield of cells at the end of the process tends to be inefficient or requires significant time to establish cultures. The use of proteolytic enzymes is the primary method used to extract primary cells. This method is applicable to large volumes of muscle tissue (Choi et al. 2021). In general, specific enzymes are used to dissociate collagen macromolecules (collagenase types I, II, III, and IV) or elements of the basal lamina and extracellular matrix (ECM), such as dispases and pronases, sometimes in combination with nonspecific digestive enzymes (e.g., trypsin).

Typically, the enzymatic digestion involves cell dilution in culture medium, filtration through 20–40 μm pore nylon filters and debris removal by using a low-speed centrifuge. The final step is to culture the cells by plating them in a basal medium supplemented with antibiotics. Both initial isolation methods may result in cultures with a low degree of purity of the isolated cells. In such cases, purification methods are required in subsequent steps (Guan et al. 2022). Examples of this type of isolation include cells from the rainbow trout (*Salmo gardneri*), which were efficiently isolated by Powell et al. in the early 1990s (Powell et al. 1989), as well as from the common carp (*Cyprinus carpio*) by Koumans et al. (1990), and from other commercially important species such as the Atlantic salmon (*Salmo salar*) (Oestbye and Ytteborg 2019) and mackerel (*Scomber scombrus*) (Saad et al. 2023). For the purification of primary cultures' enzymatic digestion, several methods have been proposed: selective precoatings, differential adhesion, centrifugation, density gradient, and selection of magnetically activated cells or fluorescence-activated cell sorting. All of these methods need to be evaluated in terms of their advantages and associated costs (Guan et al. 2022).

The selection of the biopsy site will define the profile of the isolated populations. From other biopsy sites, it is possible to obtain mesenchymal stem cells (MSCs), pre-adipocytes, adipocytes, and/or fibroblasts as sources for cultured meat production. MSCs are adult stem cells with high proliferative capacity and self-renewal and with multipotent properties to differentiate into tissues of mesenchymal origin (Bomkamp et al. 2023). Adult MSCs reside in a variety of adult tissues; in fish, they have been derived from sources such as the visceral adipose tissue of rainbow trout (Bou et al. 2017), gilthead bream (Salmerón et al. 2016), Atlantic salmon (Ytteborg et al. 2015), the vertebral column of gilthead bream (Salmerón et al. 2016), and the heart of zebrafish (Fathi et al. 2019). Despite some literature on the subject, few studies have focused on cell procurement for bioprocesses in cultured meat production. Unlike other target species such as cattle, pigs, and poultry, the scientific literature lacks information on cell acquisition, differentiation mechanisms, metabolism, and protocols for forming complex tissues from fish cells (Bomkamp et al. 2023). Table 1 provides examples of different primary cell isolation protocols.

The expression of myogenic factors and biomarkers in muscle cells is highly conserved across species and is shared by humans, mice, pigs, birds, cattle, and zebrafish. The primary biomarker is the paired box transcription factor (PAX7), which is essential for maintaining stemness and myogenic function. In terms of cell surface markers, there are a series of membrane proteins including tyrosine kinase receptor C-Met, chemokine receptor CXCR4, adhesion proteins M-cadherin and β1-integrin (CD29), and neural cell adhesion molecule (NCAM) (CD56). These have been identified predominantly in mouse and human muscle cells, but none is expressed exclusively by these cells. The combination of surface markers $CD31^-CD45^-CD56^+CD29^+$ has been successfully applied to characterize these cells in humans, cattle, and pigs but has not yet been proven for other species such as birds and fish and requires further investigation (Guan et al. 2022). Mesenchymal cells in humans, mice, and other species typically have a positive profile for the surface marker combination $CD105^+CD73^+CD90^+$ and are negative for $CD45^-CD31^-CD34^-$. Preadipocytes can be identified by the expression of mesenchymal markers such as $SCA1^+CD34^+CD29^+$ but are negative for $CD45^-CD31^-$ in

**Table 1** Isolation of primary cells in fish

| Source tissue | Type of cells | Isolation protocol | Species [References] |
|---|---|---|---|
| Lateral white epaxial muscle (above gills) | Myoblasts | 0.2% collagenase—90 min at 10 °C–13 °C | Atlantic salmon (Oestbye and Ytteborg 2019) |
| | | 0.1% collagenase type II—1 h at room temperature (RT) | Mackerel (Saad et al. 2023) |
| | | 0.2% collagenase type IV—1 h at 18 °C | Zebrafish (Froehlich et al. 2014) |
| | | 0.1% pronase + protease type XIV | Rainbow trout (Powell et al. 1989) |
| Visceral adipose tissue (white) | MSCs | 0.2% collagenase type II (130 U/mL)—1 h at 18 °C | Rainbow trout (Bou et al. 2017) |
| | | 0.125% collagenase—1 h at 11 °C | Atlantic salmon (Oestbye and Ytteborg 2019) |
| | Pre-adipocytes | 0.2% collagenase—1 h at 13 °C | Atlantic salmon (Vegusdal et al. 2003) |
| Vertebral column | MSCs | 0.125% collagenase—30 min and 90 min at 18 °C | Gilthead bream (Salmerón et al. 2016) |

mice. However, this has not been shown in a similar manner for fish species. In addition, the gene expression profile for preadipocytes varies little among vertebrates, with common expression of peroxisome proliferator-activated receptor gamma (pparγ), CCAAT/enhancer-binding protein (c/ebp), specifically c/ebpα and c/ebpβ, transgelin, and fatty acid synthase (fas) (Vegusdal et al. 2003; Bomkamp et al. 2023).

Finally, another type of primary cells used in cultured meat production are fibro-adipogenic progenitor cells (FAPs), which can be isolated from the stromal vascular fraction (SVF) obtained by digesting muscle tissue. These cells reside in the interstitial spaces of muscle and contribute to muscle development. In murine species, these cells express specific surface antigens such as CD140a, vimentin, and other factors. However, to date, FAPs have not been reported in fish specimens (Bomkamp et al. 2023).

## *2.2 Cell Sources for Cultured Chicken Production*

Tissue sampling in chicken is scarce in the literature; therefore, the main existent protocols refer to mammalian cells. The initial cell source can be retrieved from adult cells or pluripotent stem cells (Reiss et al. 2021). The latter comprise ESC and iPSC. Figure 2 shows adult stem cell types and their abilities.

The first chicken stem cells were from blastoderm of fertilized eggs at embryonic stage X and were called chicken embryonic stem cells (cESCs). From the first

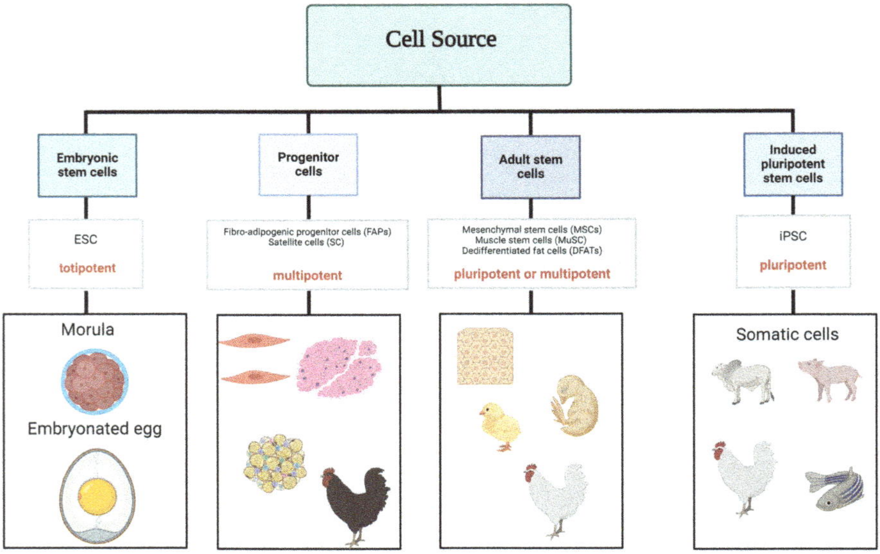


**Fig. 2** Cell types and their abilities (Created with BioRender.com)

cESCs, rapid progress was made on derivation and culture of chicken stem cells from various developmental stages (Farzaneh et al. 2017). The second population of in vitro chicken stem cells, named embryonic germ cells (EGCs), was from primordial germ cells (PGCs), while the third are called chicken-induced pluripotent stem cells (ciPSCs) being generated by in vitro reprogramming of fibroblast cells (8–12-day-old embryos) on a feeder layer. Chicken stem cells with self-renewal ability can be propagated in culture indefinitely and have the ability to differentiate into multiple cell types representing all primitive embryonic germ layers (Lavial and Pain 2010; Farzaneh et al. 2017).

MSCs can be retrieved from the bone marrow, lung, Wharton's jelly, and liver (Adhikari et al. 2019; Liu et al. 2020). Attention must be paid in order to remove blood cells while isolating MSC. MuSCs are resident cells and behave as the nuclear material source for adult muscle fiber growth (Baquero-Perez et al. 2012; He et al. 2020; Barry 2021). FAPs act as mediators of pro-myogenic signals in muscle homeostasis by excreting ECM, cytokines, ligans, and immune factors (Reggio et al. 2020).

ESCs are derived from the inner cell mass of an early embryonic structure (blastocyst), formed a few days postfertilization (Ravikumar and Powell 2023). In avian eggs, these cells can be recovered by using a paper, hair ring, or spoon or extracted from the whole embryo. In the last situation, the incubation time is crucial, once after 5 days' incubation, the embryo can be easily manipulated because it assumes a C shape (Xiong et al. 2020).

Table 2 shows some isolation protocols available in the literature regarding source tissue and type of cells found in chicken.

**Table 2** Isolation of primary cells in chicken

| Source tissue | Type of cells | Isolation protocol | References |
|---|---|---|---|
| Pectoralis muscle | MuSC | 0.2% collagenase—45 min at 37 °C | Yablonka-Reuveni et al. (1987) |
| Hindlimb muscle | MuSC | 0.1% collagenase type I—1 h at 37 °C | Kim et al. (2023b) |
| Adipose tissue | Adipose-derived progenitor cells (ADPCs) | 0.1% collagenase type I—1 h at 37 °C<br>0.2% collagenase type I—40 min at 37 °C | Gong et al. (2011)<br>Lu et al. (2014) |
| Femur bones | MSCs | Density gradient centrifugation over Ficoll–Hypaque (gradient density 1.090) | Khatri et al. (2009) |
| Bone marrow | MSCs | Two methods:<br>**(1)** Total blood adherent method<br>**(2)** density gradient centrifugation by Percoll | Bai et al. (2013) |
| Chicken embryo | ESC | Blastodermal cells taken from stage X (EG&K) embryos | Lavial and Pain (2010) |

## *2.3 Cell Sources for Cultured Mammalian Meat Production*

Mammalian cells were the first to be isolated and most protocols existent today refer to these cells. Table 3 shows some isolation protocols with respect to cell types found in some mammalian species, such as ovine, bovine, and porcine.

Pluripotent stem cells were effectively developed in many species, though the establishment of bovine-iPSC is challenging (Su et al. 2021). Methods of MuSC isolation from diverse animal species are very inconsistent, making comparisons among species hard (Baquero-Perez et al. 2012). For instance, from adipose tissue, MSC may be isolated after enzymatic reaction using collagenase (Hill et al. 2019), which is able to cleave collagen linkages. Other types of enzymes may also be used to digest muscle tissues, such as proteases, trypsin, and pronase, which may alter the ECM (Lee et al. 2021). Parameters such as enzyme load, temperature, time of incubation, among others, need to be monitored at the isolation step (Hill et al. 2019). Therefore, specific technical skills are demanded to perform isolation protocols with success.

Another protocol that might be applied to mammalian species is gradient centrifugation, for instance, to isolate pig MuSCs. Choi et al. (2021) observed that ~95% of the cells expressed the myoblast marker desmin. However, this method needs to be improved for other species, because it varies in efficiency and it has low specificity (Choi et al. 2021).

Fadel et al. (2011) evaluated seven protocols to isolate sheep cells, being three from umbilical cord blood stem cells and four regarding adipose tissue. However, only one protocol for each tissue succeeded. The authors concluded that those isolation protocols were complex and needed more assessments.

**Table 3** Isolation of primary mammalian cells

| Species | Source tissue | Type of cells | Isolation protocol | References |
|---|---|---|---|---|
| Bovine | Fetal skeletal muscle | Progenitor cells | 0.1% collagenase IV—1 h at 37 °C | Guan et al. (2017) |
| | Muscle | Satellite cells | Pronase (1 mg/mL)—50 min at 37 °C<br>0.2% collagenase II—45 min at 37 °C | Lee et al. (2007)<br>Stout et al. (2022)<br>Kim et al. (2023a) |
| | | Fibro-adipogenic progenitors (FAPs) | Collagenase D (0.75 U/ml) + 1.0 U/ml Dispase II—30 min at 37 °C<br>0.1% $_{(w/v)}$ collagenase in 0.25% trypsin–EDTA+collagenase II (1 mg/mL)—40 min at 37 °C | Ma et al. (2018)<br>Lee et al. (2023)<br>Huang et al. (2012) |
| | Mammary tissue | Primary epithelial cell | Trypsin–EDTA commercial solution—1 h at 37 °C | Jedrzejczak and Szatkowska (2014) |
| | Bovine fetal fibroblasts (BFF) | iPSC | Trypsin | Talluri et al. (2015) |
| Ovine | Umbilical cord blood and peripheral blood | MSCs | Ficoll–Hypaque ™ plus | Fadel et al. (2011)<br>Lyahyai et al. (2012) |
| | Muscle | Satellite cells | Pronase (1 mg/mL)—50 min at 37 °C | Dodson et al. (1987) |
| | Inner cell mass | Embryonic stem cells | – | Vilarino et al. (2020) |
| | Embryonic fibroblasts | iPSC | – | Liu et al. (2012) |
| Porcine | Kidney | Primary cell | 0.1% collagenase–dispase/PBS solution + DNase I (40 mg/ml)—1 h at 37 °C | Takenouchi et al. (2014) |
| | Tendon | Multipotent stem cells | Collagenase I (3 mg/ml)—1.5 h at 37 °C | Yang et al. (2018) |
| | Muscle | Satellite cells<br>FAPs | Pronase (0.8 mg/mL)—40 min at 37 °C<br>0.1% collagenase I—2 h at 37 °C | Doumit and Merkel (1992)<br>Han et al. (2017) |
| | Bone marrow | MSC | 0.2% trypsin +0.2% collagenase– 30 min at 37 °C | Bosch et al. (2006) |
| | Inner cell mass | Embryonic stem cells | 0.25% trypsin +0.2% EDTA—30 min at RT | Li et al. (2003) |
| | Embryonic fibroblasts | iPSC | 0.1% collagenase IV—5 h at 37 °C | Gao et al. (2014) |
| | Gastric epithelium | Primary cells | 200 U/mL collagenase I + 1.2 U/mL dispase II | Bautista-Amorocho et al. (2021) |

## 3 Development of Cell Lines for Cultivated Meat Production

The choice of cell line is extremely important for the development and success in cultured meat production. Cell lines with high proliferative and differentiation potential, low doubling time, genetic stability, and the ability to grow in suspension are crucial for the bioprocess (Powell et al. 1989; Specht et al. 2018; Hanga et al. 2020; Kumar et al. 2023). Cellular sources can be obtained through samples taken directly from animal tissue (primary cells) or through continuous or immortalized cell lines, which were derived from animal tissue and have acquired the ability to reproduce indefinitely (Freshney 2005; Capes-Davis et al. 2010; Specht et al. 2018; Kumar et al. 2023).

For cultured meat production, cells need to self-renew to reach high quantities, as well as differentiate into the cell types that compose meat, such as myocytes, adipocytes, fibroblasts, chondrocytes, and hematopoietic cells. Due to these requirements, stem cells are the primary source of initial cells (Reiss et al. 2021; David et al. 2023). Stem cells can be classified according to their plasticity into pluripotent stem cells and multipotent stem cells, which are present in mature tissues and conventionally referred to as adult stem cells due to their location (Sobhani et al. 2017; Thong et al. 2019).

Pluripotent stem cells include ESCs and iPSCs. Pluripotent stem cells have the capacity to undergo differentiation into all cellular lineages seen in meat, with their notable attribute of exhibiting unlimited proliferative capacity (Sobhani et al. 2017; David et al. 2023). Adult stem cells are undifferentiated progenitor cells with the function of maintaining tissue homeostasis through the repair and regeneration of damage. Undifferentiated progenitor cells represent an additional reservoir of cells that offer significant benefits in their use. Adult stem cells are characterized by their multipotency, which refers to their capacity to undergo differentiation into a limited range of cell types that are restricted to their originating germ layer (Fig. 2). This is why they are the most promising for cultured meat production, even though they have limited proliferative capacity (Reiss et al. 2021; David et al. 2023). Certain categories of adult stem cells are particularly suitable for cultured meat. Within the muscle tissue milieu, there are stem/progenitor cells feasible for cultured meat production, which are MuSCs, MSCs, FAPs, and dedifferentiated fat cells (DFATs) (Zhang et al. 2020; Reiss et al. 2021; David et al. 2023). The aforementioned progenitor cells have the capacity to undergo differentiation into many significant types of fully developed cells, including skeletal myocytes, adipocytes, chondrocytes, and fibroblasts.

Pluripotent stem cells have a remarkable capacity for self-renewal and differentiation into any type of somatic cell. However, to generate muscle-resident progenitor cells, these pluripotent stem cells must undergo differentiation into the mesodermal cell lineage. ESCs are obtained by isolating cells from the inner cell mass of the blastocyst during the embryonic stage, which requires specialized isolation procedures. These cells have the capability to generate all the cells that make up the three germ layers of an organism (mesoderm, endoderm, ectoderm). As they can differentiate into any cell type in the body, their culture demands the use of specialized culture media to inhibit unwanted differentiation and controlled conditions to maintain their desired tissue identities, which makes the cultivation expensive. Additionally, they exhibit significant genetic instability, and their use raises ethical concerns due to the need for embryo destruction (Eibl et al. 2021; Knežić et al. 2022; Kumar et al. 2023). In addition, the acquisition of ESCs is frequently challenging owing to the limited duration of the blastocyst's viability. Cell lines have been established from chicken (Lavial and Pain 2010; David et al. 2023), fish (Bomkamp et al. 2023; David et al. 2023), bovine (Bogliotti et al. 2018; David et al. 2023), sheep (Vilarino et al. 2020; Reiss et al. 2021), and pig (Li et al. 2003; Reiss et al. 2021). The cellular markers consist of Oct4, Sox2, Nanog, c-Myc, and Klf4 (Reiss et al. 2021).

iPSCs are acquired by reprogramming somatic cells using viral vectors, episomes, mRNA, or proteins, as well as through other methods such as the expression of a set of species-specific transcription factor genes (Knežić et al. 2022; Kumar et al. 2023; David et al. 2023). The major advantage of using these cells lies in their unlimited proliferative capacity and their ability to grow in suspension, forming cellular aggregates (Reiss et al. 2021). They possess epigenetic memory, characterized by reversible and heritable changes in the functional genome that do not alter the DNA nucleotide sequence (Li et al. 2003). This characteristic can be a disadvantage if the initial cells have an origin distant from the target differentiated cells. To address this issue, it is ideal that the initial cells have a close origin to the target cells (Thong et al. 2019). Cellular reprogramming can induce the emergence of phenotypic alterations and may be partial in some cases, leading to a low programming efficiency. Except for transient transfection methods (mRNA, proteins), techniques involving genome editing produce iPSCs that are regarded as genetically modified, and the resultant products must be labeled as genetically modified, triggering a distinct regulatory process (Knežić et al. 2022; David et al. 2023). The cell lines were derived from chicken (Rosselló et al. 2013; Kim et al. 2017; Reiss et al. 2021), bovine (Talluri et al. 2015; Reiss et al. 2021), pig (Gao et al. 2014; Reiss et al. 2021), and fish (Knežić et al. 2022; Bomkamp et al. 2023). They share the same pluripotency cell markers as ESCs, which are Oct4, Sox2, Nanog, c-Myc, and Klf4 (Reiss et al. 2021).

The primary and most obvious source of cells for producing cultivated meat is the muscle tissue's progenitor cells, also referred to as MuSCs. These cells are located between the basal lamina and the sarcolemma of muscle fibers. They are easily isolated from small samples of muscle tissue and require minimal efforts for differentiation into myoblasts, which give rise to myocytes, myotubes, and myofibers (Wosczyna et al. 2019; Seah et al. 2022; David et al. 2023). Although there are species- and phylum-specific variations, the mechanisms and factors associated with muscular development are highly conserved among vertebrates (Funkenstein et al. 2006; Guan et al. 2022). They have been isolated from muscle tissue of various species, such as bovine (Dodson et al. 1987; Reiss et al. 2021), chicken (Yablonka-Reuveni et al. 1987; McFarland et al. 1988; Reiss et al. 2018), sheep (Dodson et al. 1986; Reiss et al. 2021), pig (Doumit and Merkel 1992; Reiss et al. 2018), and fish

(Powell et al. 1989; Reiss et al. 2021; Knežić et al. 2022; Bomkamp et al. 2023; Saad et al. 2023). While the acquisition protocols are well established, these cells have a limited proliferation capacity of approximately 20 generations (David et al. 2023). The expansion of these cells can be conducted in static or dynamic systems that support the adherence of MuSCs (Eibl et al. 2021; David et al. 2023). They have cellular markers such as Pax7, M-cadherin, Syndecan-4, CXCR4, α-7 integrin, VCAM-1, and CD56 (Reiss et al. 2021).

MSCs are cells found in various tissues such as bone marrow, umbilical cord, skeletal muscle, and adipose tissue. They have well-established isolation protocols and the capacity to differentiate into mesodermal lineages relevant to cultivated meat, such as adipocytes, chondrocytes, and fibroblasts (Lavial and Pain 2010; Reiss et al. 2021). They have been isolated from various species, including bovines (Bosnakovski et al. 2005; Hanga et al. 2020, 2021; Reiss et al. 2021), chicken (Khatri et al. 2009; Reiss et al. 2021), sheep (Lyahyai et al. 2012; Reiss et al. 2021), pig (Bosch et al. 2006; Reiss et al. 2021), and fish (Bomkamp et al. 2023). They have limited proliferative capacity (Bomkamp et al. 2023). Cell markers include CD105, CD73, CD90, Sca-1, and PDGFRα (Reiss et al. 2021).

FAPs are cells located in the interstitial space of muscles, obtained through the isolation of muscle tissue samples (David et al. 2023). Originating from intramuscular fat, they possess a high adipogenic potential, in addition to the ability to differentiate into fibroblasts (Bomkamp et al. 2023; Kumar et al. 2023). Intramuscular fat is responsible for enhancing the palatability, texture, and flavor of meat, indicating its significant importance in cultivated meat production (Knežić et al. 2022). These cells have a limited proliferative capacity of around 30 generations (David et al. 2023). They have been isolated from a few species, such as bovines (Huang et al. 2012; Guan et al. 2017; Reiss et al. 2021) and pigs (Bosch et al. 2006; Li et al. 2020; Reiss et al. 2021). Their cell markers are Sca-1 and PDGFRα (Reiss et al. 2021).

Dedifferentiated fat cells (DFATs) are proliferative, multipotent cells similar to fibroblasts that are obtained through the dedifferentiation of mature adipocytes containing lipids (Knežić et al. 2022). Overall, it can be concluded that DFAT cells are promising for the efficient generation of adipose tissue on a large scale, owing to their notable capabilities in terms of cell proliferation and adipogenic differentiation (Knežić et al. 2022; Kumar et al. 2023). DFAT cells can be identified as distinct from other types of adipocytes based on their expression of specific stem cell markers, including CD34, Sca-1, CD90, and CD45. Additionally, DFAT cells express Oct3, Oct4, Sox2, c-Myc, and Klf4. DFAT cells exhibit multipotency under appropriate cultivation circumstances, giving rise to many cell types including adipocytes, skeletal myocytes, and smooth muscle cells. They were obtained from bovines (Knežić et al. 2022; Wei et al. 2012) and chicken (Cui et al. 2018). However, their best performance was observed in cells obtained from pigs (Peng et al. 2015; Knežić et al. 2022; Wei et al. 2013). Cell markers include CD44, CD29, and CD90 (Peng et al. 2015).

Producing cell lines relevant to food is one of the most pertinent challenges in cellular agriculture. There are few cell lines made of species and cell types suitable for cultivated meat (Soice and Johnston 2021). For large-scale production, it is necessary to increase the number of cell generations, which is limited under natural conditions as cells tend to undergo senescence over time (Freshney 2005). Cells become immortal when they lose their control pathways of the cell cycle and do not suffer senescence (Specht et al. 2018; Soice and Johnston 2021). Immortalization can occur spontaneously or be induced by the expression of the catalytic subunit of telomerase or viral gene induction that deactivates p53/p14/Rb, allowing cells to multiply without DNA damage checkpoints (Soice and Johnston 2021). In food-relevant species, such as chicken, primary preadipocytes have been immortalized by overexpressing telomerase reverse transcriptase (TERT) alone to produce immortalized chicken preadipocytes (ICP1) (Wang et al. 2017). Fish cells have a high propensity for spontaneous immortalization, unlike mammalian cells (Rubio et al. 2019; Soice and Johnston 2021; Bomkamp et al. 2023; Saad et al. 2023). Spontaneously immortalized chicken fibroblasts are being utilized by Believer Meats, as well as Upside Foods, which also employs immortalized chicken myoblasts and fibroblasts in cultivated meat production. Induced immortalization leads to genetically modified cells that require labeling and are subject to a distinct regulatory process (David et al. 2023).

The careful choice of nutrients and growth agents added to the culture medium is of utmost importance in facilitating cell proliferation and guiding cell differentiation. In terms of cellular provision, it is necessary to customize the composition of culture media in order to facilitate the growth and expansion of stem and progenitor cells, as well as their subsequent differentiation into fully developed muscle, adipose, and connective tissues.

In the first phases of cell culture, the primary objective is to optimize conditions that promote cell proliferation in order to achieve the desired quantity of cells. The bioreactor assumes a crucial function in facilitating the substantial proliferation of cells by providing the necessary capacity to accommodate the culture volume required for this large-scale development. After the desired progenitor cells achieve sufficient proliferation within the bioreactor, they become accessible and are then stimulated to undergo differentiation into fully developed cells, such as adipocytes and satellite cells.

As previously stated, the selection and characteristics of cells utilized significantly influence the proliferation and scale-up demands. In order to facilitate large-scale production, it is imperative to expand isolated cells to a substantial quantity and achieve a high cell density. Additionally, in certain scenarios, these cells need to undergo differentiation into specific cell types. This process necessitates a series of scale-up steps, commencing from the initial seed stage and culminating in large-scale production within sizable bioreactors. The expansion of the majority of mammalian cells takes place within a limited temperature range of 36.5 °C to 37.5 °C (Choi et al. 2021). Nevertheless, it has been observed that fish cells have the capacity to proliferate at reduced temperatures across a broader spectrum spanning from 15 °C to 30 °C. In addition, it has been shown that fish cells have a higher tolerance for reduced oxygen levels in comparison to human cells. Additionally, fish cells demonstrate a greater capacity for adaptation to a broader range of pH values (Fernandez et al. 1993; Rubio et al. 2019).

## 4 Cell Culture Collections for Cultivated Meat Production

Cultured meat is meat or seafood grown directly from animal cells, which can be derived from various species, including fish, cows, pigs, and others. As mentioned previously, the cultured meat industry will require the production of different cell types such as muscle, adipose, and connective tissue. Therefore, animal cell availability is crucial to advance in this field. The lack of access to cell lines continues to be a barrier to cultivated meat research and commercialization. Various organizations, companies, and initiatives are working to make cell lines more accessible for researchers and businesses involved in cultivated meat production.

In 2021, The Good Food Institute (GFI) launched one initiative to increase cell line access and sharing, which has 65 cell line records from different species, while promoting research on the development of new lines. Besides, GFI initiated a collaboration with the biotech company Kerafast (kerafast.com), where fish cell lines such as *Dicentrarchus labrax* embryonic cell line (DLEC), and *Scomber scombrus* myoblast cell line (MACK1) are available. Other companies are currently supplying animal cell lines, such as Roslin Technologies (roslintech.com) which commercializes porcine iPSC lines; Opo Bio (opobio.com), Quest Meat (questmeat.com) which provides bovine primary cell lines as well as muscle-derived and adipose-derived stem cells; PluriCells (pluricells.co.uk) and Extracellular (extracellular.com) which provide cow, lamb, and pig preadipocytes, myosatellite cells, and MSCs. Moreover, biotechnology and cell culture companies, such as ATCC, as well as academic institutions and research organizations are likely to be at the forefront of providing more cell lines for cultivated meat research. For instance, Table 4 provides examples of avian cell lines publicly available and their characteristics.

The co-cultivation of diverse cell types, such as muscle and adipose cells, has the potential to closely replicate the structural and characteristic properties of meat, poultry, or fish, including marbling. Additionally, the presence of different cell types can stimulate the secretion of factors and matrices that promote the proliferation and differentiation of other cell types (Balasubramanian et al. 2021). Several researchers employed a method wherein they arranged alternating layers of muscle and adipose cells in a stacked configuration (Pandurangan and Kim 2015; Shahin-Shamsabadi and Selvaganapathy 2022). Co-cultures can also be utilized for the development of "self-organizing" approaches in cell-based food production, serving as an alternative to scaffold-based systems. A notable obstacle in this context is the distribution of nutrients and oxygen throughout the heterogeneous tissues under development, which might potentially be addressed by the utilization of artificial blood circulation simulating principles (Bhat et al. 2015).

**Table 4** Cell lines that can be applied in 3D cell culture

| Cell line | Morphology | Organism | Tissue | References |
|---|---|---|---|---|
| SL-29 CRL-1590 | Fibroblast | *Gallus gallus*, chicken | Embryo | https://www.atcc.org/products/crl-1590 |
| LMH CRL-2117 | Epithelial | *Gallus gallus*, chicken | Liver | https://www.atcc.org/products/crl-2117 |
| DT40 CRL-2111 | Lymphoblast | *Gallus gallus*, chicken | Bursa | https://www.atcc.org/products/crl-2111 |
| UMNSAH/DF-1 CRL-12203 (spontaneously immortalized chicken cell line) | Fibroblast | *Gallus gallus*, chicken | Embryo | https://www.atcc.org/products/crl-12203 |
| Duck embryo CCL −141™ | Fibroblast | *Anas platyrhynchos domesticus*, duck, Pekin | Embryo | https://www.atcc.org/products/ccl-141 |
| LMH/2A CRL-2118™ | Epithelial | *Gallus gallus*, chicken | Liver | https://www.atcc.org/products/crl-2118 |
| DT95 CRL-2112™ | Lymphoblast | *Gallus gallus*, chicken | Bursa | https://www.atcc.org/products/crl-2112 |
| MDTC-RP19 CRL-8135™ | Lymphoblast | *Meleagris gallopavo*, Turkey | Lymph node | https://www.atcc.org/products/crl-8135 |
| QNR/D CRL-2532™ | Neuronal Schwann cell | *Coturnix coturnix japonica*, quail, Japanese | Eye, neuroretina | https://www.atcc.org/products/crl-2532 |
| QNR/K2 CRL-2533™ | Mixed | *Coturnix coturnix japonica*, quail, Japanese | Eye, neuroretina | https://www.atcc.org/products/crl-2533 |

A consistent and reproducible cell lineage is a prerequisite for all bioprocesses. In essence, as cells proliferate and replicate their DNA, the likelihood of an increased occurrence of genetic variation (single nucleotide polymorphisms, copy number variation, large insertions or deletions) also increases. The utilization of small molecules to induce differentiation may present cost benefits; however, it may also present complications if the compounds fail to obtain approval with food-safe ingredients. In this case, while still suitable for research purposes, they could pose regulatory challenges if used for commercial products. Therefore, the establishment of a biobank with cell lines (the main ingredient of cultivated meat production), which are specifically designed to supply industries and entrepreneurs, stimulates the alternative protein market. It is hard to obtain the same results experimentally in a laboratory scale and transpose them to a scale-up process at the industrial level without consistent cell lines that assure reproducibility. A considerable number of researchers who are interested in pursuing cell-based alternative protein projects are unable to do so due to a lack of access to the appropriate tissue sources or adequate protocols to initiate novel investigations in this field. A biobank should provide curation and generation of stable cell lines derived from animals that will be essential for nurturing research and innovation in order to advance in cultured meat production. The utilization of biobank has the potential to significantly reduce the need of cultivating primary cell lines repeatedly. Moreover, it will enable the cell-based industry to conduct fundamental research and product development with stable and consistent cell lines.

In order to ensure that cell lines are transferred to the production sector in accordance with prevailing legislation and provide participants with assurances regarding anticipated outcomes, measures about intellectual property management and technology transfer will be required. In case cell lines from the biobank, particularly those immortalized, are eligible for intellectual property protection, it may be necessary to establish a technical collaboration agreement. This agreement would outline the respective responsibilities and ownership rights pertaining to the technology among the involved parties. An alternative option is reaching a mutual agreement pertaining to the allocation of ownership rights between the involved parties. Finally, if the cell lines are derived from animals' specialized genetic improvement programs and possess market potential, the licensing or transfer strategy will be determined in accordance with the company's criteria.

# 5 Conservation and Quality Assessment of Cell Lines

## *5.1 Cell Line Conservation for Cultured Meat*

Cultured meat production requires efficient methods of cell conservation in order to provide a stable supply of high-quality cells. Cryopreservation is a common technique which consists in the use of very low temperatures, usually from −80 °C to −196 °C, to preserve the structure and viability of living cells over long periods of time. In the case of cultured meat production, this preservation method allows the conservation of primary, pluripotent, or immortalized cell lines, allowing the creation of working cell banks, for long-term use (and backup) of the cells without the need for continuous propagation, as well as efficient and safe cell shipping. Different strategies have been used for the cryopreservation of animal cells, such as slow freezing, vitrification, dry-state freezing, among others (Gore et al. 2022). During the freeze/thawing processes, cells are inevitably exposed to loss of viability or death, mostly due to the formation of intracellular ice or to solution effects. To improve the efficiency of cryopreservation methods and the resulting cell viability, cryoprotective agents have been used, such as dimethyl sulfoxide (DMSO). However, it may be associated with cytotoxic effects and lack of biocompatibility, thus requiring extensive washing during thawing. Additionally, controlled-rate freezing at a rate of −1 °C/min typically allows to maximize cell integrity and viability.

Traditional cryopreservation media used for mammalian cells consists of FBS containing medium, supplemented with DMSO (10%–20%). However, for cellular agriculture, the use of FBS is not a viable option and DMSO, due to its toxicity, will likely be avoided in this field. Different animal-free commercial formulations have been developed over the last years, and different cryoprotectants (such as sugar derivatives, amino acids, and others) have also been evaluated and may be explored for cellular agriculture applications (Raju et al. 2021; Weng and Beauchesne 2020). A recent study demonstrated the possibility to freeze myogenic bovine cells (Kakehi et al. 2023) for up to 1 year, at −80 °C (without liquid nitrogen), allowing their storage and shipping even to remote locations, while maintaining the proliferation and differentiation properties, but more studies are still required to optimize cryopreservation media and procedures.

## *5.2 Quality Control of Cell Lines*

The quality of the cell lines used for cellular agriculture products must be carefully and regularly assessed. Quality control is a critical step in the production process, regardless of whether the process is based on primary cells or immortalized/pluripotent lines. Primary cells are isolated from animal tissue samples and are often susceptible to contamination by microorganisms, which may potentially lead to transmission of zoonotic infectious diseases and foodborne pathogens—although with a lower probability than conventional animal food production (Treich 2021). Typical microbial contaminants may be bacteria, including *Mycoplasma* sp., animal-derived viruses, or parasites (FAO/WHO 2023). Careful and regular microbiological analysis should be performed to confirm the absence of any pathogenic microorganisms in the final product. In order to minimize microbial contamination, it is common to use antibiotics in the initial stages of cell isolation and cultivation. However, for scale-up, it is desirable to perform cultivation in the absence of antibiotics, to avoid the risk of developing antibiotic resistance, and therefore, other food preservatives may be used.

Tissue samples used for cell isolation typically consist of various cell types from the tissue of origin. Cell type selection may be performed using the techniques described in Sect. 2 and regular assessment of physicochemical properties, proliferation and differentiation potential, karyotype stability, and expression of cell markers (e.g., by flow cytometry, qRT-PCR, or immunofluorescence) should be performed to confirm cell identity and for early detection of any alteration (Ong et al. 2021). The use of cell lines also poses the risk of genetic or epigenetic drift, due to long-term cultivation, which may lead to the accumulation of mutations that may significantly alter cell characteristics (FAO/WHO 2023). To minimize this risk, apart from regular monitoring of culture behavior, cells can be continuously cultured for a defined number of passages/cell doublings, and periodically cultures are seeded again from the working cell bank.

## 6 Challenges and Future Developments

The advent of cultured meat products has prompted a broader focus on animal cell sources that may be utilized to enhance cell-based meat production. The development of multipotent and highly proliferative cell sources is a crucial technical challenge in the effort to increase cultured meat production. In addition to the chosen cell source dictating proliferative capacity and differentiation potential, it also affects the appropriate choices in production steps. Even multiple cell sources can be employed for the production of meat; nevertheless, it is crucial to take into account the senescence issue during the process.

In recent years, approaches aimed at developing pluripotent and primary cell sources from livestock species have produced significant advances. Primary cells are typically devoid of genetic alteration or mutagenesis, but they can be subjected to bioengineering techniques to enhance their ability to proliferate or slow the process of senescence, achieving a state of immortality. Therefore, advances will be needed to develop ready-to-use immortalized cell lines to achieve the scale and cost needed for commercial production and sale of cultured meat products. For example, the utilization of immortalized somatic cell lines presents several benefits in terms of enhanced proliferative capacity and more control over cell destiny. Consequently, this might potentially enhance the reliability, reproducibility, and scalability of production processes.

The use of pluripotent stem cells for cultured meat production also faces significant challenges. Although the cultivation of human or mouse pluripotent stem cells is well studied and described, for other mammalian and, particularly, nonmammalian species, the fundamental biological knowledge is still scarce. The development of cultivation strategies for these cells will likely require (1) specific culture medium formulations, able to support continuous growth and controlled differentiation; (2) optimized species-specific protocols for myogenic and adipogenic differentiation and identification of stage-specific markers; (3) molecular tools for cell analysis, such as specific antibodies, or annotated genomes.

The creation of specific biobanks to supply cell lineages for food production will also accelerate scaling-up, in addition to making this technology more accessible to new investors. Indeed, cost-reduction is also one of the most critical challenges of the field. The need to develop lower cost culture media is essential to achieve price parity with conventional meat products. The replacement of FBS by other non-animal origin and defined products is an important step in this direction. Moreover, the use of animal-original components is associated with contamination risks, and their replacement will contribute to increasing the safety of cultured meat products.

## 7 Conclusion

The production of cultured meat products has the potential to change our food system and lead to a more sustainable world and to reduce animal sacrifice. Ideal cells to produce cultivated meat must possess the self-renewing capacity and the ability to differentiate into mature cell types. Stem cells are therefore good options to be used as a starting cell source. Undifferentiated progenitor cells, e.g., MSCs and FAPs located in organs/tissues, can also differentiate into adipocytes, chondrocytes, and fibroblasts. In this chapter, we described the current knowledge regarding cell isolation and culture and discussed the already available cell lines and the challenges involved in their use for cultured meat production. The emerging field of cellular agriculture has seen impressive and fast developments over the last years, but important challenges are still ahead to develop the best cell lines and to massively implement this new food production technology worldwide.

## References

Adhikari R, Chen C, Waters E et al (2019) Isolation and differentiation of mesenchymal stem cells from broiler chicken compact bones. Front Physiol 9. https://doi.org/10.3389/fphys.2018.01892

Bai C, Hou L, Ma Y, Chen L, Zhang M, Guan W (2013) Isolation and characterization of mesenchymal stem cells from chicken bone marrow. Cell Tissue Bank 14:437–451. https://doi.org/10.1007/s10561-012-9347-8

Balasubramanian B, Liu W, Pushparaj K, Park S (2021) The epic of in vitro meat production—a fiction into reality. Foods 10:1395. https://doi.org/10.3390/foods10061395

Baquero-Perez B, Kuchipudi SV, Nelli RK, Chang KC (2012) A simplified but robust method for the isolation of avian and mammalian muscle satellite cells. BMC Cell Biol 13. https://doi.org/10.1186/1471-2121-13-16

Barry D (2021) Textbook of natural medicine, 5th edn. Elsevier, St. Louis, MO

Bautista-Amorocho H, Silva-Sayago JA, Goyeneche-Patino DA, Pérez-Cala TL, Macías-Gómez F, Arango-Viana JC, Martínez A (2021) A novel method for isolation and culture of primary swine gastric epithelial cells. BMC Mol Cell Biol 22:1–13. https://doi.org/10.1186/s12860-020-00341-7

Bentzinger CF, Wang YX, Rudnicki MA (2012) Building muscle: molecular regulation of Myogenesis. Cold Spring Harb Perspect Biol 4:a008342. https://doi.org/10.1101/cshperspect.a008342

Bhat ZF, Kumar S, Fayaz H (2015) In vitro meat production: challenges and benefits over conventional meat production. J Integr Agric 14:241–248. https://doi.org/10.1016/S2095-3119(14)60887-X

Bogliotti YS, Wu J, Vilarino M et al (2018) Efficient derivation of stable primed pluripotent embryonic stem cells from bovine blastocysts. Proc Natl Acad Sci 115:2090–2095. https://doi.org/10.1073/pnas.1716161115

Bomkamp C, Musgrove L, Marques DMC et al (2023) Differentiation and maturation of muscle and fat cells in cultivated seafood: lessons from developmental biology. Mar Biotechnol 25:1–29. https://doi.org/10.1007/s10126-022-10174-4

Bosch P, Pratt SL, Stice SL (2006) Isolation, characterization, gene modification, and nuclear reprogramming of porcine mesenchymal stem cells. Biol Reprod 74:46–57. https://doi.org/10.1095/biolreprod.105.045138

Bosnakovski D, Mizuno M, Kim G et al (2005) Isolation and multilineage differentiation of bovine bone marrow mesenchymal stem cells. Cell Tissue Res 319:243–253. https://doi.org/10.1007/s00441-004-1012-5

Bou M, Montfort J, Le Cam A et al (2017) Gene expression profile during proliferation and differentiation of rainbow trout adipocyte precursor cells. BMC Genomics 18:347. https://doi.org/10.1186/s12864-017-3728-0

Capes-Davis A, Theodosopoulos G, Atkin I et al (2010) Check your cultures! A list of cross-contaminated or misidentified cell lines. Int J Cancer 127:1–8. https://doi.org/10.1002/ijc.25242

Choi K, Yoon JW, Kim M et al (2021) Muscle stem cell isolation and in vitro culture for meat production: a methodological review. Compr Rev Food Sci Food Saf 20:429–457. https://doi.org/10.1111/1541-4337.12661

Cui HX, Guo LP, Zhao GP, Liu RR, Li QH, Zheng MQ, Wen J (2018) Method using a co-culture system with highpurity intramuscular preadipocytes and satellite cells from chicken pectoralis major muscle. Poult Sci 97:3691–3697. https://doi.org/10.3382/ps/pey023

David S, Tsukerman A, Safina D et al (2023) Co-culture approaches for cultivated meat production. Nat Rev Bioeng 1:817. https://doi.org/10.1038/s44222-023-00077-x

Dodson MV, McFarland DC, Martin EL, Brannon MA (1986) Isolation of satellite cells from ovine skeletal muscles. J Tissue Cult Methods 10:233–237. https://doi.org/10.1007/BF01404483

Dodson MV, Martin EL, Brannon MA et al (1987) Optimization of bovine satellite cell-derived myotube formation in vitro. Tissue Cell 19:159–166. https://doi.org/10.1016/0040-8166(87)90001-2

Doumit ME, Merkel RA (1992) Conditions for isolation and culture of porcine myogenic satellite cells. Tissue Cell 24:253–262. https://doi.org/10.1016/0040-8166(92)90098-R

Eibl R, Senn Y, Gubser G et al (2021) Cellular agriculture: opportunities and challenges. Annu Rev Food Sci Technol 12:51–73. https://doi.org/10.1146/annurev-food-063020-123940

Fadel L, Viana BR, Feitosa MLT et al (2011) Protocols for obtainment and isolation of two mesenchymal stem cell sources in sheep. Acta Cir Bras 26:267–273. https://doi.org/10.1590/S0102-86502011000400004

FAO/WHO (2023) Food safety aspects of cell-based food. FAO/WHO, Rome, p 146. https://doi.org/10.4060/cc4855en

Farzaneh M, Attari F, Mozdziak PE, Khoshnam SE (2017) The evolution of chicken stem cell culture methods. Br Poult Sci 58:681–686. https://doi.org/10.1080/00071668.2017.1365354

Fathi E, Farahzadi R, Sheikhzadeh N (2019) Immunophenotypic characterization, multi-lineage differentiation and aging of zebrafish heart and liver tissue-derived mesenchymal stem cells as a novel approach in stem cell-based therapy. Tissue Cell 57:15–21. https://doi.org/10.1016/j.tice.2019.01.006

Fernandez RD, Yoshimizu M, Ezura Y, Kimura T (1993) Comparative growth response of fish cell lines in different media, temperatures, and sodium chloride concentrations. Fish Pathol 28:27–34. https://doi.org/10.3147/jsfp.28.27

Feyen DAM, van den Akker F, Noort W et al (2016) Isolation of pig bone marrow-derived mesenchymal stem cells. In: Mesenchymal stem cells: methods and protocols. Springer, New York, pp 225–232

Freshney RI (2005) Subculture and cell lines. In: Culture of animal cells. Wiley
Froehlich JM, Seiliez I, Gabillard J-C, Biga PR (2014) Preparation of primary myogenic precursor cell/myoblast cultures from basal vertebrate lineages. J Vis Exp. https://doi.org/10.3791/51354
Funkenstein B, Balas V, Skopal T et al (2006) Long-term culture of muscle explants from Sparus aurata. Tissue Cell 38:399–415. https://doi.org/10.1016/j.tice.2006.09.003
Gao Y, Guo Y, Duan A et al (2014) Optimization of culture conditions for maintaining porcine induced pluripotent stem cells. DNA Cell Biol 33:1–11. https://doi.org/10.1089/dna.2013.2095
Gong X, Hou L, Bai C, Jin D, He X, Guan W, Ma Y (2011) Isolation and biological characteristics of chicken adipose-derived progenitor cells. DNA Cell Biol 30:453–460. https://doi.org/10.1089/dna.2010.1154
Gore M, Narvekar A, Bhagwat A et al (2022) Macromolecular cryoprotectants for the preservation of mammalian cell culture: lessons from crowding, overview and perspectives. J Mater Chem B 10:143–169. https://doi.org/10.1039/D1TB01449H
Guan L, Hu X, Liu L et al (2017) bta-miR-23a involves in adipogenesis of progenitor cells derived from fetal bovine skeletal muscle. Sci Rep 7:43716. https://doi.org/10.1038/srep43716
Guan X, Zhou J, Du G, Chen J (2022) Bioprocessing technology of muscle stem cells: implications for cultured meat. Trends Biotechnol 40:721–734. https://doi.org/10.1016/j.tibtech.2021.11.004
Han H, Wei W, Chu W et al (2017) Muscle conditional medium reduces intramuscular adipocyte differentiation and lipid accumulation through regulating insulin signaling. Int J Mol Sci 18:1799. https://doi.org/10.3390/ijms18081799
Hanga MP, Ali J, Moutsatsou P et al (2020) Bioprocess development for scalable production of cultivated meat. Biotechnol Bioeng 117:3029–3039. https://doi.org/10.1002/bit.27469
Hanga MP, de la Raga FA, Moutsatsou P et al (2021) Scale-up of an intensified bioprocess for the expansion of bovine adipose-derived stem cells (bASCs) in stirred tank bioreactors. Biotechnol Bioeng 118:3175–3186. https://doi.org/10.1002/bit.27842
He J, Evans NM, Liu H, Shao S (2020) A review of research on plant-based meat alternatives: driving forces, history, manufacturing, and consumer attitudes. Compr Rev Food Sci Food Saf 19:2639–2656. https://doi.org/10.1111/1541-4337.12610
Hill ABT, Bressan FF, Murphy BD, Garcia JM (2019) Applications of mesenchymal stem cell technology in bovine species. Stem Cell Res Ther 10:44. https://doi.org/10.1186/s13287-019-1145-9
Huang Y, Das AK, Yang Q-Y et al (2012) Zfp423 promotes Adipogenic differentiation of bovine stromal vascular cells. PLoS One 7:e47496. https://doi.org/10.1371/journal.pone.0047496
Jedrzejczak M, Szatkowska I (2014) Bovine mammary epithelial cell cultures for the study of mammary gland functions. In Vitro Cell Dev Biol Anim 50:389–398. https://doi.org/10.1007/s11626-013-9711-4
Kakehi R, Yoshida A, Takahashi H, Shimizu T (2023) Repeated and long-term cryopreservation of primary bovine myogenic cells to maintain quality in biomimetic cultured meat. Front Sustain Food Syst 7. https://doi.org/10.3389/fsufs.2023.1023057
Khatri M, O'Brien TD, Sharma JM (2009) Isolation and differentiation of chicken mesenchymal stem cells from bone marrow. Stem Cells Dev 18:1485–1492. https://doi.org/10.1089/scd.2008.0223
Kim YM, Park YH, Lim JM et al (2017) Technical note: induction of pluripotent stem cell-like cells from chicken feather follicle cells. J Anim Sci 95:3479–3486. https://doi.org/10.2527/jas.2017.1418
Kim B, Ko D, Choi SH, Park S (2023a) Bovine muscle satellite cells in calves and cattle: a comparative study of cellular and genetic characteristics for cultivated meat production. Curr Res Food Sci 7:100545. https://doi.org/10.1016/j.crfs.2023.100545
Kim CJ, Kim SH, Lee EY et al (2023b) Optimal temperature for culturing chicken satellite cells to enhance production yield and umami intensity of cultured meat. Food Chem Adv 2:100307. https://doi.org/10.1016/j.focha.2023.100307

Knežić T, Janjušević L, Djisalov M et al (2022) Using vertebrate stem and progenitor cells for cellular agriculture-state-of-the-art, challenges, and future perspectives. Biomolecules 12:699. https://doi.org/10.3390/biom12050699
Kong X, Wang X, Li M et al (2021) Establishment of myoblast cell line and identification of key genes regulating myoblast differentiation in a marine teleost, Sebastes schlegelii. Gene 802:145869. https://doi.org/10.1016/j.gene.2021.145869
Koumans JTM, Akster HA, Dulos GJ, Osse JWM (1990) Myosatellite cells of Cyprinus carpio (Teleostei) in vitro: isolation, recognition and differentiation. Cell Tissue Res 261:173–181. https://doi.org/10.1007/BF00329450
Kumar A, Sood A, Han SS (2023) Technological and structural aspects of scaffold manufacturing for cultured meat: recent advances, challenges, and opportunities. Crit Rev Food Sci Nutr 63:585–612. https://doi.org/10.1080/10408398.2022.2132206
Lavial F, Pain B (2010) Chicken embryonic stem cells as a non-mammalian embryonic stem cell model. Dev Growth Differ 52:101–114. https://doi.org/10.1111/j.1440-169X.2009.01152.x
Lee EJ, Choi J, Hyun JH et al (2007) Steroid effects on cell proliferation, differentiation and steroid receptor gene expression in adult bovine satellite cells. Asian Australas J Anim Sci 20:501–510. https://doi.org/10.5713/ajas.2007.501
Lee SY, Kang HJ, Lee DY et al (2021) Principal protocols for the processing of cultured meat. J Anim Sci Technol 63:673–680. https://doi.org/10.5187/jast.2021.e40
Lee JM, Lee H, Lee ST (2023) Development of an effective dissociation protocol for isolating mesenchymal stem cells from bovine intermuscular adipose tissues. J Anim Reprod Biotechnol 38:10–16. https://doi.org/10.12750/JARB.38.1.10
Li M, Zhang D, Hou Y et al (2003) Isolation and culture of embryonic stem cells from porcine blastocysts. Mol Reprod Dev 65:429–434. https://doi.org/10.1002/mrd.10301
Li X, Fu X, Yang G, Du M (2020) Review: enhancing intramuscular fat development via targeting fibro-adipogenic progenitor cells in meat animals. Animal 14:312–321. https://doi.org/10.1017/S175173111900209X
Liu J, Balehosur D, Murray B et al (2012) Generation and characterization of reprogrammed sheep induced pluripotent stem cells. Theriogenology 77:338–346.e1. https://doi.org/10.1016/j.theriogenology.2011.08.006
Liu X, Zhang S, Guan W, Zheng D (2020) Isolation and characterization of mesenchymal stem cells from chicken liver. J Biomater Tissue Eng 10:8–16. https://doi.org/10.1166/jbt.2020.2225
Lu T, Xiong H, Wang K et al (2014) Isolation and characterization of adipose-derived mesenchymal stem cells (ADSCs) from cattle. Appl Biochem Biotechnol 174:719–728. https://doi.org/10.1007/s12010-014-1128-3
Lyahyai J, Mediano DR, Ranera B et al (2012) Isolation and characterization of ovine mesenchymal stem cells derived from peripheral blood. BMC Vet Res 8:169. https://doi.org/10.1186/1746-6148-8-169
Ma YN, Wang B, Wang ZX et al (2018) Three-dimensional spheroid culture of adipose stromal vascular cells for studying adipogenesis in beef cattle. Animal 12:2123–2129. https://doi.org/10.1017/S1751731118000150
Mauro A (1961) Satellite cell of skeletal muscle fibers. J Cell Biol 9:493–495. https://doi.org/10.1083/jcb.9.2.493
McFarland DC, Doumit ME, Minshall RD (1988) The Turkey myogenic satellite cell: optimization of in vitro proliferation and differentiation. Tissue Cell 20(6):899–908. https://doi.org/10.1016/0040-8166(88)90031-6
Oestbye T, Ytteborg E (2019) Preparation and culturing of Atlantic salmon muscle cells for in vitro studies. In: Methods in molecular biology. Humana Press, Totowa, NJ, pp 319–330
Ong KJ, Johnston J, Datar I et al (2021) Food safety considerations and research priorities for the cultured meat and seafood industry. Compr Rev Food Sci Food Saf 20:5421–5448. https://doi.org/10.1111/1541-4337.12853
Pandurangan M, Kim DH (2015) A novel approach for in vitro meat production. Appl Microbiol Biotechnol 99:5391–5395. https://doi.org/10.1007/s00253-015-6671-5

Park J, Choi JK, Choi DH et al (2022) Optimization of skeletal muscle-derived fibroblast isolation and purification without the preplating method. Cell Tissue Bank 23:557–568. https://doi.org/10.1007/s10561-021-09989-7
Peng X, Song T, Hu X et al (2015) Phenotypic and functional properties of porcine dedifferentiated fat cells during the long-term culture in vitro. Biomed Res Int 2015:1–10. https://doi.org/10.1155/2015/673651
Post MJ, Levenberg S, Kaplan DL et al (2020) Scientific, sustainability and regulatory challenges of cultured meat. Nat Food 1:403–415. https://doi.org/10.1038/s43016-020-0112-z
Powell RL, Dodson MV, Cloud JG (1989) Cultivation and differentiation of satellite cells from skeletal muscle of the rainbow trout Salmo gairdneri. J Exp Zool 250:333–338. https://doi.org/10.1002/jez.1402500314
Raju R, Bryant SJ, Wilkinson BL, Bryant G (2021) The need for novel cryoprotectants and cryopreservation protocols: insights into the importance of biophysical investigation and cell permeability. Biochim Biophys Acta Gen Subj 1865:129749. https://doi.org/10.1016/j.bbagen.2020.129749
Ravikumar M, Powell D (2023) Cell line development and utilisation trends in the cultivated meat industry
Reggio A, Rosina M, Palma A et al (2020) Adipogenesis of skeletal muscle fibro/adipogenic progenitors is affected by the WNT5a/GSK3/β-catenin axis. Cell Death Differ 27:2921–2941. https://doi.org/10.1038/s41418-020-0551-y
Reiss A, Cioatera N, Chifiriuc MC et al (2018) New biologically active mixed-ligand co(II) and Ni(II) complexes of enrofloxacin: synthesis, spectral study and thermal behaviour. J Therm Anal Calorim 134:527–541. https://doi.org/10.1007/s10973-018-6994-4
Reiss J, Robertson S, Suzuki M (2021) Cell sources for cultivated meat: applications and considerations throughout the production workflow. Int J Mol Sci 22:7513. https://doi.org/10.3390/ijms22147513
Roberts RM, Yuan Y, Genovese N, Ezashi T (2015) Livestock models for exploiting the promise of pluripotent stem cells. ILAR J 56:74–82. https://doi.org/10.1093/ilar/ilv005
Rosselló RA, Chen C-C, Dai R et al (2013) Mammalian genes induce partially reprogrammed pluripotent stem cells in non-mammalian vertebrate and invertebrate species. eLife 2:e00036. https://doi.org/10.7554/eLife.00036
Rubio N, Datar I, Stachura D et al (2019) Cell-based fish: a novel approach to seafood production and an opportunity for cellular agriculture. Front Sustain Food Syst 3. https://doi.org/10.3389/fsufs.2019.00043
Saad MK, Yuen JSK, Joyce CM et al (2023) Continuous fish muscle cell line with capacity for myogenic and adipogenic-like phenotypes. Sci Rep 13:5098. https://doi.org/10.1038/s41598-023-31822-2
Salmerón C, Riera-Heredia N, Gutiérrez J et al (2016) Adipogenic gene expression in gilthead sea bream mesenchymal stem cells from different origin. Front Endocrinol 7. https://doi.org/10.3389/fendo.2016.00113
Sampaio RV, Chiaratti MR, Santos DCN et al (2015) Generation of bovine (Bos indicus) and buffalo (Bubalus bubalis) adipose tissue derived stem cells: isolation, characterization, and multipotentiality. Genet Mol Res 14:53–62. https://doi.org/10.4238/2015.January.15.7
Sänger AM, Stoiber W (2001) Muscle fiber diversity and plasticity. In: Hoar WS, Randall DJ, Farrell AP (eds) Fish physiology series. Academic Press, New York, pp 187–250
Seah JSH, Singh S, Tan LP, Choudhury D (2022) Scaffolds for the manufacture of cultured meat. Crit Rev Biotechnol 42:311–323. https://doi.org/10.1080/07388551.2021.1931803
Shahin-Shamsabadi A, Selvaganapathy PR (2022) Engineering murine adipocytes and skeletal muscle cells in meat-like constructs using self-assembled layer-by-layer biofabrication: a platform for development of cultivated meat. Cells Tissues Organs 211:304–312. https://doi.org/10.1159/000511764
Sobhani A, Khanlarkhani N, Baazm M et al (2017) Multipotent stem cell and current application. Acta Med Iran 55:6–23

Soice E, Johnston J (2021) Immortalizing cells for human consumption. Int J Mol Sci 22:11660. https://doi.org/10.3390/ijms222111660

Specht EA, Welch DR, Rees Clayton EM, Lagally CD (2018) Opportunities for applying biomedical production and manufacturing methods to the development of the clean meat industry. Biochem Eng J 132:161–168. https://doi.org/10.1016/j.bej.2018.01.015

Stout AJ, Mirliani AB, Rittenberg ML et al (2022) Simple and effective serum-free medium for sustained expansion of bovine satellite cells for cell cultured meat. Commun Biol 5:466. https://doi.org/10.1038/s42003-022-03423-8

Su Y, Wang L, Fan Z et al (2021) Establishment of bovine-induced pluripotent stem cells. Int J Mol Sci 22:10489. https://doi.org/10.3390/ijms221910489

Takenouchi T, Suzuki S, Shinkai H et al (2014) Extracellular ATP does not induce P2X7 receptor-dependent responses in cultured renal- and liver-derived swine macrophages. Results Immunol 4:62–67. https://doi.org/10.1016/j.rinim.2014.07.002

Talluri TR, Kumar D, Glage S et al (2015) Derivation and characterization of bovine induced pluripotent stem cells by transposon-mediated reprogramming. Cell Reprogram 17:131–140. https://doi.org/10.1089/cell.2014.0080

Thong T, Forté CA, Hill EM, Colacino JA (2019) Environmental exposures, stem cells, and cancer. Pharmacol Ther 204:107398. https://doi.org/10.1016/j.pharmthera.2019.107398

Treich N (2021) Cultured meat: promises and challenges. Environ Resour Econ 79:33–61. https://doi.org/10.1007/s10640-021-00551-3

Vassiliev I, Nottle MB (2013) Isolation and culture of porcine embryonic stem cells. In: Epiblast stem cells: methods and protocols. Humana Press, Totowa, NJ, pp 85–95

Vegusdal A, Sundvold H, Gjøen T, Ruyter B (2003) An in vitro method for studying the proliferation and differentiation of Atlantic salmon preadipocytes. Lipids 38:289–296. https://doi.org/10.1007/s11745-003-1063-3

Vilarino M, Alba Soto D, Soledad Bogliotti Y et al (2020) Derivation of sheep embryonic stem cells under optimized conditions. Reproduction 160:761–772. https://doi.org/10.1530/REP-19-0606

Wang W, Zhang T, Wu C, Wang S, Wang Y, Li H, Wang N (2017) Immortalization of chicken preadipocytes by retroviral transduction of chicken TERT and TR. PLoS One 12:e0177348. https://doi.org/10.1371/journal.pone.0177348

Wei S, Duarte MS, Du M, Paulino PVR, Jiang Z, Albrecht E, Fernyhough-Culver M, Zan L, Hausman GJ, Dodson MV (2012) Bovine mature adipocytes readily return to a proliferative state. Tissue Cell 44:385–390. https://doi.org/10.1016/j.tice.2012.08.001

Wei S, Duarte MS, Zan L, Du M, Jiang Z, Guan L, Chen J, Hausman GJ, Dodson MV (2013) Cellular and molecular implications of mature adipocyte dedifferentiation. J Genomics 1:5–12. https://doi.org/10.7150/jgen.3769

Weng L, Beauchesne PR (2020) Dimethyl sulfoxide-free cryopreservation for cell therapy: a review. Cryobiology 94:9–17. https://doi.org/10.1016/j.cryobiol.2020.03.012

Wosczyna MN, Konishi CT, Perez Carbajal EE et al (2019) Mesenchymal stromal cells are required for regeneration and homeostatic maintenance of skeletal muscle. Cell Rep 27:2029–2035.e5. https://doi.org/10.1016/j.celrep.2019.04.074

Xiong C, Wang M, Ling W et al (2020) Advances in isolation and culture of chicken embryonic stem cells in vitro. Cell Reprogram 22:43–54. https://doi.org/10.1089/cell.2019.0080

Yablonka-Reuveni Z, Quinn LS, Nameroff M (1987) Isolation and clonal analysis of satellite cells from chicken pectoralis muscle. Dev Biol 119:252–259. https://doi.org/10.1016/0012-1606(87)90226-0

Yang J, Zhao Q, Wang K et al (2018) Isolation, culture and biological characteristics of multipotent porcine tendon-derived stem cells. Int J Mol Med 41:3611–3619. https://doi.org/10.3892/ijmm.2018.3545

Yin H, Price F, Rudnicki MA (2013) Satellite cells and the muscle stem cell niche. Physiol Rev 93:23–67. https://doi.org/10.1152/physrev.00043.2011

Ytteborg E, Todorcevic M, Krasnov A et al (2015) Precursor cells from Atlantic salmon (Salmo salar) visceral fat holds the plasticity to differentiate into the osteogenic lineage. Biol Open 4:783–791. https://doi.org/10.1242/bio.201411338

Zhang G, Zhao X, Li X et al (2020) Challenges and possibilities for bio-manufacturing cultured meat. Trends Food Sci Technol 97:443–450. https://doi.org/10.1016/j.tifs.2020.01.026

# Culture Medium for Cultivated Meat

**Gustavo Roncoli Reigado, Izabela Daniel Sardinha Caldeira, Pedro Oliveira Marques, Mayara Santos, Felipe Santiago Chambergo Alcalde, and Viviane Abreu Nunes**

**Abstract** Cultivated meat has attracted widespread attention as a promising technology with potential benefits in food safety, sustainable development, and traceable origin. While countries worldwide invest significantly in research on cultivated meat, a fundamental challenge lies in efficiently simulating cellular growth and enabling large-scale production in bioreactors. Despite advances in tissue engineering, achieving cost-effective and technically viable large-scale cultivated meat production remains complex. Although animal cell culture media has been refined, it lacks specific adaptations to meet the unique requirements of cultivated meat. Successful production requires food-grade media with minimal cost, regulation of large-scale cell proliferation and differentiation, the creation of an acceptable product with sensory qualities, and animal-free origin, adhering to specific regulations for food-grade media. The formulation of media with these qualities draws inspiration from conventional culture media applications and principles used in optimizing large-scale microbial fermentation processes, producing ingredients for media with lower value and at a food scale. Until now, the analysis concludes that improvements in metabolic efficiency and the development of low-cost serum-free media are still insufficient to assume the substitution of conventional meat by cultivated meat.

**Keywords** Culture media · Fetal bovine serum · Serum-free media · Cell-based meat · In vitro meat · Cultured meat

G. R. Reigado · I. D. S. Caldeira · P. O. Marques · M. Santos · V. A. Nunes (✉)
Laboratory of Skin Physiology and Tissue Bioengineering, School of Arts, Sciences and Humanities, University of São Paulo (EACH-USP), São Paulo, Brazil
e-mail: vanunes@ib.usp.br

F. S. Chambergo Alcalde
Laboratory of Biochemistry and Biotechnology of Proteins, School of Arts, Sciences and Humanities, University of São Paulo, São Paulo, Brazil


C. R. Soccol et al. (eds.), *Cultivated Meat*,
https://doi.org/10.1007/978-3-031-55968-6_4

# 1 Why Think in Nontraditional Culture Media Formulations for Cultivated Meat?

As a developing technology in the world, cultured meat has provoked widespread attention due to its potential advantages of food safety, sustainable development, and traceable origin. In this sense, different countries have invested a lot of resources on research on cultured meat. Currently, the challenge of cultured meat production is to efficiently simulate cellular growth and to support large-scale production in bioreactors. Although tissue engineering has been deeply studied and achieved successful applications, it is still difficult and complex to obtain large-scale cultured meat production due to the high cost and technical requirements.

Among technical obstacles still hindering the large-scale commercialization of cultivated meat, several are related to the media that are used in the culture of the muscle, fat, and connective tissue cells. While animal cell culture media has been used and refined, it has not been specifically designed with the requirements of cultivated meat. Also, the successful production of cultivated meat requires media at food grade with minimal cost and regulation of large-scale cell proliferation and differentiation, to achieve an acceptable product with sensory qualities and animal-free origin. Strategies for achieving media formulations with these qualities can be obtained from the knowledge of conventional culture media applications and the metabolic pathways involved in myogenesis and protein synthesis. Also, the application of principles used to optimize media for large-scale microbial fermentation processes producing lower-value commodity chemicals and food ingredients can also be helpful.

This chapter will explore culture media for cell cultivation and cultured meat and the main goals for achieving an ideal culture medium for this product. In particular, it will present some of these media, their uses, costs, advantages, and obstacles to have this type of meat on our plates.

# 2 What Is a Culture Medium for In Vitro Cell Growth

Since Alexis Carrel developed a culture flask prototype, efforts have been made to ensure that a substance could provide all the required components for cell survival and proliferation while maintaining its optimal functions. In fact, many advances in animal cell culture were made. Nowadays, especially mammalian cell cultures are used for a lot of purposes in human and veterinary medicine, for new approaches to treat and prevent diseases, as well as its employment in agriculture, engineering, and in industrial fields. In addition, in its aim to collaborate with environmental demands, cell culture knowledge has been also used for finding a sustainable way to produce a cell-based meat (Carrel 1923; Yao and Asayama 2017; Cantor 2019; Okamoto et al. 2020; O'Neill et al. 2022).

In the beginning, when chicken embryonic cells were successfully cultivated with blood plasma, animal cell culture started to be made with the so-called natural media. Natural medium was essentially composed of plasma, serum, and embryonic extract. Thereafter, Margaret R. Lewis and Warren H. Lewis, in 1911, described that the glucose was important to optimize the chicken embryonic cell growth and, because of that, initiated a search for an optimal medium formulation to enhance cell culture. From there, the synthetics media emerged (Yao and Asayama 2017).

Culture media must supply all essential nutrients needed for cell synthesis, such as amino acids, fatty acids, sugars, inorganic ions, trace elements, vitamins, cofactors, and molecules that maintain the balance of the chemical environment. In 1950, medium 199 (M199) was elaborated, and in 1955, one of the first descriptions about necessary nutrients for human cell culture were made by Harry Eagle. Eagle described 27 essential nutrients for optimal HeLa (human cervical carcinoma cell line) cell growth, which he named as Basal Medium Eagle (BME). Four years later, Eagle elaborated a BME derivative, the Minimal Essential Medium (MEM), which had a greater amount of two-fold amino acids (Yao and Asayama 2017; Cantor 2019; Okamoto et al. 2020). Then, another BME derivative with higher levels of amino acids and vitamins was developed by Dulbecco and Freeman, and this improved medium was called Dulbecco's Modified Eagle Medium (DMEM). After that, another media formulated by Moore and colleagues was Roswell Park Memorial Institute Medium 1640 (RPMI 1640) (Cantor 2019). In 1963 Leibovitz's L-15 medium was developed. In addition, all these media can be supplemented with fetal bovine serum (FBS), growth factors, and hormones. However, serum-free formulations are desirable when it comes to cultured meat (Ham and McKeehan 1979). Almost a decade later, Ham's Nutrient Mixture F-12 (F-12) medium composition was described to be used as serum-free media for a Chinese Hamster Ovary (CHO) cell culture. Without serum, this medium had only the addition of albumin and fetuin purified serum proteins (Yao and Asayama 2017).

Some conditions of the culture medium are indispensable part of formulations, such as osmolality, buffering, and temperature. Most cells work well at body temperature of 37 °C and are buffered by exogenous $CO_2$ from incubators in the range of 5%–8%, although some chemicals can be used as both catalytic and physiological components. Sodium bicarbonate is added to regulate pH and osmolality and can be used as a source of carbonate as well. Along with $CO_2$ it acts as a buffer in small scales, but HEPES (4-(2-hydroxyethyl)-1-piperazineethanesulfonic acid) seems to be more efficient in larger scales (Ham and McKeehan 1979; Freshney 2010). Inorganic salts also contribute to the osmolality of the medium, being mainly those of sodium, potassium, magnesium, calcium, and chlorine (Mather 1998).

Cells reach their optimum function when all their needs are satisfied. Therefore, it is essential to know the properties of the cell and the medium that best aligns with the objective of the study in order to make the best decision. It is documented, for example, that BME and MEM media are suitable for fibroblasts, but they differ from each other due to nitrogen source. Compared to BME, MEM is formulated with increased amino acids concentrations (Mather 1998; Freshney 2010; Yao and Asayama 2017).

On the other hand, DMEM and F-12 are well suited for skeletal muscle, but they differ in proportion of vitamins and salts. F-12 media contains a higher diversity of components and includes additional trace elements, such as zinc and thymidine, that are generally not found in other media formulations. In some cases, researchers propose mixtures of DMEM and F-12 as the basis for serum-free formulation to combine the higher nutrient concentration of DMEM with the richness of F-12 (Barnes and Sato 1980; Freshney 2010).

Regarding the nutritional composition of culture media, it is necessary to mention the presence of amino acids, whether essential or nonessential. Essential amino acids must be present in the medium, while nonessential amino acids vary somewhat in concentration depending on the medium chosen. However, there is one nonessential amino acid that is present in most of the variations of the media listed above, which is L-glutamine. This amino acid is present in the blood plasma and is a great source of energy for cells, generally presenting a much higher concentration in the media than the rest of the amino acids (Freshney 2010).

DMEM medium has much higher concentrations of amino acids than the other media mentioned, but it is F-12 that has all the amino acids in its composition, albeit in much lower concentrations. Of these three media, MEM is the most lacking in amino acids and can be used in cell differentiation cultures. However, these media are supplemented with FBS or even horse serum. Thinking about media that do not receive serum as a supplement, it is possible to mention VP-SFM, a medium with a very low concentration of protein (5 μg/mL) and no substances of animal origin, designed for the growth of VERO cells for virus production and for recombinant protein production.

As previously mentioned, a very important component of culture media is the FBS, which is a mixture composed of amino acids, proteins, growth factors, hormones, vitamins, and trace elements. The concentration of serum used varies depending on the purpose of the cell culture. For proliferation purposes, a concentration of 10% in DMEM medium is generally used. For differentiation purposes, this concentration is greatly reduced and can be as low as 2% (Messmer et al. 2022).

In addition to amino acids, carbohydrates are also needed to act as a source of metabolic energy for the cells. Generally, the media contains glucose as the main source of energy and its concentration can vary. For example, there are two variations of DMEM: high glucose, with 4.5 g/L, and low glucose, with 1.0 g/L. In the case of bovine muscle cells, DMEM high glucose is used. In other cases, galactose, pyruvate, and ribose can be used as energy sources.

Finally, it is necessary to highlight the importance of the sterility of culture media and their supplements. In the case of cultured meat production, sterility conditions are a crucial factor in the approval and authorization of the final product by regulatory agencies.

Starting with culture media, there are those that are sold in a freeze-dried state and are solubilized in the laboratory. The powdered product is sterile but can be contaminated when it is dissolved. There are also media that can be autoclaved but require supplementation after solubilization, increasing the risk of contamination. There are also liquid media that pass through a 0.1 μm filter, capable of eliminating even mycoplasmas; although these media are a little more expensive, they end up offering greater safety because they require less prior handling.

It is also important to guarantee the sterility and immunological neutrality of the FBS. To guarantee sterility, the serums also pass through 0.1 μm filters. And to guarantee immunological neutrality, there are a number of techniques used to ensure this. The main one is heat inactivation, which consists of heating the serum to a temperature of approximately 56 °C in order to inactivate proteins from the complement system, which are responsible for the body's innate defense. In addition, some serums are also irradiated with 25 to 40 kGy (Hyclone Standard FBS 2023).

# 3 Culture Media for Research and Diagnostic Purposes

## *3.1 Culture Media for Animal Cells: Mammals, Avian, and Fish*

Considering all media derivatives, while the choice of medium comes from observation of what is the best formulation for each type of cells that are being cultivated, it is known that the greatest choice relies on the research purposes.

For mammalian cells, depending on cell type, the most common media used in cell culture are DMEM and RPMI-1640, followed by MEM and M199 (Phelan and May 2016; McKee and Komarova 2017; Price 2017).

In avian cell culture, according to a large amount of data, Eagle's medium and its derivatives still have been used. The basic medium described to cultivate chicken primordial germ cells is DMEM high glucose with no glutamine and calcium medium. Dehdilani et al. (2023) compared this DMEM classical medium with a DMEM enriched with lipids and albumin medium. However, in both scenarios, DMEM was the base for each preparation (Dehdilani et al. 2023). Turkey satellites cells, for example, are also cultivated in DMEM enriched with chicken serum and horse serum (Tonniges and Velleman 2019). Indeed, in avian embryonic culture, including, chicken embryo fibroblast DF-1 cells are described as cell lines that usually are cultivated in DMEM/F-12 (Lin et al. 2019; Sukparangsi et al. 2022).

When it comes to fish cell culture, the same similar media preparation was suggested to be used like Eagle's Medium, MEM, M199, and L-15. The use of those media appears to be a result of empirical analysis. Salt concentration, buffer, temperature, carbon source, and pH can be even more important to be evaluated in fish cell culture, which has distinct necessities in comparison to mammalian and avian cells. Indeed, grunt and other sea species may need additional salts into the media (Rubio et al. 2019).

Additionally, according to the literature, DMEM supplemented with FBS, insulin-like growth factor-1 (IGF-1), and insulin-like growth factor-2 (IGF-2) can be used from growth of primary trout satellite cells. DMEM and Ham's F-12 with FBS for red sea bream primary preadipocytes cell culture. L-15 medium enriched with FBS and L-glutamine, it seems to be applied for Atlantic salmon and rainbow trout preadipocytes primary cultivation (Bomkamp et al. 2023). For sturgeon germ cell culture, L-15 medium is another possibility for cell maintenance (Xie et al. 2019).

Currently, people are still using the same culture media produced many years ago. In this regard, O'Neill et al. (2022) suggested that it is necessary to investigate specific nutritional needs of different cell types, in order to elaborate new media that may be more appropriate for culturing cells from different species.

## 3.2 Serum-Free Media for In Vitro Purposes

FBS began to be used as a medium supplement in 1958 when scientists observed that in vitro cells could be cultivated for a longer period with FBS supplementation. The idea came from Burrows and Alexis Carrel, who demonstrated that medium with animal plasma improved cell culture maintenance. Since that, FBS with known and unknown factors is widely used for cell culture (Ho et al. 2021). Unfortunately, for producing few amounts of FBS, multiple calf fetuses are needed. Thus, environmental awareness and its search for animal-free products is leading science to pursue alternatives for culture media supplementation (Brindley et al. 2012; Ho et al. 2021). In parallel, few cells that could grow in the absence of serum were identified, which in some conditions can be essential for specific in vitro purposes (Karnieli et al. 2017; Ho et al. 2021). Considering all that, the idea is not just to have fewer cells that can be managed with a not enriched FBS media but also finding alternatives for the majority that cannot be maintained without serum components supplementation.

CHO is one of the few cell lines known for its capacity of growth without FBS, which was important for the production, in 1986, of the first therapeutic protein, the human plasminogen activator. After that, CHO is still chosen for recombinant protein manufacturing, since growing in a serum-free medium increases protein production safety and results in less processing variability. Moreover, the human embryonic kidney (HEK-293) cell line, which can also be maintained in a FBS-free medium and has an optimal transfection rate efficiency, is also used for therapeutic protein production (Ho et al. 2021).

In addition, it has been discussed the importance of obtaining an FBS devoid media for in vitro monoclonal antibody (mAbs) production, since elevated serum proteins impairs mAbs final product purification. The CHO cell line, used for therapeutic protein production, including antibodies, is the main tool for in vitro antibody manufacturing. There are reports that estimates that ~80% of mAbs approved in the market between 2014 and 2018 were produced in CHO cells (Du et al. 2022; Salah et al. 2022).

Considering the importance of mAbs in therapeutics approaches and in development of diagnostic tests for many human diseases, it is desirable to achieve a higher level of mAbs purity accompanied by an upscaled process. Salah et al. (2022)

recently described an alternative hybridomas secreting mAbs cultivation method, in the absence of FBS, for upscaling in vitro mAbs production.

Moreover, concerns about FBS addition in cell culture can be applied not only for producing therapeutic proteins or monoclonal antibodies but also in the use of in vitro techniques for producing vaccines. Innumerous different reasons justified the search for new alternatives, including the risk of contamination, since animal serum can be a source of pathogens like virus and mycoplasma (Butler 2015; Ho et al. 2021; Lezin et al. 2022).

Interestingly, cell culture in a medium not enriched by serum has been also employed in regenerative medicine field. It was described that even human-derived mesenchymal stem cells (hMSC) can grow and induce pluripotent stem cells and T cells in the absence of serum. Also, in human bone marrow-derived mesenchymal stem cells (BM-hMSC), serum-free media enhanced cell proliferation (Doucet et al. 2023).

In tissue-engineered skin substitutes (TESs), based in the culture of keratinocytes and fibroblasts for treatment of severe burns, it has also been considered to find serum-free alternatives for the primary development of TESs. Finding FBS substitutes for TESs will be important to avoid immunological patient's response to serum proteins (Doucet et al. 2023).

Considering all the scenarios briefly described, it is plausible to assume that drawbacks accompanied the serum media supplementation in terms of science and biosafety. Besides that, it is already known that FBS obtaining comes with ethical concerns about animals' manipulation. Finally, it is possible in the future that there will not be enough serum to support in vitro cell manipulation for medicine demands.

In this regard, in order to find ways to overcome the need of FBS in cell culture, plant-based hydrolysates emerged as an alternative source for medium supplementation. These hydrolysates can be obtained by enzymatic digestion and/or hydrolysis by acid, alkali, and fermentation process of plant products like soy, pea, rice, and rapeseed. The plant-based substitutes of FBS can represent low-cost alternatives for upscale of protein therapeutic production and for a lot of other products obtained by in vitro animal's cell culture (Ho et al. 2021).

Thus, serum-free media could be applied for many different purposes like biomarkers discovering and manufacturing, like mAbs used for human diseases screening, diagnostic, and as a therapeutic tool. Besides, media devoid of animal serum is increasingly desirable for vaccine development, skin grafts in TESs, and regenerative medicine areas.

### 3.3 Applications and Limitations for Cultivated Meat Production

The main purposes of cell-based meat production are to obtain a more sustainable food, which would be a renewable source of meat in order to preserve the environment, promote animal's welfare, and potentially could result in a safer source of protein for human consumption. However, cell-based meat still must overcome several limitations, including changes in well-established cell culture techniques and materials used in laboratories around the world, manufacturing costs and people's final product acceptance (Arshad et al. 2017; O'Neill et al. 2021; Chodkowska et al. 2022; Mateti et al. 2022; Jahir et al. 2023; Samandari et al. 2023).

In comparison with livestock farming, cell-based meat can promote a decrease in greenhouse gas emissions and, in parallel, can prevent ecosystem degradation due to feed raw demands (Chodkowska et al. 2022). Beyond all that, it is reasonable to assume that in vitro cell products can lead us to achieve a renewable source of meat that is compatible with the world's population growth rate, accompanied by less spend of natural resources like water, energy, and land (Jahir et al. 2023). Indeed, environmental awareness and conscious living are becoming increasingly popular and led many consumers to avoid animal origin products. In this regard, for in vitro meat, animal cells could be obtained safely through a biopsy, and animals would be just cell donors (Chodkowska et al. 2022; Mateti et al. 2022; Jahir et al. 2023).

In addition, cell-based meat could be applied in medicine, since cultured meat could be elaborated for specific demands. In this context, patients with a meat dietary restriction because of unwanted components in the meat from a living animal, could consume a cell-based meat without these compounds removed in the laboratory. In the same way, it could be added into the preparation of polyunsaturated fatty acids and other beneficial nutrients for specific clinical situations (Arshad et al. 2017; Chodkowska et al. 2022; Samandari et al. 2023). Additionally, in vitro meat would decrease hazards of contaminations by microorganisms like *E. coli, Campylobacter* spp. and *Salmonella* spp., as it promotes lower risk of other contaminants like dioxins, antibiotics, and would be free of hormones. Collectively, it could reduce rates of foodborne diseases and zoonoses (Chodkowska et al. 2022; Samandari et al. 2023).

On the other hand, an important limitation in cell-based meat production is to add digestible compounds that are not a risk for human health, in order to achieve organoleptic properties such as an ideal texture (Mateti et al. 2022).

Creating a reproducible environment in the laboratory that mimics tissue interactions in living animals for cultivated meat is also a challenge. There is always a limitation in overcoming the differences between in vitro cell culture and the complex in vivo tissue, considering all signaling molecules as hormones released from different organs that at some point interact with muscle cells. Moreover, media formulations can also lead to transcriptome and metabolic cell profile modification, once medium, like DMEM, which is the most used for cell maintenance, do not have the same nutrient concentrations in comparison with living systems (O'Neill et al. 2021).

The main current challenges that could be listed are as follows: (i) finding ways to cost-effectively scale up the production since muscle cells have a cell proliferation limited rate; (ii) be financially accessible and at the same time give profits; (iii) achieve consumer acceptance, since in vitro meat can be seen by the public as an unnatural product (Chodkowska et al. 2022; Mateti et al. 2022); (iv) give the response to the following question made by Samandari et al. (2023): "What are the long-term effects of cultured-meat consumption on human health and environment?"

## 4 Culture Media for Food Grade

As previously discussed, animal cell culture is a well-established technique used for both research purposes and by the pharmaceutical industry. However, seeking for a culture medium for manufacturing a food product, include other factors such as public acceptance and the organoleptic properties that the medium can give to the product. At the same time, expertise in the cultivation of animal cells for purposes other than food has provided us with a range of information about the needs and characteristics for the development of a suitable culture medium for the production of cultured meat.

Generally, these media are precisely formulated to provide essential nutrients, growth factors, and other substances for the growth and differentiation of cultured muscle cells (Ramani et al. 2021). Culture media for cultured meat are composed of a combination of amino acids, sugars, salts, and growth factors (Nikkhah et al. 2023), often obtained from vegetable sources or free from animal components to guarantee sustainability and ethical production. The objective is the same as the conventional media: to create an environment that simulates the interior of an animal's body, allowing cells to multiply and develop into muscle, adipose, vascular tissue, whatever is deemed necessary for the composition of a final meat (Kang et al. 2021; Benny et al. 2022). Furthermore, these media can be adapted to optimize the taste, texture and nutritional profile of cultured meat, making it a promising solution to address the environmental and ethical challenges associated with conventional meat production, without compromising the sensory experience. All of this must also be added to the fact that the culture media for cultivated meat must be capable of being used in bioreactors and be affordable to make the production process viable (Vergeer et al. 2021; Benny et al. 2022).

### *4.1 Serum-Free Media for Cultivated Meat Production*

As the culture medium for cells is one of the key ingredients for the production and viability of cultured meat, it also becomes one of the major bottlenecks for this product comprising ethical and economic issues.

Ethical because, as seen previously, the main culture medium for animal cells is supplemented with FBS, an additive obtained from calf fetuses in the last two thirds of pregnancy, which are discovered accidentally during the slaughter of pregnant cows in the slaughterhouse (Valk et al. 2017; van der Valk 2022), making the addition of this supplementation inconsistent with the cultured meat proposal. It is also an economical issue because replacing FBS in culture media makes its production more expensive, due to the cost of recombinant growth factors (Wood et al. 2023), which will be discussed later in this chapter.

Furthermore, the problems with FBS in traditional cell cultures also apply to meat cultivation, since the use of serum can affect growth and differentiation rates (Usta et al. 2014). Considering that it is from an in vivo source, the composition of the serum can vary drastically between different batches, which can affect the final meat product, and also brings the risk of contamination with bacteria, viruses, or prions (Park et al. 2013; Gaydhane et al. 2018; Post et al. 2020).

Thus, there is an incessant search for the development of a cheap medium, free from animal components and capable of supporting the proliferation and differentiation of cells for cultured meat. At this point, some of these ways to replace FBS in a serum-free medium will be explored.

Recently, some additives have been developed to replace FBS, which would function as an "artificial serum." The composition of these products is, in its accuracy, protected by commercial copyright, but it is known that they contain some factors also found in FBS, such as adhesion and growth factors, hormones, insulin, and vitamins (Price et al. 1998; Kadim et al. 2015). Although the idea of an easy-to-use universal replacement for FBS is the desired gold standard, these commercial products are far from that. They are expensive products, sold in small quantities, and there is no consensus on their functionality for any cell isolate (Duque et al. 2003). However, the use of one of these commercial products has shown success as a substitute for serum in a medium for the proliferation of bovine muscle cells (Messmer et al. 2022). Particularly, the Upside Foods, an American company that developed cultivated chicken meat, managed to develop an animal-component-free cell growth feed capable of being used for the production of chicken hot dogs and nuggets (Poinski 2021).

Other approaches to developing serum-free media for cultured meat production are those based on microbial fermentation to synthesize recombinant growth factors or the use of extracts of non-animal origin (Ellies-Oury et al. 2022). The production of recombinant proteins, called microbial fermentation or precision fermentation, is a strategy that has been used for years in the biotechnology sector, and it is capable of producing molecules in commercial quantities, such as the pharmaceutical insulin. The food industry also uses this technology to produce a series of proteins used in the manufacturing processes of cheese, ice cream, yogurt, sports drinks, among others (Puetz and Wurm 2019; Wood and Tavan 2022). In this context, this type of technology has been explored for the production of molecules found in the composition of FBS, allowing the replacement of serum with the exact concentrations of molecules necessary for cultured meat, by developing the so-called chemically defined media (Chen et al. 2022).

A second approach is the use of extracts of non-animal origin, a promising, cheaper, and simpler strategy (Santos et al. 2023). These products are extracts of fungi, soybeans, bacteria, or microalgae, some of which are already used in the food industry (Benjaminson et al. 2002; Gupta et al. 2018; Andreassen et al. 2020; Akcali et al. 2021; Wood and Tavan 2022). The advantages of using these components include being cheaper compared to defined media, presenting food quality that facilitates integration into the final product and the potential to enhance the flavor of the final meat. However, extracts can easily present variations in their compositions, which can affect the final product, and there is no establishment of these compounds being used in bioreactors (Batish et al. 2022; Chen et al. 2022; Yamanaka et al. 2023).

## *4.2 Applications*

Several serum-free media have been developed including both defined media and media with added extracts. However, not all of them have been studied or are currently used to produce cultured meat. In Table 1 are listed some of the media that were developed directly for the evolution of the cultured meat product and its applications.

Beefy-9 or B9 medium, an optimization of serum-free medium described by Kuo et al. (2020), was able to maintain short-term growth comparable to serum-containing medium and maintain in vitro bovine satellite cell morphology. The B9 medium is composed of a simple mixture of basal media with the addition of recombinantly produced factors and hormones (Kuo et al. 2020; Stout et al. 2022). It is a serum-free medium developed directly for the proliferation of bovine cells for the production of cultured meat. Indeed, the use of microalgae extracts also appears to be promising for the growth of primary bovine myoblasts, being recently reported the cultivation of these cells using *Chlorella vulgaris* extract in order to reducing costs and environmental impacts, as well as improving expansion of the cultured meat market (Defendi-Cho and Gould 2023; Yamanaka et al. 2023).

For growing pork, Guan et al. (2023) developed a chemically defined medium that proved to be effective in inducing the differentiation of porcine satellite cells to generate mature myotubes, already aimed at commercial application and the establishment of standards for cultured pork. Chemically defined media were also used for the culture of porcine pluripotent cells, which may have application in meat cultivation (Choi et al. 2020a, b).

One of the first in vitro meats developed was goldfish (*Carassius auratus*) by Benjaminson et al. (2002) in collaboration with the National Aeronautics and Space Administration (NASA). For that, it was produced as a serum-free medium with crude Maitake mushroom (*Grifola frondosa*) extract, which is a source of amino acids, carbohydrates, and minerals (Wu et al. 2021), in order to achieve similar increases to serum medium in the area of primary fish cell explants (Benjaminson et al. 2002). Also, for the cultivation of fish cells for in vitro meat, the use of insect and marine invertebrate hydrolysates as substitutes for serum is being studied and

**Table 1** Serum-free media for food purposes: classification, composition, and uses

| Medium | Composition | Use | Reference |
|---|---|---|---|
| Beefy-9 | DMEM/F12 in a 1:1 ratio with HEPES, 20 μg/mL insulin, 200 μg/mL L-ascorbic acid-2-phosphate, 20 μg/mL transferrin, 20 ng/mL sodium selenite, 40 ng /mL FGF-2, 0.1 ng/mL TGF-β3, 0.1 ng/mL NRG1 and 800 μg/mL human albumin | Bovine satellite cell growth | Stout AJ, Mirliani AB, Rittenberg ML, Shub M, White EC, Yuen JSK, Kaplan DL (2022) Simple and effective serum-free medium for sustained expansion of bovine satellite cells for cell cultured meat. Commun Biol 5:466. |
| Serum-free and grain-derived-nutrient-free medium using microalga-derived nutrients and mammalian cell-secreted growth factors | *Chlorella vulgaris*-derived nutrient media with RL34 cells secreto-me | Proliferation of primary bovine myoblasts | Yamanaka K, Haraguchi Y, Takahashi H, Kawashima I, Shimizu T (2023) Development of serum-free and grain-derived-nutrient-free medium using microalga-derived nutrients and mammalian cell-secreted growth factors for sustainable cultured meat production. Sci Rep 13:498. |
| Guan serum-free medium | DMEM supplemented with 0.5 mg/mL BSA, 0.5% amino acid solution, 0.2% insulin–transferrin–selenium supplement and 20 μM naringenin | Differentiation of porcine satellite cells to generate mature myotubes | Guan X, Yan Q, Ma Z, Zhou J (2023) Production of mature myotubes in vitro improves the texture and protein quality of cultured pork. Food Funct 14:3576–3587. |
| Hydrolysate insect medium | DMEM with 1% FBS and black soldier flies and 0.001-0.1 mg/mL lugworms hydrolysates | Zebrafish embryonic stem cell line ZEM2S CRL-2147 growth | Batish I, Zarei M, Nitin N, Ovissipour R (2022) Evaluating the Potential of Marine Invertebrate and Insect Protein Hydrolysates to Reduce Fetal Bovine Serum in Cell Culture Media for Cultivated Fish Production. Biomolecules 12 |
| Maitake medium | MEM-Hanks' with Maitake extract in a ratio of 1:9 | Growth of fish muscle cell explants | Benjaminson MA, Gilchriest JA, Lorenz M (2002) In vitro edible muscle protein production system (mpps): stage 1, fish. Acta Astronaut 51:879–889 |
| Yeast medium | DMEM with 10 mg/mL of yeast extract | Maintenance of primary skeletal bovine cells | Andreassen R, Pedersen M, Kristoffersen K, Rønning S (2020) Screening of by-products from the food industry as growth promoting agents in serum-free media for skeletal muscle cell culture. Food Funct 11. |

**In white: defined culture media**
In yellow: culture media with added extracts

led to a reduction of up to 90% in the use of serum with the addition of hydrolysates from black soldier flies and lugworms (Batish et al. 2022).

Yeast extract is another product already used for decades in the area of microbiology and has been tested for replacing serum in a medium able to support growth of primary bovine skeletal muscle cells. Specifically, this product was capable of recovering cell growth, metabolism, and proliferation, under reduced and serum-free conditions (Andreassen et al. 2020; Santos et al. 2023).

Although there are several alternatives to the use of serum and the development of different media for growing meat, the use of each of these additives or media comes with advantages and disadvantages for the future cultured meat industry.

## 4.3 Pros and Cons

Not everything that glitters is gold, which is why no serum-free medium has established itself as the gold standard for growing cultured meat. The need to eliminate FBS in meat farming is a consensus among the scientific community, due to all the issues related to the use of this ingredient already discussed earlier in this chapter, but the use of serum-free media is far from being well consolidated.

There are, of course, advantages of using a serum-free medium for cultured meat such as greater control, definition, and reproducibility of formulations, allowing for more convincing and consistent results (Messmer et al. 2022; Koh et al. 2023); animal welfare (Bhat et al. 2015); minimization of the risk of contamination (Jang et al. 2022); and producing media without FBS can be more sustainable, as it does not depend on animal husbandry, reducing pressure on natural resources. Thus, the advantages of using these media are all precisely linked to the fact that they do not contain FBS.

At the same time, the use of serum-free media raises some issues that still need to be resolved. The first is the optimization: satellite cells derived from different animal species have different requirements and also respond differently to certain factors used in these environments (Zidarič et al. 2020). Therefore, there is no single formulation of serum-free culture medium that can be used by a wide range of cell types, requiring optimization for each case (O'Neill et al. 2021; Jang et al. 2022). In fact, there are few serum-free media that have been developed for the cultivation of meat from aquatic animals (Batish et al. 2022).

Another issue is that none of the serum-free culture media present cell proliferation rates as good as the use of serum (Andreassen et al. 2020; Batish et al. 2022; Messmer et al. 2022; Stout et al. 2022; Yamanaka et al. 2023), even those presented here. Furthermore, chemically defined media are more expensive than serum media, because producing growth factors recombinantly, eliminating the use of additives from animal sources, still has very high costs (Vergeer et al. 2021; Wood et al. 2023).

Generally, the residual culture medium is discarded and replaced with a new one, due to the reduction of nutrients and the accumulation of waste products. This medium, requested in large volumes and then discarded, can generate environmental problems, such as the eutrophication of bodies of water, so the treatment and disposal of the culture medium for growing meat is still a hurdle to be overcome (Yamanaka et al. 2023).

Therefore, to obtain an ideal culture media and proper ways to its application in cultivated meat industry, there is still a path of research and process optimization ahead.

# 5 How to Estimate Costs of Culture Media for Cultivated Meat Production

Optimizing media for tissue engineering and cultivated meat production is difficult due to the extensive experimentation required, number of media components, non-linear and interactive responses, and the number of conflicting design objectives. In this sense, Cosenza et al. (2021) have demonstrated the capacity of a nonlinear design-of-experiments (DOE) method to predict optimal media conditions based on few experiments, which can reduce the cost for muscle cell proliferation media.

Also, when considering options for future foods, culture media to support the process has been identified as a significant contributor to the overall global warming potential (GWP) and cost of cultivated meat production. To address this issue, an artificial intelligence-based approach has been applied to simultaneously optimize the GWP, cost, and cell growth rate of a reduced-serum culture media formulation for a zebrafish (ZEM2S cell line) cultivated meat production system (Nikkhah et al. 2023). In this study, response surface methodology (RSM) was used to design the experiments with seven components—IGF, fibroblast growth factor (FGF), transforming growth factor (TGF), platelet-derived growth factor (PDGF), selenium, ascorbic acid, and serum—as independent variables. Radial basis function (RBF) neural networks and genetic algorithm (GA) were applied for prediction of dependent variables and optimization of the culture media formulation, respectively. The results indicated that the developed RBF could accurately predict the GWP, cost, and growth rate, with a model efficiency of 0.98. The suggested RSM + RBF + GA framework could be applied to sustainably optimize serum-free media development, identifying the combination of media ingredients that balances yield, environmental impact, and cost for various cultivated meat cell lines.

## 5.1 *What Parameters Contribute to the High Cost of Culture Media?*

A common observation in techno-economic analysis (TEA) for the process is that products with smaller markets should cost more. Indeed, prices for research-grade cell-culture media components, which are packaged at extremely high purity and small quantities, are liable to cost more. Even at the biopharmaceutical scale, where media components are routinely purchased in high quantities (10–1000 kg), these can cost much more than food- or feed-grade formulations of the same substance. To predict the costs of specific chemicals at scale for TEA purposes, some exponential methods have been proposed (Hart and Sommerfeld 1997; Qi et al. 2015). In general, most of the important cell-culture media components are not currently produced at scales consistent with food production. In Fig. 1, the main components of cellular culture media and those that contribute most to the cost of this product are shown.

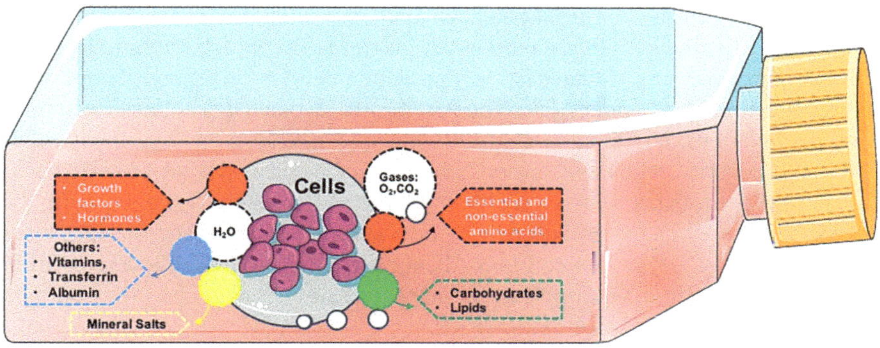


**Fig. 1** Main components found in cell culture media. In red, the components currently most expensive for media manufacturing

### 5.1.1 Macronutrients

Glucose is the primary carbon and energy source, required at 0.36 kg per kg of wet cell mass (Humbird 2021). Even though prices for glucose may range from \$0.50 to \$1.00/kg, glucose is not expected to be a bottleneck to scale-up nor a significant contributor to cost. Specifically, glucose at \$0.26/kg would only contribute \$0.24/kg wet cell mass. In the long term, 4000 kTA of dextrose would be enough to make 11,400 kTA of animal cell mass.

Amino acids, on the other hand, may be a challenge even in the short term. Each essential and nonessential amino acid is currently produced at some commercial scale (Drauz et al. 2007), for animal-feed supplements and for pharmaceutical uses including cell-culture media. However, the new demand for cultivated meat may exceed their existing market, requiring a dramatic scale-up of amino acid production.

In general, amino acids with smaller market volumes cost more, being expected that high-volume/low-cost parameters for the major amino acids reflect feed-grade formulations unsuitable for cell culture media. As an example, DL-methionine (Met) is produced by chemical synthesis as a racemic mixture that can be metabolized by humans and animals without separation; however, the same is not true for cells in culture (Zuo et al. 2019). L-threonine (Thr), the lowest-cost amino acid, is produced commercially in *E. coli* which comes together with high levels of lipopolysaccharide (LPS) endotoxin. Costs associated with LPS removal generally require the use of *E. coli* in low-volume food- or pharma-grade amino acids (Murali and Nampoothiri 2015), but LPS is not a problem for feed-grade formulations, as livestock are already exposed to high levels of *E. coli* and LPS does not cross the intestinal barrier to accumulate in tissues, for example (Wallace et al. 2016). L-lysine (Lys), on the other hand, is mainly produced by *Corynebacterium glutamicum*, which does not have an LPS problem. However, feed-grade formulations of lysine have less stringent specifications than food- or pharma-grade formulations. In fact, if potential contaminants did not directly inhibit cell growth or development in cell-culture media, they would very likely be left behind in the product.

Additionally, formulations that balance price and suitability for cell culture are, in a sense, designing a "cultured meat-grade" formulation, which seems to be a new product. This product must have its own production and purification process, quality specifications, and market size. The parameters associated with manufacture of these formulations, including yield, bioreactor size, and costs, will certainly influence the final price. For these reasons, the market volume of a novel amino acid formulation is probably a more reliable predictor of its price at scale. As such, the amino acid prices must impact decisively the cost for cultured-meat cost, since macronutrient cost is $19/kg wet cell mass, being approximately $12/kg derived from essential amino acids. Nevertheless, the nonessential ones must be provided by a variable composition of other amino acids.

### 5.1.2 Plant Protein Hydrolysates

Soybean hydrolysate has been suggested as a potential alternative source of amino acids. Especially in animal-free media formulations for cell culture, enzymatic hydrolysates, and peptones of plant proteins are sometimes used as supplements and serum replacements but only at low levels (2–20 g/L) in chemically defined media (Babcock et al. 2007). Even at low levels, Hartshorn et al. (2010) demonstrated that the hydrolysate preparation process can have a marked effect on cell growth. The hydrolysate must be ultra-filtered to remove all protein (including the enzyme) and variable-molecular weight peptides, pasteurized, and spray-dried for storage [85]. As a result, the production process can be very expensive, and the suitability of plant protein hydrolysates is still less clear.

### 5.1.3 Protein Micronutrients and the Fetal Bovine Serum

The several hormones, cytokines, vitamins, and growth factors that must be provided in cell-culture media to regulate cell growth and metabolism are responsible for the criticism and speculation among cultured meat. As previously described, at laboratory scale, growth factors are typically supplied in complex sera extracted from animal blood. Specifically, FBS is the common research serum that tends to work very well for most mammalian cell lines and therefore appears in many published laboratory protocols. FBS is indeed very expensive and has known supply-chain problems, even at research scale (Stanton 2019). Also, due to the potential for contamination by viruses and prions, FBS has largely been eliminated from large-scale biopharmaceuticals manufacturing, and it has no future chance of ever becoming an approved ingredient in anything intended for general human consumption. In this sense, cultured-meat developers aim to eliminate or replace FBS and other animal sera (Mosa Meat 2019; Rowland 2020; Watson 2020).

Some alternatives sera that have been developed are plant-based since they comprise highly processed and purified extracts of plant proteins as discussed above, and/or recombinant proteins made in *E. coli* or *Pichia pastoris* hosts grown on glucose, glycerol, or methanol, which are very expensive (Nienow 2006). In fact, commercial protein products from cell culture and fermentation have a price-volume relationship that is quite similar to that for amino acids.

In order to estimate costs for basal culture media to support muscle cells proliferation (and differentiation), Specht (2020) calculated the price for eight variations of the medium known as Essential 8 (E8) (Chen et al. 2011), with a cost per liter of $377.00, being 99% attributed to the growth factors insulin, transferrin, FGF-2, and TGF-β, where FGF-2 and TGF-β represent over 96% of the total cost of the medium. Examples of high-cost components are ascorbic acid 2-phosphate (AA2P), HEPES, D-glucose, NaCl, sodium pyruvate, L-arginine hydrochloride, L-Isoleucine, and L-Lysine hydrochloride. Finally, Specht (2020) proposed hypothetical scenarios to estimate the cost reduction, which ranged from $376,80/L to $0.24/L, when production takes place in full-scale plants, which do not still exist. As hypothetical scenarios, they have failed to correctly estimate the final cost of media for cultivated meat. In this sense, in a further publication by Vergeer et al. (2021), a correction was made in the predicted scenarios regarding the lower end price of soy hydrolysate, which was calculated for agricultural grade hydrolysate. Specifically, it was not in line with the assumptions of food-grade process inputs, showing that the problem can be more complex than initially expected, and the total costs per kg cell mass in the low-medium scenarios increased by $ 0.29. The authors reported that this represents less than 10% of the total costs; however, thinking in tons of cultivated protein to feed a population, any changes in the estimated costs can imbalance the productive food chain.

In summary, it has been shown that current cell mass production costs can be up to 10,000 times higher than values for comparable traditional meat products, depending on the requirements of medium components and its prices. Simultaneously, the reports have shown that cost reductions are feasible (Vergeer et al. 2021) but require a combination of various factors including the mandatory reduction in the production costs of growth factors and recombinant proteins.

## 6 Challenges and Future Developments

Cultured meat may be a new stage of the meat production system after hunting and livestock; however, the development of a low-cost culture medium in which all animal components should be removed from the manufacturing process is one of the top priorities for cell-based meat (Haraguchi and Shimizu 2021).

## *6.1 Low-Cost Media*

The development of the serum-free and animal component-free media is relevant for the cell-based meat process; however, it must guarantee high rates of proliferation, to be low cost, and use food-safe ingredients, as well as to meet criteria of sustainability, controllability, and reproducibility of the bioprocess, which is possible and feasible for a single cell type (Benny et al. 2022).

Cost reductions can be achieved by changing the media formulation, adopting the in-house production of components, purchasing raw material in bulk, and replacement of pharmaceutical-grade certified components for food-grade components, which is appropriate for the food end product, eliminating, for example, the need for high-performance liquid chromatography purification.

As such, the proteins (albumin, transferrin, and insulin) and the growth factors FGF-2, epidermal growth factor (EGF), IGF-1, neuregulin 1 (NRG1), TGF-β, and PDGF-B can be obtained using recombinant DNA technology, in production hosts, by several manufacturing strategies: precision fermentation in filamentous fungi or microbes such as bacteria and yeast, molecular farming in plants, cell-free production, production in insects, and in plant or animal cell cultures. In an hypothetical scenario estimated for the future, albumin would be produced at $10/kg, insulin and transferrin at $1000/kg, and growth factors at $100,000/kg (Swartz 2023).

The low-cost serum-free medium B8 was first described for human-induced pluripotent stem cells (hiPSC line 19c3); this medium contains a simple mixture of basal media (glucose, amino acids, vitamins, salts, and fatty acids), minerals, and proteins. Lab scale of B8 production was demonstrated at a cost of approximately $16/L when using in-house growth-factor production (Kuo et al. 2020). Medium B8 was adapted for bovine satellite cells through the addition of a single component, the recombinant albumin, and this new media (Beefy-9) was able to maintain growth of bovine cells (Specht 2020).

In this context, biotech companies and research groups are investing in research and development for cost reduction in each approach according to demand of the cultivated meat industry. GenScript (https://www.genscript.com/) offers recombinant proteins and growth factors (human, bovine, and salmon species-specific) for cultured meat production at kilogram level availability. Laurus Bio (https://laurus.bio/) is an integrated research-driven biomanufacturing organization with experience in precision fermentation and recombinant DNA technology that produces food-grade animal-free recombinant proteins, growth factors, and cell-culture media supplements to the cultured meat industry. Core Biogenesis (https://corebiogenesis.com/) uses the oilseed plant *Camelina sativa* as the host for the stable expression and production of growth factors and recombinant proteins. The product Core Factors® is a family of proprietary, plant-based recombinant growth factors, used in applications such as 3D cell culture, organoids production, stem cell expansion, and cellular agriculture. LenioBio (https://www.leniobio.com/) uses the cell-free protein expression (CFPE) technology that enables high yields of up to 3 g/L of growth factors and recombinant proteins. IntegriCulture (https://integriculture.

com/), a Japanese biotechnology startup, developed the IntegriCulture Minimum Essential Media (I-MEM) for cultivated meat production and proprietary cell culture technology CulNet™ system. The system is a device that uses inter-organ interactions to produce serum components and use them for cell media, artificially creating an environment similar to that inside an animal's body to grow cells. The company CellCrine (https://thespoon.tech/indiebio-startup-cellcrine-is-developing-serum-free-growth-media-that-reduces-costs-by-90/) claims that their chemically defined, serum-free media uses 90%–99% less growth factors and recombinant proteins. BioBetter™ (https://biobetter.bio/) showed a technology based on tobacco plant protein-expression platform as a solution that makes the cultivated meat industry commercially viable. Using tobacco plants as bioreactors, it produced the BioBov line of food-grade recombinant bovine growth factors (FGF-2, transferrin, insulin), and new products are in development (IGF-1, TGF-β1, Activin A, leukemia inhibitory factor). Lifeasible (https://www.lifeasible.com/) is currently developing solutions to enable the production of food-grade culture medium at the lowest cost that can regulate large-scale cell proliferation and differentiation with acceptable organoleptic quality such as microalgae and microbial-based media, platelet lysates (to serum-free culture), and serum replacement (for a single-cell type).

In addition, Venkatesan et al. (2022) presented a set of expression constructs and a simplified protocol for recombinant production of functionally bioactive growth factors (FGF-2, IGF-1, PDGF-BB and TGF-β1) from species including bovine, chicken, and fish in *E. coli*. The production cost of growth factors was $10.22/mg to $20.43/mg.

Different companies (Calysta, https://calysta.com/; Air Protein, https://www.air-protein.com/; Solar Foods, https://solarfoods.com/; Arkeon Biotechnologies, https://arkeon.bio; Farmless, https://farmless.com/) have made significant investments to make proteins using microbes and air, known as air-proteins, and created efficient fermentation platforms to producing landless proteins, with all 20 amino acids necessary for human metabolism.

Key players in the biopharma industry as Thermo Fisher (https://www.thermofisher.com/br/en/home/life-science/cell-culture/cellular-agriculture.html) and Merck (https://www.merckgroup.com/en/research/research-and-development-highlights/cultured-meat.html) offer knowledge and expertise in areas such as cell culture media and bioprocessing technologies to accelerate the emerging industry of cell-based meat.

Finally, several studies have demonstrated proof of principle for the use of size-exclusion dialysis to capture and reuse growth factors and recombinant proteins in bioprocessing applications (Nath et al. 2017; Torizal et al. 2021).

## 6.2 *Specific Regulatory Issues for Media Directed for Food Grade*

In the search for molecularly identical compounds to replace serum, growth factors, and non-animal ingredients and provide food-grade media components, peptides, peptones, small molecules, and less-refined hydrolysates from various origins can potentially provide a high protein content, suitable functional properties, and adequate amino-acid composition, while also promoting robust growth and/or differentiation of the cells.

In this context, food regulatory agencies must focus on creating guidelines or regulatory framework to monitor and promote the acceptability of new food-grade media to cell-culture meat. The safety evaluation and an objective regulatory policy of cell-based meat, as a type of brand-new meat product, must be carefully established to ensure food safety and nutrient content.

The food safety evaluation standards and policy restrictions for all products, additives and materials, with initial application in the food sector, should be developed as soon as possible to ensure the development of the cell-based meat industrial chain.

The safety concern for microbiological contamination by bacteria (including mycoplasma), fungi, and viruses can occur throughout the entire bioprocess. Mycoplasma contamination may be of particular concern, they are resistant to antibiotics, can pass through filters, and grow slowly so may not be detected in continuous cell cultures (Ong et al. 2021).

The use of hydrolysates from alternative proteins (alt-proteins) from filamentous fungus, yeast, algae, insects (flies, crickets), oysters, mussels, lugworms, and other microorganisms to replace animal serum or amino acid in the culture medium requires further study. The potential cytotoxicity and safety of alt-proteins hydrolysates has not been tested and is a safety concern. Since hydrolysate is a processed product, contaminants could be present in the raw material used or generated during the hydrolysis process. For example, (1,3)-β-glucans in the cell wall from yeast are potentially immunogenic, and hydrolysates derived from microbes (Gram-negative bacteria like *E. coli*) may contain high levels of LPS, an endotoxin that require a procedure for removal (Ho et al. 2021).

In this context, all food-grade medium components and extra additions for cell proliferation and differentiation should all be used in accordance with the applicable regulations. If sourced from upcycled or recycled streams, these ingredients must be determined safe (pathogen and chemical-residue-free) by target market regulatory authorities.

### 6.3 Future Developments

Approximately 50 L of culture media is required to produce 1 kg of cultured meat (Post 2014). The increased demand for cultured meat leads to increased culture media production. Thus, three issues associated with cultured meat must be considered: (i) the grain-derived nutrients and/or plant-based hydrolysates; (ii) use of animal sera; and (iii) mode of discard/treatment/recycling of large volumes of spent medium (media that remains after cell cultivation) generated during cultured meat.

Okamoto et al. (2020) showed that nutrients present in algae extracts (glucose from *Chlorococcum littorale* or *Arthrospira platensis* and amino acids from *Chlorella vulgaris*) can be used in the culture medium for mammalian cell cultivation. Okamoto et al. (2022) reported on the proliferation and differentiation of primary bovine myoblasts using *Chlorella vulgaris* extract (CVE) as an alternative to the conventional culture medium for sustainable production of cultured meat.

Haraguchi et al. (2022) proposed a circular cell culture (CCC) system in which microalgae (*Chlorococcum littorale*), a nutrient supply and waste-medium recycler, and growth factor-producing cells (RL34 hepatocytes) are used for the mammalian cell culture (C2C12 myoblasts). The CCC system was used, in which the rat liver epithelial RL34 cells were cultured by adding CVE to inorganic salt solution. The supernatant, containing the RL34 cell-secreted growth factors and CVE, was used as the nutrient source to primary bovine myoblast cultures (Yamanaka et al. 2023). The innovative CCC system is a resource-recycling serum-free meat production system using microalgae and mammalian cells.

Lynch and O'Connell (2022) showed that mammalian cell culture waste, used to grow CHO cells, can be recycled and used as a feed for the growth of *E. coli* cultures in secondary fermentation to produce new recombinant proteins.

Tuomisto and Teixeira de Mattos (2011) reported on the use of Cyanobacteria hydrolysate supplemented with the growth factors and vitamins to support muscle cell growth.

Other important issues to R&D are as follows: (i) the real-time monitoring of various parameters such as pH, dissolved oxygen and carbon dioxide, essential nutrient concentrations, and metabolic waste; (ii) new procedures for sterilizing the culture medium or developing components more heat stable; and (iii) the development of scaffold to efficient transport of oxygen, nutrients, and waste products to and from the cells, controlling the growing tissue's geometry and cell type distribution and contributing structure to the final product.

## 7 Conclusions

In conclusion, we have shown here the birth and improvement of culture media for cell cultivation and how this process helps in the development of culture media for the production of food products. Cultured meat, as one of these products, requires specific culture media with specific characteristics for the large-scale development and differentiation of the cells necessary for the constitution of meat. These media must be serum-free and potentially to be replaced by extracts or hydrolysates of non-animal origin or by factors produced by microbial fermentation. It has been demonstrated that suitable formulations of amino acids and protein growth factors are not currently produced at scales consistent with food production, with the exception of some recombinant proteins and growth factors as discussed. The replacement of amino acids by plant protein hydrolysates or others requires further studies. Until now, the analysis concluded that metabolic efficiency enhancements and the development of low-cost serum-free media are necessary but yet insufficient to assume the substitution of conventional meat by cultured meat.

## References

Akcali KC, Erikci E, Kiran F (2021) Microbiota-derived postbiotics: alternative supplement to fetal bovine serum for cultured meat. 13

Andreassen R, Pedersen M, Kristoffersen K, Rønning S (2020) Screening of by-products from the food industry as growth promoting agents in serum-free media for skeletal muscle cell culture. Food Funct 11:2477. https://doi.org/10.1039/C9FO02690H

Arshad MS, Javed M, Sohaib M et al (2017) Tissue engineering approaches to develop cultured meat from cells: a mini review. Cogent Food Agric 3(1). https://doi.org/10.1080/23311932.2017.1320814

Babcock J, Smith S, Huttinga H, Merrill D (2007) Enhancing performance in cell culture. GEN Genet Eng Biotechnol News 2007:20

Barnes D, Sato G (1980) Serum-free cell culture: a unifying approach. Cell 22:649–655. https://doi.org/10.1016/0092-8674(80)90540-1

Batish I, Zarei M, Nitin N, Ovissipour R (2022) Evaluating the potential of marine invertebrate and insect protein hydrolysates to reduce fetal bovine serum in cell culture media for cultivated fish production. Biomolecules 12

Benjaminson MA, Gilchriest JA, Lorenz M (2002) In vitro edible muscle protein production system (MPPS): stage 1, fish. Acta Astronaut 51:879–889. https://doi.org/10.1016/S0094-5765(02)00033-4

Benny A, Pandi K, Upadhyay R (2022) Techniques, challenges and future prospects for cell-based meat. Food Sci Biotechnol 31:1225–1242. https://doi.org/10.1007/s10068-022-01136-6

Bhat ZF, Kumar S, Fayaz H (2015) In vitro meat production: challenges and benefits over conventional meat production. J Integr Agric 14:241–248. https://doi.org/10.1016/S2095-3119(14)60887-X

Bomkamp C, Musgrove L, Marques DMC et al (2023) Differentiation and maturation of muscle and fat cells in cultivated seafood: lessons from developmental biology. Mar Biotechnol 25:1–29. https://doi.org/10.1007/s10126-022-10174-4

Brindley DA, Davie NL, Culme-Seymour EJ et al (2012) Peak serum: implications of serum supply for cell therapy manufacturing. Regen Med 7:7–13. https://doi.org/10.2217/rme.11.112

Butler M (2015) Serum and protein free media. In: Butler M (ed) Animal cell culture. Cell engineering. Springer, Cham, pp 223–236. https://doi.org/10.1007/978-3-319-10320-4_8

Cantor JR (2019) The rise of physiologic media. Trends Cell Biol 29:854–861. https://doi.org/10.1016/j.tcb.2019.08.009

Carrel A (1923) A method for the physiological study of tissues in vitro. J Exp Med 38:407–418. https://doi.org/10.1084/jem.38.4.407

Chen G, Gulbranson DR, Hou Z et al (2011) Chemically defined conditions for human iPSC derivation and culture. Nat Methods 8:424–429. https://doi.org/10.1038/nmeth.1593

Chen L, Guttieres D, Koenigsberg A et al (2022) Large-scale cultured meat production: trends, challenges and promising biomanufacturing technologies. Biomaterials 280:121274. https://doi.org/10.1016/j.biomaterials.2021.121274

Chodkowska KA, Wódz K, Wojciechowski J (2022) Sustainable future protein foods: the challenges and the future of cultivated meat. Foods 11. https://doi.org/10.3390/foods11244008

Choi K-H, Lee D-K, Oh J-N et al (2020a) Pluripotent pig embryonic stem cell lines originating from in vitro-fertilized and parthenogenetic embryos. Stem Cell Res 49:102093. https://doi.org/10.1016/j.scr.2020.102093

Choi K-H, Yoon JW, Kim M et al (2020b) Optimization of culture conditions for maintaining pig muscle stem cells *in vitro*. Food Sci Anim Resour 40:659–667. https://doi.org/10.5851/kosfa.2020.e39

Cosenza Z, Block DE, Baar K (2021) Optimization of muscle cell culture media using nonlinear design of experiments. Biotechnol J 16:2100228. https://doi.org/10.1002/biot.202100228

Defendi-Cho G, Gould TM (2023) In vitro culture of bovine fibroblasts using select serum-free media supplemented with Chlorella vulgaris extract. BMC Biotechnol 23:4. https://doi.org/10.1186/s12896-023-00774-w

Dehdilani N, Yousefi Taemeh S, Rival-Gervier S et al (2023) Enhanced cultivation of chicken primordial germ cells. Sci Rep 13:12323. https://doi.org/10.1038/s41598-023-39536-1

Doucet EJ, Cortez Ghio S, Barbier MA et al (2023) Production of tissue-engineered skin substitutes for clinical applications: elimination of serum. Int J Mol Sci 24:9–11. https://doi.org/10.3390/ijms241612537

Drauz K, Grayson I, Kleemann A et al (2007) Amino acid. In: Bohnet M (ed) Ullmann's encyclopedia of industrial chemistry, 6th edn. Wiley-VCH, Weinheim, Germany

Du Q, Zhang X, Wang T, Wang X (2022) Effects and mechanisms of animal-free hydrolysates on recombination protein yields in CHO cells. Appl Microbiol Biotechnol 106:7387–7396. https://doi.org/10.1007/s00253-022-12229-x

Duque P, Gómez E, Díaz E et al (2003) Use of two replacements of serum during bovine embryo culture in vitro. Theriogenology 59:889–899. https://doi.org/10.1016/S0093-691X(02)01134-2

Ellies-Oury M-P, Chriki S, Hocquette J-F (2022) Chapter Six - Should and will "cultured meat" become a reality in our plates? In: Wu JBT-A (ed) Emerging sources and applications of alternative proteins. Academic Press, pp 181–212

Freshney RI (2010) Culture of animal cells: manual of basic technique and specialized applications, 6th edn. Wiley, Hoboken, NJ

Gaydhane MK, Mahanta U, Sharma CS et al (2018) Cultured meat: state of the art and future. Biomanuf Rev 3:1. https://doi.org/10.1007/s40898-018-0005-1

Guan X, Yan Q, Ma Z, Zhou J (2023) Production of mature myotubes in vitro improves the texture and protein quality of cultured pork. Food Funct 14:3576–3587. https://doi.org/10.1039/D3FO00445G

Gupta S, Lee JJL, Chen WN (2018) Analysis of improved nutritional composition of potential functional food (Okara) after probiotic solid-state fermentation. J Agric Food Chem 66:5373–5381. https://doi.org/10.1021/acs.jafc.8b00971

Ham RG, McKeehan WL (1979) Media and growth requirements. In: Cell culture. Academic Press, pp 44–93

Haraguchi Y, Shimizu T (2021) Three-dimensional tissue fabrication system by co-culture of microalgae and animal cells for production of thicker and healthy cultured food. Biotechnol Lett 43:1117–1129. https://doi.org/10.1007/s10529-021-03106-0

Haraguchi Y, Okamoto Y, Shimizu T (2022) A circular cell culture system using microalgae and mammalian myoblasts for the production of sustainable cultured meat. Arch Microbiol 204:615. https://doi.org/10.1007/s00203-022-03234-9

Hart P, Sommerfeld J (1997) Cost estimation of specialty chemicals from laboratory-scale prices. Cost Eng 39:31–35

Hartshorn J, Mcnorton S, Hernandez C et al (2010) Soy hydrolysate optimization for cell culture applications. In: Cells culture. Springer, Dordrecht. https://doi.org/10.1007/978-90-481-3419-9_135

Ho YY, Lu HK, Lim ZFS et al (2021) Applications and analysis of hydrolysates in animal cell culture. Bioresour Bioprocess 8:93. https://doi.org/10.1186/s40643-021-00443-w

Humbird D (2021) Scale-up economics for cultured meat. Biotechnol Bioeng 118:3239–3250. https://doi.org/10.1002/bit.27848

Hyclone Standard FBS (2023) Hyclone Standard FBS, US Origin. In: Cytiva. https://www.cytivalifesciences.com/en/us/shop/cell-culture-and-fermentation/sera/fetal-bovine-serum/hyclone-standard-fetal-bovine-serum-fbs-u-s-origin-p-06038#related-documents%3E. Accessed 10 Sep 2023

Jahir NR, Ramakrishna S, Abdullah AAA, Vigneswari S (2023) Cultured meat in cellular agriculture: advantages, applications and challenges. Food Biosci 53:102614. https://doi.org/10.1016/j.fbio.2023.102614

Jang M, Pete E, Bruheim P (2022) The impact of serum-free culture on HEK293 cells: from the establishment of suspension and adherent serum-free adaptation cultures to the investigation of growth and metabolic profiles. Front Bioeng Biotechnol 10. https://doi.org/10.3389/fbioe.2022.964397

Kadim IT, Mahgoub O, Baqir S et al (2015) Cultured meat from muscle stem cells: a review of challenges and prospects. J Integr Agric 14:222–233. https://doi.org/10.1016/S2095-3119(14)60881-9

Kang D-H, Louis F, Liu H et al (2021) Engineered whole cut meat-like tissue by the assembly of cell fibers using tendon-gel integrated bioprinting. Nat Commun 12:5059. https://doi.org/10.1038/s41467-021-25236-9

Karnieli O, Friedner OM, Allickson JG et al (2017) A consensus introduction to serum replacements and serum-free media for cellular therapies. Cytotherapy 19:155–169. https://doi.org/10.1016/j.jcyt.2016.11.011

Koh B, Sulaiman N, Fauzi MB et al (2023) A three-dimensional Xeno-free culture condition for Wharton's Jelly-mesenchymal stem cells: the pros and cons. Int J Mol Sci 24(4):3745

Kuo H-H, Gao X, DeKeyser J-M et al (2020) Negligible-cost and weekend-free chemically defined human iPSC culture. Stem Cell Rep 14:256–270. https://doi.org/10.1016/j.stemcr.2019.12.007

Lezin C, Mauduit P, Uzan G, Abdelgawad ME (2022) An evaluation of different types of peptone as partial substitutes for animal-derived serum in Vero cell culture. Altern Lab Anim 50:339–348. https://doi.org/10.1177/02611929221122780

Lin J, Yi X, Zhuang Y (2019) Medium optimization based on comparative metabolomic analysis of chicken embryo fibroblast DF-1 cells. RSC Adv 9:27369–27377. https://doi.org/10.1039/c9ra05128g

Lynch CD, O'Connell DJ (2022) Conversion of mammalian cell culture media waste to microbial fermentation feed efficiently supports production of recombinant protein by Escherichia coli. PLoS One 17:e0266921

Mateti T, Laha A, Shenoy P (2022) Artificial meat industry: production methodology, challenges, and future. JOM 74:3428–3444. https://doi.org/10.1007/s11837-022-05316-x

Mather JP (1998) Making informed choices: medium, serum, and serum-free medium. How to choose the appropriate medium and culture system for the model you wish to create. Methods Cell Biol 57:19–30

McKee TJ, Komarova SV (2017) Is it time to reinvent basic cell culture medium? Am J Physiol Cell Physiol 312:C624–C626. https://doi.org/10.1152/ajpcell.00336.2016

Messmer T, Klevernic I, Furquim C et al (2022) A serum-free media formulation for cultured meat production supports bovine satellite cell differentiation in the absence of serum starvation. Nat Food 3:74–85. https://doi.org/10.1038/s43016-021-00419-1

Mosa Meat (2019) Growth medium without fetal bovine serum (FBS). In: Mosa Meat

Murali A, Nampoothiri KM (2015) White biotechnology for amino acids. In: Industrial biorefineries and white biotechnology. Elsevier, pp 445–471

Nath S, Nagamori E, Horie M, Kino-oka M (2017) Culture medium refinement by dialysis for the expansion of human induced pluripotent stem cells in suspension culture. Bioprocess Biosyst Eng 40:123. https://doi.org/10.1007/s00449-016-1680-z
Nienow A (2006) Reactor engineering in large scale animal cell culture. Cytotechnology 50:9–33. https://doi.org/10.1007/s10616-006-9005-8
Nikkhah A, Rohani A, Zarei M et al (2023) Toward sustainable culture media: using artificial intelligence to optimize reduced-serum formulations for cultivated meat. Sci Total Environ 894:164988. https://doi.org/10.1016/j.scitotenv.2023.164988
O'Neill EN, Cosenza ZA, Baar K, Block DE (2021) Considerations for the development of cost-effective cell culture media for cultivated meat production. Compr Rev Food Sci Food Saf 20:686–709. https://doi.org/10.1111/1541-4337.12678
O'Neill EN, Ansel JC, Kwong GA et al (2022) Spent media analysis suggests cultivated meat media will require species and cell type optimization. NPJ Sci Food 6:46. https://doi.org/10.1038/s41538-022-00157-z
Okamoto Y, Haraguchi Y, Sawamura N et al (2020) Mammalian cell cultivation using nutrients extracted from microalgae. Biotechnol Prog 36(2):e2941. https://doi.org/10.1002/btpr.2941
Okamoto Y, Haraguchi Y, Yoshida A et al (2022) Proliferation and differentiation of primary bovine myoblasts using Chlorella vulgaris extract for sustainable production of cultured meat. Biotechnol Prog 38:1–10. https://doi.org/10.1002/btpr.3239
Ong K, Johnston J, Datar I et al (2021) Food safety considerations and research priorities for the cultured Meat and seafood industry. Compr Rev Food Sci Food Saf 20(6):5421–5448
Park YH, Gong SP, Kim HY et al (2013) Development of a serum-free defined system employing growth factors for preantral follicle culture. Mol Reprod Dev 80:725–733. https://doi.org/10.1002/mrd.22204
Phelan K, May KM (2016) Basic techniques in mammalian cell tissue culture. Curr Protoc Toxicol 70:A.3B.1–A.3B.22. https://doi.org/10.1002/cptx.13
Poinski M (2021) Upside foods develops animal-free growth medium for cell-based meat. Food Dive 1. https://www.fooddive.com/news/upside-foods-develops-animal-free-growth-medium-for-cell-based-meat/611218/
Post MJ (2014) An alternative animal protein source: cultured beef. Ann N Y Acad Sci 1328:29–33. https://doi.org/10.1111/nyas.12569
Post MJ, Levenberg S, Kaplan DL et al (2020) Scientific, sustainability and regulatory challenges of cultured meat. Nat Food 1:403–415. https://doi.org/10.1038/s43016-020-0112-z
Price PJ (2017) Best practices for media selection for mammalian cells. In Vitro Cell Dev Biol Anim 53:673–681. https://doi.org/10.1007/s11626-017-0186-6
Price PJ, Goldsborough MD, Tilkins ML (1998) Embryonic stem cell serum replacement. 76
Puetz J, Wurm FM (2019) Recombinant proteins for industrial versus pharmaceutical purposes: a review of process and pricing. Processes 7
Qi W, Sathre R, Morrow WR III, Shehabi A (2015) Unit price scaling trends for chemical products. LBNL Report 189844:17
Ramani S, Ko D, Kim B et al (2021) Technical requirements for cultured meat production: a review. J Anim Sci Technol 63:681. https://doi.org/10.5187/jast.2021.e45
Rowland MP (2020) Memphis meats raises $161 million in funding, aims to bring cell-based products to consumers. In: Forbes. https://www.forbes.com/sites/michaelpellmanrowland/2020/01/22/memphis-meats-raises-161-million-series-b-funding-round-aims-to-bring-cell-based-products-to-consumers-for-the-first-time/
Rubio N, Datar I, Stachura D et al (2019) Cell-based fish: a novel approach to seafood production and an opportunity for cellular agriculture. Front Sustain Food Syst 3:1–13. https://doi.org/10.3389/fsufs.2019.00043
Salah F, Maher K, Demerdash Z et al (2022) Lab-scale production of anti-Schistosoma monoclonal antibodies by hybridoma culturing in serum free medium: enhancement of growth, cryopreservation survival rate, and diagnostic accuracy. Exp Parasitol 242:108390. https://doi.org/10.1016/j.exppara.2022.108390

Samandari M, Saeedinejad F, Quint J et al (2023) Repurposing biomedical muscle tissue engineering for cellular agriculture: challenges and opportunities. Trends Biotechnol 41:887–906. https://doi.org/10.1016/j.tibtech.2023.02.002

Santos ACA, Camarena DEM, Roncoli Reigado G et al (2023) Tissue engineering challenges for cultivated meat to meet the real demand of a global market. Int J Mol Sci 24. https://doi.org/10.3390/ijms24076033

Specht L (2020) An analysis of culture medium costs and production volumes for cultivated meat, vol 1. The Good Food Institute, Washington, DC, p 30

Stanton D (2019) Quality FBS, or just BS? Industry turning to supply chain certification. BioProcess International

Stout AJ, Mirliani AB, Rittenberg ML et al (2022) Simple and effective serum-free medium for sustained expansion of bovine satellite cells for cell cultured meat. Commun Biol 5:466. https://doi.org/10.1038/s42003-022-03423-8

Sukparangsi W, Thongphakdee A, Intarapat S (2022) Avian embryonic culture: a perspective of in ovo to ex ovo and in vitro studies. Front Physiol 13:1–9. https://doi.org/10.3389/fphys.2022.903491

Swartz E (2023) Anticipated growth factor and recombinant protein costs and volumes necessary for cost-competitive cultivated meat. The Good Food Institute, Washington, DC, p 44

Tonniges JR, Velleman SG (2019) Nutrient restriction and migration of Turkey satellite cells. Poult Sci 98:7090–7096. https://doi.org/10.3382/ps/pez349

Torizal FG, Lau QY, Ibuki M et al (2021) A miniature dialysis-culture device allows high-density human-induced pluripotent stem cells expansion from growth factor accumulation. Commun Biol 4:1316. https://doi.org/10.1038/s42003-021-02848-x

Tuomisto HL, Teixeira de Mattos MJ (2011) Environmental impacts of cultured meat production. Environ Sci Technol 45:6117–6123. https://doi.org/10.1021/es200130u

Usta SN, Scharer CD, Xu J et al (2014) Chemically defined serum-free and xeno-free media for multiple cell lineages. Ann Transl Med 2(10):97

Valk JBF, Bieback K, Buta C et al (2017) Fetal bovine serum (FBS): past – present – future. ALTEX 35:10.14573/altex.1705101

van der Valk J (2022) Fetal bovine serum—a cell culture dilemma. Science 375:143–144. https://doi.org/10.1126/science.abm1317

Venkatesan M, Semper C, Skrivergaard S et al (2022) Recombinant production of growth factors for application in cell culture. iScience 25:105054

Vergeer R, Sinke P, Odegard I (2021) TEA of cultivated meat. Future projections for different scenarios

Wallace J, Gropp J, Dierick N et al (2016) Risks associated with endotoxins in feed additives produced by fermentation. Environ Health 15:1–7. https://doi.org/10.1186/s12940-016-0087-2

Watson E (2020) Cell-based meat in focus: in conversation with meatable, finless foods, new age meats. FoodNavigator USA

Wood P, Tavan M (2022) A review of the alternative protein industry. Curr Opin Food Sci 47:100869. https://doi.org/10.1016/j.cofs.2022.100869

Wood P, Thorrez L, Hocquette J-F et al (2023) "Cellular agriculture": current gaps between facts and claims regarding "cell-based meat". Anim Front 13:68–74. https://doi.org/10.1093/af/vfac092

Wu J-Y, Siu K-C, Geng P (2021) Bioactive ingredients and medicinal values of Grifola frondosa (Maitake). Foods 10(1):95

Xie X, Li P, Pšenička M et al (2019) Optimization of in vitro culture conditions of sturgeon germ cells for purpose of surrogate production. Animals 9. https://doi.org/10.3390/ani9030106

Yamanaka K, Haraguchi Y, Takahashi H et al (2023) Development of serum-free and grain-derived-nutrient-free medium using microalga-derived nutrients and mammalian cell-secreted growth factors for sustainable cultured meat production. Sci Rep 13:498. https://doi.org/10.1038/s41598-023-27629-w

Yao T, Asayama Y (2017) Animal-cell culture media: history, characteristics, and current issues. Reprod Med Biol 16:99–117. https://doi.org/10.1002/rmb2.12024
Zidarič T, Milojević M, Vajda J et al (2020) Cultured meat: meat industry hand in hand with biomedical production methods. Food Eng Rev 12:498. https://doi.org/10.1007/s12393-020-09253-w
Zuo F, Gu Q, Li S et al (2019) Effects of different methionine sources on methionine metabolism in the IPEC-J2 cells. Biomed Res Int 2019:5464906. https://doi.org/10.1155/2019/5464906

# Microcarriers and Scaffolds in Cultivated Meat Production

**Raíssa Canova, Luiza Meurer Brand, Juliana do Canto Olegário, and Bibiana Matte**

**Abstract** The role of scaffolds in cultivated products is to provide structural support for cells as is observed in the extracellular matrix in living organs. The application of scaffolds can be at different stages of the bioprocess production of cultivated meat and can play an important role to support cell growth and differentiation, provide tissue functionality, and contribute to flavor and sensory experience. The selection of the scaffold type and process has an impact on the final product that is formulated (unstructured or structured product). Different technologies such as microcarriers, layer scaffolds, and bioprinting are being developed in order to contribute to cultivated meat processes and products. Additionally, different material types can be employed with natural or synthetic origin which contribute to attributes, advantages, and challenges for cellular agriculture products. The field of scaffold development is prominent with new materials and processes being developed with multiple requirements that can be tuned into each bioprocess. Even though the requirements can be a challenge, they are definitely potential innovative ingredients that can add value to the final product. Therefore, it is important to understand the requirements and possibilities of this field in order to develop cultivated meat products.

**Keywords** Tissue engineering · Scaffold · Microcarrier · Bioprinting · Cellular agriculture · Cultivated meat

R. Canova · B. Matte (✉)
Cellva Ingredients, Porto Alegre, Rio Grande do Sul, Brazil
e-mail: bibiana.matte@cellva.com

L. M. Brand
Cellva Ingredients, Porto Alegre, Rio Grande do Sul, Brazil

School of Dentistry, Federal University of Rio Grande do Sul, Porto Alegre, Rio Grande do Sul, Brazil

J. do Canto Olegário
Cellular Animal Science Laboratory, Federal University of Paraná, Curitiba, Paraná, Brazil



C. R. Soccol et al. (eds.), *Cultivated Meat*,
https://doi.org/10.1007/978-3-031-55968-6_5

# 1 Introduction: Applications and Background of Scaffold Technology

Meat, which is the muscle tissue of animals composed mainly of muscle fibers, connective tissue, and fat, has a hierarchy of cell organization and specific structure that needs to be mimicked in vitro to produce a product equivalent to the traditional one. This definition can be expanded to other types of animal-derived products in the sense that it is important in the cellular agriculture field to develop not only the cells that will compose the product but also the tissue structure. Therefore, scaffold development has an impact on the final product and also on the technological development of cultivated meat bioprocess.

Cell culture can be performed with various cells using different methods. The most commonly used technique is culturing cells in a monolayer. Most cell types, including those in traditional and cultivated meat, are anchorage dependent. Unlike cells from the hematopoietic system, embryonic, and tumor cells, which can grow in suspension, most cells isolated from normal tissues require support to adhere, proliferate, and differentiate (Merten 2015). Two-dimensional (2D) monolayer cultures allow cells to have access to the nutrient-rich culture medium while interacting with neighboring cells and with a rigid structure at the bottom, providing a surface for integrin attachment and creating a basal-apical orientation (Bomkamp et al. 2022). It is possible to adapt an adherent cell culture into a suspension culture. However, developing a stable suspension culture with a high proliferation rate is usually a challenging process.

In addition to depending on anchorage for proliferation, cells in living organs are adapted to a three-dimensional (3D) environment constituted by the extracellular matrix (ECM) and permeated by different components with a specific organization pattern, which is absent in 2D cultures. In order to expand and maintain their properties, normal tissue-derived cells need contact with molecules and signaling factors of the ECM, and a microenvironment similar to the naturally occurring niche is a crucial determinant of cell fate (Merten 2015). This aspect becomes especially important in cultivated meat production since primary cells are not well adapted to artificial conditions, and a 3D structure is required to mimic the architecture of animal tissues to develop a final product that closely resembles conventional meat cuts.

A critical issue in cultivated meat production is the demand for upscaling the process. Efficiency is one of the most important conditions for developing and industrializing a meat alternative (Post 2012), and culturing cells using a conventional 2D system does not enable the generation of enough cells to produce desirable quantities of food (Bodiou et al. 2020). Cultivation in bioreactors is required, and to allow the proliferation of anchorage-dependent cells in suspension, they need to be cultured in aggregates or attached to biomaterials that will serve as scaffolds. While suspension culture with aggregated cells is a simple and low-cost method for process scale-up, it can be difficult to control the size of the aggregates, which can interfere with the desired homogeneous distribution of nutrients among cells (Moritz et al. 2015).

In this context, using scaffolds, which is a technology that emerged from the tissue engineering field, in cultivated meat production can address the technical issue of providing cells with an anchoring surface while creating a 3D environment. It is also beneficial from the perspective of product functionality, helping to create a structure that mimics conventional food products, and it is a potential alternative to optimize cultivated meat production in a technical view.

## 2 Background

Tissue engineering is a multidisciplinary field that combines knowledge from engineering, biology, and medicine to create functional tissues and organs for therapeutic or research purposes. Cultured cells and biocompatible materials are used to develop 3D constructs for implantation into the body to regenerate or repair damaged structures or for application as extracorporeal systems. Tissue engineering also encompasses the creation of tissue prototypes used in diagnostic screening, drug efficacy and toxicity testing, and research studies on tissue development and functionality (Berthiaume et al. 2011). The areas of tissue engineering and regenerative medicine are collectively called "TERM" and are often used as synonyms. However, regenerative medicine is a broader field focused on restoring lost or impaired tissue function and uses various tools, including tissue engineering, to promote regeneration and recovery in damaged areas of the body (Kaul and Ventikos 2015).

While the term "tissue engineering" was coined only in 1987, the field development was initiated during the 1970s and 1980s (Berthiaume et al. 2011). In 1975, the isolation of keratinocytes from a patient's skin biopsy was described and employed to develop one of the first tissue prototypes (Rheinwald and Green 1975). At the beginning of the 1980s, Eugene Bell and colleagues created an equivalent skin organ with dermis and epidermis layers based on a collagen gel matrix seeded with fibroblasts and colonized by keratinocytes (Bell et al. 1981). Also, Ioannis Yannas and John Burke designed a porous matrix combining animal-derived biomaterials with a silicone base to create a scaffold for the proliferation of dermis cells (Yannas et al. 1982). The study of biomaterials took place in 1988 when the use of biodegradable artificial polymers as support matrices for cell culture aiming to generate functional tissues was thoroughly explored for the first time (Vacanti et al. 1988).

Regarding microcarrier technology, substantial progress in the area also happened around the 1970s and 1980s. A key issue for tissue engineering is obtaining sufficient anchorage-dependent cells while maintaining their original phenotype, and the solution relies on 3D high-density cultivation (Li et al. 2015). In 1967, the first use of microcarriers in cell culture was described by researcher Van Wezel in an attempt to adapt adherent cells to suspension culture using dextran particles to achieve large-scale production (Van Wezel 1967). In 1972, a study was conducted with a collagen substrate organized into microbundles dispersed in an aqueous medium to observe cell adhesion and 3D growth (Elsdale and Bard 1972). During this period, Van Wezel and colleagues carried out extensive research for large-scale production of vaccines using microcarrier technology, especially regarding inactivated poliomyelitis vaccines (Van Wezel et al. 1978, 1979, 1984).

Simultaneously, research on stem cells and the particularities of cell proliferation were also advancing. The existence of multipotent cells in living tissues of adult animals was validated in the 1960s (Becker et al. 1963). A decade later, studies on how tissue structure and microenvironment influence cell growth resulted in the demand for developing scaffolds compatible with the natural ECM for proper cell proliferation and functionality (Folkman and Moscona 1978; Folkman and Haudenschild 1980). Finally, in the 1990s, the use of stem cells and tissue engineering merged to create solutions to health issues arising from cellular deficiency or physical trauma (Kaul and Ventikos 2015).

Only in recent decades has the knowledge from TERM started to be applied to produce cell-based food. Muscle satellite cells were first identified in 1961 (Mauro 1961) and have been studied since then. The first cultivated burger made of bovine muscle satellite cells was successfully produced several decades later, in 2013, at Maastricht University in the Netherlands (Post 2014). Before that, in the 1990s, the first international patent describing a process for producing a three-dimensional meat product based on in vitro culture had been published (Van Eelen et al. 1999). Also, in the early 2000s, tissue engineering techniques were applied to goldfish skeletal muscle cells to produce an in vitro protein production system for consumption by professionals on flights and space stations (Benjaminson et al. 2002).

The science behind cultivated meat is based on the knowledge derived from TERM. The production of meat based on culturing animal cells shares several similarities regarding technical requirements with the construction of functional tissues for tissue regeneration purposes, including the choice of biomaterials and design aspects for scaffold development. In both scenarios, the designed scaffold should be able to recapitulate the original tissue architecture; have optimal mechanical and biological properties to favor cell adhesion, proliferation, and differentiation: and properly orchestrate molecular signaling as well as nutrient, gas, and biomolecule transport within the forming tissue (Williams 2014). However, while scaffolds used in tissue constructs for therapeutic purposes must be nontoxic, nonimmunogenic, and minimally proinflammatory for inclusion in the body (Williams 2014), these are not crucial prerequisites when developing an edible meat product. In contrast, some specific requirements not applied in medical and pharmaceutical tissue engineering constructs are important for meat production. Those include texture and thermal properties of the scaffold, edibility, nutritional profile, low cost, and scalability potential (Seah et al. 2022), which will be discussed further in this chapter.

## 3 Application of Tissue Engineering in Cultivated Meat

The terms microcarrier (MC) and scaffold are often used in tissue engineering and tissue regeneration to describe different approaches to structure support in cell and tissue culture. The materials used to develop a scaffold are similar, if not the same, to those used in MCs. These materials must be biocompatible and can be synthetic, such as some polymers and hydrogels, or natural, such as decellularized plants. The major difference between scaffold and MC is the purpose of use.

MCs are a type of scaffold consisting of small bead-like particles made of biocompatible materials used essentially for cell proliferation on a large scale. Cells can be seeded onto the MCs, forming aggregates and cell-microcarrier complexes, and suspended in a culture medium inside bioreactors, simultaneously offering a 3D culture environment and a 2D surface for cell growth (Bodiou et al. 2020). The use of MCs in the production process of cultivated meat is a prominent technology to overcome the scalability problem while providing an accessible environment for cell expansion.

Both microcarriers and scaffolds serve as a substrate upon which cells can attach and proliferate, while more complex scaffolds are a framework for cell maturation, ultimately resulting in the formation of functional tissues and structured products. Layered or multilayer scaffolds are sophisticated support structures employed to create an intricate tissue architecture, such as the one found in structured meat cuts. They can be composed of different materials with varying porosity, mechanical strength, and biochemical specificities arranged in multiple layers stacked on each other. Culturing cells in a suitable matrix is important for constructing a tissue with in vivo-like structure and desirable organoleptic properties since the intramuscular connective tissue, which consists of the extracellular matrix, influences postmortem muscle aging and meat texture (Campuzano and Pelling 2019; Bomkamp et al. 2022). It is also essential to consider the complexity of nutrient access for cells within the tissue structure.

Cell sheet stacking methods exhibit significant potential for creating adequate tissue structures (Kumar et al. 2023). The layered architecture allows control over the spatial distribution of cells, growth factors, and other molecules within the scaffold while mimicking the hierarchical organization found in native tissues. Using multilayer assembly to incorporate various layers with varying characteristics allows the creation of highly dense, textured, and complex tissues (Santos et al. 2023) and facilitates the integration of different cell types or tissue components. Layered scaffolds also address the scalability issue, enabling cost-effective and large-scale production of cultivated meat.

It is important to note that microcarriers are a type of scaffold. In this chapter, the term microcarrier will be used to refer to small bead-like scaffolds used for cultivating cells inside bioreactors for upscaling the production, and scaffold will refer to layered scaffolds employed in the differentiation and structuring process of tissues to produce meat.

The selection of suspension culture, MC, scaffold, or bioprinting techniques in bioprocess development influences the type of cultivated meat product that will reach consumers. In this context, the product type can be categorized as either nonstructured or structured. Nonstructured meat refers to products composed of small pieces or fragments of muscle tissue not presented in their whole-cut natural form. It can be found as ground meat or as the main ingredient in various food items such as burgers, sausages, and meatballs and is easier to replicate in vitro. On the other hand, structured meat is the whole muscle tissue arranged in a structured form that resembles traditional meat. It involves cultivating and arranging muscle cells to reproduce the natural organization found in conventional cuts of meat, such as steaks or filets. Graphical abstract illustrates the characteristics of scaffold types and terminology used in the chapter, as well as examples of product types.

## 4 Advantages and Disadvantages of MC and Scaffolds

Regarding the application of MCs and scaffolds in cultivated meat production, advantages and disadvantages may emerge from several aspects, including the choice of material, the manufacturing process, and their potential impact on the final product. A crucial challenge is selecting adequate biomaterials and developing cost-effective, easily scalable, biodegradable, or food-grade components capable of meeting essential operational requirements. The vascularization and nutrient perfusion characteristics and the ability to mimic the structure and organoleptic properties of conventional meat are essential for constructing a satisfactory cultivated meat product. Different materials with varying attributes and several manufacturing techniques can be employed to create MCs and scaffolds tailored for use in cultivated meat, individually presenting their own set of advantages and disadvantages.

Scaffold-free alternatives may be employed, considering the cells' capacity to secret the molecules present in the ECM and potentially develop tissue analogues without any external components (De Pieri et al. 2021). Cell sheets can be stacked on each other and held together by the secreted matrix, or culture in cell aggregates can be induced (DuRaine et al. 2015). The benefit of these self-organizing techniques in the context of cultivated meat relies on the fact that no foreign and artificial materials are used, so their inclusion in the final constructs would not pose a safety concern. However, some difficulties include not being able to generate the high number of cells required to produce a meat equivalent, frequent cell necrosis in the middle layers of the tissue due to the lack of equal oxygen and nutrient distribution, and inefficient transition from a 2D to a 3D structure (De Pieri et al. 2021). Therefore, the use of microcarriers and scaffolds for cultivated meat production does provide many advantages, including the possibility of upscaling, optimized oxygen and nutrient transport, and the creation of a complex 3D geometry that mimics the naturally occurring environment and closely resembles the characteristics of conventional meat (Bomkamp et al. 2022), as well as introducing relevant nutritional and organoleptic attributes that can add value and bring competitiveness to the final products.

For food production, it is most interesting to use edible or biodegradable materials that can be incorporated into the final product. Nonedible microcarriers made of polystyrene, polyacrylamide, or glass, for example, can be employed temporarily to support cell proliferation, after which cell removal or material degradation is necessary (Bomkamp et al. 2022). This step usually results in cell loss since the detachment process can be incomplete and inefficient (Wolf et al. 2012) and it can increase costs and raise safety concerns (Bomkamp et al. 2022). Synthetic materials used in tissue engineering for biomedical and pharmaceutical applications of scaffolds, such as polylactic acid (PLA), polylactic-co-glycolic acid (PLGA), polycaprolactone (PCL), and polyethylene glycol (PEG), are inadequate for remaining in the final meat product and can only be applied at the beginning of the production process (Bomkamp et al. 2022). Animal-derived ECM molecules such as collagen and gelatin offer a highly similar environment to in vivo tissues and are commonly employed as scaffolds for the functional reconstruction of injured tissues in regenerative medicine (Wolf et al. 2012). Nonetheless, they are unsuitable for cultivated meat production since the industry aims to prioritize using animal-free components unless fabricated through a synthetic process that results in a nontoxic ingredient (Seah et al. 2022). Plant and fungus-derived microcarriers and scaffolds are potentially interesting because they are naturally animal-free materials and can incorporate nutritional attributes into the final product (Campuzano and Pelling 2019). However, their use might require decellularization methods that do not introduce significant costs and safety issues (Bomkamp et al. 2022). The use of bacteria to produce scaffolding materials is also promising and remains to be further explored (Kolodkin-Gal et al. 2023).

The biomaterials used in cultivated meat must be either food grade or biodegradable for incorporation into the end product. Any remaining material will influence the product's organoleptic properties, including taste, scent, appearance, and texture, and it's crucial that their biodegradation results in harmless byproducts to guarantee a final construct safe for consumption (Bomkamp et al. 2022). Microcarriers, in particular, can be used temporarily to support cell expansion and remove or degrade in advanced steps for cell release, which means nonedible and nonbiodegradable materials may be applied. On the other hand, since layer scaffolds are intended to structure the tissue to create meat cuts, they are expected to be incorporated into the final product.

The scaffold's composition can have a beneficial impact on the nutritional profile of the meat product. As mentioned previously, fungus and plant-based molecules, for example, could potentially incorporate interesting dietary components naturally present in their structure, such as fiber and protein (Campuzano and Pelling 2019). Scaffolds created from plant-derived proteins, like textured soy protein (TSP) (Ben-Arye et al. 2020) and edible films of glutenin and zein (Xiang et al. 2022), might be used to improve the protein profile of the meat alternative (Singh et al. 2023). Various animal-free edible biopolymers, including polysaccharides, proteins, and lipids, already applied in conventional meat production, can potentially be employed in tissue engineering for cultivated meat. They are used as added ingredients to modulate nutrient and organoleptic aspects of meat products and could be a source of materials for microcarrier and scaffold design to regulate product stability, texture, flavor, fat, and protein content in cultivated meat (Ng and Kurisawa 2021).

The scaffold may also play a role in delivering external ingredients not part of its intrinsic structure. By incorporating nutritional compounds within the scaffold, they can be released in a controlled manner to not only support cell growth and tissue development (Zhou et al. 2016) but also enhance the nutritional content of the final product. In addition, scaffold approaches can be used to modulate fat content and distribution in cultivated meat and potentially create marbled meat. The scaffold characteristics directly impact the development and spatial distribution of adipocytes within its structure (Freeman and Kelly 2017), and layered sheets of scaffolds containing muscle and fat cells alternately stacked on top of each other may be used to create a meat product with improved and desirable taste, smell, and texture (Li et al. 2022a, b).

Also, the flavor of the meat is affected by several compounds and different reactions such as lipid and carbohydrate degradation, the Maillard reaction, and volatile compounds. To reproduce these complex sensory aspects in vitro, combining meat-associated flavor components or molecules that will lead to these reactions into edible scaffolds offers a promising approach to enhance the organoleptic properties of cultivated meat, providing a more authentic and satisfying sensory experience (Lee et al. 2022). It is important to note that, just as cooking conventional meat leads to different reactions and degradations that change its organoleptic characteristics, it could alter the properties of the materials incorporated into the end product. MC and scaffolds employed in cultivated meat production that remain in the final product should be thoroughly studied for thermal stability and degradation behavior, considering that exposure to heat can potentially impact their characteristics and functionality (Lee et al. 2022).

Altogether, specifying the process and selecting the biomaterials will have important implications in the developed product type and the technological aspects of the process. Using an MC or a scaffold can benefit the nutritional aspects, scalability of the process, and final product assembly. Therefore, these criteria regarding cultivated meat process development also make each product unique as many types of technologies and materials are being developed and employed.

## 5 Prerequisites for Biomaterials

Using biomaterials for food production requires specific criteria regarding their physical, chemical, biological, and environmental properties. The physical attributes that should be accounted for include architecture, texture, stiffness, fiber size, surface topography, and porosity. Architecture and texture are physical properties that must be considered to mimic conventional meat, mainly because they influence consumer acceptance and sensorial experience of the final product (Pajčin et al. 2022; Singh et al. 2023). These factors contribute to the final product intention. Stiffness influences cell behavior during the cultivated meat production process. For example, higher stiffness can induce muscle differentiation, while a softer composition is more favorable for cell expansion. Ideally, the stiffness for myogenic cells should be between 12 and 24 kPa, while for adipogenic cells, it should be 2–3 kPa or lower (Bomkamp et al. 2022). Morphology including fiber size, surface topography, and porosity is another set of physical properties that requires contemplation. Scaffolds with pores provide a robust structure for cell growth, and the distribution of pore sizes can be adjusted to influence cell differentiation and facilitate the alignment of muscle fibers, oxygen transport, nutrient delivery, and waste removal (Bomkamp et al. 2022; Pajčin et al. 2022). Pore sizes can range from 50 to 150 μm for myogenic cells and 40 to 400 μm for adipogenic cells. The recommended material porosity ranges from 30% to 90% to avoid compromising the mechanical properties of the scaffold (Bomkamp et al. 2022). Altogether, the physical characteristics influence cell behavior and characteristics of the final product.

The chemical properties relevant for cultivated meat production are the degradation profile of the scaffold as well as its edible, stability, and cooking properties. The degradation profile is relevant in cases where the scaffold is not designed to be edible or removed from the final construct (Pajčin et al. 2022; Bomkamp et al. 2022). If the goal is to maintain the scaffold in the final product, it's mandatory that the employed material is safe for consumption and can be digested by the human gastrointestinal system. Therefore, it must be free of toxins or harmful components, and it should be capable of breaking down and being absorbed by the body. Also, in this case, the scaffold must be stable throughout the cell culture process and maintain its shape even after cultivated meat processing and storage (Bomkamp et al. 2022). The cooking properties of the scaffold must recapitulate those of conventional meat, with the melting temperature adjusted based on the type of meat. For instance, cultivated fish meat scaffolds should have a lower thermal stability to retain the desired texture after cooking (Bomkamp et al. 2022).

As for the biological properties, the scaffold must be safe for human consumption and nontoxic, in addition to having a palatable taste, aromatic compounds, as well as cell and tissue compatibility (Seah et al. 2022; Pajčin et al. 2022; Moslemy et al. 2023). It is desirable that nutritional components, such as sugar, amino acids, and vitamins, are assessed in order to formulate a highly nutritious food product. Some materials, such as gelatin, collagen, and textured plant proteins, are inherently functional, allowing for cell adhesion and proliferation (Bodiou et al. 2020). However, the adhesion capacity of other materials like alginate is lower, requiring biochemical improvement, such as coating or incorporation of adhesion proteins (Li et al. 2022a, b). The cell adhesion rate is an important aspect, since low cell adhesion and proliferation result in cell loss and increased economic and environmental costs.

Regarding the environment, scaffolding must be sustainable. It is necessary to consider water, land, energy footprint, and greenhouse gas emissions related to production, synthesis, and processing (Post et al. 2020; Wang et al. 2023). Additionally, the cost of the product is an important issue to consider, as scaffolding must be a smaller portion of the total cost to ensure consistent production quality and cost-effectiveness, thereby enabling scalability. It is recommended that the scaffold cost represents approximately 5% of the overall production expenses (Bomkamp et al. 2022). However, substantially higher prices can be considered and accepted in the short term during the development period.

Therefore, biomaterial selection plays a significant role in the production process and composition of cultivated meat. Its properties need to be carefully considered and optimized. Not only does the scaffold provide structural support, firmness, and texture, essential for a satisfying sensory experience, but it can also serve as an additional source of nutrients and influence process scalability. Its characteristics contribute to achieving a cellular organization similar to that found in conventional meat, enhancing the overall quality and appeal of cultivated meat products.

## 6 Scaffolding Materials

Selecting the scaffold material must consider the described criteria. A variety of materials are available and have been studied to meet the requirements, and there is still room for improvement and innovation. Materials can be divided as natural or synthetic according to their origin. Box 1 describes examples of each type of material.

### *6.1 Natural Materials*

The significant advantage of natural scaffolds lies in their inherent properties of edibility and degradability, allowing them to contribute to the nutritional value and desirable textures of the final product. Plant proteins and polysaccharides, as well as decellularized plants, offer cost-effective, biocompatible, biodegradable, and non-immunogenic options. Enhancing the scaffold's performance in terms of cell adhesion to surfaces often involves cross-linking with substances such as transglutaminase, genipin, collagen, among others, to optimize mechanical characteristics. In order to achieve this on natural surfaces, functionalization is an interesting approach. For instance, in a study involving decellularized spinach leaves, fibronectin—a fibrous protein found in the extracellular matrix of vertebrates—was employed to cross-link cells with the leaf structure (Gershlak et al. 2017). Moreover, the synthesis of RGD (arginine-glycine-aspartic acid) and biomineralization, which is a process to make a scaffold more rigid, have proven to be effective methods for improving scaffold porosity and promoting cell adhesion (Fontana et al. 2017). These approaches have demonstrated promising outcomes in enhancing the scaffold properties and facilitating cell interactions. Additionally, chemical processes are employed to facilitate the functionalization of decellularized plants. In a study involving decellularized bamboo, the scaffold was functionalized through an oxidation process, resulting in enhanced cell adhesion (Aswathy et al. 2021).

Thus, certain inherent characteristics of selected plants play a crucial role in the development of cultivated meat. For example, Allan et al. (2021) selected plants based on structural shape. The elongated shape of grass leaves exhibits an organization pattern that closely resembles the natural arrangement found in muscle fibers. This highlights the potential of plant morphology to mimic the desired structure in cultivated meat production. Another application of plant proteins involves their incorporation into textured vegetable proteins (TVP). Textured soy protein (TSP) has been suggested as a scaffold for cultivated meat production due to its advantageous properties such as porosity and malleability. Ben-Arye et al. (2020) used two types of TSP, differing in the concentrations of proteins, fibers, and fat. They observed that the tested TSPs were able to support cell adhesion and proliferation. In a separate study, Lee et al. (2022) functionalize the TSP by mixing it with fish gelatin and agar, improving cell adhesion and proliferation. Wei et al. (2023) conducted a study utilizing a porous scaffold composed of soy protein amyloid fibrils. Within this innovative scaffold, mouse muscle cells (C2C12) were cultured, leading to remarkable findings. The research demonstrated the scaffold's capacity to support cell proliferation and differentiation, showcasing its potential for various applications in tissue engineering and regenerative medicine. Additionally, a viable alternative to natural scaffolds is the utilization of fungi-derived scaffolds. These scaffolds offer several advantages, including low production costs, sustainability, and scalability (Levi et al. 2022). Their inherent properties make them a promising candidate in cellular agriculture applications.

Plants can also be utilized in the production of microcarriers, as demonstrated in a study by Kong and Huang (2023). In this study, proteins extracted from different seeds were employed to create MCs by cross-linking them with alginate spheres using genipin as a binding agent. Notably, pumpkin protein exhibited remarkable cell adhesion to these MCs, highlighting its suitability for such applications. Moreover, broccoli florets have emerged as a promising food-grade candidate for MCs in the production of cultivated meat, further expanding the range of options available in this field (Thyden et al. 2022).

The utilization of substances that naturally constitute the extracellular matrix of animals, such as gelatin and collagen, presents prospects for the functionalization of MCs and scaffolds. Kong et al. (2022) explored this concept by enhancing the cell affinity through the incorporation of gelatin derived from various animal sources like porcine, bovine, or fish into alginate microbeads. However, it is crucial to remember that gelatin and collagen are sourced from animals, and since the fundamental principle of cellular agriculture is to eliminate all inputs of animal origin, except the cells of donor animals, the substitution of these inputs with nonanimal-derived alternatives or their production via recombinant protein technology becomes imperative. By adopting such approaches, we can uphold the ethical and sustainable objectives of cellular agriculture while advancing the development of animal-free solutions for various applications.

## *6.2 Synthetic Materials*

The synthetic materials used as scaffolds and MCs find widespread application in regenerative medicine for tissue and organ production, as well as in the pharmaceutical industry for the manufacture of vaccines, antibodies, proteins, and more. However, just like plant scaffolds, they need to be functionalized to enhance their surface properties for improved cell adhesion. Examples of such materials include polystyrene, lactic and glycolic acid (PLGA), and polyethylene glycol (PEG). It is important to note that these materials are often not classified as food grade, posing challenges for their application in cellular agriculture.

The polystyrene MCs are usually spherical particles covered with substances such as gelatin and collagen that facilitate cell adhesion. Some commercial MCs made of polystyrene are Synthemax II, Plastic Plus, MicroHex, Hillex CT, ProNectin, and SphereCol that are specifically designed for cell growth (Rafiq et al. 2016). In a study utilizing Synthemax, cellBIND, and Cytodex I MCs, the ability of bovine myoblasts to adhere and migrate from one bead to another was observed (Verbruggen et al. 2018). Additionally, in their 2023 study, Tzimorotas and colleagues employed Cytodex 1 and 3 microcarriers to investigate cell adaptation in both spinner flasks and bioreactors. Remarkably, their observations revealed that the cells maintained adherence for a 38-day duration within the bioreactor setting. The polylactic acid (PLA) and polyglycolic acid (PGA) MC are widely used in biomedical applications due to their biocompatibility and biodegradability. Moreover, their size, porosity, and degradation rate can be tailored to suit various cell types (Rodrigues et al. 2019; Kankala et al. 2019). Kankala et al. (2019) developed highly open porous microspheres based on PLGA for skeletal muscle cell proliferation. The cells encountered a favorable microenvironment for cellular proliferation and differentiation. Polyethylene glycol (PEG) MCs are synthetic particles coated with PEG polymers. A key advantage is their nonimmunogenic nature, making them suitable for biomedical tissue engineering applications (Hume et al. 2012). Additionally, they exhibit protein-repellent properties, meaning they do not absorb protein from the culture medium. Moreover, PEG MCs can be functionalized to incorporate substances that enhance cell adhesion. While these MCs effectively facilitate mammalian cell adhesion for cultivated meat production, it is crucial to highlight that such MCs are nonedible and should be either removed from the final biomass or degraded during the production process to ensure no traces of MCs remain in the final product.

In regenerative medicine, the utilization of synthetic scaffolds is prevalent, particularly because the ultimate purpose of the tissue produced is not for consumption. For instance, a study from Zeltinger et al. (2004) employing scaffolds based on polylactic acid (L-PLA) and tested with three different cell lines demonstrated remarkable cell uniformity within the scaffold. Moreover, to further advance clinical applications, Ellis and Chaudhuri (2006) successfully developed a PLGA-based scaffold capable of supporting cell culture, promoting pseudovascularitazion, and facilitating channels for angiogenesis. Considering cellular agriculture, it becomes indispensable that the scaffold meets food-grade standards to ensure its safety and suitability for human consumption. Striving for food-grade scaffolds is crucial to aligning the principles of cellular agriculture with the production of sustainable and safe cell-based food products.

**Box 1 Description and Examples of Materials Used and Type of Scaffolds Exploited for Cultivated Product Development**

| Name | Material | Surface | References |
|---|---|---|---|
| Natural microcarriers | | | |
| Broccoli floret | Broccoli floret celullose | – | Thyden et al. (2022) |
| *Aspergillus oryzae* | Fungal cell wall | – | Ogawa et al. (2022) |
| Pectin and Cardosin A | Pectin and Cardosin A | – | Marga (2015) (patent) |
| Alginate | | – | Lucena-Thomas et al. (2020) |
| Natural scaffolds | | | |
| Green onion bulb | Cellulose | Coated with fibronectin | Cheng et al. (2020) |
| Grass | Cellulose | – | Allan et al. (2021) |
| Carrots, apple, and celery | Cellulose | Poly-L-lysine | Contessi et al. (2020) |
| Bamboo | Cellulose | Oxidation | Aswathy et al. (2021) |
| Spinach | Cellulose | – | Gershlak et al. (2017) |
| Textured soy protein (TSP) | Protein | – | Ben-Arye et al. (2020) |
| Synthetic microcarriers | | | |
| Cytopore 1 | Cellulose | Positive charge | Spearman et al. (2005) |
| Cytodex 1 | Cross-linked dextran | Positive charge | Metiner et al. (2019) |
| Cytodex 3 | Cross-linked dextran | Gelatin coated | Lupberger et al. (2006) |
| Hillex | Polystyrene dextran matrix | Positive charge amine | Knibbs et al. (2003) |
| DM-SII (Synthemax ® II) | Polystyrene | RGD sequence of human vitronectin | Verbruggen et al. (2018) |
| Synthetic scaffolds | | | |
| PLGA-SF-COL | PLGA-silk fibroin-collagen | Fiber scaffold | Wang et al. (2011) |
| PEG hydrogel | PEG | Hydrogel | Hume et al. (2012) |
| Disc-shaped L-PLA scaffold | L-PLA | Porous scaffold | Zeltinger et al. (2004) |

## 7 Microcarriers for Cultivated Meat Production in Bioreactors

MCs offer a 3D culture environment, while the cells grow on a 2D surface suspended in a medium (Koh et al. 2020). MCs typically have a spherical structure and are composed of porous materials that provide a large surface area for the cell to adhere and proliferate. MCs are commonly used for large-scale cell growth in suspension within a stirred bioreactor since it offers a large surface/volume ratio. This suspended system provides better nutrient intake and gas exchange and also a better control over shear stress. The great advantage of cell culture in MC is that the surface given for the cells to grow can be increased simply by adding more MCs in the culture, causing the cells to migrate from MC to MC (Bodiou et al. 2020). This characteristic of the cell migrating to another MC can occur in two ways: by detaching the cell from the MC and going to populate another MC or by making "bridges" between MCs through contact between two MCs (Bodiou et al. 2020). This type of methodology using MCs has been established for vaccine production, monoclonal antibodies, and fermentation. However, these processes were only focused on metabolites released by the cells; for cellular agriculture, the product of interest is the cells itself.

Cell adhesion in the MC is the main prerequisite for the development of a good MC as it influences the entire process of cell expansion. Cell adhesion occurs through receptors of the integrin family. In order to increase cell adhesion, there are different strategies to optimize the modification of the properties of MCs, among them, covering the MC with proteins that contain, for example, an amino acid sequence called RGD (Bodiou et al. 2020). In addition, chemical modifications, such as hydrophilicity and changing the charge of the MC, can be employed to increase cell adhesion. For the development of cultivated meat, it is important to emphasize that three scenarios can be used: (1) The MC can be a temporary support for cell proliferation, and then the cells will be removed and processed in the sequence. (2) The MC can be temporary, but it will be dissolved releasing the cell. (3) The MC can be part of the final product and therefore must be edible (Bomkamp et al. 2022).

Cell removal from the MC can be designed in different ways such as chemical, mechanical, or thermal principles. The main purpose here is to remove the MC while keeping cell viability. The most well-studied approach is using an enzymatic process to dissociate cells from MC. However, the use of proteases can affect cell stability and viability. Therefore, other processes such as the usage of mechanical, thermal, electric, and magnetic responsive materials are being studied. After cells are detached from MCs, the next step is to separate cells from them. This is an added step that must be performed with this type of material. When complete separation is concluded, the MCs could be re-used in the process accordingly (Bomkamp et al. 2022).

The MCs might also be dissolved in order to release cells as a final process step. This type of MC eliminates the step of cell-MC separation. There are a few options of materials like PGA, PLGA, dextran, and chitosan that may be employed. The MC degradation can be accomplished by thermal, chemical, mechanical, photo, and biological processes (Handral et al. 2023). The study of different materials and degradation solutions are being studied and applied to different processes. The goal would be to dissolve the MC, wash the solution, and then be able to collect the cells released as the final product. It is important to conduct analysis on the cells to guarantee that the degradation process did not alter its viability and characteristics.

In these two scenarios, the material used will be considered a food contact material, so they must be sufficiently inert to not affect consumer health or food quality. If the MC remains in the final product, then the edible MC must comply with regulations to be used as a food ingredient or additive. In addition, the edible MC will contribute to the sensorial experience of the product and can impact taste, color, nutritional value, and texture. Recent work has been conducted in order to develop edible MCs. There are different identified materials as plant components, polysaccharides, proteins, and fungi. When studying food-grade ingredients, it is fundamental that the materials are also adapted to promote cell attachment. For example, one patent described the use of pectin coupled with cardosin A because cardosin naturally contains a RGD polypeptide as an option for edible MC (Marga 2015). The process with an edible MC is more challenging from a development point of view. However, it does not require an additional step as in the process with dissociation, degradation, and separation techniques. In addition, the presence of the MC in the final product can incorporate organoleptic and nutritional benefits. As a recent field with abundant possibilities, it is mandatory to conjugate different knowledge in order to obtain abundant, cheap, food-safe materials that can accelerate cultivated meat scalability.

## 8 Layer Scaffolds for Tissue Maturation

In order to achieve a structured product, for example a whole-cut beef, different approaches to develop layer scaffolds are being conducted. With this type of scaffold, it is possible to create a 3D structure that more closely resembles the natural environment of the tissue, where it is possible to control stiffness, porosity, chemical composition, and controlled release of bioactive factors. As a main advantage to resemble the whole tissue containing cells and the extracellular matrix, the scalability and the process design of this type of scaffold are still challenging and are mostly under development. Layer scaffolds can be developed with a variety of materials and can be categorized as porous scaffolds, fiber scaffolds, and hydrogels.

Porous scaffolds have a spongelike structure with pores on the magnitude of hundreds of microns (Bomkamp et al. 2022). It provides mechanical stability for the cells to form tissue and can also deposit extracellular matrix. The textured soy protein developed by Ben-Arye et al. (2020) is an example of porous scaffold. It was demonstrated that cells can differentiate into this scaffold in myotubules. Decellularized spinach leaf is another example of porous scaffold (Jones et al.

2021). There are possibilities on tuning mechanical and organoleptic properties, for example, in this type of scaffold. Additionally, specific bioreactor or large-scale equipment must be developed in order to produce cultivated meat with this type of technology (Bomkamp et al. 2022).

Fibrous scaffolds are composed of long and thin fibers usually produced by electrospinning techniques (Bomkamp et al. 2022). The fibers' alignment can support cell adhesion and maturation and can be produced with a variety of materials such as PLA, PLGA, fibronectin, and gelatin. There are several approaches in order to also add biological cues into the fibers that can promote cell proliferation and differentiation, for example (Uehara et al. 2020). Technologies to produce fiber scaffolds in scale are being developed that would accelerate the incorporation into the cultivated meat space. The different materials to produce fibers can also be tuned to provide benefits in food quality. Also, fibers are produced in the nanoscale if needed, which can support oxygen and nutrient transport (Wei 2008). The fibrous scaffold can be stacked in order to conjugate and form a structured cultivated meat product.

Hydrogels are composed of a hydrophilic polymer matrix that is cross-linked by physical or chemical process and has a large water-holding capacity (Bomkamp et al. 2022). There are several materials that can be employed such as collagen, gelatin, and carrageenan that are food-grade and derived from edible seaweed. In this type of scaffold, the gelation step is essential to form the hydrogels, and cells can be incorporated into the hydrogel before or after the gelation step (MacQueen et al. 2019). Hydrogels are often used in biomedical applications as they can mimic the structure of naturally occurring ECM, and this can also be an important feature in the cultivated meat industry (Bomkamp et al. 2022).

## 9 3D Bioprinting

3D bioprinting technology is emerging as a potential solution to the challenges encountered in the production of cultivated meat within the food industry. This innovative technology involves the precise arrangement of cells and acellular components, such as biomaterials, to create a complex 3D tissue structure. There are three main bioprinting techniques: inkjet printing, light-assisted printing, and extrusion printing—the extrusion technique being the most commonly employed method (An et al. 2015). The extrusion bioprinters work based on pneumatic pressure or mechanical screw plunger to dispense the filament-forming bioink as per the customization of the predefined 3D design of layers, pores, and cell distribution. This technique allows working with a range of transfers, materials, and printheads, allowing the simultaneous printing of several bioinks (Tibrewal et al. 2023).

Bioink is the base material used to produce a living tissue with 3D bioprinting, and it's composed of cells and biomaterials. While various materials have been studied for bioink development, hydrogels, particularly based on alginate, gelatin, and collagen, have emerged as significant bioink options for cultivated meat applications. Plant-based strategies for coating hydrogels or as a biomaterial are also being explored in the field of bioprinting (Ianovici et al. 2022). In the context of cultivated meat production, it is crucial for the biomaterial to be safe and edible or capable of degrading during the growing period (Bomkamp et al. 2022). Additionally, the cost-effectiveness and the sensory quality provided by the biomaterial must be considered.

Bioinks employed in 3D extrusion printing necessitate specific rheological properties. They should exhibit low viscosity and yield stress during the extrusion process. However, once extruded, these materials must possess adequate viscosity and yield strength to retain their shape, provide support, and facilitate connection with subsequent layers (Dick et al. 2019). Mechanical support is also important for maintaining structural stability and self-support within the printed construct. Furthermore, the biomaterial must offer sufficient surface area and porosity to support cell viability, adhesion, and differentiation (Bomkamp et al. 2022) It should exhibit favorable gelling properties and a desirable level of softness to ensure optimal performance.

The utilization of bioinks in 3D bioprinting enables the production of personalized food products, without preprocessing and complicated cutting, thus reducing waste (Guo et al. 2023). The personalization of bioprinted cultivated meat is a promising advantage of the technique since it can be designed according to nutritional needs, such as the addition of vitamins, and the individual's personal preference, for example, defining the texture, amount, and location of fat in the meat according to the consumer's desire (Handral et al. 2022) The automation and precision of the technique allows the controlled distribution and interconnectivity of the pores in the biomaterial, important for the transport of oxygen and nutrients to the cellular components. In addition, the customization of the 3D design allows for an adequate architecture with the induction of myotube and creation of vascular-like channels (Wang et al. 2023).

On the other hand, there are still challenges in the choice of biomaterials, the gelation strategy, and the feasibility, because the rheological properties of the material, such as the shear dilution behavior, viscosity, and yield stress, are critical to determine whether it can be printed with success and ensure product quality (Guo et al. 2023). The unique fibrous structure of the flesh adds to the challenge of printing, requiring the bioink to smoothly exit the nozzle and maintain stable self-support and formability. Most current bioprinters are single or dual printheads, and the feed tube can only hold a small amount of bioink, so for the production of cultivated meat on a larger scale, customization of the bioprinter is necessary, resulting in a high investment and very large equipment. Therefore, large-scale production of cultivated meat through bioprinting remains a challenge that needs to be overcome (Handral et al. 2022).

In 2021, Osaka University made a cultivated meat bioprinted with muscle, fat, and blood vessels that mimics the arrangement of a conventional steak. When cut perpendicularly, the meat had a marbling structure similar to a Wagyu beef (Kang et al. 2021). Also, since 2019, Steakholder Food has been developing technology to produce whole cuts of meat and fish, using animal cell cultivation and 3D bioprinting. The company has developed its proprietary technology to print meat on a scale of tons of meat per day. The possibilities of bioprinting food with animal cells in the desired quantities can be of great advantage for all consumers.

## 10 Challenges and Future Developments

As the cellular agriculture field has been quickly advancing in the past decade combining technologies from medical, pharmaceutical, and food fields, there is an increase in knowledge that has been exploited. The different possibilities of material and cell combination and application will deliver novel solutions for food products development.

Along with material design and developments to ensure satisfactory cell attachment and proliferation, scalability with the materials is fundamental. Thus, bioreactors and bioprocesses must be developed incorporating the biomaterials for cultivated meat production. This encompasses the equipment design and upstream and downstream process with optimizations in order to deliver the food product. The usage of scaffolds is definitely an interesting approach to reduce and optimize production costs. However, in order to validate this, the materials used for scaffold development must also have a compatible cost.

The cultivated meat industry evolved to be a sustainable solution for food production. This main goal has to be incorporated into the whole process. Therefore, a cyclic process with reduced waste products is an interesting path to follow. In the specific case of scaffolds, the biomaterials can be produced from residues of other industries, for example. There are a variety of options that can be exploited in order to fulfill a sustainable process for the cultivated meat.

Scientific progress is visible for the cultivated meat sector to enable scalable cell production. This is a fundamental need for cultivated products to become a consolidated part of the food industry. However, the ultimate goal is that the food tastes delicious for the consumers. The addition of flavor and texture can be accomplished by tuning scaffold characteristics. The same concept also applies to the nutritional value of the product. There are many possibilities to increase and modify nutritional aspects by developing materials and new blends of scaffolds. With this, there are great opportunities for the industry but mostly for the benefit of consumers.

There is a broad list of requirements in order to find a suitable material and process for cultivated meat scaffolds, which makes this an open field of development and innovation possibilities. The search and creation of new biomaterials or repurposed materials will definitely have an impact on new and better products for the cultivated meat industry and consumers. Within the concepts aligned and in combination with new findings, innovative products will emerge and benefit the society.

## References

Allan SJ, Ellis MJ, Bank PA (2021) Decellularized grass as a sustainable scaffold for skeletal muscle tissue engineering. J Biomed Mater Res A 109(12):2471–2482. https://doi.org/10.1002/jbm.a.37241

An J, Teoh JEM, Suntornnood R et al (2015) Design and 3D printing of scaffolds and tissues. Engineering 1(2):261–268. https://doi.org/10.15302/J-ENG-2015061

Aswathy SH, Chandini C, Unnikrishnann PS et al (2021) Decellularization and oxidation process of bamboo stem enhance biodegradation and osteogenic differentiation. Mater Sci Eng C Mater Biol Appl 119:111500. https://doi.org/10.1016/j.msec.2020.111500

Becker A, McCulloch E, Till J (1963) Cytological demonstration of the clonal nature of spleen colonies derived from transplanted mouse marrow cells. Nature 197:452–454. https://doi.org/10.1038/197452a0

Bell E, Ehrlich HP, Buttle DJ et al (1981) Living tissue formed in vitro and accepted as skin-equivalent tissue of full thickness. Science 211(4486):1052–1054. https://doi.org/10.1126/science.7008197

Ben-Arye T, Shandalov Y, Ben-Shaul S et al (2020) Textured soy protein scaffolds enable the generation of three-dimensional bovine skeletal muscle tissue for cell-based meat. Nat Food 1:210–220. https://doi.org/10.1038/s43016-020-0046-5

Benjaminson MA, Gilchriest JA, Lorenz M (2002) In vitro edible muscle protein production system (MPPS): stage 1, fish. Acta Astronaut 51:879–889. https://doi.org/10.1016/S0094-5765(02)00033-4

Berthiaume F, Maguire TJ, Yarmush ML (2011) Tissue engineering and regenerative medicine: history, progress, and challenges. Annu Rev Chem Biomol Eng 2:403–430. https://doi.org/10.1146/annurev-chembioeng-061010-114257

Bodiou V, Moutsatsou P, Post MJ (2020) Microcarriers for upscaling cultured meat production. Front Nutr 7:10. https://doi.org/10.3389/fnut.2020.00010

Bomkamp C, Skaalure SC, Fernando GF et al (2022) Scaffolding biomaterials for 3D cultivated meat: prospects and challenges. Adv Sci 9:2102908. https://doi.org/10.1002/advs.202102908

Campuzano S, Pelling AE (2019) Scaffolds for 3D cell culture and cellular agriculture applications derived from non-animal sources. Front Sustain Food Syst 3:38. https://doi.org/10.3389/fsufs.2019.00038

Cheng YW, Shiwarski DJ, Ball RL et al (2020) (2020) engineering aligned skeletal muscle tissue using decellularized plant-derived scaffolds. ACS Biomater Sci Eng 6(5):3046–3054. https://doi.org/10.1021/acsbiomaterials.0c00058

Contessi NN, Toffoletto N, Farè S et al (2020) Plant tissues as 3D natural scaffolds for adipose, bone and tendon tissue regeneration. Front Bioeng Biotechnol 8:723. https://doi.org/10.3389/fbioe.2020.00723

De Pieri A, Rochev Y, Zeugolis DI (2021) Scaffold-free cell-based tissue engineering therapies: advances, shortfalls and forecast. NPJ Regen Med 6:18. https://doi.org/10.1038/s41536-021-00133-3

Dick A, Bhandari B, Prakash S (2019) 3D printing of meat. Meat Sci 153:35–44. https://doi.org/10.1016/j.meatsci.2019.03.005

DuRaine GD, Brown WE, Hu JC, Athanasiou KA (2015) Emergence of Scaffold-free approaches for tissue engineering musculoskeletal cartilages. Ann Biomed Eng 43:543–554. https://doi.org/10.1007/s10439-014-1161-y

Ellis MJ, Chaudhuri JB (2006) Poly(lactic-co-glycolic acid) hollow fibre membranes for use as a tissue engineering scaffold. Biotechnol Bioeng 1:177–187. https://doi.org/10.1002/bit.21093

Elsdale T, Bard J (1972) Collagen substrata for studies on cell behavior. J Cell Biol 54(3):626–637. https://doi.org/10.1083/jcb.54.3.626

Folkman J, Haudenschild C (1980) Angiogenesis in vitro. Nature 288:551–556. https://doi.org/10.1038/288551a0

Folkman J, Moscona A (1978) Role of cell shape in growth control. Nature 273:345–349. https://doi.org/10.1038/273345a0

Fontana G, Gershlak J, Adamski M et al (2017) Biofunctionalized plants as diverse biomaterials for human cell culture. Adv Healthc Mater 160:1225. https://doi.org/10.1002/adhm.201601225

Freeman FE, Kelly DJ (2017) Tuning alginate bioink stiffness and composition for controlled growth factor delivery and to spatially direct MSC fate within bioprinted tissues. Sci Rep 7:17042. https://doi.org/10.1038/s41598-017-17286-1

Gershlak J, Hernandez S, Fontana G et al (2017) Crossing kingdoms: using decellularized plants as Perfusable tissue engineering scaffolds. Biomaterials 125:13–22. https://doi.org/10.1016/j.biomaterials.2017.02.011

Guo X, Wang D, He B et al (2023) 3D bioprinting of cultured meat: a promising avenue of meat production. Food Bioprocess Technol. https://doi.org/10.1007/s11947-023-03195-x

Handral H, Hua Tay S, Wan Chan W et al (2022) 3D printing of cultured meat products. Crit Rev Food Sci Nutr 62(1):272–281. https://doi.org/10.1080/10408398.2020.1815172

Handral HK, Wyrobnik TA, Luam AT (2023) Emerging Trends in Biodegradable Microcarriers for Therapeutic Application. Polymers 15:1487. https://doi.org/10.3390/polym15061487

Hume SL, Hoyt SM, Walker JS et al (2012) Alignment of multi-layered muscle cells within three-dimensional hydrogel macrochannels. Acta Biomater 6:2193–2202. https://doi.org/10.1016/j.actbio.2012.02.001

Ianovici I, Zagury Y, Redenski I et al (2022) 3D-printable plat protein-enriched scaffolds for cultivated meat development. Biomaterials 284:121487. https://doi.org/10.1016/j.biomaterials.2022.121487

Jones J, Rebello A, Gaudette G (2021) Decellularized spinach: an edible scaffold for laboratory-grown meat. Food Biosci 41:100986. https://doi.org/10.1016/j.fbio.2021.100986

Kang DH, Louis F, Liu H et al (2021) Engineering whole cut meat-like tissue by the assembly of cells fibers using tendon-gel integrated bioprinting. Nat Commun 12:5059. https://doi.org/10.1038/s41467-021-25236-9

Kankala RK, Zhao J, Liu CG et al (2019) Highly porous microcarriers for minimally invasive in situ skeletal muscle cell delivery. Nano Micro Small 25:1901397. https://doi.org/10.1002/smll.201901397

Kaul H, Ventikos Y (2015) On the genealogy of tissue engineering and regenerative medicine. Tissue Eng Part B Rev 21(2):203–217. https://doi.org/10.1089/ten.TEB.2014.0285

Knibbs RN, Dame M, Allen MR et al (2003) Sustained high-yield production of recombinant proteins in transiently transfected COS-7 cells grown on trimethylamine-coated (hillex) microcarrier beads. Biotechnol Prog 19(1):9–13. https://doi.org/10.1021/bp020092r

Koh B, Sulaiman N, Fauzi MB et al (2020) Three dimensional microcarrier system in mesenchymal stem cell culture: a systematic review. Cell Biosci 10:75. https://doi.org/10.1186/s13578-020-00438-8

Kolodkin-Gal I, Dash O, Rak R (2023) Probiotic cultivated meat: bacterial-based scaffolds and products to improve cultivated meat. Trends Biotechnol. https://doi.org/10.1016/j.tibtech.2023.09.002

Kong Y, Huang D (2023) Pumpkin seed proteins rival animal gelatin in increasing the cytoaffinity of edible microbeads for cell-based meat culture. Food Res Int 168:0963–9969. https://doi.org/10.1016/j.foodres.2023.112750

Kong Y, Ong S, Liu MH et al (2022) Functional composite microbeads for cell-based meat culture: effect of animal gelatin coating on cell proliferation and differentiation. J Phys D Appl Phys 55:345401. https://doi.org/10.1088/1361-6463/ac7011

Kumar A, Sood A, Han SS (2023) Technological and structural aspects of scaffold manufacturing for cultured meat: recent advances, challenges, and opportunities. Crit Rev Food Sci Nutr 63(5):585–612. https://doi.org/10.1080/10408398.2022.2132206

Lee M, Park S, Choi B et al (2022) Tailoring a gelatin/agar matrix for the synergistic effect with cells to produce high-quality cultured meat. ACS Appl Mater Interfaces 14:38235–38245. https://doi.org/10.1021/acsami.2c10988

Levi S, Yen FC, Baruch L et al (2022) Scaffolding technologies for the engineering of cultured meat: towards a safe, sustainable, and scalable production. Trends Food Sci Technol 126:13–25. https://doi.org/10.1016/j.tifs.2022.05.011

Li B, Wang X, Wang Y et al (2015) Past, present, and future of microcarrier-based tissue engineering. J Orthop Translat 3(2):51–57. https://doi.org/10.1016/j.jot.2015.02.003

Li L, Chen L, Chen X et al (2022a) Chitosan-sodium alginate-collagen/gelatin three-dimensional edible scaffolds for building a structured model for cell cultured meat. Int J Biol Macromol 209(Pt A):668–679. https://doi.org/10.1016/j.ijbiomac.2022.04.052

Li CH, Yang IH, Ke CJ et al (2022b) The production of fat-containing cultured meat by stacking aligned muscle layers and adipose layers formed from gelatin-soymilk scaffold. Front Bioeng Biotechnol 10:875069. https://doi.org/10.3389/fbioe.2022.875069

Lucena-Thomas JP, Boonprasirt P, Luetchford K et al (2020) Bed expansion properties of tissue engineering particles in a fluidised bed bioreactor. Biochem Eng J 107632:1369–703X. https://doi.org/10.1016/j.bej.2020.107632

Lupberger J, Mund A, Kock J, Hildt E (2006) Cultivation of HepG2.2.15 on Cytodex-3: higher yield of hepatitis B virus and less subviral particles compared to conventional culture methods. J Hepatol 45(4):547–552. https://doi.org/10.1016/j.jhep.2006.05.012

MacQueen LA, Alver CG, Chantre CO et al (2019) Muscle tissue engineering in fibrous gelatin: implications for meat analogs. NPJ Sci Food 3:20. https://doi.org/10.1038/s41538-019-0054-8

Marga F (2015) Edible and animal-product-free microcarriers for engineered meat. Patent number WO 2015/038988A1

Mauro A (1961) Satellite cell of the skeletal muscle fibers. J Biophys Biochem Cytol 9:493–495. https://doi.org/10.1083/JCB.9.2.493

Merten OW (2015) Advances in cell culture: anchorage dependence. Philos Trans R Soc B 370:20140040. https://doi.org/10.1098/rstb.2014.0040

Metiner SP, Can H, Ayyıldız TD, Karakavuk M, Kımız GI et al (2019) The use of toxoplasma gondii tachyzoites produced in HeLa cells adhered to Cytodex 1 microcarriers as antigen in serological assays: an application of microcarrier technology. Cytotechnology 71(1):91–105. https://doi.org/10.1007/s10616-018-0269-6

Moritz MSM, Verbruggen SEL, Post MJ (2015) Alternatives for large-scale production of cultured beef: a review. J Integr Agric 14(2):208–216. https://doi.org/10.1016/S2095-3119(14)60889-3

Moslemy N, Sharifi E, Asadi-Eydivand M, Abolfathi N (2023) Review in edible materials for sustainable cultured meat: scaffolds and microcarriers production. Int J Food Sci Technol 58:6182. https://doi.org/10.1111/ijfs.16703

Ng S, Kurisawa M (2021) Integrating biomaterials and food biopolymers for cultured meat production. Acta Biomater 124:108–129. https://doi.org/10.1016/j.actbio.2021.01.017

Ogawa M, García M, Nitin J et al (2022) Assessing edible filamentous fungal carriers as cell supports for growth of yeast and cultivated meat. Foods 11:3142. https://doi.org/10.3390/foods11193142

Pajčin I, Knežić T, Savic Azoulay I et al (2022) Bioengineering outlook on cultivated meat production. Micromachines 13(3):402. https://doi.org/10.3390/mi13030402

Post MJ (2012) Cultured meat from stem cells: challenges and prospects. Meat Sci 92(3):297–301. https://doi.org/10.1016/j.meatsci.2012.04.008

Post MJ (2014) Cultured beef: medical technology to produce food. J Sci Food Agric 94:1039–1041. https://doi.org/10.1002/JSFA.6474

Post MJ, Levenberg S, Kaplan DL et al (2020) Scientific, sustainability and regulatory challenges of cultured meat. Nat Food 1:403–415. https://doi.org/10.1038/s43016-020-0112-z

Rafiq QA, Coopman K, Nienow AW et al (2016 Mar) Systematic microcarrier screening and agitated culture conditions improves human mesenchymal stem cell yield in bioreactors. Biotechnol J 11(4):473–486. https://doi.org/10.1002/biot.201400862

Rheinwald JG, Green H (1975) Serial cultivation of strains of human epidermal keratinocytes: the formation of keratinizing colonies from single cells. Cell 6(3):331–343. https://doi.org/10.1016/s0092-8674(75)80001-8

Rodrigues AL, Rodrigues CAV, Gomes AA et al (2019 Apr) Dissolvable microcarriers allow scalable expansion and harvesting of human induced pluripotent stem cells under xeno-free conditions. Biotechnol J 14(4):e1800461. https://doi.org/10.1002/biot.201800461

Santos ACA, Camarena DEM, Reigado GR et al (2023) Tissue engineering challenges for cultivated meat to meet the real demand of a global market. Int J Mol Sci 24:6033. https://doi.org/10.3390/ijms24076033

Seah JSH, Singh S, Tan LP, Choudhury D (2022) Scaffolds for the manufacture of cultured meat. Crit Rev Biotechnol 42:311–323. https://doi.org/10.1080/07388551.2021.1931803

Singh A, Kumar V, Singh SK et al (2023) Recent advances in bioengineered scaffold for in vitro meat production. Cell Tissue Res 391:235–247. https://doi.org/10.1007/s00441-022-03718-6

Spearman M, Rodriguez J, Huzel N, Butler M (2005) Production and glycosylation of recombinant beta-interferon in suspension and cytopore microcarrier cultures of CHO cells. Biotechnol Prog 21(1):31–39. https://doi.org/10.1021/bp0498084

Thyden R, Perreault LR, Jones JD et al (2022) An edible, decellularized plant derived cell carrier for lab grown meat. Appl Sci 12:5155. https://doi.org/10.3390/app12105155

Tibrewal K, Dandekar P, Jain R (2023) Extrusion-based sustainable 3D bioprinting of meat & its analogues: a review. Bioprinting 29:e00256. https://doi.org/10.1016/j.bprint.2022.e00256

Tzimorotas D, Solberg NT, Andreassen RC et al (2023) Expansion of bovine skeletal muscle stem cells from spinner flasks to benchtop stirred-tank bioreactors for up to 38 days. Front Nutr 2023(10):1192365. https://doi.org/10.3389/fnut.2023.1192365

Uehara TM, Piano IMM, Santos FA et al (2020) Fabrication of randon and aligned eletrospun nanofibers containing graphene oxide for skeletal muscle cells scaffold. Polym Adv Technol 31:1437–1443. https://doi.org/10.1002/pat.4874

Vacanti JP, Morse MA, Saltzman WM et al (1988) Selective cell transplantation using bioabsorbable artificial polymers as matrices. J Pediatr Surg 23(1):3–9. https://doi.org/10.1016/S0022-3468(88)80529-3

Van Eelen WF, Van Kooten WJ, Westerhof W (1999) Industrial scale production of meat from in vitro cell cultures. World Intellectual Property Organisation. https://patentscope.wipo.int/search/en/WO1999031222

Van Wezel A (1967) Growth of cell-strains and primary cells on micro-carriers in homogeneous culture. Nature 216:64–65. https://doi.org/10.1038/216064a0

Van Wezel AL, Van Steenis G, Hannik CA et al (1978) New approach to the production of concentrated and purified inactivated polio and rabies tissue culture vaccines. Dev Biol Stand 41:159–168

Van Wezel AL, Van Herwaarden JA, Van de Heuvel-de Rijk EW (1979) Large-scale concentration and purification of virus suspension from microcarrier culture for the preparation of inactivated virus vaccines. Dev Biol Stand 42:65–69

Van Wezel AL, Van Steenis G, Van der Marel P et al (1984) Inactivated poliovirus vaccine: current production methods and new developments. Rev Infect Dis 6(Suppl 2):S335–S340. https://doi.org/10.1093/clinids/6.supplement_2.s335

Verbruggen S, Luining D, Essen AV et al (2018) Bovine myoblast cell production in a microcarriers-based system. Cytotechnology 70:503–512. https://doi.org/10.1007/s10616-017-0101-8

Wang G, Hu X, Lin W et al (2011) Electrospun PLGA–silk fibroin–collagen nanofibrous scaffolds for nerve tissue engineering. In Vitro Cell Dev Biol Anim 47:234–240. https://doi.org/10.1007/s11626-010-9381-4

Wang Y, Zou L, Liu W, Chen X (2023) An overview of recent progress in engineering three-dimensional scaffolds for cultured meat production. Foods 12(13):2614. https://doi.org/10.3390/foods12132614

Wei G (2008) Nanostructured biomaterials for regeneration. Adv Funct Mater 18(22):3566–3582. https://doi.org/10.1002/adfm.200800662

Wei Z, Dai S, Huang J, Hu X et al (2023) Soy protein amyloid fibril scaffold for cultivated meat application. ACS Appl Mater Interfaces 15(12):15108–15119. https://doi.org/10.1021/acsami.2c21702

Williams DF (2014) The biomaterials conundrum in tissue engineering. Tissue Eng Part A 20(7–8):1129–1131. https://doi.org/10.1089/ten.tea.2013.0769

Wolf MT, Daly KA, Reing JE, Badylak SF (2012) Biologic scaffold composed of skeletal muscle extracellular matrix. Biomaterials 33:2916–2925. https://doi.org/10.1016/j.biomaterials.2011.12.055

Xiang N, Yao Y, Yuen JSK et al (2022) Edible films for cultivated meat production. Biomaterials 287:121659. https://doi.org/10.1016/j.biomaterials.2022.121659

Yannas IV, Burke JF, Orgill DP et al (1982) Wound tissue can utilize a polymeric template to synthesize a functional extension of skin. Science 215(4529):174–176. https://doi.org/10.1126/science.7031899

Zeltinger J, Sherwood JK, Graham DA et al (2004) Effect of pore size and void fraction on cellular adhesion, proliferation, and matrix deposition. Tissue Eng 7(5):557. https://doi.org/10.1089/107632701753213183

Zhou A, Chen S, He B et al (2016) Controlled release of TGF-beta 1 from RADA self-assembling peptide hydrogel scaffolds. Drug Des Devel Ther 10:3043–3051. https://doi.org/10.2147/DDDT.S109545

# Bioreactors for Cultivated Meat Production

**Luciana Porto de Souza Vandenberghe, Ariane Fátima Murawski de Mello, Giuliana Biagini, Patricia Beatriz Gruening de Mattos, Igor Negreiros Piazenski, João Pedro Manica Candelario, and Carlos Ricardo Soccol**

**Abstract** The interest in cultivated meat production technology is continually and exponentially growing, with the objective of providing new solutions for protein production and consumption globally. These processes involve the cultivation of animal cells, the peculiarities, and specificities of which demand controlled conditions to maintain viability and enable efficient growth. Cell cultivation is carried out in various types of bioreactors, spanning from traditional to newly developed models, each with its own characteristics and possibilities. Stirred tank reactors (mechanically agitated) and airlift/bubble column bioreactors (pneumatically agitated) are the most commonly employed, with adaptations and/or modifications to facilitate satisfactory cultured meat production. Microcarriers/scaffolds are imperative to aid cell adhesion and multiplication. Other bioreactor models, such as hollow fibers and packed bed bioreactors, have also been studied. Single-use technology is gaining traction due to its advantages such as easier cleaning and sterilization and shorter operation times, but it comes with the associated environmental impact of generated plastic waste. In addition to describing various bioreactor models, this chapter also presents some information about instrumentation and control. The goal is to present the main reported bioreactor models currently used for cultivated meat production, highlighting their characteristics, advantages, disadvantages, and technological barriers to overcome.

**Keywords** Stirred tank reactors · Airlift bioreactors · Hollow fibers bioreactors · Packed-bed bioreactors · Single-use bioreactors · Shear stress · Agitation · Animal cells · Scaffolds

---

L. Porto de Souza Vandenberghe (✉) · A. F. Murawski de Mello · G. Biagini
P. B. Gruening de Mattos · I. Negreiros Piazenski · J. P. Manica Candelario · C. R. Soccol
Bioprocess Engineering and Biotechnology Department, Federal University of Paraná, Curitiba, Paraná, Brazil
e-mail: lvandenberghe@ufpr.br


C. R. Soccol et al. (eds.), *Cultivated Meat*,
https://doi.org/10.1007/978-3-031-55968-6_6

## 1 Introduction

A bioreactor, commonly referred to as "fermenter," is a device wherein biotransformations catalyzed by enzymes or whole cells take place. Depending on the ultimate goal of the bioprocess, the model and design of the bioreactor are modified to meet the specific demands of the cells, the enzymes, or the microorganisms being employed. Consequently, decisions made regarding the design aspect of the bioreactor can have an impact on the efficiency of the bioprocess. In this context, a fundamental understanding of the process kinetics, the necessary agitation, energy, and mass equilibrium, as well as the heat and mass transfer, proves to be imperative. For cultivated meat (CM), the purpose of the bioreactor can be to expand the cells on the initial step of the process or to allow an adequate environment for differentiation (Allan et al. 2019).

Animal cell culture in bioreactors presents additional challenges, due to the absence of a cell wall, unlike bacteria, fungi, and plants. This characteristic makes animal cells more sensitive to shear rates (Bellani et al. 2020), influencing the choice of agitators. Furthermore, in the context of meat production, the majority of cells utilized for this purpose are anchorage-dependent, requiring physical support for growth and development (Bellani et al. 2020). Therefore, in an ideal scenario, these cells should not be cultured in suspension, and the bioreactor model should facilitate the incorporation of microcarriers or scaffolds.

These cells also exhibit heightened sensitivity to temperature and pH variations. Since $CO_2$ plays a crucial role in maintaining the medium's pH (Dubey et al. 2021; Lim et al. 2022), its supplementation demands greater caution and control compared to microbial cultures. Additionally, due to the adherence of the cells in CM production, assessing cell viability and growth during bioreactor culture is not straightforward. Protocols exist for detaching cells from the microcarriers for analysis, but they can stress the cells, potentially influencing results (Soccol et al. 2022). The unique characteristics of animal cell culture impact not only the design of the bioreactor but also the choice of sensors. Ideally, in addition to standard temperature, pH, and dissolved oxygen (DO) sensors, sensors for cell viability and dissolved $CO_2$ sensors should be incorporated (Chopda et al. 2020). However, their high cost may render their use unfeasible depending on the final product's value. Ultimately, when designing a bioreactor and selecting accessories for CM production, additional variables should be considered to adequately address the cells' nutritional and physical requirements.

Hence, this chapter provides information on bioreactor models that are commonly employed in CM production, detailing their characteristics, modifications, technologies, and challenges tailored to these innovative bioprocesses.

## 2 Animal Cells' Cultivation and Differentiation

Industrial process involving the cultivation and differentiation of animal cells still appear to be relatively uncommon, primarily due to intricate mechanisms involved in growth, maintenance, and differentiation of these cells. However, in the past decade, there has been an increased exploration of various types of animal cells, such as those from bovine, fish, pigs, and chickens, particularly for ex situ cultivation to obtain CM. These cells are typically procured through animal biopsies, yielding samples containing both stem cells and differentiated cells. Both primary cells, which exhibit limited ex vivo proliferation, and immortalized cell lineages have been investigated in this context (Ng and Kurisawa 2021).

The stem cells are desirable due to functional multilineage differentiation capacity, enabling the development of complex structures akin to muscle tissues. However, precisely controlling differentiation along this specific path poses a significant challenge (Mizrahi and Lazar 1988; Warner 2019; Zhang et al. 2020; Jairath et al. 2021). An appealing alternative involves the use of adult stem cells instead of embryonic ones, as they are present in various tissues but exhibit limited differentiation ability (Ulloa-Montoya et al. 2005; Letti et al. 2021). For CM process, various cell types, including myosatellites, myoblasts, myocytes (myotubes and myofibers), adipose-derived stem cells, adipocytes, fibroblasts, and induced pluripotent stem cells (iPSCs), are under consideration (Allan et al. 2019; O'Neill et al. 2021). These distinct cell lineages possess specific properties, such as culture medium requirements, dry and wet mass, protein and water content, specific growth rate, anchorage dependence, and methods for proliferation and differentiation (Allan et al. 2019).

The proliferation and differentiation represent the primary stages in the large-scale production of CM. The cell proliferation or expansion stage is the simplest due to ease of scaling-up in a semicontinuous or fed-batch process. This approach not only maximizes cell proliferation but also helps prevent contaminations and reduces the need of cell handling. Stirred-tank bioreactors (STR) are the most used bioreactor in this step, followed by wave bioreactor, packed and fluidized bed, and airlift and bubble column bioreactors (Ng and Kurisawa 2021). For the large-scale cultivation of stem cells for food, suspension cultures in bioreactors are essential. This is particularity true because muscle cells necessitate adhesion and the formation of aggregates for their growth and proliferation. To achieve high-density culture suspensions and cultivation on microcarriers or in aggregated form, such as cell aggregates, can be employed (Moritz et al. 2015; Steiner et al. 2010; Abbasalizadeh et al. 2012).

Under these conditions, it is possible to achieve a cell density ranging from $10^5$ to $10^6$ cells/mL with various cell lines. Recent studies have demonstrated that changing the medium every 2–3 days allows cells to reach confluence and promotes better growth, resulting in higher-density levels (Hanga et al. 2021; Bunnell et al. 2008; Verbruggen et al. 2018; Andreassen et al. 2022; Stout et al. 2022). Furthermore, in cases where microcarriers are employed, the addition of new beads can enhance cell transfer and colonization, mitigating the formation of large clumps that might impede proper growth (Moritz et al. 2015; Stout et al. 2022; Hanga et al. 2021). In a specific instance, a production of $5.053 \times 10^5$ cells/mL of bovine adipose-derived stem cells (bASCs) was accomplished using a 1 L stirred tank bioreactor with SoloHill Plastic microcarriers in fed-batch mode (Hanga et al. 2021).

The cell differentiation stage requires the adaptation of microcarriers or scaffolds in bioreactors to support higher cell densities ($10^8$–$10^9$ cells/mL) resulting from cell expansion and specialization. Scaffolds or microcarriers are porous matrixes obtained from animal, plant, microbial, or synthetic sources that act as templates for the development of muscle, fat, and connective tissues through cell adhesion (Seah et al. 2022). Compared to monolayer culture, they offer a larger surface-to-volume ratio, allowing better proliferation and differentiation. The microcarrier/scaffold must possess characteristics as low cost and adequate chemical, physical, and mechanical properties including bead-to-bead transference biocompatibility, removability, edibility, and biodegradability. For CM production, examples include SoloHill, Cellbind, Cytodex 1, Cytodex-3, Synthemax II, Biosilon, labs glass-coated polymer, poly-L-lactic acid (PLLA), polypropylene (PP), polyethersulfone (PES), polysulfone (PS), and cellulose triacetate hollow fibers (Allan et al. 2019). Alternative microcarriers, such as food-grade microcarriers, have been studied to improve nutritional and organoleptic properties of CM, seeking similarity with traditional meat and avoiding the need to separate cells (Allan et al. 2019; Ben-Arye et al. 2020; Messmer et al. 2022). A recent example is the textured soy protein scaffold, enabling the generation of three-dimensional bovine skeletal muscle tissue with good taste and improved texture (Ben-Arye et al. 2020).

Bioreactors such as hollow fiber bioreactors and fixed bed bioreactors can be efficiently employed in this stage. A high cell density of $5.1 \times 10^8$/mL was reported to be achieved with packed bed bioreactors (Park and Stephanopoulos 1993). The fixed bed allows continuous circulation of oxygen and nutrient supply, providing a large surface-to-volume ratio for cell attachment and expansion (Park and Stephanopoulos 1993; Rafiq et al. 2016). For hollow fiber bioreactors, a cell density on the order of $1.0 \times 10^9$/mL was reported due to the flow operation, known as intracapillary inoculation with extracapillary perfusion (Martin and Vermette 2005; Roberts et al. 2012). This design allows automated monitoring and control of metabolite concentration to maintain process consistency with efficient nutrient supply, removal of waste, and oxygen exchange (Martin and Vermette 2005; Godara et al. 2008; Kumar and Starly 2015).

The differentiation stage begins with the addition of inducing substances to the culture medium, a composition that varies according to the cell type within the culture. Accomplishing this often necessitates a reduction in the agitation of the medium, with controlled mixing rates to avoid cell damage. Similar to proliferation stage, it is imperative to change the medium every 2–3 days to sustain the proper rate of muscle tissue formation (Hanga et al. 2021; Stout et al. 2022). Tissue perfusion bioreactors prove to be optimal for cells to mature and form tissue resembling meat. These bioreactors facilitate the continuous flow of medium through tissues already structured in 3D, mirroring the vascular system (Chen et al. 2022).

In both stages, agitation is a parameter of great importance. Altering this parameter allows the control of the aggregate size and the reduction of shear stress impact on the cells. Numerous studies demonstrate the necessity of maintaining agitation at an appropriate level for effective cell suspension while simultaneously preventing aggregate rupture, typically between 60 and 80 rpm (Hanga et al. 2021; Allan et al. 2019; Stout et al. 2022). Oscillating platform bioreactors with fluid movement resembling waves, such as air transport, have been explored. These systems promote limited shear stress while maintaining homogeneity, enabling large-scale culture, for instance, >20,000 L (Merchuk 1990; Li et al. 2020). A 300,000 L airlift bioreactor design has been evaluated using computational fluid dynamics, presenting adequate mass transfer conditions, with promising results estimated to meet the demand of meat supply (Li et al. 2020).

Moreover, bioprocess conditions, including dissolved oxygen, carbon dioxide, pH, and temperature must be meticulously controlled to enhance yields. Animal cells exhibit a high demand for oxygen, which is solubilized in the culture medium through natural or forced aeration (Allan et al. 2019). Studies indicate that saturation with air supplemented with 5% $CO_2$ is optimal for meeting cellular demands (Hanga et al. 2021; Stout et al. 2022). Temperature, along with pH, are variables requiring control around 37 °C and a neutral pH range of 7.0–7.5 (Hanga et al. 2021; Bunnell et al. 2008; Verbruggen et al. 2018; Moritz et al. 2015; Stout et al. 2022).

Recently, Upside Foods introduced a perfusion apparatus technology that supports cell proliferation and differentiation. This cost-efficient system is designed for growing cell sheets and includes a separator for recovering the meat product from the substrate (Leung et al. 2021).

# 3 Bioreactor Models for Animal Cell Cultivation

## 3.1 *Mechanically Agitated Bioreactors*

Despite the wide variety of bioreactors available today, the most widely used in the bioprocessing industry are the mechanically stirred tanks (STR), also known as mixing reactors (Schirmer et al. 2021). This is attributed to the fact that STRs can provide the necessary conditions for culturing a large array of microorganisms and cells used in industrial applications (Schirmer et al. 2021). Given that they are the most commonly employed type of bioreactor and have been used for an extended period, their structures and mechanisms are well understood, facilitating the scaling-up process compared to other bioreactor types. STRs can operate in batch, fed-batch, and continuous modes, making them an easily adaptable tool for various bioprocesses (Spier et al. 2011). For these reasons, the STR has become the default choice for culturing mammalian cells and, consequently, CM (Bellani et al. 2020). Indeed, some reports detail successful cultures of bovine-derived cells in small-scale STRs (Hanga et al. 2021; Tzimorotas et al. 2023).

Mechanical agitation is necessary for the homogeneous distribution of nutrients and also contributes to heat transfer in the bioreactor (Spier et al. 2011). Typically, this is achieved by employing three turbine impellers. The most common models of STR impellers are the Rushton impeller, pitched blade impeller, and marine blade impeller. However, mechanical agitation creates a high shear rate environment that can be damaging to animal cells due to not only their lack of cell wall but also because it can detach them from the microcarrier, potentially leading to cell death

(Allan et al. 2019). Hence, when considering animal cell culture, extra attention should be paid to the choice of the impeller.

The Rushton impeller provides radial flow and more vigorous agitation (Grenville et al. 2017; Schirmer et al. 2021). Consequently, the resulting shear stress is high, making this type of impeller typically suitable for low-viscosity media and processes involving more resistant cells, such as microbial cells (Schirmer et al. 2021). For CM production, the Rushton impeller may be deemed inadequate. However, it is noteworthy that shear stress in bioreactors is not solely related to the speed of the agitation and the impeller used; it also involves the bubbles created due to the vortex generated by the agitation process, which can occur even at low rpm. Hu et al. (2011) observed that hybridoma cell viability was maintained even at a speed of 1500 rpm with a Rushton impeller. They attribute this result to the presence of baffles in the bioreactor and emphasize that, in comparison to those without baffles, animal cells can be damaged at agitation speeds as low as 200 rpm due to the breakage of bubbles from the vortex. Moreover, for animal cell culture, shear stress is a crucial aspect to be considered because it can influence certain differentiation pathways for stem cells and affect the mechanical properties of the final product (Bellani et al. 2020; Lim et al. 2022). This consideration is especially significant for CM applications, as the production of meat involves different cell types (fat and muscle cells) and the mechanical properties, particularly softness, play a vital role in its quality and similarity to common meat.

Despite conflicting and incomplete data in the literature regarding animal shear stress tolerance (Nienow 2010; Sieblist et al. 2016), it is established that myocyte precursors can be more fragile. Given their anchorage-dependent nature, it is advisable to ensure a low-shear rate environment for CM production (Allan et al. 2019; Nienow 2010). In STRs, this can be accomplished by employing the pitched blade or the marine impeller. Both options provide gentler agitation of the system, but the pitched blade, in particular, allows for both axial and radial flow, enhancing more efficient nutrient and oxygen diffusion in the culture (Grenville et al. 2017).

## *3.2 Pneumatically Agitated Bioreactors*

Although innovative bioreactor systems have gained interest in recent years, conventional pneumatically agitated systems remain among the most familiar and widely utilized bioreactors (Mutaf and Oncel 2023). Pneumatically agitated bioreactors (Fig. 1) have been the focus of research for many years in the field of animal cell culture. This type of bioreactor is generally categorized into two main groups: bubble column bioreactors and airlift bioreactors.

Bubble columns are typically cylindrical containers initially filled with liquid medium, and gas bubbles are continuously introduced at the bottom of these columns (Lubbert 2023). In a standard airlift bioreactor designed for mammalian cells, a single internal baffle is utilized to establish a vertical division within the reactor.

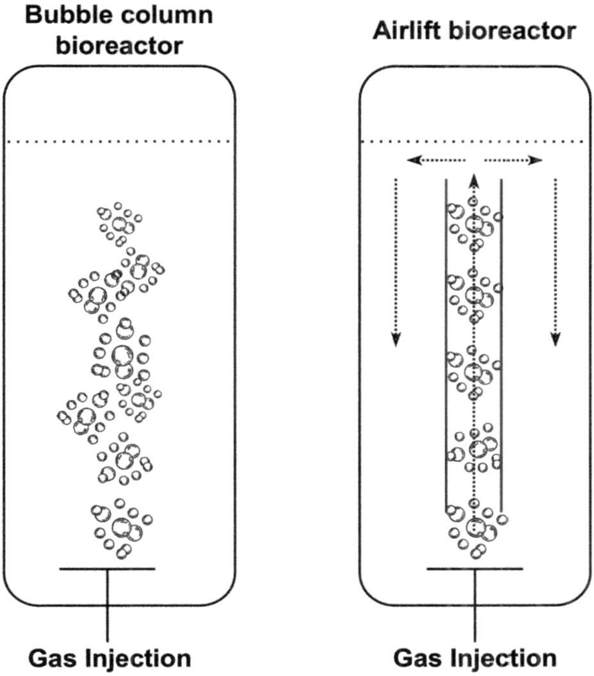


**Fig. 1** Pneumatically agitated bioreactors: Schematics of bubble column and airlift bioreactors

This division separates the reactor into an upward-flow riser section and a downward-flow downcomer section (Kuystermans and Al-Rubeai 2011).

Stirred tank reactors are typically a suitable choice for working with volumes under 20 $m^3$, offering advantages such as effective mixing and efficient energy consumption. However, when larger volumes are required, pneumatically agitated bioreactors become a more suitable option (Li et al. 2020). Both bubble columns and airlift systems lack moving parts, resulting in minimal heat generation and contamination issues (Letti et al. 2021). Additionally, they can provide uniform shear stress, a crucial factor in the design of animal cell reactors. These combined benefits make it feasible to achieve high cell densities (Li et al. 2020).

On the other hand, a significant challenge hinders the progress in the utilization of this reactor type. Despite pneumatic agitated bioreactors being capable of providing low-shear stress, animal cells can be harmed by the bursting of bubbles at the liquid surface (Nienow 2010). Cells in suspension have a tendency to adhere to the outer surface of rising bubbles or be carried along in the wake of the bubble. If the cells are in close proximity to the bubble when it eventually bursts at the top of the bioreactor, they experience significant shear forces (Humbird 2021). Due to their relatively larger size and the absence of a rigid cell wall, animal cells are especially vulnerable to these forces, potentially leading to physiological alterations, including cell death (Al-Rubeai et al. 1995). The extent of this damage varies depending on the specific cell line, the medium used, the operating conditions, and whether the cells are freely suspended or grown on microcarriers (Li et al. 2020; Neunstoecklin et al. 2015).

Another important consideration is that smaller bubbles are preferable for microcarrier-based cell cultures (Li et al. 2020; Neunstoecklin et al. 2015). In high-density animal cell cultures, it is common practice to avoid sparging altogether, and instead, oxygen is supplied through the headspace of the bioreactor (Humbird 2021). Gently sparged STRs are indeed the preferred choice, and there is limited research and work that specifically concentrate on the use of pneumatically agitated bioreactors in the field of CM production (Humbird 2021).

## *3.3 Other Models*

### 3.3.1 Wave or Rocking Motion Bioreactors

The wave bioreactor, also known as the rocking motion bioreactor, is a unit capable of achieving effective mixing, gas transfer, and off-bottom anchored cell suspension without the buildup of shear stress (Zhan et al. 2019; Kahan et al. 2020). The system (Fig. 2) comprises a single-use cell bag with ports for gas and utilities on a heated incubation platform, featuring a wavelike motion mechanism that agitates the cells back and forth, allowing gentle mixing and gas transfer via the bioreactor head-space. Novel systems can provide motion with higher degrees of freedom, allowing increased mass transfer compared to standard wave-type bioreactors (e.g., 2 degrees

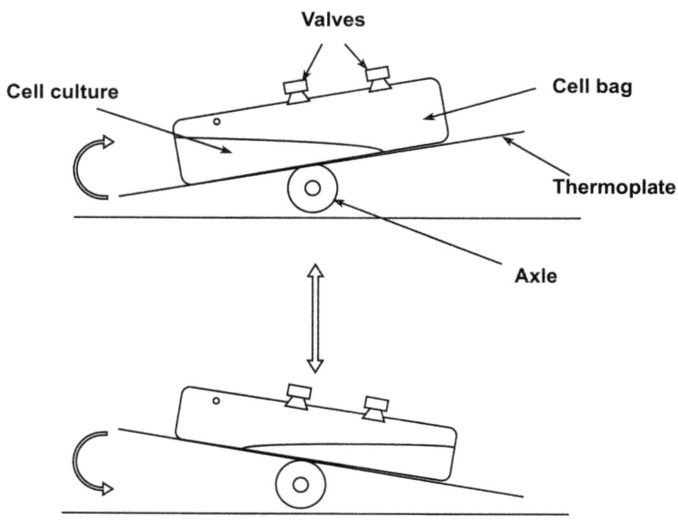


**Fig. 2** Schematics for wave or rocking motion bioreactor

of freedom involving translation along the X-axis and rotation along the Y-axis) (Seidel et al. 2022; Singh 1999).

The cell bag is designed to provide mechanical strength and maintain inert fluid contact within the bioreactor. The bag ports are crucial for its operation, as they are used for process monitoring with sensors, including oxygen, pH, and temperature probes; sampling valves; inoculation and harvest lines; and inlet and outlet air. Furthermore, additional configurations can be achieved through supplementary ports, allowing operation in batch, fed-batch, repeated fed-batch, as well as perfusion modes (Bellani et al. 2020; GE Healthcare Bio-Sciences AB 2015; Singh 1999).

This approach offers advantages in terms of safe operation and sterility, primarily due to the system's ease of handling. Additionally, thanks to its low-shear stress and efficient mass transfer capabilities, it enables the attainment of high cell densities, making it a commonly employed method for expanding mammalian cell lines (Bellani et al. 2020).

However, for CM production, relying solely on this system as the primary strategy can pose challenges. This system operates at capacities of up to 500 L, but scaling beyond 100 L can be labor-intensive and demand significant space. While it may be unfeasible for large-scale bioreactors, it remains a valuable tool with essential applications in inoculum buildup and the scaling-up process, owing to its inherent advantages (Bellani et al. 2020; Kahan et al. 2020).

### 3.3.2 Fluidized and Fixed-Bed Bioreactors

The cultivation of anchorage-dependent cell lines is extensively explored in CM production, involving the use of microcarriers and scaffolds. Notably, fluidized and fixed-bed bioreactors exhibit characteristics that are compatible with this technology (Kuystermans and Al-Rubeai 2011).

Fluidized bed bioreactors (Fig. 3) typically consist of a column-type bioreactor wherein the fluids flow upward through a bed of microcarriers, creating a constant circulation loop that suspends them, inducing a fluidlike behavior while minimizing the formation of oxygen and temperature gradients. In this configuration, cell viability tends to remain high; however, the available working volumes are still very limited, since spatial heterogeneity can become more pronounced when dealing with large manufacturing scales (Mendonça da Silva et al. 2020; Rodrigues et al. 2016).

Fixed-bed reactors (Fig. 4), well-researched and promising in animal cell culture due to their potential for high cell density in a low shear-stress environment, are typically utilized in perfusion mode. In this mode, cells are seeded within a fixed matrix, while the medium is continually pumped through, ensuring a constant supply of oxygen and nutrients for cell growth (Ng and Kurisawa 2021).

This setup enables the use of porous scaffolds/fibers, extending the possibilities for structured animal cell culture, thanks to the large surface-to-volume ratio that supports cell attachment, expansion, and differentiation. Besides that, the

Fig. 3 Schematics for fluidized bed bioreactor. Adapted from Kong et al. (1999)

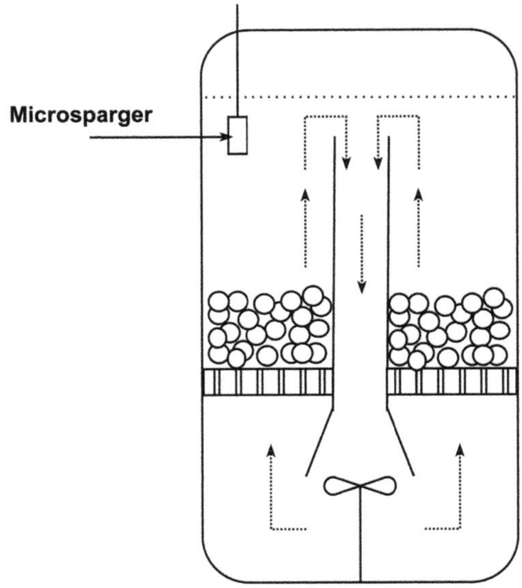


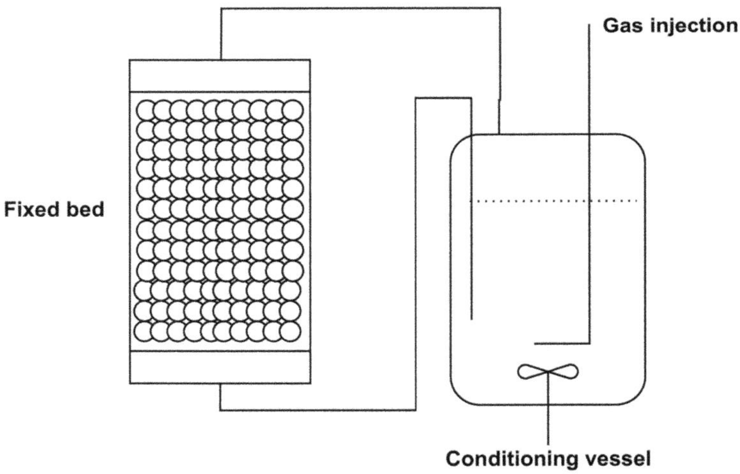


**Fig. 4** Schematics for fixed bed bioreactor. Adapted from Pörtner and Faschian (2019)

microenvironments offered by the scaffold tend to be more stable for the cells (Bellani et al. 2020; Kuystermans and Al-Rubeai 2011; Pörtner and Faschian 2019).

On the other hand, in this operation, higher perfusion rates are required, and the structural aspects may lead to the formation of axial and radial concentration gradients, particularly on a large scale. Additionally, cell harvesting can be challenging due to the high cell densities and the limited accessibility of detachment supplements to the cell culture. Packed bed bioreactors have only achieved a maximum reported working volume of 30 L, making them valuable for inoculum buildup but not yet suitable for large-scale production (Bellani et al. 2020; Kuystermans and Al-Rubeai 2011).

### 3.3.3 Hollow Fiber Bioreactor

Hollow fiber bioreactors also have extensive use in perfusion mode, and, similar to fixed-bed reactors, they can provide a significant surface-to-volume ratio, thereby allowing for high cell density in a low-shear environment (Kuystermans and Al-Rubeai 2011).

These bioreactors consist of a cylindrical chamber stacked with semipermeable hollow fibers (Fig. 5). During operation, cells can be cultivated within the hollow fibers, on their external surface, or in both locations, with the culture medium being circulated through the fibers. This setup allows the delivery of nutrients and oxygen to the cells by diffusion, while metabolites flow out and can be subsequently removed (Bellani et al. 2020; Kuystermans and Al-Rubeai 2011; Kahan et al. 2020).

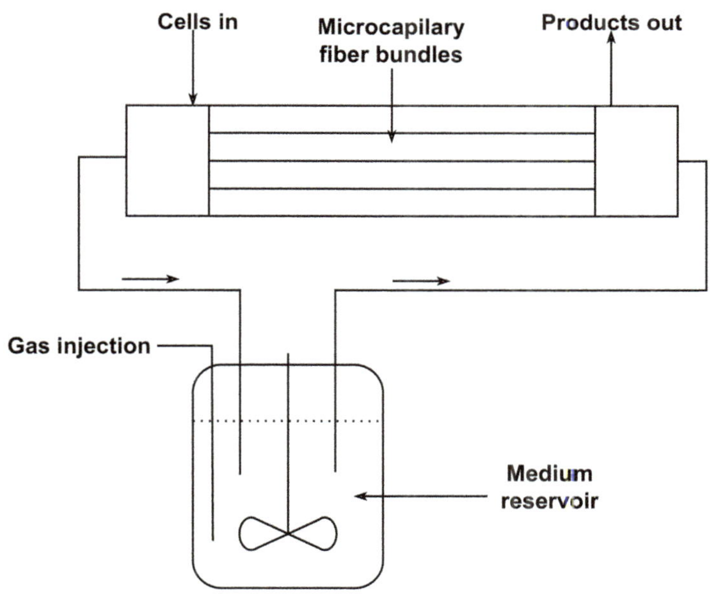


**Fig. 5** Schematics for hollow fiber bioreactor. Adapted from Rodrigues et al. (2009)

The major disadvantages linked to this type of reactor include challenges in scaling-up, high per-unit costs, common issues with membrane fouling, operational difficulties, such as mechanical failures, the potential for concentration gradients, and the difficulty of ensuring full access for cell harvesting components (Bellani et al. 2020; Kuystermans and Al-Rubeai 2011; Rodrigues et al. 2009).

It is suggested that hollow fiber could be useful in the cell differentiation stage of CM production, where the scaffold is designed to be edible without requiring separation from differentiated cells (Ng and Kurisawa 2021). Some works employ hollow fiber bioreactors and make use of bench-scale experimental data for proliferation and differentiation of $C_2C1_2$ cell (Tuomisto et al. 2022; Nie et al. 2022) and other cell lineages.

### 3.3.4 Single-Use Technology

The adoption of larger-scale single-use technology (SUT) has gained significant acceptance in the biopharmaceutical industry (Bellani et al. 2020). Often, this technology is regarded as time and cost-saving since it eliminates the need for cleaning and sterilization processes, while shortening setup, and turn-around time between runs, making it easy to operate and validate (Rodrigues et al. 2009; Jossen et al. 2017).

Typically, these presterilized bioreactors are either rigid or flexible plastic cultivation containers that can operate in various modes depending on the intended use. They can be manufactured as stirred tanks, wave bags, or even with perfusion modules (Seidel et al. 2022). Nonetheless, a few studies already showed that the plastic bags in single-use bioreactors can release substances that are possibly be harmful for mammalian cell and halt their growth, which will inevitably affect the process yield and the safety of the final product.

The primary factor limiting the size is the sterilization method, specifically, the depth of penetration of gamma radiation, which is approximately 30 cm. This imposes a 2000 L scale limit on this technology. Consequently, SUTs are primarily utilized for smaller-scale cultivations, including seed culture expansion and inoculation of large conventional bioreactors (Humbird 2021; Rodrigues et al. 2009).

Other limitations of SUTs include the assembly time with manual unpacking of individual bags, risk of leaks, the potential introduction of leachables and extractables into the final product, and the constraints associated with using single-use sensors for instrumentation. Moreover, the environmental impact of single-use items should be taken into account, as the plastic waste can end up in landfills or the sea, posing recycling challenges due to mixed plastic content (including daily media-preparation bags, filters, hoses, and sensors) and contamination (Allan et al. 2019).

Table 1 shows different types of bioreactors that can be employed for proliferation and differentiation steps of CM, with the description of their advantages and disadvantages, applied volumes, and reached final cells' concentration.

**Table 1** Bioreactors used in the proliferation and differentiation steps of cultivated meat

| Bioreactor type | Cell cultivation stage | Max. practical capacity per unit | Advantages | Disadvantages | Final cell concentration (cells/mL) | References |
|---|---|---|---|---|---|---|
| Bubble column/airlift | Proliferation | 100 L–2000 L | – Low-shear stress<br>– Low-heat generation<br>– Effective mixing<br>– Low contamination | – Bubble bursting cell damage<br>– Foaming formation<br>– Requires microcarriers | $10^5$–$10^6$ | Bellani et al. (2020); Letti et al. (2021) Li et al. (2020) |
| Stirred tank | Proliferation | 2000 L–20,000 L | – Easier to scale-up<br>– Great mass transfer<br>– Long-term sterility | – Require microcarriers<br>– High-shear stress<br>– Requires optimal cell line<br>– Aggregate dissociation<br>– Large volumes required | $10^5$–$10^8$ | Allan et al. (2019) Bellani et al. (2020) Humbird (2021) Kuystermans and Al-Rubeai (2011) |
| Wave/rocking platform | Proliferation | 100 L–500 L | – Effective mixing, suspension, and gas transfer<br>– Disposable/sterility<br>– Inoculumn buildup<br>– Low contamination | – Scale-up limitations<br>– Generates too much residues<br>– Low cell density | $10^6$–$10^7$ | Allan et al. (2019) Bellani et al. (2020) Humbird (2021) Kuystermans and Al-Rubeai (2011) Chen et al. (2022) Slivac et al. (2006) |
| Fluidized/packed bed | Proliferation/differentiation | 30 L | – Low shear<br>– Scaffolding compatibility<br>– High cell density | – Low volume<br>– Harvesting issues<br>– Concentration gradient | $10^6$–$10^8$ | Allan et al. (2019) Bellani et al. (2020) Humbird (2021) Kuystermans and Al-Rubeai (2011) |
| Hollow fiber | Differentiation | – | – High cell density<br>– Scaffolding compatibility<br>– High surface-to-volume ratio | – Harvesting issues<br>– Concentration gradient | $10^8$–$10^9$ | Allan et al. (2019) Bellani et al. (2020) Humbird (2021) |

## 4 Bioreactor Instruments and Control

Given the sensitive nature of animal cells, monitoring and controlling the parameters of the production process is a mandatory task that requires meticulous attention. Monitoring must be systematically organized and should provide clear, unambiguous results. This data is essential for assessing the physiological and morphological conditions of the cells as well as for identifying their stages of cellular growth, which are critical to the success of cultivation. As a result, gathering of this cultivation data assists in optimizing the process and facilitating its scaling (Moutsatsou et al. 2019; Matthews et al. 2016), creating a vital link between data collection and process efficiency.

In line with this, to guarantee efficiency in bioreactors, it is essential to use suitable instrumentation, complemented by rapid and accurate data analysis. Moreover, it is ideal that all actions directed toward monitoring and control be automated to avoid any human handling (Moutsatsou et al. 2019). This combination is essential for effectively guiding the biotransformation process at its peak, avoiding contaminations or in-process variations.

Ensuring optimal cell development requires monitoring a broad spectrum of parameters (Table 2). These include specific parameters of the cells, such as morphology, density, viability, and size; physical and mechanical parameters such as agitation, shear stress, partial gas pressures, temperature and density; along with physicochemical parameters, which include the medium's pH, a factor that is collectively influenced by the concentration of nutrients, salts, and metabolites in the media.

To address the need for consistent and reliable monitoring, three categories of sensing methods are utilized: in-line sensors, featuring sterilizable probes immersed in the culture fluid or scaffold matrix; noncontact sensors, positioned outside the

**Table 2** Important parameters to monitor and control in cultivated meat production

| Parameter | Optimal values | Details | References |
|---|---|---|---|
| Temperature | 37 °C | Depends on the cell type | Humbird (2020) |
| pH | 6.7–7.8 | Depends on the cell type | Pajčin et al. (2022) |
| Dissolved oxygen (DO) | 20%–60% | Depends on the cell type | Pawar et al. (2023), Djisalov et al. (2021) |
| Carbon dioxide partial pressure | 40–100 mbar | Above 130 mbar is inhibitory | Humbird (2020) |
| Ammonia | < 2–10 mM | Depends on the cell type. In excess promotes growth inhibition | Humbird 2021 |
| Lactate | <10–22 mM | Depends on the cell type. In excess promotes growth inhibition | Hassell et al. (1991) |
| Agitation | – | Effective mixing and low-shear rate are recommended | – |
| Cell morphology | – | According to the matrix, cell type, and differentiation stage | – |

bioreactor, which employ techniques like spectrophotometry or ultrasound for monitoring; and indirect sensing, comprising methods such as parameters correlations and offline analysis, where samples are collected and subsequently assessed in a laboratory setting (Djisalov et al. 2021).

## *4.1 Physical and Mechanical Parameters*

Physical and mechanical parameters are key factors in cellular growth, as they directly influence the environment where the cells develop. For instance, agitation and shear stress can affect cell morphology and viability (Humbird 2021), while factors like minor deviations in temperature and gas partial pressure impact the metabolic rate and the overall efficiency of the cultivation process (Pawar et al. 2023).

The ideal internal temperature of the bioreactor will depend on the cells being cultivated. Generally, sensors should be capable of reading in the range of 10 °C–40 °C with a maximum reading deviation of 0.5 °C for optimal performance. As the bioreactor volume increases, it is recommended to implement more sensor points for uniform readings. To achieve this, in-line resistance temperature detectors (RTDs) are used for cell temperature monitoring due to their high accuracy and fast response time (Pawar et al. 2023; Djisalov et al. 2021). However, thermistors (non-contact sensors) and thermocouples are cheaper and may be of interest for single-use technologies (SUT). Still, depending on the model, they may be less sensitive (Djisalov et al. 2021).

Effective supply of $O_2$-enhanced air is imperative due to the cells' dependency on oxygen to fulfill their metabolic requirements (Humbird 2021). To identify the partial pressure of $O_2$, a close correlation can be done with dissolved oxygen (DO), the variable commonly employed for monitoring purposes. Additionally, the monitoring of dissolved carbon dioxide holds equal significance as it is one of the primary metabolites and directly affects the pH of the extra- and intracellular environment (Pawar et al. 2023). Dissolved ammonia should also be monitored as it has a strong inhibitory effect at low concentrations (Humbird 2021). Furthermore, it can be generated both as a metabolic by-product and through nonenzymatic glutamine degradation, a crucial amino acid present in the culture medium (Xie and Wang 2000). In general, for monitoring dissolved gases, two main types of sensors are used. First, in-line electrochemical sensors, which can employ a variety of electrodes. These sensors are fast, offering quick monitoring of different molecules with good sensitivity. Second, in-line optical sensors, although more expensive and slightly slower than electrochemical ones, are gaining prominence. They stand out for their high anti-interference capabilities and lower maintenance needs (Djisalov et al. 2021; Pawar et al. 2023; Wei et al. 2019). Besides these in-line methods, offline analysis is also an option. Alternatively, spectrometers or analyzers specifically designed for the parameter of interest can be coupled to reactors.

The mechanical state of cells is a critical factor in both monitoring and control processes within bioreactors, with its importance varying depending on the specific reactor and growth matrix utilized. This is particularly relevant in scenarios where cells are prone to shear stress, which can impact their morphology and viability. Agitation, a crucial factor in many reactors, is typically a fixed variable optimized during the scale-up process. As such, it demands rigorous control to ensure the preservation of cell integrity and function. For instance, in airlift reactors or bubble columns, it is essential to control the aeration rate and bubble size to maintain optimal conditions. In stirred tank reactors (STRs), attention must be paid not only to aeration but also to the agitation rate, particularly focusing on the design and efficiency of the impeller. Additionally, ongoing monitoring of cell morphology is necessary to assess the impact of these mechanical factors. Further details on the monitoring of cell morphology will be discussed subsequently.

## *4.2 Physicochemical Parameters*

The pH of the medium provides valuable information about the process condition, and an analysis combined with other variables helps to better understand the metabolic behavior of the cells. Just like dissolved gases, pH is also monitored by electrochemical and optical sensors.

As mentioned, cell metabolism correlates with changes in the pH of the medium. These metabolic pathways generate different responses depending on the concentration and consumption of glucose and amino acids, which result in by-products of cell growth, increasing the concentrations of inhibitory components such as lactate, ammonia, and carbon dioxide. It has been demonstrated that a controlled feeding of glucose and amino acids can help maintain a balanced metabolic environment, slow metabolite accumulation, and improve cell growth and productivity (Domján et al. 2022; Matthews et al. 2016; Xie and Wang 2000).

Monitoring the concentration of reagents and products in bioprocesses is vital for understanding and controlling pH and cell development, leading to further process optimization. Besides the use of electrochemical sensors (Pawar et al. 2023; Djisalov et al. 2021), various coupled equipment techniques are used for this purpose. For instance, Raman spectroscopy allows for the online monitoring of lactate, glucose, and up to 20 amino acids (Domján et al. 2022; Matthews et al. 2016). UV-visible spectroscopy can detect metabolites and nutrients at different wavelengths, while mass spectrometers are useful for identifying volatile organic compounds (Pawar et al. 2023). Offline analyses are also feasible using these and other methods such as liquid and gas chromatography.

While some of these methods can be expensive and challenging to implement on a large scale, they can be cost-effective by enabling controlled feeding of the culture medium, thereby minimizing waste. This balance between cost and efficiency is crucial in optimizing bioprocessing.

### *4.3 Cell Parameters*

In cultivated meat production, where cells are the product, precise monitoring of their conditions is essential. Tracking cell viability and physiological states ensures product quality and safety. For in-depth morphological analysis, microscopic examination of samples from the reactor is crucial, providing insights into cell structure and development. However, manual procedures for assessing cell count, viability, density, and size can be time-consuming and subject to human error. To overcome these limitations, technologies like flow cytometry and cell counters have been developed. These advanced methods automate the analysis process, reducing time and bias (Akbari et al. 2022).

Additionally, routine assessments can be efficiently conducted using automated techniques integrated with the reactors. Raman spectroscopy, for instance, effectively monitors cell density and concentration, providing real-time data crucial for process optimization. The dielectric spectroscopy sensor is utilized for assessing biomass viability, offering key insights into cell conditions. Moreover, the Doppler ultrasound technique, known for its noninvasive approach, is proposed for measuring both cell concentration and viability (Pawar et al. 2023; Akbari et al. 2022).

## 5 Challenges and Future Developments in Bioreactor Design and Scale-Up

Despite being a promising alternative to conventional meat, mainly in the sustainability and environmental aspects, the CM production process still faces several challenges for its proper industrialization and commercialization. As an example, the cost associated with the cultivation media applied for both process steps (proliferation and differentiation) (Soccol et al. 2022). For the proper culture of eukaryotic cells, the medium has to be nutrient-rich in order to meet the requirements of energy and cofactors needed for cell development. In comparison to the media applied for cultivation of microorganisms, for example, the media for eukaryotic cells need significantly more components, specifically amino acids, vitamins, and trace elements (Humbird 2021; Jairath et al. 2021). There are some commercial formulated media that provide all the necessary nutrients. However, they tend to be animal based (using either bovine or horse derived serum), which contradicts the CM being a slaughter-free product, along with the high cost of these components (Jairath et al. 2021; Soccol et al. 2022). Besides, there is the need of using specific factors that will induce either cell multiplication or differentiation in each step of the process. These specific components can be obtained by recombinant technologies (Soccol et al. 2022). Finally, nutrient-rich media are more susceptible to contamination. Antibiotics can be added to the media to prevent contamination; however, regulations are usually strict about the indiscriminate use of these compounds as resistant bacteria can be developed, and the use of one more component in the cultivation medium can make it even more costly (Letti et al. 2021). Another challenge regarding contamination is the proper equipment sterilization, mainly the bioreactors. This process needs to be well conducted in order to guarantee minimal risks of contamination. Clean-in-place (CiP) and sterilization-in-place (SiP) are usually the strategies applied, as they are automated systems that do not require manual manipulation. However, significant amounts of water and energy are used in these systems (Pajčin et al. 2022). Another alternative is the use of wave bioreactors or single-use STRs as the bags are already sterile, being disposable when the process ends. Yet, there is an environmental issue regarding the use of these systems as more plastic waste is generated (Post et al. 2020).

The CM production usually involves the culture of muscle cells in the bioreactors. However, meat is a complex structure of muscle, blood, adipose, and skeletal cells (Letti et al. 2021). The co-culture of different cell types in an efficient manner is not yet a reality, despite the recent scientific efforts that have been made (Pajčin et al. 2022; Reiss et al. 2021). Therefore, minced meat or processed products, such as hamburgers, nuggets, and meatballs, are usually the ones obtained. The obtainment of a "steak" or "steak-like" product is not yet possible, and a great research effort is still needed for it. The advantage of cultured meat, however, is the possibility of modulating or enhancing the nutritional value and even protein content as certain safe additives can be used. Besides, there is the possibility of genetic engineering of the cells, even though this raises the discussion about safety of genetically modified organisms. As the cells applied for the cultivated meat obtainment are anchorage dependent, a solid matrix is usually needed in both of the production steps for their proper development (Allan et al. 2019). Therefore, scaffolds are applied, being biopolymers that will serve as a matrix for the cells to attach along the process. These materials need to mandatorily follow food regulation minimal standards and be either biocompatible or biodegradable in order to be present in the product or easily disassociated previously to commercialization (Reiss et al. 2021). Industrial scale production for CM scaffolds is not yet a widespread reality, and, therefore, the cost of these biomaterials will naturally increase the cost of the process. Besides, the required scaffolds for each step (proliferation and differentiation) are usually different, which will also lead to the addition in the final cost. Scaffolds, therefore, need to be well chosen, being compatible with the applied bioreactors, cheap, with a scalable production, and must come from a safe and animal-free source.

The eukaryotic cells required for the CM production are also known to be more sensible to process conditions than microorganisms usually are (Allan et al. 2019). Therefore, scaling-up and obtaining high cell densities can become a challenge. Process control of basic conditions such as temperature, pH, dissolved oxygen, carbon source consumption, and cell development is crucial. Some bioreactor systems (e.g., STR) can be easily controlled with well-established strategies. However, other types of bioreactors can require refined strategies since controlling is difficult, mainly those with a solid matrix that immobilizes cells (e.g., hollow fibers and packed bed) (Pajčin et al. 2022). The cells are also shear sensitive, that being, either low agitation rates or specific agitation strategies need to be applied in order to guarantee the cell integrity. Nonetheless, poor agitation can lead to poor mixing, leading to gradients inside the bioreactor. Low aeration rates, when applied, can lead to poor mass transfer and even to high $CO_2$ concentration inside the bioreactor as a product of the cellular respiration. High $CO_2$ levels will decrease the medium's pH. These consequences reveal the complicated scenario of controlling the conditions inside an eukaryotic cell culture (Zhang et al. 2020). The scale-up of some bioreactors is still a challenge as the models and relations have not yet been fully established and developed. As a general rule, geometric relations of the bioreactors' models between two scales (lab and pilot or pilot and industrial) need to be maintained constant in order to guarantee the proper reproducibility of the process when scaling-up. Besides, some process parameters mainly related to agitation and aeration—guaranteeing therefore the minimal conditions for cell development—can be kept constant among scales. Bench-scale bioreactors and pilot scale can be used as models and tests of different conditions regarding process parameters for further industrial application.

As the scale-up of some models applied to the cultivated meat production is not yet well established, some other strategies can be used in order to achieve the required final product concentration. This is the scale-out and/or parallelization strategy in which several smaller bioreactors are applied rather than one unique large-scale bioreactor. This can reduce the impact that scaling-up has on the shear-induced stress and the need to develop a scaffold that can fit into a large-scale bioreactor. However, this is a more labor-intensive strategy than a single larger bioreactor (Pajčin et al. 2022; Stephens et al. 2018). Another strategy to facilitate scale-up is the use of computational fluid dynamics (CFD), which can be a useful tool for simulating bioreactors' conditions, helping to determine the optimal conditions for cell development, as analysis about fluid behavior, characteristics, and interactions can be performed with CFD. This kind of analysis can be helpful in predicting parameters and giving useful insights about the bioprocess through simulation, making the further optimization and scale-up easier (Bellani et al. 2020; Rivera and Chen 2022).

The fabrication of bioreactors for CM production needs to be customized as each process will require certain conditions. Therefore, there is the need for an open dialogue with bioreactors' manufacturers in order to find the best possible design for the proposed process. A disruptive innovation is the use of 3D printing as a step for cultivated meat production, consisting in constructing a meat product by printing a bioink in a layer-by-layer manner through extrusion (Handral et al. 2022; Tibrewal et al. 2023). Generally, a cell expansion step is still required previously to the printing; however, not necessarily the bioreactor for the differentiation is needed. The final obtained product is, theoretically, a structured 3D product, rather than the unstructured meat that is obtained without the bioprinting. Post-processing may change some characteristics of the final product, being important to modulate it before commercialization (Handral et al. 2022). Despite being a disruptive technology with huge potential for CM production, it is still a nascent and costly technique and, therefore, requires deep research and understanding prior to application (Handral et al. 2022; Ramachandraiah 2021; Tibrewal et al. 2023).

## 6 Conclusions and Future Perspective

The CM technology is experiencing significant growth, necessitating substantial advancements in the development of new and/or adapted bioreactors' models for the proper and efficient cultivation of cells. The unique characteristics of various types of mammalian cells employed for this purpose require the research and development of novel devices and conditions, encompassing the composition of media, agitation, mass transfer, structure, and composition of microcarriers that serve as essential supports. Nevertheless, traditional bioreactor types find application in CM production with certain adaptations, illustrating, for instance, that STRs and airlift bioreactors remain viable and effective options for culturing these cells with proper control and instrumentation. Other bioreactors options, including hollow fibers, packed bed, and single-use bioreactors, are also worth considering. Consequently, the design and manufacture of bioreactors for CM production demand in-depth discussions among researchers and manufactures to arrive at efficient solutions.

## References

Abbasalizadeh S, Larijani MR, Samadian A, Baharvand H (2012) Bioprocess development for mass production of size-controlled human pluripotent stem cell aggregates in stirred suspension bioreactor. Tissue Eng Part C Methods 18(11):831–851. https://doi.org/10.1089/ten.tec.2012.0161

Akbari S, Anderson P, Zang H et al (2022) Non-invasive real-time monitoring of cell concentration and viability using Doppler ultrasound. SLAS Technol 27:368–375. https://doi.org/10.1016/j.slast.2022.09.003

Allan SJ, De Bank PA, Ellis MJ (2019) Bioprocess design considerations for cultured meat production with a focus on the expansion bioreactor. Front Sustain Food Syst 3(June). https://doi.org/10.3389/fsufs.2019.00044

Al-Rubeai M, Singh RP, Goldman MH, Emery AN (1995) Death mechanisms of animal cells in conditions of intensive agitation. Biotechnol Bioeng 45(6):463–472. https://doi.org/10.1002/bit.260450602

Andreassen RC, Rønning SB, Solberg NT, Grønlien KG, Kristoffersen KA, Høst V, Kolset SO, Pedersen ME (2022) Production of food-grade microcarriers based on by-products from the food industry to facilitate the expansion of bovine skeletal muscle satellite cells for cultured meat production. Biomaterials 286. https://doi.org/10.1016/j.biomaterials.2022.121602

Bellani CF, Ajeian J, Duffy L, Miotto M, Groenewegen L, Connon CJ (2020) Scale-up technologies for the manufacture of adherent cells. Front Nutr 7(November). https://doi.org/10.3389/fnut.2020.575146

Ben-Arye T, Shandalov Y, Ben-Shaul S, Landau S, Zagury Y, Ianovici I, Lavon N, Levenberg S (2020) Textured soy protein scaffolds enable the generation of three-dimensional bovine skeletal muscle tissue for cell-based meat. Nat Food 1(4):210–220. https://doi.org/10.1038/s43016-020-0046-5

Bunnell BA, Flaat M, Gagliardi C, Patel B, Ripoll C (2008) Adipose-derived stem cells: isolation, expansion and differentiation. Methods 45(2):115–120. https://doi.org/10.1016/j.ymeth.2008.03.006

Chen L, Guttieres D, Koenigsberg A, Barone PW, Sinskey AJ, Springs SL (2022) Large-scale cultured meat production: trends, challenges and promising biomanufacturing technologies. Biomaterials 280(July 2021):121274. https://doi.org/10.1016/j.biomaterials.2021.121274

Chopda VR, Holzberg T, Ge X, Folio B, Tolosa M, Kostov Y, Tolosa L, Rao G (2020) Real-time dissolved carbon dioxide monitoring I: application of a novel in situ sensor for CO2 monitoring and control. Biotechnol Bioeng 117:981–991. https://doi.org/10.1002/bit.27253

Djisalov M, Knežić T, Podunavac I et al (2021) Cultivating multidisciplinarity: manufacturing and sensing challenges in cultured meat production. Biology (Basel) 10:204. https://doi.org/10.3390/biology10030204

Domján J, Pantea E, Gyürkés M et al (2022) Real-time amino acid and glucose monitoring system for the automatic control of nutrient feeding in CHO cell culture using Raman spectroscopy. Biotechnol J 17:2100395. https://doi.org/10.1002/biot.202100395

Dubey A, Lavanya L, Sadananda D, Gouthami K, Elfansu K, Singh A, Singh A (2021) Inferences of carbon dioxide in present-day cell culture systems: an unacknowledged problem and perspectives. Austin Ther 6(1):1033. https://doi.org/10.26420/austintherapeutics.2021.1033

GE Healthcare Bio-Sciences AB (2015) Disposable CellbagTM bioreactors for WAVE BioreactorTM systems

Godara P, McFarland CD, Nordon RE (2008) Design of bioreactors for mesenchymal stem cell tissue engineering. J Chem Technol Biotechnol 83(4):408–420. https://doi.org/10.1002/jctb.1918

Grenville RK, Giacomelli JJ, Padron G, Brown DAR (2017) Impeller performance in stirred tanks. Chem Eng J 2017:46–55

Handral HK, Hua Tay S, Wan Chan W, Choudhury D (2022) 3D printing of cultured meat products. Crit Rev Food Sci Nutr 62(1):272–281. https://doi.org/10.1080/10408398.2020.1815172

Hanga MP, de la Raga FA, Moutsatsou P, Hewitt CJ, Nienow AW, Wall I (2021) Scale-up of an intensified bioprocess for the expansion of bovine adipose-derived stem cells (bASCs) in stirred tank bioreactors. Biotechnol Bioeng 118(8):3175–3186. https://doi.org/10.1002/bit.27842

Hassell T, Gleave S, Butler M (1991) Growth inhibition in animal cell culture: the effect of lactate and ammonia. Appl Biochem Biotechnol 30:29–41. https://doi.org/10.1007/BF02922022

Hu W, Berdugo C, Chalmers JJ (2011) The potential hydrodynamic damage to animal cells of industrial relevance: current understanding. Cytotechnology 63:445–460. https://doi.org/10.1007/s10616-011-9368-3

Humbird D (2020) Scale-up economics for cultured meat: techno-economic analysis and due diligence. engrXiv. https://doi.org/10.31224/osf.io/795su

Humbird D (2021) Scale-up economics for cultured meat. Biotechnol Bioeng 118(8):3239–3250. https://doi.org/10.1002/bit.27848

Jairath G, Mal G, Gopinath D, Singh B (2021) A holistic approach to access the viability of cultured meat: a review. Trends Food Sci Technol 110:700–710. https://doi.org/10.1016/j.tifs.2021.02.024

Jossen V, Eibl-Schindler R, Pörtner R, Kraume M, Eibl D (2017) Stirred bioreactors: current state and developments, with special emphasis on biopharmaceutical production processes. In: Current developments in biotechnology and bioengineering: human and animal health applications. Elsevier, New York, pp 179–215

Kahan S, Camphuijsen J, Cannistra C, Potter G, Consenza Z, Shmulevich I (2020) Cultivated meat modeling consortium: Inaugural meeting whitepaper. Authorea. https://doi.org/10.22541/au.158057683.31004563

Kong D, Cardak S, Chen M, Gentz R, Zhang J (1999) High cell density and productivity culture of Chinese hamster ovary cells in a fluidized bed bioreactor. Cytotechnology 29(3):215–220. https://doi.org/10.1023/A:1008064217040

Kumar A, Starly B (2015) Large scale industrialized cell expansion: producing the critical raw material for biofabrication processes. Biofabrication 7(4):044103. https://doi.org/10.1088/1758-5090/7/4/044103

Kuystermans D, Al-Rubeai M (2011) Bioreactor systems for producing antibody from mammalian cells. In: Al-Rubeai M (ed) Antibody expression and production. Cell engineering. Springer, Dordrecht

Letti LAJ, Karp SG, Molento CFM, Colonia BSO, Boschero RA, Soccol VT, Herrmann LW, de Penha RO, Woiciechowski AL, Soccol CR (2021) Cultivated meat: recent technological developments, current market and future challenges. Biotechnol Res Innov 5(1):e2021001. https://doi.org/10.4322/biori.202101

Leung M, Warner M, Vanderpol RE, Hsiu TP-J, Carswell K (2021) Apparatuses and systems for preparing a meat product. Patent No. US20210145031A1. https://uspto.report/patent/app/20210145031

Li X, Zhang G, Zhao X, Zhou J, Du G, Chen J (2020) A conceptual air-lift reactor design for large scale animal cell cultivation in the context of in vitro meat production. Chem Eng Sci 211:115269. https://doi.org/10.1016/j.ces.2019.115269

Lim D, Renteria ES, Sime DS, Ju YM, Kim JH, Criswell T, Shupe TD, Atala A, Marini FC, Gurcan MN, Soker S, Hunsberger J, Yoo JJ (2022) Bioreactor design and validation for manufacturing strategies in tissue engineering. Biodes Manuf 5:43–63. https://doi.org/10.1007/s42242-021-00154-3

Lubbert A (2023) Bubble columns and airlift bioreactors. Encyclopedia of Life Support Systems (EOLSS)

Martin Y, Vermette P (2005) Bioreactors for tissue mass culture: design, characterization, and recent advances. Biomaterials 26(35):7481–7503. https://doi.org/10.1016/j.biomaterials.2005.05.057

Matthews TE, Berry BN, Smelko J et al (2016) Closed loop control of lactate concentration in mammalian cell culture by Raman spectroscopy leads to improved cell density, viability, and biopharmaceutical protein production. Biotechnol Bioeng 113:2416–2424. https://doi.org/10.1002/bit.26018

Mendonça da Silva J, Erro E, Awan M, Chalmers S-A, Fuller B, Selden C (2020) Small-scale fluidized bed bioreactor for long-term dynamic culture of 3D cell constructs and in vitro testing. Front Bioeng Biotechnol 8. https://doi.org/10.3389/fbioe.2020.00895

Merchuk JC (1990) Why use air-lift bioreactors? Trends Biotechnol 8(C):66–71. https://doi.org/10.1016/0167-7799(90)90138-N

Messmer T, Klevernic I, Furquim C, Ovchinnikova E, Dogan A, Cruz H, Post MJ, Flack JE (2022) A serum-free media formulation for cultured meat production supports bovine satellite cell differentiation in the absence of serum starvation. Nat Food 3(1):74–85. https://doi.org/10.1038/s43016-021-00419-1

Mizrahi A, Lazar A (1988) Media for cultivation of animal cells: an overview. Cytotechnology 1(3):199–214. https://doi.org/10.1007/BF00145023

Moritz MSM, Verbruggen SEL, Post MJ (2015) Alternatives for large-scale production of cultured beef: a review. J Integr Agric 14(2):208–216. https://doi.org/10.1016/S2095-3119(14)60889-3

Moutsatsou P, Ochs J, Schmitt RH et al (2019) Automation in cell and gene therapy manufacturing: from past to future. Biotechnol Lett 41:1245–1253. https://doi.org/10.1007/s10529-019-02732-z

Mutaf T, Oncel SS (2023) Bubble column and airlift bioreactor systems for animal cell culture applications. Asia-Pacific J Chem Eng 18(1). https://doi.org/10.1002/apj.2872

Neunstoecklin B, Stettler M, Solacroup T, Broly H, Morbidelli M, Soos M (2015) Determination of the maximum operating range of hydrodynamic stress in mammalian cell culture. J Biotechnol 194:100–109. https://doi.org/10.1016/j.jbiotec.2014.12.003

Ng S, Kurisawa M (2021) Integrating biomaterials and food biopolymers for cultured meat production. Acta Biomater 124:108–129. https://doi.org/10.1016/j.actbio.2021.01.017

Nie M, Shima A, Takeuchi S (2022) Hollow fiber bioreactor with micro anchor arrays for the biofabrication of skeletal muscle tissues. In: IEEE 35th International Conference on Micro Electro Mechanical Systems Conference (MEMS), pp 287–288

Nienow AW (2010) Impeller selection for animal cell culture. In: Flickinger MC (ed) Encyclopedia of industrial biotechnology: bioprocess, bioseparation, and cell technology. Wiley
O'Neill EN, Cosenza ZA, Baar K, Block DE (2021) Considerations for the development of cost-effective cell culture media for cultivated meat production. Compr Rev Food Sci Food Saf 20(1):686–709. https://doi.org/10.1111/1541-4337.12678
Pajčin I, Knežić T, Azoulay IS, Vlajkov V, Djisalov M, Janjušević L, Grahovac J, Gadjanski I (2022) Bioengineering outlook on cultivated meat production. Micromachines 13(3):1–44. https://doi.org/10.3390/mi13030402
Park S, Stephanopoulos G (1993) Packed bed bioreactor with porous ceramic beads for animal cell culture. Biotechnol Bioeng 41(1):25–34. https://doi.org/10.1002/bit.260410105
Pawar D, Lo Presti D, Silvestri S et al (2023) Current and future technologies for monitoring cultured meat: a review. Food Res Int 173:113464. https://doi.org/10.1016/j.foodres.2023.113464
Pörtner R, Faschian R (2019) Design and operation of fixed-bed bioreactors for immobilized bacterial culture. In: Mishra M (ed) Growing and handling of bacterial cultures. IntechOpen
Post MJ, Levenberg S, Kaplan DL, Genovese N, Fu J, Bryant CJ, Negowetti N, Verzijden K, Moutsatsou P (2020) Scientific, sustainability and regulatory challenges of cultured meat. Nat Food 1(7):403–415. https://doi.org/10.1038/s43016-020-0112-z
Rafiq QA, Heathman TRJ, Coopman K, Nienow AW, Hewitt CJ (2016) Scalable manufacture for cell therapy needs. In: Bioreactors. Wiley, pp 113–146
Ramachandraiah K (2021) Potential development of sustainable 3D-printed meat analogues: a review. Sustainability 13(2):1–20. https://doi.org/10.3390/su13020938
Reiss J, Robertson S, Suzuki M (2021) Cell sources for cultivated meat: applications and considerations throughout the production workflow. J Mol Sci 22(14):7513. https://doi.org/10.3390/ijms22147513
Rivera FJC, Chen J (2022) Computational fluid dynamics modeling of cell cultures in bioreactors and its potential for cultivated meat production—a mini-review. Future Foods 6:100195. https://doi.org/10.1016/j.fufo.2022.100195
Roberts I, Baila S, Rice RB, Janssens ME, Nguyen K, Moens N, Ruban L, Hernandez D, Coffey P, Mason C (2012) Scale-up of human embryonic stem cell culture using a hollow fibre bioreactor. Biotechnol Lett 34(12):2307–2315. https://doi.org/10.1007/s10529-012-1033-1
Rodrigues ME, Costa AR, Henriques M, Azeredo J, Oliveira R (2009) Technological progresses in monoclonal antibody production systems. Biotechnol Prog 26:332–351. https://doi.org/10.5772/intechopen.87944
Rodrigues C, Vandenberghe LPS, Woiciechowski AL, de Oliveira J, Letti LAJ, Soccol CR (2016) Production and application of lactic acid
Schirmer C, Maschke RW, Pörtner R, Eibl D (2021) An overview of drive systems and sealing types in stirred bioreactors used in biotechnological processes. Appl Microbiol Biotechnol 105:2225–2242. https://doi.org/10.1007/s00253-021-11180-7
Seah JSH, Singh S, Tan LP, Choudhury D (2022) Scaffolds for the manufacture of cultured meat. Crit Rev Biotechnol 42(2):311–323. https://doi.org/10.1080/07388551.2021.1931803
Seidel S, Maschke RW, Kraume M, Eibl R, Eibl D (2022) CFD modelling of a wave-mixed bioreactor with complex geometry and two degrees of freedom motion. Front Chem Eng 4:1–19. https://doi.org/10.3389/fceng.2022.1021416
Sieblist C, Jenzsch M, Pohlscheidt M (2016) Equipment characterization to mitigate risks during transfers of cell culture manufacturing process. Cytotechnology 68:1381–1401. https://doi.org/10.1007/s10616-015-9899-0
Singh V (1999) Disposable bioreactor for cell culture using wave-induced agitation. Cytotechnology 30(1/3):149–158. https://doi.org/10.1023/A:1008025016272
Slivac I, Srček VG, RadoŠević K, Kmetič I, Kniewald Z (2006) Aujeszky's disease virus production in disposable bioreactor. J Biosci 31(3):363–368. https://doi.org/10.1007/BF02704109
Soccol CR, Karp SG, Letti LAJ, Biagini G, Balla G, Boligon AP, Candelario JPM, Faria TF, de Mattos PBG, Piazenski IN, Raymundo LCK, Valeixo DARR, Soccol VT (2022) Is the devel-

opment of low-cost media one of the greatest challenges to produce cultivated meat on an industrial scale? Biotechnol Res Innov 6(2):e2022005. https://doi.org/10.4322/biori.20226201

Spier MR, Vandenberghe LPDS, Medeiros ABP, Soccol CR (2011) Application of different types of bioreactors in bioprocesses

Steiner D, Khaner H, Cohen M, Even-Ram S, Gil Y, Itsykson P, Turetsky T, Idelson M, Aizenman E, Ram R, Berman-Zaken Y, Reubinoff B (2010) Derivation, propagation and controlled differentiation of human embryonic stem cells in suspension. Nat Biotechnol 28(4):361–364. https://doi.org/10.1038/nbt.1616

Stephens N, Di Silvio L, Dunsford I, Ellis M, Glencross A, Sexton A (2018) Bringing cultured meat to market: technical, socio-political, and regulatory challenges in cellular agriculture. Trends Food Sci Technol 78:155–166. https://doi.org/10.1016/j.tifs.2018.04.010

Stout AJ, Mirliani AB, Rittenberg ML, Shub M, White EC, Yuen JSK, Kaplan DL (2022) Simple and effective serum-free medium for sustained expansion of bovine satellite cells for cell cultured meat. Commun Biol 5(1). https://doi.org/10.1038/s42003-022-03423-8

Tibrewal K, Dandekar P, Jain R (2023) Extrusion-based sustainable 3D bioprinting of meat & its analogues: a review. Bioprinting 2023:e00256. https://doi.org/10.1016/j.bprint.2022.e00256

Tuomisto HL, Allan SJ, Ellis MJ (2022) Prospective life cycle assessment of a bioprocess design for cultured meat production in hollow fiber bioreactors. Sci Total Environ 851:158051. https://doi.org/10.1016/j.scitotenv.2022.158051

Tzimorotas D, Solberg N, Andreassen RC, Moutsatsou P, Bodiou V, Pedersen ME, Rønning SB (2023) Expansion of bovine skeletal muscle stem cells from spinner flasks to benchtop stirred-tank bioreactors for up to 38 days. Front Nutr. https://doi.org/10.3389/fnut.2023.1192365

Ulloa-Montoya F, Verfaillie CM, Hu WS (2005) Culture systems for pluripotent stem cells. J Biosci Bioeng 100(1):12–27. https://doi.org/10.1263/jbb.100.12

Verbruggen S, Luining D, van Essen A, Post MJ (2018) Bovine myoblast cell production in a microcarriers-based system. Cytotechnology 70(2):503–512. https://doi.org/10.1007/s10616-017-0101-8

Warner RD (2019) Review: analysis of the process and drivers for cellular meat production. Animal 13(12):3041–3058. https://doi.org/10.1017/S1751731119001897

Wei Y, Jiao Y, An D et al (2019) Review of dissolved oxygen detection technology: from laboratory analysis to online intelligent detection. Sensors 19:3995. https://doi.org/10.3390/s19183995

Xie L, Wang DIC (2000) High cell density and high monoclonal antibody production through medium design and rational control in a bioreactor. Biotechnol Bioeng 51:725–729. https://doi.org/10.1002/(SICI)1097-0290(19960920)51:6<725::AID-BIT12>3.0.CO;2-C

Zhan C, Hagrot E, Brandt L, Chotteau V (2019) Study of hydrodynamics in wave bioreactors by computational fluid dynamics reveals a resonance phenomenon. Chem Eng Sci 193:53–65. https://doi.org/10.1016/j.ces.2018.08.017

Zhang G, Zhao X, Li X, Du G, Zhou J, Chen J (2020) Challenges and possibilities for biomanufacturing cultured meat. Trends Food Sci Technol 97:443–450. https://doi.org/10.1016/j.tifs.2020.01.026

# Modeling and Simulation of Cell Cultures in Cultivated Meat Production

**Luiz Alberto Junior Letti, Andressa Tedesco Andretta, Walter José Martinez Burgos, Fernando Enrique Rosas Vega, and Carlos Ricardo Soccol**

**Abstract** The production of cultured meat or in vitro meat is a process developed in several stages, including the proliferation of mother cells and ex situ differentiation. This chapter will discuss the different mathematical models used to describe the stages of proliferation and differentiation in cultured meat production. In the first stages of the process, the objective is to facilitate cell proliferation and obtain the largest quantities of biomass possible. For modeling this stage, there are different mathematical models that can predict the size and dynamics of cell growth with great certainty, including the Gompertz model, logistic model, Richards model, among others. This last model included a new parameter where the deviation from symmetric dynamics is considered. In the case of the differentiation stage, where stem cells become muscle cells, it turns out to be a more complex process to model since, in addition to being a long process, it also depends on a set of physical and chemical factors. One of the best-known models to evaluate this stage is the mesenchymal stem cell (MSC), which takes into account the density of cell types and the volume in the process at this stage. Other models take into account chemical conditions and cellular maturity, such as the model for mechanical stimuli. Therefore, this chapter aims to enhance the comprehension of the intricate field of cultivated meat production by employing mathematical models to describe crucial steps of cell growth.

**Keywords** Cultivated meat · Mathematical modeling · Differentiation · Proliferation · Cell growth kinetics

L. A. J. Letti · A. T. Andretta · W. J. M. Burgos · F. E. R. Vega · C. R. Soccol (✉)
Bioprocess Engineering and Biotechnology Department, Federal University of Paraná, Curitiba, Paraná, Brazil
e-mail: soccol@ufpr.br


C. R. Soccol et al. (eds.), *Cultivated Meat*,
https://doi.org/10.1007/978-3-031-55968-6_7

# 1 Introduction

Cultivated meat, also known as in vitro meat, is an emerging alternative to address the growing demand for sustainable and ethical food sources. Cell culture development involves various stages, encompassing the proliferation of mother cells and their subsequent differentiation. There are several challenges to develop this technology, and this chapter will deal with the modeling and simulation of cell cultures, which is essential for developing efficient and scalable solutions.

Understanding and manipulating the stages of cell proliferation and differentiation are essential to achieving the desired outcomes in terms of biomass yield and product quality. In the initial phases, the primary objective is to facilitate robust cell proliferation, laying the basis for substantial biomass production. Mathematical models play a crucial role for a better qualitative and quantitative description of this stage, providing predictive insights into the size and dynamics of cell growth. Among these models, the Gompertz model, logistic model, and Richards model stand out, each offering particular perspectives on cell proliferation dynamics.

The subsequent differentiation stage, where stem cells transform into specialized muscle cells, introduces a new layer of complexity. This phase, characterized by its duration and dependence on a myriad of physical and chemical factors, necessitates sophisticated modeling approaches. One prominent model in this context is the mesenchymal stem cell (MSC) model, which incorporates considerations of cell-type density and volume in the differentiation process. Additionally, other models delve into the intricate interplay of chemical conditions and cellular maturity, exemplified by models addressing the influence of mechanical stimuli.

This chapter details these mathematical models, trying to elucidate the complexity of proliferation and differentiation steps in the context of cultivated meat production. By exploring the nuances of these models, we aim to provide a comprehensive understanding of these particular systems and help to provide a deeper appreciation of this interdisciplinary field.

# 2 Modeling and Simulation of Stem Cell Proliferation in Cultivated Meat Production

The production of in vitro meat is a multistep process that involves obtaining muscle stem cells, proliferating and differentiating them ex situ, to produce edible tissues that mimic conventional meat (Datar and Betti 2009; Guan et al. 2021; Jaques et al. 2021; Post 2014). During the early stages of cell culture, the main priority is to facilitate cell proliferation to obtain the large quantity of cells necessary for creating cultivated meat products at a commercial scale (Jaques et al. 2021; Reiss et al. 2021). For this reason, a mathematical model that could represent the size and the growth dynamics of a population of stem cells would be highly useful (Deasy et al. 2003; Tabatabai et al. 2011). This is the first step in the development of predictable and reproducible bioreactor processes, as they can act as predictive tools for the expansion potential of stem cells (Deasy et al. 2003). In the following paragraphs, we present a review of some existing growth models applied to the area of stem cell proliferation.

The method most commonly used to quantitate the proliferative capacity of cells applies the exponential growth equation $N = N_0 \cdot 2^{\frac{t}{DT}}$, which allows for estimating the population doubling time (DT) (Sherley et al. 1995). This method assumes that all cells in the studied population divide and give rise to two daughter cells that also both divide, making the fraction of daughter cells that divide ($F_d$) equal to 1. However, not all populations have an $F_d$ equal to 1, usually it is significantly lower, which makes the determined DT being in an average value since the exponential growth equation is inappropriately applied (Sherley et al. 1995). Taking this into consideration, Sherley et al. (1995) described a more general equation that has a broader application to the growth of cells in culture. Their equation (Eq. 1) defines cell growth in terms of the variable $F_d$, providing a sensitive method for detecting either exponential or nonexponential division dynamics (Sherley et al. 1995).

$$P(t) = P_0 \left[ 0.5 + \frac{1-(2F_d)^{\frac{t}{DT}+1}}{2(1-2F_d)} \right] \quad (1)$$

The nonexponential dynamic of stem cells is a result of a proliferative heterogeneity, where the nondividing nature of some cells can be attributed to quiescence, differentiation, or cell death (Deasy et al. 2003). Thus, the population size at time $t$ depends directly on the birth rate, the mitotically active fraction, and the death rate. Mathematical models designed to describe nonexponential growth are the result of variations on these three themes. Taking this into account, Deasy et al. (2003) took as a base the Sherley model described below and incorporated a term to account for cell loss by death (Eq. 2).

$$P(t) = P_0 \left[ 0.5 + \frac{1-(2F_d)^{\frac{t}{DT}+1}}{2(1-2F_d)} \right] - M \quad (2)$$

Tumors, described as a cell mass, also exhibit a nonexponential growth pattern (Tatro 2018). Constructed with differential equations, the Gompertz model (Eq. 3) accurately predicts tumor growth curves (Benzekry et al. 2014), considering an exponential decay since the growth rate of tumors diminishes in a nonlinear way, while its mass increases (Domingues 2012). Even though the Gompertz equation had been proposed to study the development of solids tumors, it has proved useful for a wide range of growth curves (Tabatabai et al. 2005), fitting to everything, from plant to cell growth (Tjørve and Tjørve 2017). The Gompertz model was, for example, applied by Tabatabai et al. (2011) to analyze the growth of stem cells.

$$\frac{dN}{dt} = r\,Nln\left(\frac{K}{N}\right) \tag{3}$$

Another model widely used in several areas, like statistics, economics, medicine, and in biology, is the logistic model proposed by the mathematician Pierre Verhulst (Omer 2018). The logistic model suggests that the rate of increase in population growth may decrease as the population near the saturation point of the environment, which can be caused by interacting factors, nutrient availability, crowding, or competition (Wachenheim et al. 2003). The logistic function (Eq. 4) can be derived for the coefficients have a specific biological meaning (Omer 2018; Wachenheim et al. 2003).

$$\frac{dN(t)}{dt} = rN(t)\left[1 - \frac{N(t)}{K}\right] \tag{4}$$

Richards model (Eq. 5), also known as the generalized logistic model, covered the features of the previous model and included a new factor (Wang et al. 2016). The new parameter included is used to capture the deviation of the symmetric S-shaped dynamics of the simple logistic model (Zreiq et al. 2020). However, most of the works found in the literature relies on the logistic or the Gompertz models due to the difficulties to estimate the parameters of the Richards model (Loibel et al. 2010).

$$\frac{dN(t)}{dt} = rN(t)\left[1 - \left(\frac{N(t)}{K}\right)^{a}\right] \tag{5}$$

To provide more accuracy and flexibility in the growth rate as the population reaches its carrying capacity, the hyperbolastic model H3 (Eq. 6) was recently introduced. This model has been demonstrated to be highly accurate in modeling biological growth. Tabatabai et al. (2011) used it to analyze the stem cell proliferation, applied both to embryonic and adult mesenchymal stem cells, having comparable results to the other models cited above (Tabatabai et al. 2011).

$$\frac{dN(t)}{dt} = \left(L - N(t)\right)\left(\delta\gamma t^{\gamma-1} + \frac{\theta}{\sqrt{1+\theta^2 t^2}}\right) \tag{6}$$

All the cited models are prevalent in modeling population growth, and for a better understanding of the progression and regression of a population size, the choice of an accurate growth model is important (Tabatabai et al. 2011). A comparison between the mean absolute relative error of all the models here described was performed in a study of stem cell data by Tabatabai et al. (2011).

A resume of the equations and the parameters meaning for each model is presented in Table 1.

**Table 1** Resume of the models with the respective equations and parameters meanings

| Model | Equation | Parameters | References |
|---|---|---|---|
| Sherley model | $P(t) = P_0\left[0.5 + \frac{1-(2F_d)^{\frac{t}{DT}+1}}{2(1-2F_d)}\right]$ | $P(t)$ = population time at time $t$<br>$P_0$ = initial cell number<br>$F_d$ = fraction of daugther cells that divide<br>$DT$ = division time<br>$M$ = number of dead cells | Sherley et al. (1995) |
| Deasy model | $P(t) = P_0\left[0.5 + \frac{1-(2F_d)^{\frac{t}{DT}+1}}{2(1-2F_d)}\right] - M$ | | Deasy et al. (2003) |
| Gompertz model | $\frac{dN}{dt} = r\,N ln\left(\frac{K}{N}\right)$ | $N$ = cell population<br>$r$ = intrinsic rate of growth<br>$K$ = carrying capacity of the population<br>$a$ = deviation of the logistic model | Domingues (2012) |
| Logistic model | $\frac{dN(t)}{dt} = rN(t)\left[1-\frac{N(t)}{K}\right]$ | | Omer (2018) |
| Richards model | $\frac{dN(t)}{dt} = rN(t)\left[1-\left(\frac{N(t)}{K}\right)^a\right]$ | | Zreiq et al. (2020) |
| Hyperbolastic H3 model | $\frac{dN(t)}{dt} = (L-N(t))\left(\delta\gamma t^{\gamma-1} + \frac{\theta}{\sqrt{1+\theta^2 t^2}}\right)$ | $N$ = cell population<br>$\delta$ = intrinsic rate of growth<br>$\gamma$ = allomeric constant<br>$L$= carrying capacity | Tabatabai et al. (2011) |

## 3 Modeling and Simulation of Stem Cell Differentiation in Cultivated Meat Production

The second part of obtaining muscle cells, and for mesenchymal stem cell (MSC) cultures to become muscle cells and form cultured meat tissues, they have to undergo a cell differentiation process. This process is long and complicated and involves many important factors, as these cells can differentiate not only into muscle tissue cells but also into adipose tissue cells, chondrocytes, osteoblasts, cardiac muscle cells, smooth muscle cells, and skeletal muscle cells. All of this is possible, thanks to the multipotency of these cells.

These cultures, also known as adult stem cell cultures, are found in various parts of developed organisms, such as the bone marrow, trabecular bone, adipose tissue, skeletal muscle, and other locations (Baksh et al. 2004).

Cell differentiation begins with the presence or absence of certain important chemical compounds in the environment. Depending on the maturity of each cell, they will secrete different proteins into the extracellular environment. These proteins, in turn, trigger reactions in neighboring cells to also undergo a process of phenotypic differentiation.

Cell differentiation can also be influenced by the surface topology it is located on and the elasticity conditions of its matrix, as Even-Ram et al. (2006) mention that stem cells that differentiate in a 1 kPa matrix become nerve cells, in a 10 kPa matrix we will observe the formation of muscle tissue, which is the main goal of our modeling, and in a matrix with an elasticity of 100 kPa, we will find differentiated cells that form bone tissue.

Similarly, the differentiation of MSCs can be affected by their motility. These cultures can move at a rate of 30 μm/h, according to Zhao et al. (2011). In his research, "Directed migration of human bone marrow mesenchymal stem cells in a physiological direct current electric field," another example of the analysis of MSC motility for cell differentiation can be found in the study conducted by Bertolo et al. (2015) where he relates the mobility of MSCs to differentiation properties, stating that this in vitro cell mobility is a useful tool for the characterization and classification of the differentiation potential of MSCs.

Naturally, the differentiation of stem cells into muscle tissue cells follows the pattern of myogenesis. These cells will go through a gene expression sequence, genes that will activate the processes of delamination, migration, proliferation, determination, differentiation, and specific muscle formation (Hollway and Currie 2003). However, even when there are fully mature and developed cells, there will still be satellite cells that serve for the constant regeneration of muscle tissue. Nevertheless, for this modeling, gene expressions that occur during the cell differentiation phase will be taken into account, including genes like myogenin, MCF2, MyoD, and Myf6 (Hollway and Currie 2003).

Myogenin: This is a muscle-specific transcription factor that can induce myogenesis in a variety of cell types and tissue cultures. It belongs to the family of proteins related by sequence homology, the helix-loop-helix (HLH) proteins. Activation of this gene is essential for the development of functional skeletal muscle (Heerema-Mckenney et al. 2008).

MCF2L: This gene regulates cell migration in neurotrophin-3-induced cells, which are members of the neurotrophin family. It is also associated with osteoarthritis when it is overexpressed (Day-Williams et al. 2011; Panoutsopoulou and Zeggini 2013). The product of this gene is a signaling protein in the process of chondrocyte hypertrophy and can also activate proteins related to invasive properties of fibroblast-like synoviocytes (Wang and Beier 2005).

MyoD: It is a myogenic differentiation protein 1 that plays a crucial role in regulating muscle tissue differentiation. It belongs to the family of proteins known as myogenic regulatory factors (MRFs), and its action is related to other MRF family members such as Myf5, myogenin, and MRF4. Its main function is to direct mesodermal cells to differentiate into muscle tissue repair cells. It also prevents cells from continuing the cell cycle by enhancing transcription and is inhibited by cyclin-dependent kinases (Olson et al. 2020).

MYF6: It is a member of the myogenic factor (MRF) family of transcription factors that regulate skeletal muscle myogenesis and muscle regeneration. Myogenic factors are basic helix-loop-helix (bHLH) transcription factors. MYF6 is a gene that encodes a protein involved in the regulation of myogenesis. The precise role of Myf6 in myogenesis is unclear, although in mice, it can initiate myogenesis in the absence of Myf5 and MyoD, two other MRFs. The protein integral to myogenesis regulation requires the conserved basic helix-loop-helix (bHLH) domain shared by all the genes in the MRF family. MYF6 is expressed exclusively in skeletal muscle, with higher levels of expression in adult skeletal muscle compared to other MRF family genes. In mice, Myf6 exhibits two-phase expression, being transiently expressed in the somites during the early stages of myogenesis along with Myf-5 and more noticeably expressed postnatally. This suggests that it plays an important role in the maintenance and repair of adult skeletal muscle (Moretti et al. 2016).

Satellite cells can be considered as quiescent myoblasts. These cells re-enter a proliferation cycle and supply newly differentiated cells to the tissue through asymmetric division. However, there are different viewpoints on the location and role of satellite cells in muscle tissues. Shefer et al. (2004) report that these cells are capable of differentiating into muscle tissue, bone tissue, and fat tissue; on the other hand, Starkey et al. (2011) demonstrate that this may only be possible for satellite cells of the same lineage.

These satellite cells acted as stem cells, first migrating to the muscle tissues where the injury was located and initiating a new differentiation process for the regeneration or healing of the damaged muscle tissue.

As previously mentioned, the differentiation process begins with cell-to-cell communication through certain proteins called cytokines or other growth factors. When it comes to muscle cells, we refer to the cytokines produced by myosins. Within these secreted proteins, there is a group that is released after the proliferation process, which has autocrine, paracrine, or endocrine effects.

Most models use the balance between masses for simulating the differentiation of cell phenotypes. In these models, one cell type can directly transition to another phenotype through the use of a simulation in a differential equation if and only if the phenotypes developed in that cell have not been completed.

The mathematical model for MSC differentiation (Eq. 7) produced by Prokharau et al. (2014) for differentiation is considered a gradual process, and modeling takes into account the density of a cell type per unit of maturity and unit of volume over time (t) and the differentiation space (a). By defining a set of boundary conditions, this model can alter the cellular population composition and, consequently, the development of different cell tissues. The main equation used for modeling MSC differentiation is:

$$\frac{\partial c_i}{\partial t} = \nabla\cdot\left(D_c \nabla c_t\right) - \frac{\partial}{\partial a}\left(u_i c_i\right) + A_c\left(1 - c_{\text{tot}} - f - b\right)c_i, i = b, f \tag{7}$$

where $c_{\text{tot}}$ is considered as the normalized total density of MSC (Eq. 8) was formulated for finding the total concentration of an MSC at a specific time and Ac is the proliferation rate of the culture. It is well-known that the value of Ac will remain constant when the total density of MSC reaches its limit, which in this case will be equal to one.

For the normalized total density, we will use the following equation:

$$c_{\text{tot}}\left(x,t\right) = \int_0^1 (c_b + c_f)da \tag{8}$$

Replacing the value of "∇" in the first equation, the physical space or development matrix will be represented, and the constant $D_c$ is the cell mobility coefficient.

The differentiation process is newly modeled in the second term of the equation "$-\frac{\partial}{\partial a}\left(u_i c_i\right)$" where the flow $u_i c_i$ originates from the differentiation state domain.

The baseline parameter values necessary for this modeling to work are described in the following Table 2. This table provides descriptions of the units used in the mathematical modeling of muscle cell differentiation from MSCs, based on units found in the articles (Bailón-Plaza and Van Der Meulen 2001; Harrevelt 2015; Prokharau et al. 2014), where we can find formulas for the mathematical modeling of MSC differentiation.

Furthermore, a more developed mathematical model (Eq. 9) to evaluate the influence of mechanical and chemical conditions was extracted from Harrevelt (2015), a more complex modeling, in which the influence of mechanical and chemical stimuli on cellular differentiation could be better evaluated for more effective results.

$$\frac{\partial c_k}{\partial t} + \nabla\cdot\left(v_m c_k\right) - D_k \Delta c_k = \sum_{i=1}^{C}\sum_{j=1}^{L} \gamma_k^i\left(t\right)\delta\left(x - x_j^i\left(t\right)\right)\Delta\Gamma_j^i\left(t\right), \quad x\epsilon\,\Omega, \tag{9}$$

$$c_k\left(x,0\right) = 0, x\epsilon\,\Omega,$$

$$D_k\frac{\partial c_k}{\partial n} + \kappa c_k = 0, x\epsilon\,\partial\Omega,$$

In this work, the new values of $k$ will be 1 or 2 depending on the type of cell to be differentiated. The modeling in his work was focused on simulating the

**Table 2** Parameters and descriptions of the units used for the mathematical modeling of muscle cell differentiation from MSCs, taken from works of other authors, referenced in the description column, adapted from Prokharau et al. (2014)

| Parameter values | Description |
|---|---|
| $C = 10^6$ cells/mm$^2$ | Saturated cell density used as a scaling coefficient for the normalized cell densities (Bailón-Plaza and Van Der Meulen 2001) |
| $D_c = 0.133$ mm$^2$/day | MSC diffusion coefficient (Moreo et al. 2009) |
| $D_f = 0.5$ mm$^2$/day | Fibroblast diffusion coefficient (Andreykiv et al. 2008) |
| $D_k = 0.81$ mm$^2$/day | Coefficient of diffusion ($k = 1$, myokines; $k = 2$, adipokines) (Goodhill 1998) |
| $A_c = 0.5$ day$^{-1}$ | MSC proliferation coefficient (Moreo et al. 2009) |
| $A_b = 0.5$ day$^{-1}$ | Osteoblast proliferation coefficient (Andreykiv et al. 2008) |
| $A_f = 0.1$ day$^{-1}$ | Fibroblast proliferation coefficient (Andreykiv et al. 2008) |
| $\alpha_w = 0.1$ day$^{-1}$ | Woven bone production coefficient (Moreo et al. 2009) |
| $\gamma = 0.06$ day$^{-1}$ | Bone remodeling coefficient (Moreo et al. 2009) |
| $\alpha_f = 0.06$ day$^{-1}$ | Fibrous tissue production coefficient (Andreykiv et al. 2008) |
| $\psi_1 = 0.008$ | Mechanical stimulus threshold corresponding to the maximum rate of differentiation of MSCs into osteoblast |
| $\psi_2 = 0.05$ | Maximum mechanical stimulus for which MSCs differentiate into osteoblasts (Isaksson et al. 2006) |
| $\psi_3 = 0.07$ | Mechanical stimulus threshold corresponding to the maximum rate of differentiation of MSCs into fibroblasts |
| $\delta_\psi = 10^{-4}$ | Half-width of the "transition region" for the function $U(\psi)$ |
| $U_b^0 = 0.1\,\text{day}^{-1}$ | Rate of differentiation of MSCs into osteoblasts for a zero mechanical loading |
| $U_b^{\max} = 0.2\,\text{day}^{-1}$ | Maximum rate of differentiation of MSCs into osteoblasts mass transfer coefficient in the liquid state |
| $U^{\min} = 0.0143$ day$^{-1}$ | Threshold for the MSC differentiation rate for which MSCs are able to complete differentiation within 63 days |
| $U_f^{\max} = 0.1\,\text{day}^{-1}$ | Maximum rate of differentiation of MSCs into fibroblasts |
| $v_m = 0.1$ day$^{-1}$ | Maturation rate |
| $U_c^i = 0.1\,\text{day}^{-1}$ | Differentiation rate in response to chemical stimuli |
| $U_M^i = 0.1\,\text{day}^{-1}$ | Differentiation rate in response to mechanical stimuli |
| $L = 23$ | Number of nodules traced around the cell (Harrevelt 2015) |
| $\gamma_k^0 = 0.5\,\text{day}^{-1}$ | Cell formation rate ($k = 1$, myocytes; $k = 2$, adipocytes) (Harrevelt 2015) |
| $\varepsilon_G = 5.4\%$ | Gas volume fraction or gas retention |
| $v_L = 1.31$ mm/s | Liquid superficial velocity |
| $k_La = 1.5$ day$^{-1}$ | Mass transfer coefficient in the liquid state |

[a]The data were standardized to the same units

differentiation that cultures of MSCs would undergo into myocytes (1) or adipocytes (2); other variables are explained and standardized in Table 2.

The modeling focused on the degree of maturity of MSCs (Eq. 10) is carried out taking into account the chemical and mechanical stimuli of the environment (Harrevelt, 2015), using as a basis for this study the phenotypes expressed by the cells. The model determines the cellular maturity ($m^i(t)$) of the satellite cells ($i$).

$$\frac{d}{dt}m^i(t) = \frac{1}{2}\left(U_c^i(t) + U_M^i(t)\right), i = 1,\ldots,C, \tag{10}$$

$$m(0) \sim U(0,1)$$

In this modeling, the differentiation is based solely on two terms, the first, $U_c^i(t)$ representing the chemical part of the cell maturation rate (Eq. 11) was extracted from Harrevelt (2015), and the second, $U_M^i(t)$ representing the mechanical part of the cell (Eq. 12) considering the influence of mechanical stimuli. Both growth stimuli are determined by the chemical and mechanical cues that a cell perceives through extracellular communications via its boundary nodes.

The maturation velocity generated by the chemical stimuli is explained in the following equation:

$$U_c^i(t) = U_{\max} \tan h\left(2\frac{\psi_c^i(t)}{\xi(t)}\right) I^i\left(l_c^i, t\right), i = 1,\ldots,C, \tag{11}$$

In this equation, the hyperbolic tangent has been used to limit the rate of chemical aging through the constant $U_{\max}$ which should be equal to 1. The term within the hyperbolic tangent is a scaled version of the chemical stimulus (Eq. 8), where $\xi(t)$ also depends on the chemical stimulus. The table describes the values and unit descriptions found in Eq. (11).

Furthermore, the model for mechanical stimuli is represented by the following formula:

$$U_M^i(t) = U_{\max} \tan h\left(2\frac{\psi_M^i(t)}{\eta(t)}\right) I^i\left(l_c^i, t\right), i = 1,\ldots,C, \tag{12}$$

Again, the hyperbolic tangent has been used to limit the rate of mechanical aging. Additionally (Eqs. 13–14), the values of $\eta(t)$, $\xi(t)$ are used to adjust the values that mechanical ($\psi_M^i(t)$) and chemical ($\psi_C^i(t)$) stimuli have when influencing differentiation.

However, these parameters associated with chemical and mechanical maturation are challenging to determine, and their biological relevance may be negligible. They are primarily mathematical concepts used to model behavior that is considered appropriate. This is how we define it:

$$\eta(t) = \max_{I\epsilon[1,C]} \left\|\psi_M^i(t)\right\| \tag{13}$$

$$\xi(t) = \max_{I\epsilon[1,C]} \left\| \psi_c^i(t) \right\| \tag{14}$$

The reason for choosing these parameters within the model is the simplicity of their implementation and their scalability with the cell population size. In the case of $\xi(t)$, it is known that the choice depends on the relative chemical concentration detected by the cell population at a given time "$t$" which contrasts with the natural process of cell differentiation. But from a biological standpoint, all of this sequence is being regulated by the absolute quantity of biomolecules that are part of chemical reactions, which, in turn, obey the laws of mass and energy conservation. Therefore, equating that value to the maximum population size that cell cultures can reach may not be appropriate unless it signifies an external formation of by-products.

Furthermore, Li et al. (2020) in their research "A conceptual air-lift reactor design for large-scale animal cell cultivation in the context of in vitro meat production" evaluate the oxygen concentration to determine the oxygen concentration in the culture medium. Cells must be added to ensure complete cellular differentiation. Eq. (15) was extracted from Li et al. (2020). This leads us to the following modeling.

$$\frac{\partial C}{\partial t} = -v_L \frac{\partial C}{\partial z} + D_L \frac{\partial^2 C}{\partial z^2} - r_{0_2} + \frac{k_L a}{1 - \varepsilon_G}\left(C^* - C\right) \tag{15}$$

where $C^*$, $C$ are the concentrations (mol/m$^3$) of dissolved oxygen and the concentration of oxygen dissolved at saturation, respectively. The variable $v_L$ would be in units of (m/s) as it represents the liquid velocity and $D_L$ is the oxygen dispersion coefficient in the liquid.

The units of $r_{0_2}$ would be in (mol*s/m$^3$) as it is the rate of oxygen consumption. Its value in the exponential growth phase is constant with a value of $10^{-17}$mols/cell; regardless of the dissolved oxygen concentration, this value is typical for animal cell models.

The variables, normalized parameters, and mathematical models are explained in Table 2. These were developed under controlled conditions using simulation software.

## 4 Challenges and Future Developments in Modeling and Simulation of Cultivated Meat Production

This chapter has underscored the nuanced nature of crucial steps for mathematical modeling of cell proliferation and differentiation for cultivated meat production.

However, in acknowledging the progress made, it is imperative to recognize that the models presented are not exhaustive or infallible. While they provide valuable insights, the dynamic nature of cellular processes demands an openness to diverse modeling approaches. Future research may well unveil novel models that capture subtleties overlooked by existing frameworks, thereby refining our understanding of cell proliferation and differentiation in the context of cultivated meat production.

Looking ahead, mathematical modeling faces additional challenges that extend beyond the intricacies of cellular behavior. Expanding the scope to incorporate a broader set of parameters, such as environmental factors, mechanical stimuli, and cell-cell interactions, presents an avenue for more comprehensive models. The complexity of the bioreactor systems integral to large-scale cultivation also beckons attention. Modeling these systems, considering spatial gradients, fluid dynamics, and nutrient distribution, will be paramount for optimizing the efficiency and scalability of cultivated meat production.

## 5 Conclusions

An array of mathematical models elucidating the intricate processes of cell proliferation and differentiation in cultivated meat production have been shown in this chapter: from the Gompertz and logistic models capturing the dynamics of cell growth to the Richards model introducing parameters for asymmetric dynamics, the landscape of mathematical modeling has been revealed as a versatile toolkit for understanding and predicting cellular behavior. Recognizing the pivotal role of modeling in scaling-up processes, developing economically viable methods, and ensuring environmental sustainability underscores the imperative for ongoing interdisciplinary collaboration. However, it is crucial to recognize that these models, while valuable, represent a snapshot of our current understanding, and the ever-evolving nature of scientific inquiry beckons further exploration and refinement.

## References

Andreykiv A, Van Keulen F, Prendergast PJ (2008) Simulation of fracture healing incorporating mechanoregulation of tissue differentiation and dispersal/proliferation of cells. Biomech Model Mechanobiol 7(6):443–461. https://doi.org/10.1007/S10237-007-0108-8/METRICS

Bailón-Plaza A, Van Der Meulen MCH (2001) A mathematical framework to study the effects of growth factor influences on fracture healing. J Theor Biol 212(2):191–209. https://doi.org/10.1006/JTBI.2001.2372

Baksh D, Song L, Tuan RS (2004) Adult mesenchymal stem cells: characterization, differentiation, and application in cell and gene therapy. J Cell Mol Med 8(3):301–316

Benzekry S, Lamont C, Beheshti A, Tracz A, Ebos JML, Hlatky L, Hahnfeldt P (2014) Classical mathematical models for description and prediction of experimental tumor growth. PLoS Comput Biol 10(8):e1003800. https://doi.org/10.1371/JOURNAL.PCBI.1003800

Bertolo A, Gemperli A, Gruber M, Gantenbein B, Baur M, Pötzel T, Stoyanov J (2015) In vitro cell motility as a potential mesenchymal stem cell marker for multipotency. Stem Cells Transl Med 4(1):84–90. https://doi.org/10.5966/SCTM.2014-0156

Datar I, Betti M (2009) Possibilities for an in vitro meat production system. Innov Food Sci Emerg Technol. https://doi.org/10.1016/j.ifset.2009.10.007

Day-Williams AG, Southam L, Panoutsopoulou K, Rayner NW, Esko T, Estrada K, Helgadottir HT, Hofman A, Ingvarsson T, Jonsson H, Keis A, Kerkhof HJM, Thorleifsson G, Arden NK, Carr A, Chapman K, Deloukas P, Loughlin J, McCaskie A et al (2011) A variant in MCF2L is associated with osteoarthritis. Am J Hum Genet 89(3):446. https://doi.org/10.1016/J.AJHG.2011.08.001

Deasy BM, Jankowski RJ, Payne TR, Cao B, Goff JP, Greenberger JS, Huard J, Huard J (2003) Modeling stem cell population growth: incorporating terms for proliferative heterogeneity. Stem Cells 21(5):536–545. https://doi.org/10.1634/STEMCELLS.21-5-536

Domingues JS (2012). Gompertz model: resolution and analysis for tumors

Even-Ram S, Artym V, Yamada KM (2006) Matrix control of stem cell fate. Cell 126(4):645–647. https://doi.org/10.1016/J.CELL.2006.08.008

Goodhill GJ (1998) Diffusion of molecules and cytokines. https://personal.math.ubc.ca/~ais/website/status/diffuse.html

Guan X, Lei Q, Yan Q, Li X, Zhou J, Du G, Chen J (2021) Trends and ideas in technology, regulation and public acceptance of cultured meat. Future Foods 3:100032. https://doi.org/10.1016/j.fufo.2021.100032

Harrevelt SD (2015) Simulating the differentia-tion of stem cells using a 2D and 3D mathematical model. http://repository.tudelft.nl/

Heerema-Mckenney A, Wijnaendts LCD, Pulliam JF, Lopez-Terrada D, McKenney JK, Zhu S, Montgomery K, Mitchell J, Marinelli RJ, Hart AAM, Van De Rijn M, Linn SC (2008) Diffuse myogenin expression by immunohistochemistry is an independent marker of poor survival in pediatric rhabdomyosarcoma: a tissue microarray study of 71 primary tumors including correlation with molecular phenotype. Am J Surg Pathol 32(10):1513–1522. https://doi.org/10.1097/PAS.0B013E31817A909A

Hollway GE, Currie PD (2003) Myotome meanderings. Cellular morphogenesis and the making of muscle. EMBO Rep 4(9):855. https://doi.org/10.1038/SJ.EMBOR.EMBOR920

Isaksson H, Wilson W, van Donkelaar CC, Huiskes R, Ito K (2006) Comparison of biophysical stimuli for mechano-regulation of tissue differentiation during fracture healing. J Biomech 39(8):1507–1516. https://doi.org/10.1016/J.JBIOMECH.2005.01.037

Jaques A, Sánchez E, Orellana N, Enrione J, Acevedo CA (2021) Modelling the growth of in-vitro meat on microstructured edible films. J Food Eng 307:110662. https://doi.org/10.1016/J.JFOODENG.2021.110662

Li X, Zhang G, Zhao X, Zhou J, Du G, Chen J (2020) A conceptual air-lift reactor design for large scale animal cell cultivation in the context of in vitro meat production. Chem Eng Sci 211:115269. https://doi.org/10.1016/J.CES.2019.115269

Loibel S, Andrade MG, do Val JBR, de Freitas AR (2010) Richards growth model and viability indicators for populations subject to interventions. An Acad Bras Cienc 82(4):1107–1126. https://doi.org/10.1590/S0001-37652010000400028

Moreo P, García-Aznar JM, Doblaré M (2009) Bone ingrowth on the surface of endosseous implants. Part 1: mathematical model. J Theor Biol 260(1):1–12. https://doi.org/10.1016/J.JTBI.2008.07.040

Moretti I, Ciciliot S, Dyar KA, Abraham R, Murgia M, Agatea L, Akimoto T, Bicciato S, Forcato M, Pierre P, Uhlenhaut NH, Rigby PWJ, Carvajal JJ, Blaauw B, Calabria E, Schiaffino S (2016) MRF4 negatively regulates adult skeletal muscle growth by repressing MEF2 activity. Nat Commun 7. https://doi.org/10.1038/NCOMMS12397

Olson NJ, Fritchie KJ, Torres-Mora J, Folpe AL (2020) MyoD1 expression in fibroepithelial stromal polyps. Hum Pathol 99:75–79. https://doi.org/10.1016/J.HUMPATH.2020.03.006

Omer TA (2018) Analysis of bacterial population growth using extended logistic growth model with distributed delay. https://arxiv.org/abs/1807.09108v1

Panoutsopoulou K, Zeggini E (2013) Advances in osteoarthritis genetics. J Med Genet 50:715–724. https://doi.org/10.1136/jmedgenet-2013-101754

Post MJ (2014) Cultured beef: medical technology to produce food. J Sci Food Agric 94(6):1039–1041. https://doi.org/10.1002/JSFA.6474

Prokharau PA, Vermolen FJ, García-Aznar JM (2014) A mathematical model for cell differentiation, as an evolutionary and regulated process. Comput Methods Biomech Biomed Eng 17(10):1051–1070. https://doi.org/10.1080/10255842.2012.736503

Reiss J, Robertson S, Suzuki M (2021) Cell sources for cultivated meat: applications and considerations throughout the production workflow. Int J Mol Sci 22(14):7513. https://doi.org/10.3390/IJMS22147513

Shefer G, Wleklinski-Lee M, Yablonka-Reuveni Z (2004) Skeletal muscle satellite cells can spontaneously enter an alternative mesenchymal pathway. J Cell Sci 117(22):5393–5404. https://doi.org/10.1242/JCS.01419

Sherley JL, Stadler PB, Stadler JS (1995) A quantitative method for the analysis of mammalian cell proliferation in culture in terms of dividing and non-dividing cells. Cell Prolif 28(3):137–144. https://doi.org/10.1111/J.1365-2184.1995.TB00062.X

Starkey JD, Yamamoto M, Yamamoto S, Goldhamer DJ (2011) Skeletal muscle satellite cells are committed to myogenesis and do not spontaneously adopt nonmyogenic fates. J Histochem Cytochem 59(1):33. https://doi.org/10.1369/JHC.2010.956995

Tabatabai M, Williams DK, Bursac Z (2005) Hyperbolastic growth models: theory and application. Theor Biol Med Model 2(1):1–13. https://doi.org/10.1186/1742-4682-2-14/TABLES/5

Tabatabai MA, Bursac Z, Eby WM, Singh KP (2011) Mathematical modeling of stem cell proliferation. Med Biol Eng Comput 49(3):253–262. https://doi.org/10.1007/S11517-010-0686-Y/FIGURES/6

Tatro D (2018) The mathematics of cancer: fitting the Gompertz equation to tumor growth

Tjørve KMC, Tjørve E (2017) The use of Gompertz models in growth analyses, and new Gompertz-model approach: an addition to the unified-Richards family. PLoS One 12(6):e0178691. https://doi.org/10.1371/JOURNAL.PONE.0178691

Wachenheim DE, Patterson JA, Ladisch MR (2003) Analysis of the logistic function model: derivation and applications specific to batch cultured microorganisms. Bioresour Technol 86(2):157–164. https://doi.org/10.1016/S0960-8524(02)00149-9

Wang G, Beier F (2005) Rac1/Cdc42 and RhoA GTPases antagonistically regulate chondrocyte proliferation, hypertrophy, and apoptosis. J Bone Miner Res 20(6):1022–1031. https://doi.org/10.1359/JBMR.050113

Wang X, Liu S, Huang Y (2016) A study on the rapid parameter estimation and the grey prediction in Richards model. J Syst Sci Inf 4(3):223–234. https://doi.org/10.21078/JSSI-2016-223-12

Zhao Z, Watt C, Karystinou A, Roelofs AJ, McCaig CD, Gibson IR, De Bari C (2011) Directed migration of human bone marrow mesenchymal stem cells in a physiological direct current electric field. Eur Cell Mater 22:344–358. https://doi.org/10.22203/ECM.V022A26

Zreiq R, Kamel S, Boubaker S, Al-Shammary AA, Algahtani FD, Alshammari F (2020) Generalized Richards model for predicting COVID-19 dynamics in Saudi Arabia based on particle swarm optimization algorithm. AIMS Public Health 7(4):828–843. https://doi.org/10.3934/PUBLICHEALTH.2020064

# Downstream Processes for Cultivated Meat

**Julio Cesar de Carvalho, Susan Grace Karp, Luis Daniel Goyzueta Mamani, Giuliana Biagini, Gabriela dos Santos Costa, Leonardo Wedderhoff Herrmann, and Carlos Ricardo Soccol**

**Abstract** Cultivated meat production is being developed worldwide and is gaining traction. Whether the technology becomes a staple in the food industry, becomes a niche application, or takes an unforeseen turn remains to be seen. Early research in the field showed the feasibility of small-scale culturing cells and transforming them into meat-like products. These lab-scale proofs of concept relied on the manual separation of cells using simple operations such as straining and centrifuging. However, industrial processing requires working with large culture volumes and recovering cells with minimal damage. This chapter presents the basics of cellular meat production from the point of view of downstream processing, overviewing culture propagation, cell differentiation, and suspension composition and discusses cell separation aspects. Although there are no industrial-scale cultivated meat processes to serve as examples, translational research from other cell culturing processes hints at centrifugation and membrane processes as starting points for cell recovery. The final step of forming meat-like products using 3D bioprinting is also discussed.

**Keywords** Centrifugation · Tangential flow filtration · Shear stress · Spent media analysis · Bioprinting

## 1 Introduction

Cultivated meat (CM), or cultured, lab-grown, cell-based, or clean meat, refers to meat products created by cultivating animal cells in a laboratory environment. Unlike conventional meat production, which requires the rearing and slaughtering

J. C. de Carvalho (✉) · S. G. Karp · L. D. G. Mamani · G. Biagini · G. dos Santos Costa · L. W. Herrmann · C. R. Soccol
Bioprocess Engineering and Biotechnology Department, Federal University of Paraná, Curitiba, Paraná, Brazil
e-mail: jccarvalho@ufpr.br


C. R. Soccol et al. (eds.), *Cultivated Meat*,
https://doi.org/10.1007/978-3-031-55968-6_8

of animals, cultivated meat is produced without animal husbandry (Chriki and Hocquette 2020). The procedure begins with collecting a small sample of animal cells, typically through a biopsy or another noninvasive technique. Then, these cells are cultivated in a nutrient-rich medium that promotes their growth and multiplication. As the cells multiply, they become muscle tissue and ultimately form edible meat (Bryant 2020).

Cultivated meat addresses several issues associated with conventional meat production, including environmental sustainability, animal welfare, and resource consumption. It is considered a more sustainable and ethical alternative to conventional livestock farming because it can potentially reduce the environmental impact of animal agriculture, including greenhouse gas emissions, land use, and water consumption (Post 2012).

While cultivated meat production technology is still in its early stages, ongoing research and development efforts are aimed at making the production process more scalable and affordable. The goal is to make cultivated meat commercially viable and easily accessible to consumers, thereby providing a sustainable and ethical option for meeting the global demand for meat products (Heathman et al. 2019). It is anticipated that cultivated meat will play a significant role in shaping the future of food production and contributing to more sustainable and humane food systems as the field advances (Tuomisto and Teixeira de Mattos 2011).

Producing CM commerciality means extrapolating from small-scale processes that rely on sophisticated lab equipment and much manual labor, to large factories with robust processes. Today's world production of beef, chicken, and pork is about 300 million tons per year (FAOSTAT-Database 2022). Using current technology, a biofactory capable of supplying *one-millionth* of the world's demand would have to process 9 million liters of culture media per year or 32 $m^3$/day. This is why cell lines, bioreactor technology, and culture media composition catch most of the research attention today—these are the first technological bottlenecks in CM production. However, complete process development requires considering other aspects, such as separating large amounts of fragile cells from suspension and transforming them into a meat-like product. The following sections present fundamental aspects of CM production, focusing on cell separation and formulation of a final product and the challenges and opportunities ahead.

## 1.1 Basics of the Production Process

Producing cultivated meat is a meticulously regulated scientific process. The procedure can be divided into the following stages:

**Cell Line Development** Animal cells, including muscle and fat cells, are primary cells isolated from tissue via biopsies or immortalized cell lines. Optimal cell lines must have robust proliferative and differentiating capabilities and adaptability to function effectively in a bioreactor environment, whether in suspension or scaffolds.

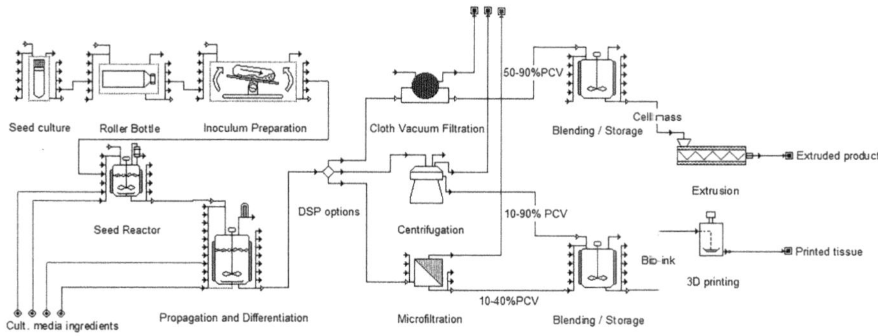


**Fig. 1** General process for cultivated meat production, from seed cultures to final product. After one or more steps of propagation, and differentiation, a cell suspension with 1–10% PCV (packed cell volume) can be filtered, centrifuged, or microfiltrated. The choice of operation depends on cell type and next formulation steps. In this conceptual flowchart, utilities, sterile media and air, and residual streams are not detailed

Additionally, they must be resistant to mechanical forces, manage the accumulation of metabolic waste, and endure nutrient deprivation (Scarfone et al. 2020) (Fig. 1).

**Cell Culture and Scaffold Integration** Isolated cells are cultivated in a nutrient-rich culture medium that provides essential nutrients, growth factors, and hormones to promote their growth and multiplication. The culture media can be evaluated according to the minimum working volume needed for bioreactor efficiency (cells/mL) and cell consumption (Campuzano and Pelling 2019).

The design and structure of scaffolds must be included in the overall bioprocess design. The relationship between surface area-to-volume ratio and the resulting bioreactor size requirements highlights the significance of scaffold selection. This choice significantly impacts bioreactor seeding efficiency, system passaging requirements, downstream processing, bioreactor fluid dynamics, mass transfer, and overall cost considerations. It is possible to classify scaffolds as edible or nonedible (Bellani et al. 2020).

**Cell Growth** On an industrial scale, cultivated meat is produced in bioreactors that meticulously control critical parameters such as temperature, pH, dissolved oxygen, and carbon dioxide. The process begins with cell growth in bioreactors. Subsequently, these cells are attached to scaffolds or organized into the final 3D configuration before being transferred to perfusion bioreactors for complete tissue maturation. The bioreactors used for cultivated meat production are specifically designed to reduce shear stress, maintain a uniform environment, allow for extensive cell growth at an exponential rate, and operate with minimal energy and consumable resources (Allan et al. 2019).

**Scaffolding and Cell Differentiation** After expansion, undifferentiated cells are induced to differentiate into specific cell types, such as myoblasts or myosatellites, adipocytes, and connective tissue cells, mimicking the composition of conventional meat. To achieve this, scaffolds are required as a platform enabling cell attachment and the formation of tissue exhibiting a 3D cytoarchitecture resembling a natural meat structure (Allan et al. 2019). The scaffold material must meet strict criteria for industrial use: food-safe, either edible or biodegradable, and derived from nonanimal sources. It is critical that scaffolds exhibit favorable biocompatibility and effectively facilitate cell adherence (Enrione et al. 2017).

**Maturation and Harvesting** The assembled tissue is subjected to a controlled maturation process that allows it to develop its texture and flavor by applying the appropriate environmental conditions and simulating stress. When it is necessary to remove cells from the scaffold, several dissociation techniques are available. While enzymatic dissociation with trypsin is common in laboratory-scale cultures, mechanical or shear-induced dissociation is also an option (Nienow et al. 2014).

To ensure the safety and quality of the final cultivated meat product, it is essential to eliminate specific components from the culture medium throughout the production process. These include waste metabolite byproducts of cellular metabolism, such as lactate, which, if allowed to accumulate, can inhibit cell growth and compromise the quality of the final product. Removing excess growth factors and unused nutrients is essential to prevent disruptions in CM composition and flavor profile (Yang et al. 2023).

## *1.2 Cells, Scaffolds, and 3D Structures*

For the creation of 3D meat tissue, various scaffolds, including natural, synthetic, and composite biomaterials, are considered. Raw and natural materials with high biocompatibility and degradability include collagen, fibrin, hyaluronic acid, and various plant-derived materials such as decellularized plants, cellulose, alginate, and chitosan (from crustaceans, yeast, or fungi), as well as fungal mycelium. Animal-derived scaffolds are naturally functionalized with native cell adhesion motifs, making them beneficial for various tissue engineering applications (Dhandayuthapani et al. 2011; Seah et al. 2022).

Since the early stages of cultivated meat development, several startups used animal-derived scaffolds, such as gelatine-based. The search for nonanimal scaffolds capable of promoting the maturation of three-dimensional meat tissue is ongoing. Matrix Meats, for example, has developed electro-spun nanofiber scaffolds that are adaptable to the requirements of various cultivated meat development processes (Levi et al. 2022). These scaffolds are composed of polymers, including polycarbonate, polylactic acid, polyethylene, and natural polymers, chosen based on the desired mechanical, degradative, and biocompatible properties. However, comprehensive data on the performance and cost-effectiveness of these scaffolds for producing large quantities of cultivated meat are still lacking (Kumar et al. 2023).

Protein or polysaccharide matrices derived from plants or microbes could be cost-effective and environmentally friendly scaffolding alternatives (Wang et al. 2023). These plant-derived options have positive characteristics, including scalability, affordability, and adaptability. The vascular network of decellularized plant tissues can resemble that of animal tissues. Decellularized spinach leaves, for instance, can serve as edible scaffolds that effectively support the growth and differentiation of bovine satellite cells into meat tissue, highlighting their potential for large-scale cultivated meat production. Textured soy protein (TSP), a byproduct of soybean oil processing, also show potential for scaffold production (Young and Skrivergaard 2020).

In contrast to natural polymers, synthetic polymers lack bioactivity and are generally less compatible with biological systems. A hybrid scaffold composed of natural and synthetic components could be a viable strategy for meeting all the criteria for an effective scaffold (Levi et al. 2022).

### *1.3 Nonmeat Adjuvants and Ingredients*

Numerous protein sources, including vegetable proteins such as gluten, cottonseed protein, and sesame protein, are already available as meat substitutes. Due to their economic viability as meat alternatives, soybeans currently dominate production. New protein sources like fungi and bacteria have unique technical properties and nutritional values, allowing for improved and diverse meat substitutes. These proteins, already accepted by consumers, can be used as benchmarks for CM presentation forms and as components of hybrid structured, meat-like foods. Common meat substitutes include the following:

**Protein-Rich Precursors** These include gluten from cereals, defatted oilseeds, flour from cereals or beans, and meat-based precursors such as byproducts of mechanical meat disintegration, fish meal, and dry egg white which are widely used in the production of meat analogs (van der Weele et al. 2019).

**Microalgae** Long-chain polyunsaturated fatty acids, proteins, vitamins, carotenoids, minerals, and phenolic compounds are prevalent in microalgae such as *Chlorella* and *Spirulina*. Spirulina's high protein content makes it an ideal ingredient for developing protein-rich foods (Sathasivam et al. 2019). *Tofu*, derived from soybeans, is a popular meat substitute high in protein and minerals. Its production is distinguished by its simplicity, cleanliness, comfort, and manageability (Wang et al. 2018); *seitan*, derived from wheat gluten, is also used to produce vegetarian sausages, burgers, nuggets, and ground meat. Adding reducing agents or proteases modifies their physical properties (Schmidinger 2012).

**Mushrooms** The fibrous nature of mushrooms improves the chewability of meat substitutes. They are abundant in polyunsaturated fatty acids, dietary fibers, and proteins. *Fusarium graminearum* is an edible filamentous fungus that imparts a meat-like flavor to commercial products such as burger patties and sausages (Rousta et al. 2021).

**Texturized Vegetable Proteins (TVP)** TVP is a protein derived from legumes, nuts, and oilseeds that is fibrous and porous. Due to their high protein content, soybeans are frequently used to make textured soy protein (TSP) or soy-meat, characterized by its chewy texture and robust flavor.

**Single-Cell Protein** Bacteria are viable candidates for single-cell protein (SCP) production owing to their rapid growth and ability to utilize various raw materials. *Brevibacterium*, *Acinetobacter*, and *Bacillus* are frequently employed in SCP production (Lee et al. 2020).

Cultivated meat can show specific characteristics, such as the color of developing muscle fibers appearing darker due to the suppression of myoglobin in ambient oxygen during cell culture. However, studies have shown that myoglobin expression in bovine muscle fibers can be enhanced under low-oxygen conditions (Simsa et al. 2019). The distinctive flavor of meat stems from a complex blend of proteins, sugars, and aromatic compounds. Incorporating adipose tissue enriches the flavor and enhances the product's nutritional value (Fraeye et al. 2020). This can be achieved using various cell types, including mesenchymal stem cells, adipose tissues, and myoblasts. To ensure the desired texture, it is essential to simulate the maturation of white adipose tissue, which takes significantly longer than muscle development. The sensory experience, texture, and mouthfeel of meat are influenced by factors such as tissue composition, both at the microscopic and macroscopic levels, as well as the meat's ability to retain moisture and fat (Post and Hocquette 2017; Paredes et al. 2022).

Replicating the texture of small meat pieces, such as those in hamburgers or processed meats, requires the self-assembly of muscle cells and fibers within a specific matrix or scaffold.

## *1.4 Downstream Processing as a Part of CM Development*

The fact that CM production is in its infancy is reflected in the lack of data on downstream processing. It is not that cells *are not* separated, but that in the small scale used so far, gentle centrifugation or straining is enough for cell recovery. On the laboratory scale, cell separation is not seen as a problem, and even in life cycle analyses exclude details of equipment for downstream processing (Tuomisto et al. 2022). The imagined process for suspended cells is usually an unspecified centrifugation step (Sinke et al. 2023), and for cells in microcarriers, a centrifugation or a dead-end filtration step (Schnitzler et al. 2016). However, as the challenges in cell cultivation are solved and the scale increases, the downstream constraints may become significant.

# 2 Cell Proliferation, Differentiation, and Post-Processing in Bioreactors

## *2.1 From Cell Lines to Inocula*

The downstream processing of CM is directly affected by the type of cell, the type and phase of culture, and the final cell concentration. Small-scale, proof-of-concept processes in the past have even prepared meat-like tissues by surface culture and manual stacking of cell layers. That is uneconomical for industrial production, which requires an approach that allows the production and separation of large amounts of cells. However, it all starts with an inoculum.

Typically, a specific cell line is selected for a particular process. From this cell line, a first inoculum, also called a starter culture, is prepared, maintained under carefully controlled conditions, and used to inoculate sterile plates or liquid medium in shake flasks. After these cells multiply, the preculture is used to inoculate a pilot bioreactor. Generally, industrial fermenters have large volumes (150–250 $m^3$), so the inoculum is built through several successive stages, up to 5–10% of the working volume of the production fermenter. This strategy ensures optimal use of the production fermenter while minimizing batch time in this vessel. In animal cell culture, the inoculum size is generally selected to provide an initial cell count of 2–4 × $10^5$ cells/ml (Ammam et al. 2010).

## *2.2 Cultivation of Cells on Surfaces Versus Cell Suspension*

Cultivated cells can appear as adherent cells or non-adherent cells. Adherent cells originate from hard tissues, which require anchoring, that is, they require adhesion to a contact surface to proliferate. On the other hand, non-adherent cells can be cultivated in suspension in the medium, as they come from tissues that do not need an anchoring structure to proliferate and survive. This capacity is limited to hematopoietic, transformed, and cancer cells (Alves and Guimarães 2010).

In 2D cultures, in most cases, plastic culture surfaces coated with proteins or adhesion molecules are used to aid cell attachment. In 3D cultivation, cells that require adhesion to proliferate and differentiate can be cultivated on structures such as scaffolds or microcarriers. Scaffolds provide a favorable structure for growth as they mimic the extracellular matrix. Cells that do not require an attachment surface to grow can be cultivated in suspension. In this case, microcarriers can be used to suspend the culture. Suspension culture can be achieved by coating the surfaces of microspheres with an extracellular matrix and suspending them in the culture medium. A suspension culture can be used for cells, such as CHO cells, which are often used for antibody production and do not require cell adhesion (Horie et al. 2022).

Cultivating animal cells in suspension requires constant agitation so that there is homogeneity within the bioreactor and the cells do not form aggregates. However, animal cells are sensitive to the shear stress generated by mechanical agitation and, therefore, need to be agitated delicately to avoid stress and disintegration of the cells. Cell aggregates can be dissociating enzymatic solutions, but their use induces considerable loss of cells due to sensitivity to physical stress and single-cell dissociation (Singh et al. 2010; Otsuji et al. 2014). Some studies report using polymers such as gellan gum and mesh filters to replace agitation (Otsuji et al. 2014) or reduce shear stress (Horiguchi et al. 2021).

## 2.3 *Scaffolds: Concept and Importance in Cell Cultivation*

The scaffold concept has been used in tissue engineering for at least 20 years (Chng and Wan 2023). Scaffolds are highly porous structures that provide support for cell adhesion and growth. These structures mimic the extracellular matrix, enabling cell attachment, proliferation, and differentiation (Seah et al. 2022). The raw material for scaffold production can come from animal, vegetable, and microbial sources. The choice of material must consider the final application of the scaffold and criteria such as biocompatibility, biodegradability, edibility, low cost, availability, and physical, chemical, and mechanical properties (Letti et al. 2021).

Scaffolds can be produced from natural or synthetic polymers. Natural polymers are preferred when the scaffold is required to be edible, for example, in cultivated meat production. Since the scaffold is edible, it does not need to be removed in the downstream process. The FDA (USA Food and Drug Administration) approves polymers such as pectin, chitosan, gluten, gellan gum, cellulose, gelatine, collagen, isolated soy protein, starch, glucomannan, and alginate as edible (Ali and Ahmed 2018). Some natural polymers used for scaffolds are silk, elastin, keratin, actin, and myosin. Polysaccharides such as amylose, dextran, chitin, and glycosaminoglycans can also be used (Stevens 2008). Synthetic polymers include polyethylene glycol (PEG), polyglycolic acid, and polyacrylamide.

The preparation of scaffolds has been carried out using various methods, such as decellularization, solvent casting, 3D bioprinting, electrospinning, and freeze-drying (Singh et al. 2023; Rao et al. 2023). The production methodology must be chosen by evaluating the cost of reagents and operation and the purpose of the scaffold. The most reported methods for cultivated meat are electrospinning, decellularization, and freeze-drying (Levi et al. 2022).

## 2.4 *Proliferation*

The cell proliferation phase aims to produce the maximum possible cell density so these cells can go through differentiation (Post 2012). For an effective proliferation process, there must be support to ensure the fixation and adhesion of the cells.

Proliferation depends on a combination of signals that promote cell division (Alberts 2017). There are signaling pathways that affect cell proliferation, differentiation, and metabolism. One is the PI3K/AKT/mTOR pathway, an important pathway induced by growth factors that promote anabolic pathways and cell proliferation in multiple contexts. This pathway stimulates glucose import and glycolysis by regulating glucose transporters and glycolytic enzymes, promoting fatty acid synthesis and protein synthesis (Agathocleous and Harris 2013).

Growth factors are essential components that must be added to the medium when culturing animal cells. These factors are proteins that bind to cell surface receptors and result in proliferation or differentiation. Adding growth factors to the medium ensures a favorable environment for cell development. Some growth factors used are FGF2, belonging to the family of fibroblast growth factors; neuregulin 1 (NRG1), a factor of the tyrosine kinase ligand family; and TGF-β3, a regulator of adhesion molecules and extracellular matrix formation (Taya et al. 1999; Teo et al. 2016).

## 2.5 *Differentiation*

Cellular differentiation constitutes a complex and vital biological process, as it regulates the expression of many genes linked to tissue-specific functions and controls cell proliferation (Naves and Moreno 2000). Stem and progenitor cells divide to produce different types of cells, a process that requires extensive cellular rearrangement. In differentiation, proliferated cells differentiate and mature into specialized tissues.

Stem cells are cells with a high proliferative capacity and can produce several copies of themselves, also having a high rate of specialization and being able to differentiate into several cell types (Post et al. 2020; Ng and Kurisawa 2021). Due to this capacity and because it is found in different types, some stem cell lines can be applied to produce cultivated meat (Kadim et al. 2015).

The production of cultivated meat depends on the proliferation of mononucleated myoblasts that will differentiate into multinucleated cells without the ability to proliferate. This process generates the muscle tissue analogous to beef that we know. Muscle satellite cells are the most recommended type of cells in cultivated meat production since they can differentiate into a specific tissue, the muscular tissue where they originated (Ye et al. 2022).

As well as for proliferation, some growth factors induce the differentiation stage, such as insulin-like growth factor 1 (IGF-1) and epidermal growth factor (EGF) 9 (Stout et al. 2022). IGF-1 is one of the most potent natural activators of the AKT signaling pathway, a stimulator of cell growth and proliferation, and a potent inhibitor of programmed cell death. Epidermal growth factor (EGF) is a potent growth factor that stimulates various epidermal and epithelial cell proliferation (Ahmad et al. 2023).

# 3 Downstream Processes for Cultivated Meat Production

After cultivation, the cells that will make up the CM products must be separated from the culture medium, with the minimum impact possible on the cells, to be used as raw materials or mounted into meat-like assemblies. Besides concentrating the cells so that they may get closer to the concentrations found in meat tissues, several compounds are present in the culture medium, and its separation is necessary because they cannot be present in the final product (e.g., antibiotics) or because they can be recycled (e.g., glucose). Separation of animal cells is routine work in producing monoclonal antibodies and viruses and, on a smaller scale, in cell therapy. Separation on a large scale has not yet been done but is essential for future cultivated meat production. To design a CM downstream process, it is essential to understand the composition of the suspension exiting the reactors and the properties of the suspended particles—cells or microcarriers. Another challenge is that animal cells are less robust than microbial cells routinely produced in bioprocesses. So, existing processes must be adapted or developed.

## *3.1 Cell Centration and Spent Media Composition*

The final concentration of cells after proliferation and differentiation is reported to range between $10^7$ and $2 \cdot 10^8$ cells/mL, which translates to 35–700 g/L of wet cells, considering an average wet weight of 3.5 ng/cell (calculated from data presented by Myers et al. 2023). Cells must be separated from the culture media to increase cell concentration, giving a cell paste that can be processed into CM, while nutrients can, in theory, be recovered. The analysis of spent media can be informative both for media development (showing what is unconsumed and what is depleted, possibly limiting nutrients) and for evaluating recovery opportunities. Two such analyses were done by Haraguchi and Shimizu (2021) and O'Neill et al. (2022), showing that although glucose and glutamine from DMDM culture media are reduced by 70–99%, many amino acids remain, and there is a significant generation of lactate and ammonia. Table 1 shows the average composition of the spent medium after cultivation of attached murine cells (C2C12 myoblasts).

The concentration of minerals remained approximately constant, except for phosphate (reduced by 15%) and ammonia (increased by 33%), as determined by Haraguchi and Shimizu (2021), while in the study by O'Neill et al. (2022), the same trend was observed for $Na^+$, $K^+$, $Ca^{2+}$, and $Mg^{2+}$. Both studies observed net maintenance of vitamins, detailed in the study by O'Neill, with the reduction of thiamine and riboflavin, a slight reduction in pyridoxine and folic acid, a slight increase in niacinamide, and an increase in pantothenate, biotin, and cyanocobalamin—but never exhaustion of the vitamins.

These results show that the presence of many components, but not necessarily their complete consumption, may be a requirement for cell growth. However, this also means that after cell removal, the supernatant has a considerable nutrient concentration that calls for recycling. Also, since the analytical techniques used are tailored to look for specific substances, the concentration of cell debris from lysed cells and minor metabolites is not yet known. These could build up upon recycling. Therefore, the best way to reuse the nutrients with today's technology is indirectly using leftover culture media as raw material for another bioprocess, such as anaerobic digestion or microalgae cultivation. In particular, Myers et al. (2023) estimate a nitrogen loss of 91 gN/kg of CM and a generation of 1.57 kg COD/kg of CM; these figures show the opportunity and economic need for recycling nitrogen and reducing COD.

## 3.2 *Solid-Liquid Separation*

Typical processes for cell separation in mammalian cell cultures are dead-end filtration, centrifugation, and tangential flow filtration in the microfiltration mode. However, most of the industrial experience with these processes is with from

**Table 1** Average concentration of spent culture media after cultivation of C2C12 murine myoblast cells in surface cultures

| Compound | Residual concentration, mM | |
|---|---|---|
| | 3 days culture in DMEM[a] | 7 days in 40% DMEM + 40% Ham F10 + 20% FBS[b] |
| Glucose | 0.32 | 3.70 |
| Lactate | n.d. | 25.51 |
| Glutamine | 0.62 | 0.55 |
| Amino acids, except glutamine | 4.4 | 3.82[c] |
| Pyruvate | 0.82 | n.d. |
| Vitamins | 45 | n.d. |
| $Na^+$ | 160 | 159.42 |
| $K^+$ | 6.3 | 5.89 |
| $Ca^{2+}$ | 2.1 | 1.32 |
| $HPO_4^{2-}$ | 0.93 | n.d. |
| $Mg^{2+}$ | 0.92 | 0.79 |
| $NH_4^+$ | 0.76 | n.d. |

Obs: [a]Haraguchi and Shimizu (2021), [b]Estimated from charts in O'Neill et al. (2022), [c]From the sum of 21 amino acids, using an average molar mass of 110 g/mol. Ammonium is not an original component of DMEM, but is present from the start due to aminoacid degradation.

clarification of the supernatant to recover monoclonal antibodies and other recombinant molecules (Shekhawat et al. 2018), not for recovery of cells—although this is also increasing in small-scale, highly controlled processes, because of the development of cell therapies.

### 3.3 Particle Sedimentation

Microcarriers can be separated by filtration, centrifugation, and, less commonly, by magnetism (Bodiou et al. 2020) and inertia (Moloudi et al. 2018). However, magnetism is impractical for CM because of the need for magnetic (inedible) microcarriers, and inertia is difficult to scale up.

Sedimentation of small particles follows Eq. (1) (Belter et al. 1987; De Carvalho et al. 2013), which is derived from Stokes' law and the force balance in a particle suspended in a fluid. For example, with relatively large scaffolds of 100–300 μm in diameter, it would seem that sedimentation could be simple. However, hydrogel scaffolds have a low density—1.04 g/mL for Cytodex® dextran microspheres and 1.02–1.04 g/mL for Cultispher® gelatine microspheres—making it almost buoyant with culture media, which is suitable for culture in bioreactors. DMEM+G media, for example, has been shown to have a density of 1.006–1.009 and a viscosity of 0.9598–1.011 Pa·s (or centipoise) (Hinderliter et al. 2010).

$$V = g \cdot d^2 \left(\rho_\mathrm{p} - \rho_\mathrm{m}\right) / 18 \cdot \mu \tag{1}$$

Or

$$V = d^2 \left(\rho_\mathrm{p} - \rho_\mathrm{m}\right) / 5000 \cdot \mu \tag{2}$$

In Eq. (1), any unit can be used. In Eq. (2), $V$ (the sedimentation velocity) is found in cm/h, $d$ (the particle diameter) is used in μm, and $\rho_\mathrm{p} - \rho_\mathrm{m}$ (the density difference between particles and culture medium) is used in g/mL or g/cm$^3$. For microcarriers with a density of 1.04 g/mL, diameter of 175 μm, and a culture medium with a density of 1.007 g/mL and viscosity of 1 Pa·s, the velocity would be 0.20 cm/h, not practical for sedimenting large volumes.

For cells with about 15 μm, sedimentation is even slower, but centrifugation is a challenging alternative because of shear stress. There are commercial systems for filtration, but these are for small scale and mostly thought of for cell therapy (Schnitzler et al. 2016), even consisting of disposable systems that can be impractical in terms of cost and large-scale plastic residue production. Tangential filtration also has a pre-sterilization problem (Schnitzler et al. 2016).

### 3.4 Centrifugation and Shear

Cell separation is typically done by centrifugation at low relative centrifugal force for small-scale setups. Cell culture protocols state that cell lines should be centrifuged at 125×g for 10 min (ATCC 2014) or 100–200×g for 5–10 min for reculturing; for the concentration of cells, it is possible to use 800×g for 10 min, but exact values are highly dependent of the cell line (Invitrogen 2010). Some cell lines are very sensitive to shear.

Shear stress toward cells comes from the velocity of cells relative to the liquid. Although that is frequently cited as of some concern in centrifugation, the shear from cell movement during centrifugation (except in the acceleration and ejection zones of the centrifuge) is not as crucial as the shear from agitation and pumping. However, the apparent weight of the cell is another story, which can cause some deformation stress due to centrifugation. Consider the simple example of a spherical cell at the bottom of a tube, with a density slightly above that of the liquid; the force balance means that a small force is applied by the cell on the wall, and that will cause a slight tension that the cell cytoskeleton must support. However, in centrifugation, that force is directly proportional to the relative centrifugal force and can cause the cell's deformation. Eventually, it is lysed if the deformation overcomes the cell Young's modulus. A similar problem may happen during filtration, where the filter cake can be deformed because of the pressure, leading to increased resistance to filtration and reduced filtration flux. These considerations are on par with research for *D. salina*, a microalga (not an animal cell, but similarly devoid of a cell wall) that was found to be more affected by the liquid head and pellet pressure than by increased cell weight and viscous drag at 9000×g, but resisting damage at 3000×g (Xu et al. 2015).

Much of the stress in cell centrifugation may not arise from cell movement but from cell acceleration from a low-velocity fluid in the inlet to the high-speed rotation of the bowl; it is akin to jumping *into* a train in movement. A centrifuge bowl rotating at 3000 rpm, which is ordinary for disk-stack centrifuges, has a tangential velocity of 3.1 m/s at just a 1 cm radius from the axis. The acceleration is not a problem for small and strong cells such as bacteria or yeast but can be a problem for animal cells and requires a careful design in the feeding section of centrifuges.

### 3.5 Using Existing Centrifuges

Centrifuge makers such as GEA, Alfa Laval, and Westfalia do not yet offer solutions for the CM market, but it is possible that existing equipment can be adapted. Most process centrifuges were developed to clarify fermented broths, beverages, and cell cultures, having RCFs of 3000–3500×g for decanters (e.g., Foodec® line, Alfa Laval) and above 2000 up to 11,000×g for disc-stack centrifuges with solids ejection (e.g., MBUX® and Culturefuge® lines, Alfa Laval). It is possible that these centrifuges, operating at lower rotational speeds, could separate CM cells from broths. However, the throughput of the equipment would be significantly lower. Surimi, a protein product from minced fish meat, can be processed using Flottweg decanter centrifuges, operating at 3500–4100×g for capacities of 0.6–2.2 $m^3/h$, depending on the model.

Using concentrations in the same range imagined for myoblasts produced in bioreactors (6.6%–7.3% packed cell volume), Hundley and Königsson (2020) reached product yields of up to 96%, with packed cell volumes of 97%, but with 15% lysis and some residual turbidity in the supernatant, using a CF100 disk-stack centrifuge. Rotational speeds are not disclosed. In further developments, the same authors reached similar yields with lower lysis (2–7%) by changing the bowl design and using lower load factors (0.075 $L/h/m^2$), presumably by reducing rpm and relative centrifugal force.

## 3.6 Direct (Dead-End) Filtration

Cellular suspensions are routinely filtered on a small scale, both for preparation of cell suspensions from tissue homogenates (mesh filtration) and for clarification of supernatants (fine filter media, usually membrane filtration). However, direct filtration can be challenging for the large volumes necessary in a CM production process. Using a membrane filter would lead to cell buildup (the filter *cake*) with a quick increase in resistance for the filtration. Using a mesh filter in direct filtration would initially be feasible, but again, the accumulated cells would create resistance. Although animal cells are relatively large, making initial cake resistance low, the cake is deformable—hence the buildup in resistance. However, it is conceivable that a mesh filter mounted in a belt or rotary vacuum filtration system could work, provided the biomass is blown out rather than removed with the usual advancing knife—a setup yet to be tested.

## 3.7 Tangential Filtration

Tangential filtration operated as microfiltration can be used to separate cells—this is regularly done in small-scale perfusion reactors, where membranes with pore sizes of 0.1–1 μm (e.g., Cytiva CFP-6-D-4X2MA at 0.65 μm or Sartorius Green Line Explorer at 0.45 μm) are used to retain cells while removing unwanted solutes. However, the apparent viscosity of suspensions increases exponentially with the packed volume; while increasing the PCV of coarse spherical particle suspensions from 5% to 25% only doubles the viscosity, further increasing PCV to 50% increases viscosity another five to ten times (Haruki et al. 2011; Pal 2023). Although this was determined for rigid spheres, the same trend is observed in fermentation broths. Even if the fluids show shear thinning, as is the case for blood (Fedde and Wideman Jr 1996; Windberger et al. 2019), the higher viscosity puts a practical limit to tangential flow filtration since suspensions must be pumped at high velocities to avoid membrane clogging, and higher viscosities may imply more significant shear stress. However, because of the possibility of selectively removing leftover solutes, TFF could be used for diafiltration or concentration before cell recovery through filtration or centrifugation.

## 4 3D Bioprinting in Cultivated Meat Production

Harvested cells are not the final product: they must still be formulated into a tissue-like structure to be called CM. A common strategy for manufacturing three-dimensional (3D) matrices for tissue engineering is 3D printing or bioprinting. In general, 3D printing allows the development of solid and porous structures by laying polymeric filaments onto a surface. The dimensions and structures of these matrices can be easily tuned and controlled according to the goal at hand.

A variation of 3D printing is 3D bioprinting, which involves directly printing cells in variable formulations (bioinks) to manufacture organs and tissues. The 3D bioprinting technique was first developed in the early 2000s when researcher Thomas Boland modified a conventional printer to print live bovine cells immersed in culture medium (Wilson Jr and Boland 2003). In theory, 3D bioprinting would allow printing organs and tissues in their original size and structure, but it is still considered an evolving technique that requires improvement.

The biggest challenge 3D bioprinting faces is the bioink formulation because each cell demands a distinct combination of factors to reproduce its biological environment and guarantee its adequate development. The ideal bioink for bioprinting follows three main roles. First, it must maintain cellular viability, protecting cells against mechanical stress and allowing good diffusion of nutrients and biomolecules. Second, it must be mechanically compatible with the bioprinting technique, yielding the ideal rheology, especially in terms of viscosity. Finally, the bioink must direct specific cellular behavior, enabling cellular differentiation through the combination of mechanical properties for mechanobiological signaling and cellular mobility along the matrix so that cells can migrate, interact, and remodel the environment appropriately (Morgan et al. 2020).

Bioprinting meat involves recreating the skeletal muscle tissue, which is composed of a complex structure comprising the muscle fiber, connective tissue (endomysium, perimysium, and epimysium), and intramuscular fat. All these components play a vital role in the organoleptic characteristic of the meat (Bomkamp et al. 2022), underscoring the importance of considering them in the formulation of the bioink. The vasculature and nerves of the muscle tissue influence the meat's texture. They are crucial for tissue maturation and maintaining the adequate function of the cells (Bomkamp et al. 2022) to obtain a CM closer to conventional meat. Ideally, they should also be recreated in bioprinting.

The most widely explored bioink formulation is based on hydrogels, decellularized tissues, synthetic polymers, composites, and bioactive molecules (Chimene et al. 2020). Among these options, hydrogel bioinks seem to be the preferred option for tissue engineering, including CM. Hydrogels are usually made up of polysaccharides, polymers, or proteins, and their advantage lies in the fact that they offer a biochemical environment like the extracellular matrix and can easily be combined with biological factors to create a richer structure (Schätzlein and Blaeser 2022).

In the case of CM, it is also important to note that the bioink formulation will affect the meat's properties, such as taste, tenderness, and nutritional value (Wollschlaeger et al. 2022). Consequently, there are a myriad of proposed bioink formulations in literature. For example, Kang et al. (2021) described a bioink made up of fibrinogen and Matrigel®. In contrast, Dutta et al. (2022) used a combination of alginate and gelatine supplemented with plant and edible insect hydrolysates. In both cases, the authors obtained a product that is structurally quite similar to conventional meat.

Lee et al. (2023) also tested an alginate-based formulation. However, in this case, the authors highlight that natural alginate has low bioprintability and offers low cellular function. Hence, it was necessary to adapt the alginate structure by conjugating it with tyramine. The need to adapt bioinks is a common scenario in 3D bioprinting, given that there is still no ideal formulation that yields the three main rules (bioprintability, maintenance of cellular viability, and signaling for specific cellular behaviors).

Even though hydrogel seems to be the preferred base of the bioinks for CM, its mechanical strength and stiffness are low, which hinders the control of the shape of the printed tissue (Schätzlein and Blaeser 2022). However, this challenge presents an opportunity for improvement by exploring various gelation processes with different cross-linking techniques. By applying innovative cross-linking methods, such as chemical, physical, enzymatic, or hybrid approaches, it is possible to tune and possibly enhance the structural integrity and mechanical properties of hydrogel-based bioinks (Schätzlein and Blaeser 2022). Another aspect that can affect the bioprintability of the hydrogel is the concentration of protein applied, which changes the gelation point. This impacts the hydrogel's elastic and shear moduli and viscosity (Wollschlaeger et al. 2022).

Besides the bioink, the success of the printed tissue also depends on the bioprinters used. There are three main types of bioprinters available: inkjet (thermal and piezoelectric), extrusion-based (piston, pneumatic, and screw-based), and light-based (laser-induced, stereolithography, and matrix-assisted pulsed laser evaporation) (Jamieson et al. 2021). The choice of the equipment will depend on the bioink and the scaffold that is going to be printed. For example, extrusion-based bioprinters are more adequate for printing viscous and dense bioinks. In contrast, stereolithography bioprinters are envisaged for bioinks capable of photopolymerizing, and despite being more expensive, they provide a higher resolution (Jamieson et al. 2021).

# 5 Challenges and Future Developments

Every step is challenging when scaling complex processes. The scale-up of existing animal cell cultivation and downstream processes into a wholly new industry such as CM is comparable to the development of the automobile or the computer industry, beginning with bulky and intricate systems and now offering much more sophisticated but practical and user-friendly mass products.

We are living the days of relatively low titers, and no large volumes processes exist to look to as a benchmark. Cell concentration after proliferation typically reaches $10^5$–$10^6$ cells/mL in bioreactors with free media, and higher concentrations in more complex setups (Letti et al. 2021), such as perfusion systems. However, even considering a concentration of $2 \cdot 10^7$ cells/mL, which translates to 70 g/L of wet cells (calculated from data presented by Myers et al. 2023), a volume of 28 liters of medium would be necessary for producing 1 kg of cellular meat paste, i.e., a mass with a packed cell volume (PCV) of 95–100%. For producing a bioink, a lower PCV could be enough; however, to avoid leftover solutes, maybe the way forward would still be concentration to high PCVs followed by paste dilution.

There are plenty of challenges, but a clear path ahead: scale must be increased. From the results of pilot-scale processes, more information will allow the simulation and development of high-throughput systems. Four aspects are more likely to drive the downstream process research for CM production in the coming years: cell line development and culture, raw material prices and media recovery, equipment for cell separation and bioprinting, and quality assurance.

## 5.1 *Cell Line Development*

Current CM production uses cell lines developed for other uses, such as the production of monoclonal antibodies or in vitro studies of toxicity. New cell lines must be selected for their processing and food product qualities, such as growth rate, resistance to shear, nutritional requirements, and resistance to bioprinting and integration into a product. Lower lysis and higher resistance warrant process intensification, with higher flow rates and higher throughput in equipment.

## 5.2 *Raw Material and Recovery*

The ideal CM must be both accessible and environmentally friendly. This can only happen due to the efficient use of raw materials and recovery processes. The rational development of culture media tailored for cell lines and production phases impacts the amount of non-product substances that must be removed from cells in the downstream process; at the same time, it is essential to recover macro- and micronutrients. Because of the complexity of recycling those to the process, the first step would be to make a convoluted recovery, using leftover nutrients to produce microbial biomass for proteins and recover energy. Next, water recovery through reverse osmosis and nutrient molecules through ultra- and nanofiltration can be incorporated into CM processing and ion exchange resins for mineral recovery. Both techno-economical and life cycle analysis can be used to compare the innumerable operational possibilities.

### 5.3 Equipment for Cell Separation

The industrial production of CM must involve processing large volumes of cell suspensions and pastes. Plenty of large-scale separation equipment has been developed to separate other suspended cells, such as bacteria and yeast. These processes can separate animal cells, but not yet efficiently, and have not yet been tested on a large scale. As soon as bioreactors allow the production of pilot-scale volumes of cellular suspensions, it will be possible to apply basic downstream processing principles to determine cake resistances, settling velocities, fluid properties such as apparent viscosities, fouling characteristics, and strategies for reduction of stress on cells in different setups. From these initial results, new designs shall appear.

### 5.4 Quality Assurance

Like any food, meat can lose nutritional and organoleptic qualities if improperly processed. Also, like any food, it can be contaminated or spoiled by microorganisms. Unlike most foods, however, CM is an entirely new processed food. CM must resemble meat, but it is not developed in an organism, which poses new challenges to the industry, such as cell line stability, virus clearance, and proper cell development. Downstream processes must be developed with quality by design in mind. All processes must adhere to strict sanitary practices, which affect the construction materials and equipment design, and the process must ensure that large volumes of healthy cell suspensions are converted into residue-free, highly concentrated cell pastes and 3D products. The pharma industry is resorting to disposable parts for cell products such as monoclonal antibodies, but that is not an option for low-cost and sustainable CM production. Therefore, materials and designs must be robust and allow containment, sterilization, and minimal manipulation.

## 6 Conclusion

Most processes described for CM production are small-scale, relying on processes that cannot be scaled up, such as direct filtration and manual cell stacking. Even the patents, techno-economic, and life cycle analyses describe the downstream step as an unspecified cell separation. There is no real-world data, probably because there is no real-world large-scale cultivation of animal cells yet. Preliminary research and comparison with other industries indicate that the downstream processing of cultivated meat can be done using existing technologies—tangential flow filtration and disk centrifuges. However, the yields are modest, and the loss of cells to lysis is significant. The impact of mechanical cues from processing on the final cell state is unknown. To increase throughput and efficiency compatible with the industrial production of CM, it is essential to determine process properties for the cells in pilot-scale studies and use these results to choose the best processing approaches and improve equipment design. Downstream processing requires research, with a significant impact on the process economy.

## References


Agathocleous M, Harris WA (2013) Metabolism in physiological cell proliferation and differentiation. Trends Cell Biol 23:484–492. https://doi.org/10.1016/J.TCB.2013.05.004

Ahmad SS, Chun HJ, Ahmad K, Shaikh S, Lim JH, Ali S, Han SS, Hur SJ, Sohn JH, Lee EJ, Choi I (2023) The roles of growth factors and hormones in the regulation of muscle satellite cells for cultured meat production. Journal of Animal Science and Technology 65(1):16

Alberts B (2017) Biologia molecular da Célula, 6th edn. Taylor & Francis

Ali A, Ahmed S (2018) Recent advances in edible polymer based hydrogels as a sustainable alternative to conventional polymers. J Agric Food Chem 66:6940–6967

Allan SJ, De Bank PA, Ellis MJ (2019) Bioprocess design considerations for cultured meat production with a focus on the expansion bioreactor. Front Sustain Food Syst 3:44

Alves EA, Guimarães ACR (2010) Capítulo 5 - Cultivo Celular. In: Conceitos e métodos para a formação de profissionais em laboratórios de saúde. EPSJV, Rio de Janeiro, pp 215–253

Ammam K, Bornscheuer U, Böttcher D et al (2010) Industrial biotechnology: sustainable growth and economic success. Wiley-VCH Verlag GmbH, Weinheim

ATCC (2014) Animal cell culture guide, vol 39. ATCC

Bellani CF, Ajeian J, Duffy L et al (2020) Scale-up technologies for the manufacture of adherent cells. Front Nutr 7. https://doi.org/10.3389/fnut.2020.575146

Belter PA, Cussler EL, Hu W (1987) Bioseparations: downstream processing for biotechnology. Wiley, New York

Bodiou V, Moutsatsou P, Post MJ (2020) Microcarriers for upscaling cultured meat production. Front Nutr 7:506214. https://doi.org/10.3389/FNUT.2020.00010/BIBTEX

Bomkamp C, Skaalure SC, Fernando GF et al (2022) Scaffolding biomaterials for 3D cultivated meat: prospects and challenges. Adv Sci 9:2102908

Bryant CJ (2020) Culture, meat, and cultured meat. J Anim Sci 98:skaa172

Campuzano S, Pelling AE (2019) Scaffolds for 3D cell culture and cellular agriculture applications derived from non-animal sources. Front Sustain Food Syst 3:38

Chimene D, Kaunas R, Gaharwar AK (2020) Hydrogel bioink reinforcement for additive manufacturing: a focused review of emerging strategies. Adv Mater 32:1902026

Chng VJY, Wan ACA (2023) The scaffold concept for alternative proteins. J Food Eng 357:111622. https://doi.org/10.1016/J.JFOODENG.2023.111622

Chriki S, Hocquette J-F (2020) The myth of cultured meat: a review. Front Nutr 7:7

De Carvalho JC, Medeiros ABP, Rodríguez-Fernández DE et al (2013) Downstream operations of fermented products. In: Fermentation processes engineering in the food industry. Taylor & Francis, pp 201–236. https://doi.org/10.1201/b14070

Dhandayuthapani B, Yoshida Y, Maekawa T, Kumar DS (2011) Polymeric scaffolds in tissue engineering application: a review. Int J Polym Sci 2011:1

Dutta SD, Ganguly K, Jeong M-S et al (2022) Bioengineered lab-grown meat-like constructs through 3D bioprinting of antioxidative protein hydrolysates. ACS Appl Mater Interfaces 14:34513–34526

Enrione J, Blaker JJ, Brown DI et al (2017) Edible scaffolds based on non-mammalian biopolymers for myoblast growth. Materials 10:1404

FAOSTAT-Database (2022) Food and Agriculture Organization of the United Nations. FAO, Rome, Italy. http://www.fao.org/faostat/es/#data/QC. Retrieved 28 October 2022

Fedde MR, Wideman RF Jr (1996) Blood viscosity in broilers: influence on pulmonary hypertension syndrome. Poult Sci 75:1261–1267

Fraeye I, Kratka M, Vandenburgh H, Thorrez L (2020) Sensorial and nutritional aspects of cultured meat in comparison to traditional meat: much to be inferred. Front Nutr 7. https://doi.org/10.3389/fnut.2020.00035

Haraguchi Y, Shimizu T (2021) Microalgal culture in animal cell waste medium for sustainable 'cultured food' production. Arch Microbiol 203:5525–5532

Haruki S, Sukenaga S, Saito N, Nakashima K (2011) Viscosity estimation of spherical particles dispersed suspension. High Temp Mater Processes 30:405–409

Heathman TRJ, Nienow AW, Rafiq QA et al (2019) Development of a process control strategy for the serum-free microcarrier expansion of human mesenchymal stem cells towards cost-effective and commercially viable manufacturing. Biochem Eng J 141:200–209. https://doi.org/10.1016/j.bej.2018.10.018

Hinderliter PM, Minard KR, Orr G et al (2010) ISDD: a computational model of particle sedimentation, diffusion and target cell dosimetry for in vitro toxicity studies. Part Fibre Toxicol 7:36. https://doi.org/10.1186/1743-8977-7-36

Horie M, Yamano-Adachi N, Kawabe Y et al (2022) Recent advances in animal cell technologies for industrial and medical applications. J Biosci Bioeng 133:509–514. https://doi.org/10.1016/J.JBIOSC.2022.03.005

Horiguchi I, Torizal FG, Nagate H et al (2021) Protection of human induced pluripotent stem cells against shear stress in suspension culture by Bingham plastic fluid. Biotechnol Prog 37:e3100. https://doi.org/10.1002/btpr.3100

Hundley R, Königsson S (2020) Development of a single-use hermetic centrifuge system. Bioprocess Int

Invitrogen (2010) GIBCO® Cell culture basics handbook

Jamieson C, Keenan P, Kirkwood D et al (2021) A review of recent advances in 3D bioprinting with an eye on future regenerative therapies in veterinary medicine. Front Vet Sci 7:584193

Kadim IT, Mahgoub O, Baqir S et al (2015) Cultured meat from muscle stem cells: a review of challenges and prospects. J Integr Agric 14:222–233

Kang D-H, Louis F, Liu H et al (2021) Engineered whole cut meat-like tissue by the assembly of cell fibers using tendon-gel integrated bioprinting. Nat Commun 12:5059

Kumar A, Sood A, Han SS (2023) Technological and structural aspects of scaffold manufacturing for cultured meat: recent advances, challenges, and opportunities. Crit Rev Food Sci Nutr 63:585–612. https://doi.org/10.1080/10408398.2022.2132206

Lee HJ, Yong HI, Kim M et al (2020) Status of meat alternatives and their potential role in the future meat market - a review. Asian Australas J Anim Sci 33:1533–1543. https://doi.org/10.5713/ajas.20.0419

Lee S, Choi G, Yang YJ et al (2023) Visible light-crosslinkable tyramine-conjugated alginate-based microgel bioink for multiple cell-laden 3D artificial organ. Carbohydr Polym 313:120895

Letti LAJ, Karp SG, Molento CFM et al (2021) Cultivated meat: recent technological developments, current market and future challenges. Biotechnol Res Innov 5:e2021001. https://doi.org/10.4322/biori.202101

Levi S, Yen FC, Baruch L, Machluf M (2022) Scaffolding technologies for the engineering of cultured meat: towards a safe, sustainable, and scalable production. Trends Food Sci Technol 126:13–25. https://doi.org/10.1016/J.TIFS.2022.05.011
Moloudi R, Oh S, Yang C et al (2018) Inertial-based filtration method for removal of microcarriers from mesenchymal stem cell suspensions. Sci Rep 8:1–10. https://doi.org/10.1038/s41598-018-31019-y
Morgan FLC, Moroni L, Baker MB (2020) Dynamic bioinks to advance bioprinting. Adv Healthc Mater 9:1901798
Myers GM, Jaros KA, Andersen DS, Raman DR (2023) Nutrient recovery in cultured meat systems: impacts on cost and sustainability metrics. Front Nutr 10:1151801
Naves MMV, Moreno FS (2000) Diferenciação celular: importância na hepatocarcinogênese e papel modulador do ß-caroteno/Cellular differentiation: importance in hepatocarcinogenesis and modulator role of ß-carotene
Ng S, Kurisawa M (2021) Integrating biomaterials and food biopolymers for cultured meat production. Acta Biomater 124:108–129
Nienow AW, Rafiq QA, Coopman K, Hewitt CJ (2014) A potentially scalable method for the harvesting of hMSCs from microcarriers. Biochem Eng J 85:79–88
O'Neill EN, Ansel JC, Kwong GA et al (2022) Spent media analysis suggests cultivated meat media will require species and cell type optimization. NPJ Sci Food 6:46
Otsuji TG, Bin J, Yoshimura A et al (2014) A 3D sphere culture system containing functional polymers for large-scale human pluripotent stem cell production. Stem Cell Rep 2:734–745. https://doi.org/10.1016/J.STEMCR.2014.03.012
Pal R (2023) Recent progress in the viscosity modeling of concentrated suspensions of unimodal hard spheres. ChemEngineering 7:70
Paredes J, Cortizo-Lacalle D, Imaz AM et al (2022) Application of texture analysis methods for the characterization of cultured meat. Sci Rep 12:3898. https://doi.org/10.1038/s41598-022-07785-1
Post MJ (2012) Cultured meat from stem cells: challenges and prospects. Meat Sci 92:297–301
Post MJ, Hocquette J-F (2017) New sources of animal proteins: cultured meat. In: New aspects of meat quality. Elsevier, New York, pp 425–441
Post MJ, Levenberg S, Kaplan DL et al (2020) Scientific, sustainability and regulatory challenges of cultured meat. Nat Food 1:403–415
Rao KM, Choi SM, Han SS (2023) A review on directional muscle cell growth in scaffolding biomaterials with aligned porous structures for cultivated meat production. Food Res Int 168:112755
Rousta N, Hellwig C, Wainaina S et al (2021) Filamentous fungus aspergillus oryzae for food: from submerged cultivation to fungal burgers and their sensory evaluation—a pilot study. Foods 10:2774
Sathasivam R, Radhakrishnan R, Hashem A, Abd Allah EF (2019) Microalgae metabolites: a rich source for food and medicine. Saudi J Biol Sci 26:709–722
Scarfone RA, Pena SM, Russell KA et al (2020) The use of induced pluripotent stem cells in domestic animals: a narrative review. BMC Vet Res 16:477. https://doi.org/10.1186/s12917-020-02696-7
Schätzlein E, Blaeser A (2022) Recent trends in bioartificial muscle engineering and their applications in cultured meat, biorobotic systems and biohybrid implants. Commun Biol 5:737
Schmidinger K (2012) Worldwide alternatives to animal derived foods – overview and evaluation models. Solution to global problems caused by livestock. University of Natural Resources and Life Sciences, Vienna, Austria
Schnitzler AC, Verma A, Kehoe DE et al (2016) Bioprocessing of human mesenchymal stem/stromal cells for therapeutic use: current technologies and challenges. Biochem Eng J 108:3–13
Seah JSH, Singh S, Tan LP, Choudhury D (2022) Scaffolds for the manufacture of cultured meat. Crit Rev Biotechnol 42:311–323. https://doi.org/10.1080/07388551.2021.1931803

Shekhawat LK, Sarkar J, Gupta R et al (2018) Application of CFD in bioprocessing: separation of mammalian cells using disc stack centrifuge during production of biotherapeutics. J Biotechnol 267:1–11
Simsa R, Yuen J, Stout A et al (2019) Extracellular Heme proteins influence bovine myosatellite cell proliferation and the color of cell-based meat. Foods 8
Singh H, Mok P, Balakrishnan T et al (2010) Up-scaling single cell-inoculated suspension culture of human embryonic stem cells. Stem Cell Res 4:165–179. https://doi.org/10.1016/J.SCR.2010.03.001
Singh A, Kumar V, Singh SK et al (2023) Recent advances in bioengineered scaffold for in vitro meat production. Cell Tissue Res 391:235–247
Sinke P, Swartz E, Sanctorum H et al (2023) Ex-ante life cycle assessment of commercial-scale cultivated meat production in 2030. Int J Life Cycle Assess 28:234–254
Stevens MM (2008) Biomaterials for bone tissue engineering. Mater Today 11(5):18–25
Stout AJ, Mirliani AB, Rittenberg ML et al (2022) Simple and effective serum-free medium for sustained expansion of bovine satellite cells for cell cultured meat. Commun Biol 5:466. https://doi.org/10.1038/s42003-022-03423-8
Taya Y, O'Kane S, Ferguson MWJ (1999) Pathogenesis of cleft palate in TGF-β3 knockout mice. Development 126:3869–3879
Teo JCM, Boularaoui SM, AlWahab NSA, Christoforou N (2016) Mending the heart through in situ cardiac regeneration. In: In situ tissue regeneration: host cell recruitment and biomaterial design. Elsevier, New York, pp 313–344. https://doi.org/10.1016/B978-0-12-802225-2.00017-9
Tuomisto HL, Teixeira de Mattos MJ (2011) Environmental impacts of cultured meat production. Environ Sci Technol 45:6117–6123
Tuomisto HL, Allan SJ, Ellis MJ (2022) Prospective life cycle assessment of a bioprocess design for cultured meat production in hollow fiber bioreactors. Sci Total Environ 851:158051
van der Weele C, Feindt P, Jan van der Goot A et al (2019) Meat alternatives: an integrative comparison. Trends Food Sci Technol 88:505–512. https://doi.org/10.1016/j.tifs.2019.04.018
Wang T, Xu P, Chen Z et al (2018) Alteration of the structure of rice proteins by their interaction with soy protein isolates to design novel protein composites. Food Funct 9:4282–4291
Wang Y, Zou L, Liu W, Chen X (2023) An overview of recent Progress in engineering three-dimensional scaffolds for cultured meat production. Foods 12:2614
Wilson WC Jr, Boland T (2003) Cell and organ printing 1: protein and cell printers. Anat Rec A Discov Mol Cell Evol Biol 272:491–496
Windberger U, Auer R, Seltenhammer M et al (2019) Near-Newtonian blood behavior – is it good to be a camel? Front Physiol 10:906
Wollschlaeger JO, Maatz R, Albrecht FB et al (2022) Scaffolds for cultured meat on the basis of polysaccharide hydrogels enriched with plant-based proteins. Gels 8:94
Xu Y, Milledge JJ, Abubakar A et al (2015) Effects of centrifugal stress on cell disruption and glycerol leakage from Dunaliella salina. Microalgae Biotechnol 1. https://doi.org/10.1515/MICBI-2015-0003
Yang M, Wang Q, Zhu Y et al (2023) Cell culture medium cycling in cultured meat: key factors and potential strategies. Trends Food Sci Technol 138:564
Ye Y, Zhou J, Guan X, Sun X (2022) Commercialization of cultured meat products: current status, challenges, and strategic prospects. Future Foods 6:100177. https://doi.org/10.1016/J.FUFO.2022.100177
Young JF, Skrivergaard S (2020) Cultured meat on a plant-based frame. Nat Food 1:195. https://doi.org/10.1038/s43016-020-0053-6

# Ingredients and Formulation of Cultivated Meat Products

**Paula Toshimi Matumoto-Pintro, Fernando Antônio Anjo, and Bianka Rocha Saraiva**

**Abstract** The cultivated meat production chain has been developing in recent years. For consumers to adopt the consumption of meat products derived from cultivated meat, it is essential that they have similar sensory characteristics to meat products from animal production. The development of the cultivated meat production chain requires ingredients and additives to provide traditional characteristics while also meeting the criteria of sustainability and healthiness. This chapter explores the importance of ingredients and additives in formulating cultivated meat products that meet traditional meat standards for texture, flavor, and nutritional value. Ingredients and additives play a crucial role in creating cultivated meat products that resemble conventional meat. This includes proteins derived from meat, fats, and other plant components and agro-industrial by-products. The use of natural ingredients, such as antioxidants and preservatives from plants, as well as by-products, offers a healthier and more sustainable alternative than synthetic additives. Food safety, regulation, development of effective cell cultivation technologies, production of safe ingredients and additives, and consumer awareness about the benefits of this alternative are challenges in this production chain. Challenges that can be solved through collaboration between industries, research centers, and regulatory government entities. Cultivated meat represents an opportunity to address global challenges of food security, sustainability, and animal welfare.

**Keywords** Meat products · Additives · Technological properties · Sensory characteristics · Sustainability · Healthiness

---

P. T. Matumoto-Pintro (✉)
Department of Agronomy, Functional Food Research Group, State University of Maringá, Maringá, Paraná, Brazil
e-mail: ptmpintro@uem.br

F. A. Anjo · B. R. Saraiva
Functional Food Research Group, State University of Maringá, Maringá, Paraná, Brazil



C. R. Soccol et al. (eds.), *Cultivated Meat*,
https://doi.org/10.1007/978-3-031-55968-6_9

# 1 Introduction

Almost 400 years ago, when ordered rectangular structures on a piece of cork were visualized under a primitive microscope, one would never have imagined that this was the first step towards the development of cell biology (Lodish et al. 2013). Since then, several studies around the world have revealed the complexity of cells and the possibility of studying them in a simulated environment. Cells could be maintained in the laboratory under conditions that allow their survival and growth (cell culture). Cell cultivation in laboratories allowed the study of their activities, differentiation, and proliferation. Cell culture has become the primary method for studying their operational mechanisms, being used in various knowledge areas such as medicine, biotechnology, and, more recently, in food production.

In recent years, in the field of food science and technology, cell culture has been studied with a focus on microorganisms of interest in food production. Currently, this technique, in which certain cells are kept alive and in full development in an appropriate environment that simulates natural conditions, has been adapted for food production, the cultivated meat production.

The animal cell's cultivation for food production is an emerging technique that, in the last seven years, has been the subject of many studies around the world. The interest in this type of alternative food production is valid, as meat is an important source of nutrients for the population, and the accelerated growth of the world population stimulates the search for alternative food sources. Meat production involves the slaughter of animals, an unacceptable practice for some social and ethnic groups, and the meat production chain uses large amounts of natural resources (Letti et al. 2021; Galland and Pacheco 2022).

Cultivated meat production involves the collection of cells from cattle, poultry, pigs, fish, seafood, or other animals, their cultivation/proliferation in bioreactors, differentiation into various cell types (muscle, fat, connective tissue, or other cell types), structuring, and, finally, the collection for use in the processing of meat products (Fasano and Michael 2023). Knowledge on food, tissue, and bioprocess engineering, as well as cell biology, biochemistry, and genetics, is required for processing cultured meat.

Cultured cellular meat can be obtained in unstructured form, which is more easily produced and used in products such as burgers, meatballs, kibbehs, breaded meats, and, in structured form, as whole cuts such as steaks, chicken filets, fish, and other desired cuts. These meat cuts obtained by cell culture may be subjected to the application of preservation techniques, the same ones already applied to traditional meat, such as sterilization, pasteurization, smoking, fermentation, and use of additives (conventional, produced by fermentation, or obtained from vegetables). Unstructured meat products can be subjected to appropriate technological processes, obtaining processed meat products such as restructured meat products (burgers, meatballs, and nuggets) and emulsified meat products (mortadella, sausage, and pâtés), among others (Kieliszek and Misiewicz 2014; Ong et al. 2021; Galland and Pacheco 2022). In the latter case, the use of cultivated meat for the processing of meat derivatives will require the use of specific ingredients and additives capable of interacting with the cultured cells to structure the meat product, giving it technological properties identical to those of traditional meat products. This chapter aims to present the possible ingredients and food additives that can be used in the preparation of meat products made with cultivated meat.

## 2 Ingredients in the Formulation of Cultivated Meat Products

The food industry is constantly improving, and these changes do not exclude the already established meat products industry. Current challenges involve the production of food that meets the demands of an increasingly discerning consumer, the development of nutritious and safe products with sensory properties that appeal to the consumer's palate, the high demand for food, sustainable production, and the development of products that contain lower amounts of fats, carbohydrates, and sodium in their formulation (Tarté 2009). All these challenges have led to technological advancements that have culminated in the development of alternative meat production technologies, such as cultivated meat.

Cellular meat, a cultivated product, can be processed like traditional meat. The use of cultivated meat products to create meat derivatives becomes advantageous, as one of the major challenges in cultivated meat production is the structuring of these muscle cells to enable the production of a steak (Guan et al. 2021).

All meat products are produced by grinding meat. Therefore, the production of cultured ground meat, for example, would be faster and more economically advantageous, as the cell structuring stage would not require the use of such specific scaffolding. Several companies worldwide produce ground cultivated meat (Good Food Institute 2023).

The elaboration of meat products derived from cultivated meat will require ingredients and food additives that, similar to the processing of products made from traditional meat, enhance technological properties (texture modulation), sensory properties (flavor, aroma, and color), nutritional aspects (proteins and lipids), and product preservation (bioactive compounds) (Tarté 2009; Ng and Kurisawa 2021). Protein ingredients derived from meat (Table 1) are widely used in the preparation of meat products, not only to enhance product quality but also often to maximize the use of the raw materials (the animal). These ingredients are by-products obtained from the slaughter of the animal and the meat itself (Tarté 2009).

Since the cultivated meat industry initially focuses on producing muscle cells, these ingredients obtained from by-products are still relatively unexplored. Thus, it requires the development of technologies that allow the specific cultivation of these cells to obtain these ingredients or that these ingredients can be replaced by others, obtained, for example, from plant sources (Kieliszek and Misiewicz 2014; Ong et al. 2021; Galland and Pacheco 2022).

**Table 1** Some protein ingredients derived from meat (Tarté 2009)

| Source | Ingredients |
|---|---|
| Blood | Blood plasma (liquid, frozen, dried)<br>Whole blood (liquid and dried)<br>Red cell protein (decolorized)<br>Plasma transglutaminase |
| Bone | Gelatin (type B)<br>Edible bone collagen (ossein)<br>Bone collagen hydrolysates (stocks and broths)<br>Edible bone phosphate<br>Edible fat |
| Pig skin | Gelatin (type A)<br>Stocks and broth |
| Beef hides | Gelatin (type B) |
| Poultry skin (chicken, turkey) | Concentrated collagen<br>Stocks and broths<br>Edible fat |
| Collagen-rich tissues | Concentrated collagen<br>Collagen hydrolysates |

These protein ingredients obtained from meat have been used for generations in the production of meat products, e.g., traditional products such as Thuringian blood sausage, a typical German dish that uses pig head meat, fatty trimmings, skin, glands, and pig blood in its formulation; tongue and blood sausage from Austria, made with pig tongue, skin, fat, and blood; and black pudding from England, which generally does not contain any muscle tissue, being made only with blood, cubed pork fat, cooked barley, and toast (Feiner 2006).

Some popular meat products also have added protein ingredients derived from meat. In the case of chicken nuggets, the percentage of skin used can reach 35%. The skin has technological and sensorial importance for the product, as it is rich in connective tissue, increases juiciness, adds flavor, and is a relatively inexpensive ingredient, which reduces production costs (Feiner 2006). The replacement of chicken skin with other ingredients that can provide the same quality attributes has been carried out in nuggets produced with traditional meat: the replacement of chicken skin with canola oil, wheat, and chia flour (*Salvia hispanica* L.) (Kim et al. 2015; Barros et al. 2018; Seó et al. 2024). These studies provide important information so that these alternative ingredients can also be used in the preparation of cultivated meat nuggets. Skin can also be cultured, just like muscle cells. Technologies such as 3D bioprinting, which is already used in the production of skin for medical applications, can be adapted to produce skin from poultry, pigs, and cattle, enabling its use as an ingredient in food production (Ramachandraiah 2021; Gao et al. 2021).

The preparation of structured meat products (burgers and meatballs), emulsified products (mortadella and pâtés), and salami involves the use of fat from pork or beef as an ingredient. The fat added to these products contributes to shine, flavor, aroma, texture, and juiciness (Marianski and Marianski 2010; Iida et al. 2015; Frank et al. 2017). Assuming the use of cultivated meat for the production of these meat products, it is important to investigate not only the cultivation of muscle cells but also the cultivation of adipose tissue cells.

Cultured fat cells can be isolated, produced, used, and incorporated as food constituents, improving the quality of cultivated meat. The challenges of cultivating fat cells are the same as those of cultivating muscle cells: the development of cell lines, optimization of cell proliferation and differentiation, serum-free and food-grade cell culture media, supply of appropriate biomaterials and structures, design of bioreactors, scaled manufacturing, and process development (Warner 2019). However, the cultivation of these cells allows, using personalized culture media, the production of fat cells with a lower content of saturated lipids and a higher concentration of unsaturated lipids. In this way, it is possible to maintain the sensory characteristics and increase the concentration of omega-3 unsaturated fatty acids, which confer health benefits (Sugii et al. 2023).

In addition to the cultivation of fat cells, another option would be the use of fat substitutes, which are already used in meat products obtained from traditional meat. Compounds capable of reducing calories and providing succulence similar to that of fat, due to their interaction with water molecules, and having rheological properties similar to fat, such as starches, hydrocolloids, fibers, and proteins, have been used (Feiner 2006; Tarté 2009; Ng and Kurisawa 2021). Konjac starch, isolated soy protein, and modified starch, ethylcellulose oleogels, polysaccharide gels, microcrystalline cellulose, and brewing residue have been used as fat substitutes in meat products (Hoffman and Mellett 2003; Kao and Lin 2006; Krawczyk et al. 2009; Zetzl et al. 2012; Herrero et al. 2014; Oh et al. 2019; Saraiva et al. 2019). New alternatives to fat include algae, insects, and fungi (Asgar et al. 2010; Rubio et al. 2019).

## 3 Food Additives in the Formulation of Cultivated Meat Products

Every food manufacturing process demands the integration of several operations, ranging from the production of raw materials to the distribution of the final product. These integrated operations form the concept of a production chain, responsible for supporting the production of a given product (Drahokoupil 2023). A factory that produces processed meat products, such as hamburgers, for example, requires suppliers of ingredients (meat, fat, condiments, and seasonings), additives (texturizers and preservatives), and packaging that will support the manufacturing of the product.

Since the production of cultivated meat has been developing over recent years, it is extremely important for industries to emerge that supply meat products obtained from cultivated meat with inputs that make up this production chain. In this way, an entire production chain is formed around the cultivated meat industry: suppliers of cell lines, cultivation media, bioprocess design and bioreactor manufacturing, optimization of biotechnological processes, development of scaffolds for structuring, and the production of protein itself through cell cultivation (Galland and Pacheco 2022; Garcia et al. 2023). These companies will be responsible for developing, mainly, food additives that contribute to technological, sensory, and nutritional quality to the development of meat products from cultivated meat.

Food additives can be used to improve and maintain quality, enhance nutritional value, provide and improve functionality, facilitate processing, and enhance consumer acceptance. Authorization for additive use occurs through their toxicological assessment and proven technological need. Furthermore, the use of food additives is prohibited when there is evidence or suspicion that the additive has real or potential toxicity, significantly and unfavorably interferes with the nutritional value of the food, serves to cover up flaws in processing and food handling techniques, conceals changes to the raw material or product already produced, misleads the consumer, deceives or confuses, and does not meet each country's food additive legislation (Feiner 2006).

Most regulatory institutions around the world seek to limit the use of additives in food, especially synthetic additives, which can have harmful effects on consumers' health (Feiner 2006). Limiting the use of additives is also a desire of consumers, who, when adopting new eating habits, increasingly seek well-being associated with food. As a consequence of this consumption behavior, the segment of clean label, green, and organic foods grows every day, and the development and production of natural food additives from vegetables is inseparable from this new way of producing food (Zang et al. 2023).

When proposing a new method of meat production, such as cellular cultivation, resistance from consumers is observed—a phenomenon known as food neophobia (Faccio and Guiotto Nai Fovino 2019; Kantono et al. 2022). They believe that the production of meat through cell culture may require the use of various additives, which, in their conception, will have harmful effects on consumer health (Siegrist and Sütterlin 2017). It is important to highlight that the development of cell cultivation techniques aimed at meat production is based on sustainability and healthiness, making it impossible to dissociate these two concepts (Lin-Hi et al. 2022; Kantono et al. 2022). Therefore, even if the production of cultivated meat or its derivatives requires the use of additives, researchers, industries, and governments are committed to developing production technologies and additives that promote technological functions and beneficial effects on consumers' health.

The processed meat industry derived from traditional meat has been using various additives for many years. There are several classes of additives used and the functionalities they provide (Fig. 1). These additives are well established in conventional meat products development. However, the trend towards sustainable food production, combined with consumer knowledge and concern about the healthiness and sustainability of the foods they consume, leads to the search for ways to reduce the use and consumption of synthetic chemical additives, foods with high levels of carbohydrates, sodium, and lipids (Tarté 2009). On the other hand, there are healthier, more sustainable, and safer alternatives, such as natural plant-based additives

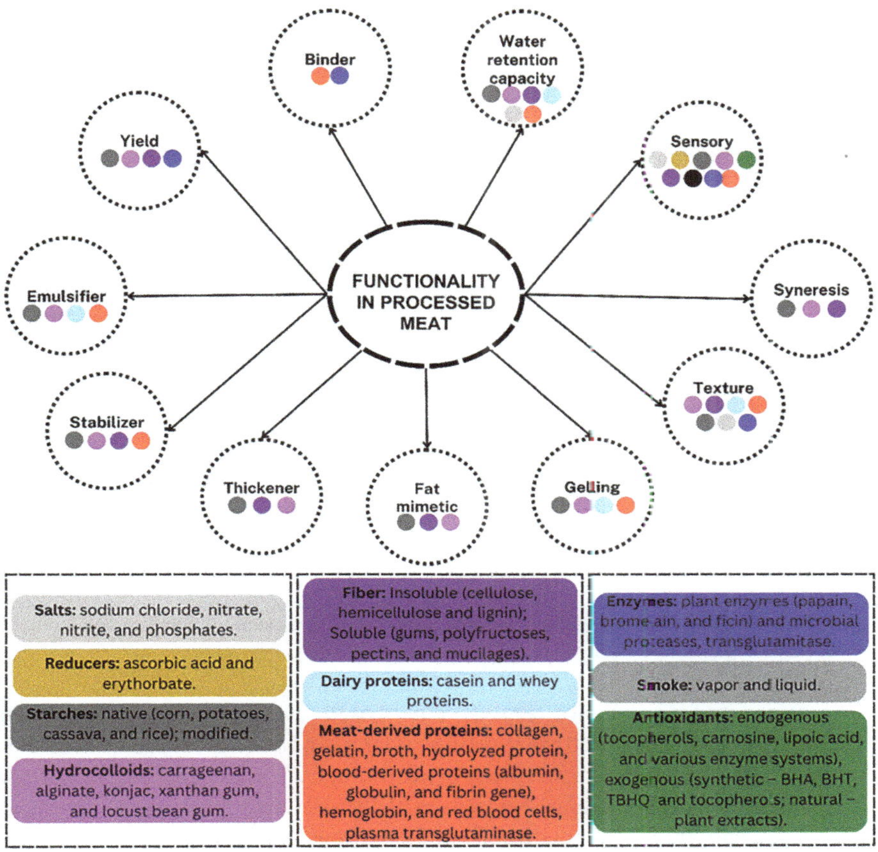


**Fig. 1** Main additives used in the meat derivatives industry (based on theoretical information provided by Feiner 2006 and Tarté 2009)

(Zang et al. 2023), which can be used both to replace synthetic additives already used in conventional meat products and in plant-based meat analogues and primarily to develop cultivated meat products.

Natural compounds extracted from plants can be used as additives with various functionalities, such as preservation and providing aroma, color, texture, and nutritional enhancements, among other functions (Carocho et al. 2015). Antioxidant components extracted from plants can be used as preservatives and antimicrobials in foods (Zang et al. 2023), largely due to the presence of bioactive compounds, especially phenolics (Aziz and Karboune 2016), as well carotenoids, vitamins, and peptides (Li et al. 2014; Makris and Boskou 2015; Aziz and Karboune 2016). Plant-based additives that have been used as antioxidants, antimicrobials, sweeteners, and colors in meat products are shown in Fig. 2, along with the species from which they were extracted.

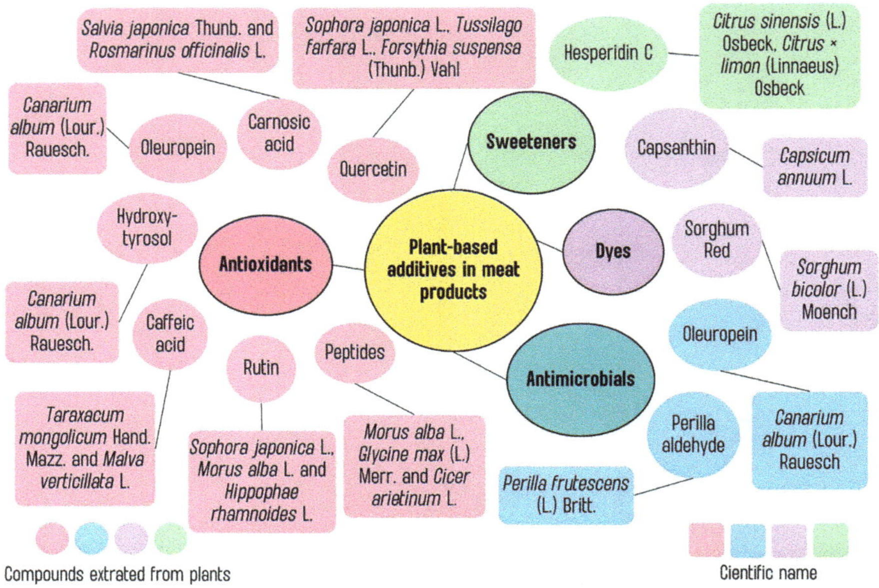


**Fig. 2** Plant-based food additives extracted from plants used in meat products (based on theoretical information provided by Zang et al. 2023)

As demonstrated in studies (Fig. 2), plant-based additives have been explored over time and have shown efficient results in terms of their antioxidant effect. For example, carnosic acid extracted from dried rosemary leaves was used in buffalo meat patties and chicken patties (Naveena et al. 2013). Oleuropein, hydroxytyrosol, caffeic acid, and rutin were obtained from the wastewater of the pomace from olive oil processing, aqueous extraction of red grape, and green tea extract and were used in cooked beef and pork (DeJong and Lanari 2009). In addition to olive processing by-products, olive leaf extracts were also used by Alirezalu et al. (2017) in Frankfurter-type sausage; and the green tea, which has also been studied for its catechin compounds (Guo et al. 2017), has been used in burgers as antioxidant source (Fachinello et al. 2018). Plant extracts from cloudberry, willow herb, and beetroot extracts have also been used as antioxidants in cooked pork patties and were shown to be potent antioxidants, with results comparable to those observed for pure quercetin (Rey et al. 2005). Peptides from the enzymatic hydrolysis of partially defatted chia flour and chia protein concentrate were used as antioxidants in ground beef (Silveira Coelho et al. 2019). Fruits can also be sources of phenolic compounds with antioxidant potential in burgers, such as goji berries (Cardoso et al. 2023).

In addition to compounds extracted from plants and fruits, even more sustainable alternatives can be found in agro-industrial by-products. They are often low- or no-cost materials that are generally discarded or underused for composting or animal feed but can be used as raw materials in the development of alternative additives. The use of by-products also helps to reduce the environmental impacts caused by the inadequate disposal of these materials, such as the high biological load discarded directly into water bodies or the burning or even burying of these by-products, which are practices that occur in industries and in the field (Koul et al. 2022). Demonstrating the possibilities of adding value to these materials, despite the additional cost and effort, can lead to lower environmental impacts.

These additives are also of natural and vegetable origin and generally contain many nutrients and compounds of interest in their composition, such as phenolic compounds, antioxidants, and proteins. Malt bagasse, a by-product from brewing, was used in burgers to replace fat and improve their cooking performance, resulting in less cooking loss, a smaller reduction in diameter, and less shrinkage (Rubio et al. 2019). By-products from pomegranate, tomato, grape, and olive were studied for their potential antioxidant and antimicrobial properties in chilled stored raw lamb patties (Andrés et al. 2017). Phenolic extracts from Colombian mango by-products from "Sugar Mango" and "Tommy Atkins" cultivars were used as antioxidants in cooked beef samples (Castro-Vargas et al. 2019). Passion fruit peel powder extract has also been studied for its antibacterial and antioxidant activity in preserved meat products (Ramli et al. 2020).

Furthermore, other by-products, initially used in dairy products, can also be applied to other food matrices, such as cultivated meat products. For example, *Acácia mearnsii* gum, a by-product from tannin production, obtained in the exudate form, was used as a food stabilizer (Anjo et al. 2021, 2023). The co-product of the bitterness extraction from trub, obtained from beer production, was used as a natural antioxidant in processed cheese (Saraiva et al. 2023). Additionally, okara and grape juice residues were used as antioxidant sources in milk and ice cream, respectively (Vital et al. 2018a, b).

These results demonstrate that natural additives based on plants, fruits, and residues have been studied for reducing lipid oxidation, extending shelf life, and improving other properties of meat products, as well as dairy products. The additives used in different food matrices have great potential for use in cultivated meat products, especially those already studied in this same food matrix, as cultivated meat has an identical composition to animal meat (Ramli et al. 2020; Stout et al. 2020; Hossain et al. 2022). Although research into cultivated meat is recent, the production of carotenoids by engineering primary bovine and murine muscle cells immortalized with endogenous prokaryotic enzymes has been studied as a way to produce phytonutrients in muscle cells to then improve the nutrition of cultivated meat (Stout et al. 2020). More research in this direction is needed to complement the supply of additives for the processing of cultivated meat.

## 4 Challenges and Future Developments

The production of cultured meat, as well as the additives and ingredients necessary to develop processed products, still presents many challenges. It is necessary to conduct studies that evaluate the existing bottlenecks in this emerging production chain so that viable alternatives and alternative technologies can be proposed to overcome these challenges.

The cultivated meat industry is still taking its first steps, and it is a rising market. Every year, the value of investments and the number of startups, industries, and research on cultivated meat increase (Bioinformant 2023). As previously mentioned, these cultivated meat processing companies need a well-established production chain to compete effectively in this highly competitive market.

The challenges of this new production chain are numerous, starting with the lack of regulation for the production processes of cells, additives, ingredients, and subsequently derived products (Ye et al. 2022). The question is, how can we regulate a completely new production chain that is still in its early stages of development? The key to solidifying this new market lies in partnerships between companies (industries), research centers, and government entities. Both must collaborate mutually to ensure that the development of the cultivated meat production chain aligns with the pillars that were the reasons for its emergence: sustainability, healthiness, animal protection, and fight against hunger.

Cultivated meat production also demands the development of new technological processes, novel tissue sampling techniques, cell culture and fermentation methods, 3D printing applications, the creation of natural ingredients and additives (with properties similar to synthetic ones), and cultivated meat processing improvements. New technologies must be studied to enhance techniques for obtaining, extracting, concentrating, isolating, and preserving ingredients and additives, especially those of plant origin. Sustainable methods for producing ingredients and additives to meet the industry's demand for cultivated meat products, such as utilizing industrial by-products, should be explored. It's essential to propose research and development partnerships between industries generating by-products and those producing ingredients and additives.

The production costs of cultivated meat and consequently its processed products are a concern, and one of the alternatives both to reduce costs and to increase the supply and diversity of foods is the production of hybrid products. These are products developed with cultured meat and vegetable proteins/ingredients in similar portions, ranging from 25 to 50% vegetable ingredients in their formulation (Grasso and Jaworska 2020). Currently, hybrid products have been commercialized, but with meat of animal origin and vegetable ingredients, such as sausages, burgers, meatballs, and minced meat, showing greater sensory acceptance from consumers than analogous vegetable-based meat products (Grasso 2020; Grasso et al. 2022; Grasso and Goksen 2023). As shown in the previous sections, plant-origin ingredients and additives are already used in the production of meat products. Their use is justified not only by technological contributions (gelation, fat emulsification, and water binding) but also by the reduction of production costs. In hybrid meat products, the use of vegetable ingredients in large proportions can also accelerate the supply of these products to the market and consequently to the consumer, as it will be necessary to use less cultured meat compared to a 100% cultured meat product, which, although estimates point to the optimization of the process, still takes a long production time. The hybrid product can also serve groups of consumers who do not consume or consume little meat of animal origin due to taste restrictions, but wish to consume it for nutritional reasons. It is crucial to emphasize that these new ingredients and additives must be safe, meaning they should not have any harmful effects on consumers' health. Numerous tests and years of research are needed for a food additive to gain approval. Once again, regulation is necessary to determine the safety level of these inputs.

At the end of the production chain, these new ingredients and additives play a pivotal role in the development of meat products derived from cultivated meat. Therefore, alongside the advancement of cell cultivation technologies, there is a simultaneous development of ingredients and additives. Another challenge in this production chain is the need for professionals from various fields who are capable of working not only in technical areas (research and development) but also in administration, policy, and marketing of these products (Morais-da-Silva et al. 2022). Marketing professionals, in particular, are of great importance, especially in addressing food neophobia and in demystifying production processes to educate consumers about the benefits of the cultivated meat production chain.

## 5 Conclusion

The global population is growing, and challenges related to food production are, therefore, increasing. The search for alternative and sustainable forms of food production is constant, and cultivated meat production offers a sustainable and ethical alternative for alternative food production. There are several challenges for the development of the cultivated meat production chain to consolidate. The structuring of products and their sensory characteristics are some of the major challenges of this type of food production. This chapter addressed the crucial role of ingredients and additives in the development of cultivated meat products. These inputs are obtained from various sources, including animal by-products and plant extracts, and offer a range of functionalities, from texture to preservation, effectively contributing to the improvement of the technological, sensory, and nutritional properties of cultivated meat products.

It is important to remember that consumers' new dietary habits need to be considered in the development of additives and ingredients. The use of natural and sustainable sources, such as plants, fruits, and agro-industrial by-products, aligns with consumers' growing preference for clean labels and sustainable foods. These inputs obtained from natural sources or agro-industrial by-products contribute to the quality and safety of cultivated meat products and are in line with the fundamental principles of sustainability and health that underpin the cultivated meat industry. The development of the cultivated meat production chain also faces a lack of regulation. Only through the collaboration of industries, research centers, and government entities can the regulation and production of safe cultured foods be possible. The harmony between these sectors should drive development, making it continuous, so that it is possible to develop cell culture technologies, create new ingredients and additives, and raise consumer awareness about the benefits of cultivated meat.

**Acknowledgments** We thank the Fundação Araucária (FA/PR – Brazil) for supporting the research and the Conselho Nacional de Desenvolvimento Científico e Tecnológico (CNPq/Brazil) for the scholarship.

## References

Alirezalu K, Hesari J, Eskandari MH et al (2017) Effect of green tea, stinging nettle and olive leaves extracts on the quality and shelf life stability of Frankfurter type sausage. J Food Process Preserv 41:e13100. https://doi.org/10.1111/jfpp.13100

Andrés AI, Petrón MJ, Adámez JD et al (2017) Food by-products as potential antioxidant and antimicrobial additives in chill stored raw lamb patties. Meat Sci 129:62–70. https://doi.org/10.1016/j.meatsci.2017.02.013

Anjo FA, Saraiva BR, Da Silva JB et al (2021) *Acacia mearnsii* gum: a residue as an alternative gum Arabic for food stabilizer. Food Chem 344:128640. https://doi.org/10.1016/j.foodchem.2020.128640

Anjo FA, Saraiva BR, da Silva JB et al (2023) A new food stabilizer in technological properties of low-fat processed cheese. Eur Food Res Technol 249:597–606. https://doi.org/10.1007/s00217-022-04155-z

Asgar MA, Fazilah A, Huda N et al (2010) Nonmeat protein alternatives as meat extenders and meat analogs. Compr Rev Food Sci Food Saf 9:513–529. https://doi.org/10.1111/j.1541-4337.2010.00124.x

Aziz M, Karboune S (2016) Natural antimicrobial/antioxidant agents in meat and poultry products as well as fruits and vegetables: a review. Crit Rev Food Sci Nutr 1–26:486. https://doi.org/10.1080/10408398.2016.1194256

Barros JC, Munekata PES, Pires MA et al (2018) Omega-3- and fibre-enriched chicken nuggets by replacement of chicken skin with chia (*Salvia hispanica* L.) flour. LWT 90:283–289. https://doi.org/10.1016/j.lwt.2017.12.041

Bioinformant (2023) The global market for cultured meat - market size, trends, competitors, and forecasts (2023). BioInformant

Cardoso MAP, Vital ACP, Medeiros A et al (2023) Goji berry effects on hamburger quality during refrigerated display time. Food Sci Technol 43. https://doi.org/10.1590/fst.68322

Carocho M, Morales P, Ferreira ICFR (2015) Natural food additives: quo vadis? Trends Food Sci Technol 45:284–295. https://doi.org/10.1016/j.tifs.2015.06.007

Castro-Vargas H, Ballesteros Vivas D, Ortega Barbosa J et al (2019) Bioactive phenolic compounds from the agroindustrial waste of Colombian mango cultivars 'sugar mango' and 'Tommy Atkins'—an alternative for their use and valorization. Antioxidants 8:41. https://doi.org/10.3390/antiox8020041

DeJong S, Lanari MC (2009) Extracts of olive polyphenols improve lipid stability in cooked beef and pork: contribution of individual phenolics to the antioxidant activity of the extract. Food Chem 116:892–897. https://doi.org/10.1016/j.foodchem.2009.03.053

Drahokoupil J (2023) Production chain. In: Britannica Money. https://www.britannica.com/money/production-chain. Accessed 26 Sep 2023

Faccio E, Guiotto Nai Fovino L (2019) Food neophobia or distrust of novelties? Exploring consumers' attitudes toward GMOs, insects and cultured meat. Appl Sci 9:4440. https://doi.org/10.3390/app9204440

Fachinello MR, Vital ACP, Chambo APS et al (2018) Effect of freeze-dried green tea added in hamburgers as source of antioxidant during freezing storage. J Food Process Preserv 42:e13780. https://doi.org/10.1111/jfpp.13780
Fasano J, Michael M (2023) Animal cell-culture food technology: a new regulatory frontier. Food Safety Magazine
Feiner G (2006) Meat products handbook: practical science and technology. CRC Press, Boca Raton
Frank D, Kaczmarska K, Paterson J et al (2017) Effect of marbling on volatile generation, oral breakdown and in mouth flavor release of grilled beef. Meat Sci 133:61–68 https://doi.org/10.1016/j.meatsci.2017.06.006
Galland FAB, Pacheco MTB (2022) Série tecnológica das proteínas alternativas: carne cultivada. Tiki Books: The Good Food Institute Brasil, São Paulo
Gao C, Lu C, Jian Z et al (2021) 3D bioprinting for fabricating artificial skin tissue. Colloids Surf B Biointerfaces 208:112041. https://doi.org/10.1016/j.colsurfb.2021.112041
Garcia EEC, Berto MI, Padula M et al (2023) Brasil Food Safety Trends 2030: transformações, tendências e desafios para a governança da e gestão da segurança dos alimentos, 1st edn. Instituto de Tecnologia de Alimentos - ITAL, Campinas
Good Food Institute (2023) Alternative protein company database. In: Alternative protein company database. https://gfi.org/resource/alternative-protein-company-database/. Accessed 25 Sep 2023
Grasso S (2020) Hybrid meat. Food Sci Technol 34:48–51. https://doi.org/10.1002/fsat.3403_12.x
Grasso S, Goksen G (2023) The best of both worlds? Challenges and opportunities in the development of hybrid meat products from the last 3 years. LWT 173:114235. https://doi.org/10.1016/j.lwt.2022.114235
Grasso S, Jaworska S (2020) Part meat and part plant: are hybrid meat products fad or future? Foods 9:1888. https://doi.org/10.3390/foods9121888
Grasso S, Rondoni A, Bari R et al (2022) Effect of information on consumers' sensory evaluation of beef, plant-based and hybrid beef burgers. Food Qual Prefer 96:104417. https://doi.org/10.1016/j.foodqual.2021.104417
Guan X, Lei Q, Yan Q et al (2021) Trends and ideas in technology, regulation and public acceptance of cultured meat. Future Foods 3:100032. https://doi.org/10.1016/j.fufo.2021.100032
Guo X, Cheng M, Zhang X et al (2017) Green tea polyphenols reduce obesity in high-fat diet-induced mice by modulating intestinal microbiota composition. Int J Food Sci Technol 52:1723–1730. https://doi.org/10.1111/ijfs.13479
Herrero AM, Carmona P, Jiménez-Colmenero F, Ruiz-Capillas C (2014) Polysaccharide gels as oil bulking agents: technological and structural properties. Food Hydrocoll 36:374–381. https://doi.org/10.1016/j.foodhyd.2013.08.008
Hoffman LC, Mellett FD (2003) Quality characteristics of low fat ostrich meat patties formulated with either pork lard or modified corn starch, soya isolate and water. Meat Sci 65:869–875. https://doi.org/10.1016/S0309-1740(02)00293-0
Hossain A, Yeo J, Dave D, Shahidi F (2022) Phenolic compounds and antioxidant capacity of sea cucumber (*Cucumaria frondosa*) processing discards as affected by high-pressure processing (HPP). Antioxidants 11:337. https://doi.org/10.3390/antiox11020337
Iida F, Saitou K, Kawamura T et al (2015) Effect of fat content on sensory characteristics of marbled beef from Japanese Black steers. Anim Sci J 86:707–715. https://doi.org/10.1111/asj.12342
Kantono K, Hamid N, Malavalli MM et al (2022) Consumer acceptance and production of *in vitro* meat: a review. Sustainability 14:4910. https://doi.org/10.3390/su14094910
Kao WT, Lin KW (2006) Quality of reduced-fat Frankfurter modified by Konjac-starch mixed gels. J Food Sci 71:S326–S332. https://doi.org/10.1111/j.1750-3841.2006.00003.x
Kieliszek M, Misiewicz A (2014) Microbial transglutaminase and its application in the food industry. A review. Folia Microbiol (Praha) 59:241–250. https://doi.org/10.1007/s12223-013-0287-x

Kim H-Y, Kim K-J, Lee J-W et al (2015) Quality evaluation of chicken nugget formulated with various contents of chicken skin and wheat fiber mixture. Korean J Food Sci Anim Resour 35:19–26. https://doi.org/10.5851/kosfa.2015.35.1.19

Koul B, Yakoob M, Shah MP (2022) Agricultural waste management strategies for environmental sustainability. Environ Res 206:112285. https://doi.org/10.1016/j.envres.2021.112285

Krawczyk G, Venables A, Tuason D (2009) Microcrystalline cellulose. In: Handbook of hydrocolloids. Elsevier, pp 740–759

Letti LAJ, Karp SG, Molento CFM et al (2021) Cultivated meat: recent technological developments, current market and future challenges. Biotechnol Res Innov 5:e2021001. https://doi.org/10.4322/biori.202101

Li S, Chen G, Zhang C et al (2014) Research progress of natural antioxidants in foods for the treatment of diseases. Food Sci Human Wellness 3:110–116. https://doi.org/10.1016/j.fshw.2014.11.002

Lin-Hi N, Schäfer K, Blumberg I, Hollands L (2022) The omnivore's paradox and consumer acceptance of cultured meat: an experimental investigation into the role of perceived organizational competence and excitement. J Clean Prod 338:130593. https://doi.org/10.1016/j.jclepro.2022.130593

Lodish HF, Berk A, Zipursky SL et al (2013) Molecular cell biology, 7th edn. H. Freeman and Company, New York

Makris DP, Boskou D (2015) Plant-derived antioxidants as food additives. In: Plants as a source of natural antioxidants. CABI, London, UK, pp 169–190

Marianski S, Marianski A (2010) Home production of quality meats and sausages. Bookmagic, Seminole

Morais-da-Silva RL, Villar EG, Reis GG et al (2022) The expected impact of cultivated and plant-based meats on jobs: the views of experts from Brazil, the United States and Europe. Humanit Soc Sci Commun 9:297. https://doi.org/10.1057/s41599-022-01316-z

Naveena BM, Vaithiyanathan S, Muthukumar M et al (2013) Relationship between the solubility, dosage and antioxidant capacity of carnosic acid in raw and cooked ground buffalo meat patties and chicken patties. Meat Sci 95:195–202. https://doi.org/10.1016/j.meatsci.2013.04.043

Ng S, Kurisawa M (2021) Integrating biomaterials and food biopolymers for cultured meat production. Acta Biomater 124:108–129. https://doi.org/10.1016/j.actbio.2021.01.017

Oh I, Lee J, Lee HG, Lee S (2019) Feasibility of hydroxypropyl methylcellulose oleogel as an animal fat replacer for meat patties. Food Res Int 122:566–572. https://doi.org/10.1016/j.foodres.2019.01.012

Ong KJ, Johnston J, Datar I et al (2021) Food safety considerations and research priorities for the cultured meat and seafood industry. Compr Rev Food Sci Food Saf 20:5421–5448. https://doi.org/10.1111/1541-4337.12853

Ramachandraiah K (2021) Potential development of sustainable 3D-printed meat analogues: a review. Sustainability 13:938. https://doi.org/10.3390/su13020938

Ramli ANM, Manap NWA, Bhuyar P, Azelee NIW (2020) Passion fruit (*Passiflora edulis*) peel powder extract and its application towards antibacterial and antioxidant activity on the preserved meat products. SN Appl Sci 2:1748. https://doi.org/10.1007/s42452-020-03550-z

Rey AI, Hopia A, Kivikari R, Kahkonen M (2005) Use of natural food/plant extracts: cloudberry (*Rubus Chamaemorus*), beetroot (*Beta vulgaris* "vulgaris") or willow herb (*Epilobium angustifolium*) to reduce lipid oxidation of cooked pork patties. LWT Food Sci Technol 38:363–370. https://doi.org/10.1016/j.lwt.2004.06.010

Rubio NR, Fish KD, Trimmer BA, Kaplan DL (2019) Possibilities for engineered insect tissue as a food source. Front Sustain Food Syst 3. https://doi.org/10.3389/fsufs.2019.00024

Saraiva BR, Agustinho BC, Vital ACP et al (2019) Effect of brewing waste (malt bagasse) addition on the physicochemical properties of hamburgers. J Food Process Preserv 43. https://doi.org/10.1111/jfpp.14135

Saraiva BR, Cegudo ACM, Gibin MS et al (2023) Co-product from debittering process of trub (brewing by-product) as natural antioxidant in processed cheese. Int J Food Sci Technol 58:6752. https://doi.org/10.1111/ijfs.16643

Seó ST, de Souza XR, Scabora MH et al (2024) Replacement of chicken skin with canola oil in chicken nuggets. Ciência Rural 54. https://doi.org/10.1590/0103-8478cr20220460

Siegrist M, Sütterlin B (2017) Importance of perceived naturalness for acceptance of food additives and cultured meat. Appetite 113:320–326. https://doi.org/10.1016/j.appet.2017.03.019

Silveira Coelho M, de Araujo AS, Machado Latorres J, de las Mercedes Salas-Mellado M (2019) In vitro and in vivo antioxidant capacity of chia protein hydrolysates and peptides. Food Hydrocoll 91:19–25. https://doi.org/10.1016/j.foodhyd.2019.01.018

Stout AJ, Mirliani AB, Soule-Albridge EL et al (2020) Engineering carotenoid production in mammalian cells for nutritionally enhanced cell-cultured foods. Metab Eng 62:126–137. https://doi.org/10.1016/j.ymben.2020.07.011

Sugii S, Wong CYQ, Lwin AKO, Chew LJM (2023) Alternative fat: redefining adipocytes for biomanufacturing cultivated meat. Trends Biotechnol 41:686–700. https://doi.org/10.1016/j.tibtech.2022.08.005

Tarté R (2009) Ingredients in meat products. Springer, New York

Vital ACP, Croge C, da Silva DF et al (2018a) Okara residue as source of antioxidants against lipid oxidation in milk enriched with omega-3 and bioavailability of bioactive compounds after in vitro gastrointestinal digestion. J Food Sci Technol 55:1518–1524. https://doi.org/10.1007/s13197-018-3069-2

Vital ACP, Santos NW, Matumoto-Pintro PT et al (2018b) Ice cream supplemented with grape juice residue as a source of antioxidants. Int J Dairy Technol 71:183–189. https://doi.org/10.1111/1471-0307.12412

Warner RD (2019) Review: analysis of the process and drivers for cellular meat production. Animal 13:3041–3058. https://doi.org/10.1017/S1751731119001897

Ye Y, Zhou J, Guan X, Sun X (2022) Commercialization of cultured meat products: current status, challenges, and strategic prospects. Future Foods 6:100177. https://doi.org/10.1016/j.fufo.2022.100177

Zang E, Jiang L, Cui H et al (2023) Only plant-based food additives: an overview on application, safety, and key challenges in the food industry. Food Rev Int 39:5132–5163. https://doi.org/10.1080/87559129.2022.2062764

Zetzl AK, Marangoni AG, Barbut S (2012) Mechanical properties of ethylcellulose oleogels and their potential for saturated fat reduction in frankfurters. Food Funct 3:327. https://doi.org/10.1039/c2fo10202a

# Packaging, Conservation, and Shelf Life of Cultivated Meat

**Débora Gonçalves Bortolini, Luane de Oliveira Maior, Gustavo Henrique Couto, Marjorie Benegra, Maria Giovana Binder Pagnoncelli, and Luiz Gustavo Lacerda**

**Abstract** The global demand for food is on the rise and undoubtedly poses a challenge for the coming decades, especially considering the finite resources of the world. Laboratory-grown meat, also referred to as cultivated or cell-based meat, presents an innovative and sustainable alternative to traditional meat production while addressing concerns regarding animal welfare. As this technology continues to advance and gain popularity, the packaging of cultured meat products becomes paramount in ensuring their safety, quality, and marketability. Cultured meat is inherently highly perishable due to its composition; therefore, protective measures are essential to minimize adverse effects during transportation and storage. Consequently, strategies must be devised to preserve its unique characteristics and safety, a necessity that certainly extends to lab-grown meat. Food packaging holds a central position in the contemporary food industry. While it plays a pivotal role in preserving the quality of food products, it can also raise concerns about food safety. This chapter explores significant aspects of packaging, preservation, and shelf life concerning cultured meat. Additionally, it delves into the challenges facing this burgeoning industrial sector.

---

D. G. Bortolini
Environmental Science and Technology Program, Department of Chemistry and Biology, Federal University of Technology - Paraná, Curitiba, Paraná, Brazil

L. de Oliveira Maior · L. G. Lacerda (✉)
Food Science and Technology Graduation Program, Department of Food Engineering, State University of Ponta Grossa, Ponta Grossa, Paraná, Brazil
e-mail: lglacerda@uepg.br

G. H. Couto · M. G. B. Pagnoncelli
Biotechnology Program, Department of Chemistry and Biology, Federal University of Technology - Paraná, Curitiba, Paraná, Brazil

M. Benegra
Biomedical Engineering Program, Department of Mechanical Engineering, Federal University of Technology - Paraná, Curitiba, PR, Brazil


C. R. Soccol et al. (eds.), *Cultivated Meat*,
https://doi.org/10.1007/978-3-031-55968-6_10

**Keywords** Cultivated meat preservation · Cultivated meat supply chain · Shelf-life extension

## 1 Introduction

The increase in meat consumption and production is poised to impact the global environment in the years to come, underscoring the need for more effective alternatives to conventional meat. While plant-based offerings hold significant potential as an alternative, they currently fall short of fully replicating the characteristics of meat, a challenge that many consumers find problematic (Escribano et al. 2021). In this regard, cultured meat presents a viable option as the emerging trend of lab-grown meat gains momentum as a potential substitute for traditional livestock in future global markets (Siddiqui et al. 2022). Figure 1 illustrates the entire process of cultivated meat from its source to the consumer.

The quality of cultured meat must be meticulously crafted to match that of regular or traditional meat. The consumption of lab-grown meat has become a tangible reality in certain nations. Cultured meat has the ability to closely approximate conventional meat in almost every aspect, including its visual appearance, aroma, texture, flavor, and, naturally, its nutritional profile (Cintoni et al. 2023). It is possible to consider that the challenges and opportunities associated with packaging cultivated meat are manifold. Preserving freshness is paramount, given the product's inherent perishability due to its unique cellular composition (Datar and Betti 2010).

Packaging must provide a protective barrier for foods, including meat and its related products, throughout their life cycle, shielding them against various

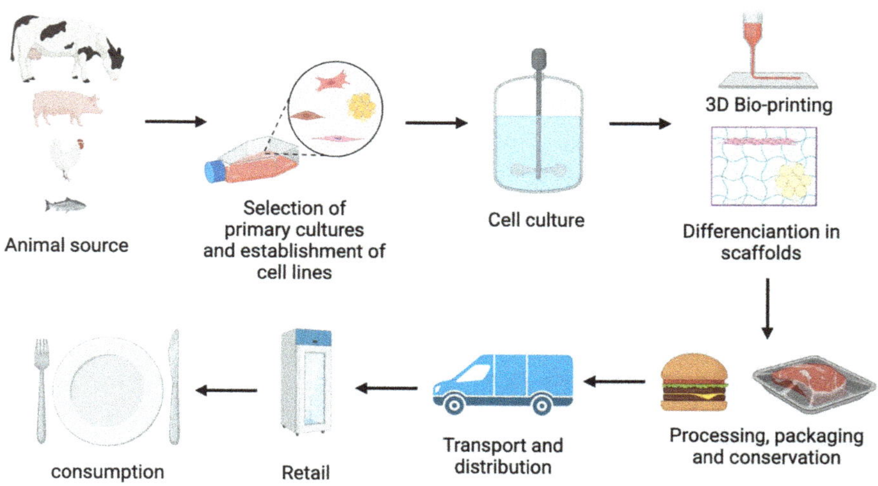


**Fig. 1** The supply chain of cultivated meat (Created with BioRender.com)

potential hazards such as physical harm, chemical exposure, and biological threats (Cenci-Goga et al. 2020). Furthermore, the use of modified atmosphere packaging (MAP) can create an environment characterized by reduced oxygen (O2). This condition serves to inhibit oxidation reactions, prevent microbial spoilage of food, minimize food shrinkage, and preserve sensory quality (Qu et al. 2022).

Consumer awareness and preferences are also shaping packaging strategies, and they are increasingly conscious of their food's origin and production methods, making clear and informative labeling a critical component (Bryant and Barnett 2018). Sustainability is another key consideration, with eco-friendly materials and waste reduction playing an integral role (Wilks et al. 2019).

Sustainable packaging aligns with the core motivation behind cultivated meat production: reducing the environmental impact of meat consumption (Tuomisto and Teixeira De Mattos 2011). Convenience remains a priority for consumers, driving the demand for user-friendly and resealable packaging options (Verbeke et al. 2015). Indeed, cultured meat products are not yet widely available in the market, and the extent of consumer acceptance for these products is still under investigation (Siddiqui et al. 2022). In this comprehensive exploration of cultivated meat packaging, this chapter delves into the multifaceted aspects shaping this critical industry component. From preserving freshness to enhancing sustainability and food safety, this examination provides insights into the challenges and innovations that will define the future of packaging within the cultivated meat sector.

## 2 Packaging of Cultivated Meat Products

Cultured meat offers an innovative approach with a reduced environmental impact and improved animal welfare. Packaging for cultured meat must be designed to encompass aspects related to meat preservation, shelf life, and consumer appeal, given the existing resistance to consuming such products (Siddiqui et al. 2022).

Currently, consumer acceptance represents a significant challenge for the large-scale adoption of this new technology (Siegrist and Hartmann 2020a, b). In this context, packaging characteristics play a vital role in shaping the consumer's perception of the product. In addition to ensuring safety, quality, and shelf life and maintaining the product's appearance, which are conventional requirements in traditional livestock farming, packaging for cultured meats requires additional elements to enhance its appeal. Regarding consumer acceptance, factors such as color, labeling, and nomenclature are significant elements that must be carefully considered in packaging development (Califano et al. 2023).

Furthermore, the packaging needs to ensure the product's integrity and be tamper-evident, guaranteeing quality and conveying safety to the consumer. Environmental considerations should also be taken into account, with the choice of sustainable materials and an attractive design that facilitates opening during consumption in compliance with regulations. Some technological processes can be employed to develop intelligent packaging that interacts actively with the product.

The following subtopics discuss how packaging and labeling influence consumer acceptance of cultured meat and the primary types of food packaging that can be used to preserve cultured meat.

## 2.1 *Influence of Cultured Meat Packaging on Consumer's Acceptance*

Color is one of the first attributes perceived by consumers. Studies have shown that specific colors evoke different emotions and associations (Muller et al. 2019). A color palette that conveys freshness and naturalness may be more attractive for cultured meat. However, foods marketed in predominantly green packaging tend to be perceived as natural and healthy (Dantec et al. 2022). In the case of cultured meat, the color red can be used to associate with beef, for example. Although cultured meat is of animal origin and not a plant-based substitute, there is still no consensus between using green, which refers to sustainability and healthiness, or red, which reinforces the similarity with traditional meat (Califano et al. 2023). On the other hand, transparent packaging can show consumers how cultivated meat is similar to traditional ones, increasing the credibility of these products and the likelihood of purchase (Guzman-Puyol et al. 2022). In any case, choosing the appropriate color helps establish a visual connection between the product and consumer expectations (Sucapane et al. 2021).

Consumers receive essential product information from the label. Providing clear and accurate information on the product's cultivation method, place of origin, and nutritional qualities is crucial. Additionally, sustainable and animal welfare-related labels and certifications can enhance consumer confidence in products made from cultured meat. The product name, contents, nutritional information, allergen warnings, and suitable storage and handling directions must be included on the packaging. However, it is equally crucial that the label is carefully planned to prevent overwhelming customers with excess or irrelevant information (Siddiqui et al. 2022).

The name chosen for cultured meat can directly impact consumer acceptance. Terms such as "cultured meat" or "clean meat" highlight the environmental and ethical benefits. In contrast, others such as "in vitro meat," "synthetic meat," "artificial meat," "lab meat," and "grown meat" can generate uncertainty or distrust (Bryant and Barnett 2020). In October 2019, federal legislation in the United States was enacted to restrict the widespread use of the term "meat" on package labels. Several countries have approved this restriction, mandating that all products that do not contain "real" meat must be labeled as "imitation" (Santo et al. 2020). However, there are differing opinions about the term "imitation." This designation is considered inaccurate by Silva and Semprebon (2021) since cell-based meats have equivalent levels of all essential elements of conventional meat.

While marketing appeal is vital for product acceptance, food safety must be considered foremost. The packaging material must not contaminate the meat or leave undesirable residues that compromise flavor and odor. Cultured meats are sterile, and maintaining this sterility is crucial, so packaging must prevent cross-contamination and serve as a microbial barrier (Furuhashi et al. 2021). The development of cultured meat packaging is based on its origin and properties. The packaging method depends on fat and muscle content, which influences the quality of cultured meat, including proteolysis and oxidation (Siddiqui et al. 2022). Thus, cultured meat packaging can be designed to prevent or reduce biochemical changes that lead to cultured deterioration. The following subtopic discusses the primary potential packaging for preserving cultured meat.

## 2.2 *Cultured Meat Packaging Technology*

Various types of packaging can be employed to mitigate significant deterioration in fresh meats. Several factors contribute to the degradation and reduction in shelf life of fresh meats, the primary factors being the presence of deteriorating microorganisms, lipid oxidation, endogenous enzymes, meat pH, temperature, humidity, and light. Properly implementing the cold chain and using suitable packaging can effectively retard the deterioration process.

Given that cultured meat shares the same physicochemical characteristics as traditional meat, the technologies employed in producing their packaging can also be analogous. Table 1 outlines the key types of packaging traditionally used for meats, which have potential applications in the context of cultured meats.

### 2.2.1 Modified Atmosphere Packaging (MAP)

Modified atmosphere packaging (MAP) involves altering the atmosphere within the package to a controlled combination of artificial gases, thus extending the shelf life and preserving the quality of meat products (Wang et al. 2022). The gas composition in a modified atmosphere plays a critical role in the preservation process. For instance, carbon dioxide inhibits the proliferation of aerobic bacteria, oxygen prevents the growth of anaerobic bacteria and maintains the appealing color of meat, while nitrogen is essential for delaying fat oxidation and prevents package collapse (Lei et al. 2023). Lai et al. (2023) observed that both MAP (78.8% $O_2$ + 18.8% $CO_2$ + 2.4% $N_2$) and $HiCO_2MAP$ (60% $CO_2$ + 40% $N_2$), when maintained at 4 °C, extended the shelf life and retained higher water content in beef steaks. However, HiCO2MAP exhibited higher oxidation levels and resulted in an unattractive color.

**Table 1** Potential packaging technologies for cultured meat

| Quality properties | Influencing factor | Proposed packaging material | Packaging type | Shelf life | Main results | References |
|---|---|---|---|---|---|---|
| Microbial growth, oxidative parameters | Oxidation, deterioration, and microbial growth | Kappa-carrageenan-poly(vinyl alcohol) electrospun fiber mats encapsulated with *Prunus domestica* anthocyanins and epigallocatechin gallate | Smart packaging/ antioxidant packaging | 15 days | Electrospun fibers rich in anthocyanins showed a color change when subjected to different pH buffer solutions, indicating alterations in minced beef. The packaging enriched with anthocyanins and epigallocatechin gallate showed antioxidant and antimicrobial activities, preserving the minced beef meat for 15 days | Goudarzi et al. (2023) |
| Oxidative parameters | Lipid profile and oxidative stability | Plastic bags of polyethylene (thickness 70 μm) | Vacuum packaging under longtime freezing | 360 days | Fallow deer (FD), wild boar (WB), and pig meat (PM) were vacuum-packed and freeze-stored for 360 days. The monitoring of meat quality was carried out through lipid oxidation levels. During meat storage, long-term vacuum packaging showed lower lipid oxidation. The duration of the storage period affected the lipid oxidation levels | Reitznerová et al. (2023) |
| Sensory parameters | Color | MAP (80% $O_2$+ 20% $CO_2$; 60% $O_2$ + 20% $CO_2$ + 20% $N_2$; 40% $O_2$ + 20% $CO_2$ + 40% $N_2$) | Modified atmosphere packaging (MAP) | 0–14 days | The study was conducted in DFD (dark, form, and dry) beef. DFD beef ($n$ = 180) was previously treated with high hydrostatic pressure (400 MPa). After pressuring, beef was removed from vacuum packages and repacked in modified atmosphere packaging. HPP (400 MPa) combined with MAP improved DFD beef color parameters and delayed discoloration. Total plate counts for DFD beef stakes placed in 60% $O_2$-MAP showed ≤6 $\log_{10}$ CFU/g after 14 days of storage | Mao et al. (2023) |

| Quality properties | Influencing factor | Proposed packaging material | Packaging type | Shelf life | Main results | References |
|---|---|---|---|---|---|---|
| Microbial growth | Bacteria growth | Pad was manufactured using polyurethane (PU) nanofibers as the top hydrophobic layer, added with polyhexamethylene biguanide hydrochloride (PHMB) as a cationic antimicrobial agent. Polyvinyl alcohol (PVA) nanofibers were used to build the bottom absorbent layer, added with anthocyanin as a pH indicator | Antimicrobial pad packaging | 7 days | PHMB showed antimicrobial properties in PU-PVA pads because of its capacity to keep the antibacterial ability during meat spoilage that increases pH. The juice of the meat was successfully absorbed into the bottom layer. Thus, the top layer remained dry, reducing the multiplication of bacteria. Anthocyanins added to the bottom layer worked as a pH indicator. The new antibacterial indicator absorbent pad successfully inhibited bacterial growth, prolonging the shelf life of fresh meat | Jiao et al. (2023) |
| Sensory parameters and microbial growth | Color and deteriorative bacteria growth | Chitosan edible films added with flavonoids (catechin, quercetin, and luteolin) | Antioxidant edible films | 14 days | Chitosan-flavonoid edible films inhibited the growth of pathogenic strains (*L. monocytogenes*, *S. typhimurium*, *E. coli*, and *S. aureus*). During the storage, the color of the beef was maintained, and the aerobic plate count remained low. Therefore, chitosan composite films prepared with flavonoids could be used to improve beef's shelf life and color quality without adding artificial additives | Sutharsan et al. (2023) |

### 2.2.2 Vacuum Packaging

Vacuum packaging is a technological process that removes air from the packaging, creating a vacuum environment that significantly enhances meat preservation. The preservation effects of vacuum packaging are primarily attributed to the removal of oxygen, thus preventing spoilage, fat oxidation, and microbial growth on the meat (Zhao et al. 2022). Rovira et al. (2023) compared two materials with different oxygen permeabilities, one with low $O_2$ permeability and another with high $O_2$ permeability and antimicrobial properties, used for vacuum packaging of beef meat under refrigeration (0–1.15 °C; 120 days) and refrigeration followed by freezing storage (28 days at 0–1.5 °C and 92 days at −20 °C). The results suggested that meat subjected to vacuum packaging and treated with refrigeration before freezing exhibited better microbiological indicators. Thus, vacuum packaging is an excellent alternative for reducing microbial counts in meat products.

### 2.2.3 Edible Film Packaging

Edible films are made from Generally Recognized As Safe (GRAS) materials, forming thin and flexible sheets or coatings for food protection. Edible films can be based on carbohydrates, proteins, lipids, and waxes and can be incorporated with additives, antioxidants, antimicrobials, and flavor compounds (Gaspar and Braga 2023). In the realm of meat products, edible films are a growing trend. Xiang et al. (2022) assessed various potential biomaterials for manufacturing edible films, with a focus on their application in cultured meat. The authors concluded that protein-based films, particularly glutenin and zein, show promise as biomaterials due to their characteristics related to proliferation, adherence, differentiation, mechanisms, and material availability. Edible films incorporated with bioactive compounds can give rise to active packages with antioxidant properties (Sutharsan et al. 2023), antimicrobial properties, and smart functions (Manzoor et al. 2023), as demonstrated.

### 2.2.4 Antioxidant Packaging

Bioactive compounds, especially phenolic compounds, exhibit antioxidant properties owing to the presence of one or more hydroxyl radicals attached to an aromatic ring, capable of donating either an atom of hydrogen or a single electron, thereby curbing free radical reactions (Yan et al. 2020). Excessive free radicals in the human body can lead to various illnesses, such as cardiovascular diseases, cancer, and metabolic disorders (Dresen et al. 2023). Phenolic compounds are found in almost all vegetables and their by-products (Benvenutti et al. 2019; Rossetto et al. 2020; Maroldi et al. 2022). Regular consumption of bioactive compounds from natural sources can potentially offer health benefits, reducing the risk of developing such conditions (Rathod et al. 2023). Incorporating bioactive compounds into food packaging can have technological benefits. Antioxidant compounds can delay the oxidation process of meats, thereby extending their shelf life. Sutharsan et al. (2023) coated beef with chitosan-flavonoid edible films and found that the color of the beef was maintained, and the aerobic plate count remained low during storage.

Moreover, the edible film on the meat inhibited the growth of pathogenic bacteria strains, including *L. monocytogenes*, *S. typhimurium*, *E. coli*, and *S. aureus*. Thus, chitosan-based edible films demonstrate the potential to preserve beef meat, with the incorporation of flavonoids having a positive effect against bacteria strains. Another technological property of bioactive compounds in food packaging is the creation of smart packages (Zainal Arifin et al. 2023; Jiang et al. 2023). These compounds can also exhibit antimicrobial properties, as discussed later.

### 2.2.5 Smart Packaging

Anthocyanins, which are phenolic compounds belonging to the flavonoid class, are water-soluble pigments that demonstrate pH-sensitive behavior. Consequently, their inclusion in food packaging can indicate pH changes, giving rise to smart packaging. Jiang et al. (2023) developed a film using sweet potato starch, κ-carrageenan, and *Oxalis triangularis* extract (a plant rich in anthocyanins). The product exhibited different colors in response to varying pH environments. Hence, this material was applied to fresh beef meat, showing significant color changes during storage. Thus, this biomaterial can be utilized as smart packaging for meats.

### 2.2.6 Antimicrobial Packaging

As previously discussed, some microorganisms are sensitive to bioactive compounds. Phenolic compounds can possess antimicrobial properties, and their incorporation into food packaging can inhibit bacterial growth, thereby improving shelf life. Jiao et al. (2023) produced a pad using polyurethane (PU) nanofibers as the top hydrophobic layer, augmented with polyhexamethylene biguanide hydrochloride (PHMB) as a cationic antimicrobial agent. Polyvinyl alcohol (PVA) nanofibers were used to create the bottom absorbent layer, supplemented with anthocyanin as a pH indicator. PHMB serves as an antimicrobial agent in PU-PVA pads because it maintains antibacterial efficacy during meat spoilage, which leads to an increase in pH. The meat juice is absorbed into the bottom layer, leaving the top layer dry and reducing the multiplication of bacteria. Anthocyanins incorporated into the bottom layer serve as a pH indicator. This innovative antibacterial indicator absorbent pad successfully inhibits bacterial growth, extending the shelf life of fresh meat.

As explored in this section, proper meat packaging is vital for preserving cultured meat. Nonetheless, meats, in general, are highly susceptible to spoilage. Therefore, it may be necessary to combine other preservation methods with meat packaging to achieve better results. The following section describes the main potential preservation methods for cultured meat.

# 3 Conservation and Shelf Life of Cultivated Meat Products

The application of technologies in the production of cultured meat has led to products that closely resemble traditional meat in terms of composition, texture, and appearance. However, cultured meats are produced within sterile environments to prevent contamination by microorganisms during cell and tissue growth. Nonetheless, it is imperative to meticulously develop other unitary operations involved in producing these foods to prevent contamination of cultivated meat. Despite the increased attention to cultured meat production and consumption, there is limited existing literature, and specific research on the preservation of cultured meat remains scarce. Consequently, conservation techniques employed to maintain the freshness of cultured meat align with those used in traditional meat preservation (Siddiqui et al. 2022). Figure 2 illustrates various preservation techniques that can be listed to preserve cultured meat.

Meat comprises essential amino acids, fatty acids, vitamins, and minerals. These nutrients, coupled with high moisture content and a neutral pH, render meat susceptible to microbial growth. Microorganisms play a pivotal role in meat deterioration and the occurrence of foodborne diseases (Ren et al. 2022). Hence, it is essential to shield fresh meat with appropriate packaging barriers and to store it under refrigeration or in a frozen state (Nastasijević et al. 2017). Therefore, packaging and the maintenance of the cold chain, acknowledged as the ultimate defensive strategies against harmful agents (Nethra et al. 2023), are pivotal factors that require strict control to ensure the preservation and extension of the shelf life of cultured meat. Furthermore, the implementation of quality management systems is essential for ensuring the safe production of cultured meat (Siddiqui et al. 2022).

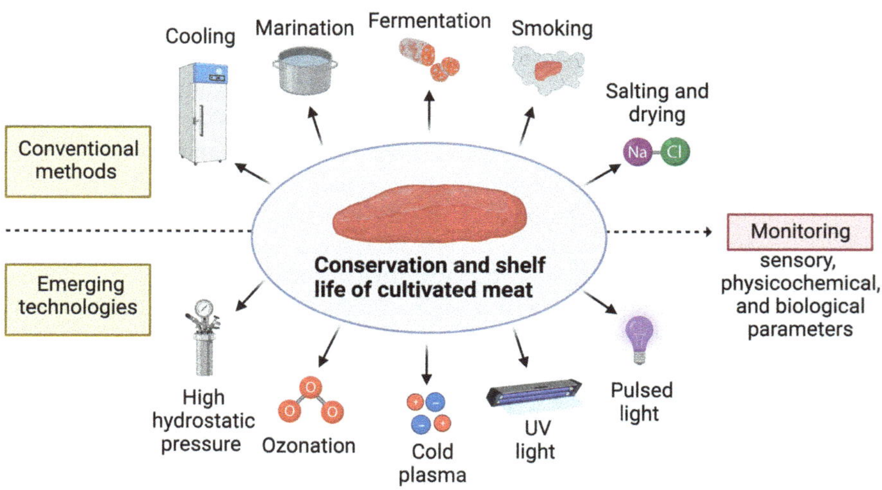


**Fig. 2** Conventional and emerging methods of cultivated meat conservation (Created with BioRender.com)

The shelf life of meat can be monitored through the evaluation of sensory, physicochemical, and biological parameters. Key sensory indicators include color, tenderness, juiciness, and flavor, which are assessed by trained panelists and subjected to statistical analysis. Physicochemical indicators comprise parameters such as water-holding capacity, pH, shear force, TVB-N content, and protein content, which must align with predefined ranges. Finally, biological indicators can be evaluated through the total bacteria count and the presence of pathogenic microorganisms (Ren et al. 2022).

As previously discussed, a diverse range of food packaging options exists, varying in terms of materials used, technological properties (such as microorganism inhibition, reduction of oxidation rates, maintenance of water activity, and sensory properties), and their desirable effects on meat products, including the preservation of color and moisture (Gupta and Dudeja 2017). Consequently, various packaging methods have been widely employed for the preservation of fresh meat, including modified atmosphere packaging (Mao et al. 2023), vacuum packaging (Reitznerová et al. 2023), antioxidant packaging, antimicrobial packaging, pad packaging (Jiao et al. 2023), and edible films (Goudarzi et al. 2023). It's worth noting that these packaging techniques hold potential for conserving fresh cultured meat as well (Siddiqui et al. 2022).

In addition to suitable packaging, the processing and storage temperature of fresh meat are crucial factors in extending its shelf life. The cold chain encompasses multiple phases of handling, commencing with the chilling of carcasses following slaughter, followed by the transportation of meat to distribution points, storage in retail establishments, the management and transportation of meat cuts within these establishments, delivery to local markets, placement on store shelves, and, ultimately, refrigeration at consumers' homes, managed by consumers themselves (Nethra et al. 2023). Due to temperature fluctuations that may occur during these phases, physicochemical and microbiological changes can take place, potentially reducing the shelf life of meat products. Maintaining temperatures within the range of 4–12 °C can effectively curtail microbial growth and enzymatic activity, as these activities tend to cease at freezing temperatures (−18 °C). However, it's important to note that some microorganisms may rebound after the meat is defrosted, so it is not advisable to consume meat that has been exposed to room temperature for more than 2 h (Nethra et al. 2023).

The combination of cooling and appropriate packaging constitutes concurrent methods for meat preservation. These methods can be effectively applied to both fresh and processed meat. While it's true that these methods have not yet been extensively studied for cultured meat, the findings from existing literature encourage the belief that they may also be effective in extending the shelf life of cultured meat. Cultivated meat, as a relatively new technology, is currently treated as a part of fresh meat production. However, meat preservation technology extends far beyond these conventional methods. As a result, one can envision a future where cultured meat undergoes traditional preservation processes and is used as a raw material in the production of various processed meat products. This could lead to the creation of different products, such as marinated beef (Osaili et al. 2023), smoked beef products (Borges et al. 2023), cured beef (Zhu et al. 2023; He et al. 2023), fermented beef (Yu et al. 2020), and more.

Fermentation represents one of the oldest technologies employed for meat preservation. Traditionally, meat fermentation occurs spontaneously due to the growth of natural microbiota. However, in industrial settings, meat products are inoculated with a starter culture to facilitate controlled fermentation. The process begins with the conversion of sugars into lactic acid, accompanied by the migration of water from the center to the surface of the meat, aiding in evaporation. As sugars are consumed, other microorganisms like yeasts, molds, and staphylococci come into play, feeding on alternative energy sources and producing the characteristic aromas and flavors associated with fermented meat products (Toldrá and Hui 2007).

Smoking is another preservation technique that involves exposing meat to compounds resulting from pyrolysis. Hot smoking entails hanging meat over an open flame, which serves to reduce water activity, expose the product to antioxidant and antimicrobial agents found in the smoke, and enhance product preservation. However, it's important to note that hot smoking can lead to the formation of polycyclic aromatic hydrocarbons (PAHs), which may have adverse health effects. To mitigate the formation of PAHs, cold smoking (at temperatures of 20–25 °C) and the use of liquid smoke, a concentrated liquid containing key compounds associated with the smoked flavor, have been widely adopted as alternatives for flavoring meats (Xin et al. 2021).

Marination is a process that involves submerging raw meat in a marinade composed of sugar, salts, and phosphates to enhance its texture and flavor. During marination, the marinade permeates the interior of the meat piece until reaching a dynamic equilibrium (Shi et al. 2023). In addition to its impact on sensory characteristics, marinades can influence meat preservation due to their low pH and the presence of potential antimicrobial ingredients such as salt, herbs, and spices. For example, Zawani et al. (2022) examined the inhibitory effect of cinnamon extract on bacteria from the genus *Pseudomonas* and the quality of marinated meat stored at different temperatures. Their study revealed that cinnamon extract exhibited antimicrobial activity, with the pure extract achieving a bactericidal concentration of 50% and an inhibitory concentration of 25%. Meat marinated with 50% cinnamon extract and stored at temperatures of 5, 10, 15, and 25 °C for 48 h experienced a reduction in *Pseudomonas* counts. However, the effectiveness of marinated meat preservation was influenced by temperature, with better results observed at 5 and 10 °C. Comparing the marinated meat with the control sample, significant sensory differences were noted in terms of color, texture, and overall acceptance. Consequently, the use of cinnamon extract was effective in reducing *Pseudomonas* counts and extending the shelf life of marinated meat when stored between 5 and 10 °C.

The salting method, often combined with drying, serves as a primary preservation process for cured beef. Sodium nitrite is a widely utilized ingredient to prevent the growth of *Clostridium botulinum*, reduce oxidation, and enhance the sensory properties of meat products (Mirzazadeh et al. 2021). Essentially, nitric oxide reacts with myoglobin, a naturally occurring pigment in meat, to form nitrosomyoglobin, which, when heated, produces nitrosylhemochrome—a compound responsible for imparting color to cured meats. However, this reaction can also lead to the formation of N-nitrosamines, which are potentially carcinogenic and mutagenic (Wójciak et al. 2019).

The process of curing meat gives rise to various regional meat products, such as cecina, a dry-cured beef from Spain (Revilla et al. 2020); Yunnan dry-cured beef from China (Yu et al. 2020); and carne-de-sol, charque, and jerked beef, traditional salted and dry-cured meats from Brazil (Freixo et al. 2022). During the curing process, meat undergoes biochemical reactions including oxidation, proteolysis, and lipolysis, which result in the production of volatile compound metabolites, as well as amino acids, free fatty acids, creatine, and nucleotides (Tian et al. 2020; Freixo et al. 2022; Martín-Gómez et al. 2022).

These transformations ultimately impact the sensory characteristics of the product (Fu et al. 2022). While salt is a crucial component in the manufacture of cured meats, studies have been conducted to replace a portion of sodium chloride with potassium chloride, L-lysine, and L-histidine in an effort to reduce sodium concentration, which can be detrimental to human health when consumed in excess (Liu et al. 2019a, b, 2020). Additionally, research has been focused on minimizing the formation of nitrosamines (Mirzazadeh et al. 2021). These conventional meat preservation processes, as mentioned above, effectively extend the shelf life of meat. However, they also introduce sensory changes to the raw products, resulting in alterations in flavor, texture, and color. Emerging technologies provide an alternative means of conserving meats while minimizing sensory changes. Notably, these new technologies offer potential applications in the preservation of cultured meat. Within this context, emerging thermal technologies such as pulsed light (Duma-Kocan et al. 2023), ultraviolet light (Soro et al. 2022), and nonthermal technologies like high hydrostatic pressure (Zhang et al. 2023), ozonation (Khanashyam et al. 2022), and the application of cold plasma (Abdel-Naeem et al. 2022) have been prominently employed in meat preservation. Further examples of nonconventional technologies for meat preservation are provided in Table 2.

High hydrostatic pressure (HHP) technology entails submerging appropriately packaged food underwater and exposing it to pressures ranging from 100 to 1000 MPa at temperatures between 0 and 120 °C. Short exposure times (<5 min) can achieve results similar to commercial pasteurization, while temperatures exceeding 100 °C are necessary to deactivate spores. HHP technology is applicable to meat, among other food products, and has been successfully used to preserve fruits, vegetables, and beverages (Aganovic et al. 2021). HHP is considered a nonthermal method for meat preservation (Roobab et al. 2022) and can be employed to minimize sensory alterations in meats. However, sensory attributes, particularly color and texture profiles, are the most sensitive parameters, as observed in the research presented in Table 2.

Cap et al. (2020) investigated the effects of HHP on frozen chicken breasts, focusing on *Salmonella* spp. inactivation and sensory quality. The researchers confirmed that salmonella bacteria could be inactivated by subjecting the chicken breasts to 500 MPa for 1 min or 400 MPa for 5 min. They also observed a reduction of 2 log CFU/g of spoilage microorganisms. Chromatic analysis and texture profiles

**Table 2** Trends and emerging technologies for meat preservation

| Conservation methods | Meat product | Main results | References |
|---|---|---|---|
| High hydrostatic pressure (HHP) | Frozen chicken breast | The authors evaluated the inactivation of *salmonella* spp. through applying HHP and its effects on the sensory quality of frozen chicken breast meat. Treatments with 500 MPa/1 min and 400 MPa/5 min were sufficient to inactivate bacteria of the *Salmonella* genus. Furthermore, for spoilage microorganisms, a reduction of 2 log CFU/g was observed, which indicates a possible improvement in the conservation of the product. The application of HHP influenced the chromatic parameters and texture profile. The treatment with 500 MPa/1 min showed fewer sensory changes, indicating that such characteristics are better preserved in shorter exposure to high pressures | Cap et al. (2020) |
| High hydrostatic pressure (HHP) | Marinated ground chicken breast | HHP (200–600 MPa /5–15 min) was applied on ground chicken breast. A decrease in texture springiness parameter and L* value was detected. However, aroma, aftertaste, and hardness were not modified. The 600 MPa/15 min application provided inactivation on the initial total mesophilic aerobic bacteria and total mold and yeast counts. The initial *Salmonella* and total coliform counts were reduced to undetectable levels by 400 MPa/15 min and 600 MPa for 5, 10, and 15 min, respectively. Thus, HHP can be a nonthermal alternative for meat conservation | Evrendilek (2022) |
| Pulsed light | Poultry | The researchers investigated the use of pulsed light for the control of *campylobacter* on poultry. Inoculated and control samples were submitted to five pulses/s of 300 ms, 1 Hz, and 1 $J/cm^2$. Pulsed light treatment significantly reduced the presence of *Enterobacteriaceae* in poultry meat. However, the applied treatments did not lower *Campylobacter* counts by one log per gram of poultry meat. Additionally, the pulsed light-treated poultry meat exhibited a slightly lighter, redder, and more yellowish appearance than untreated meat. Pulsed light treatment effectively decreased the proportion of aldehydes in the overall volatile compounds found in the meat. Thus, pulsed light can be used for microorganisms' growth control in poultry; however, higher energy doses of pulsed light might be tested to improve poultry conservation | Baptista et al. (2022) |

**Table 2** (continued)

| Conservation methods | Meat product | Main results | References |
|---|---|---|---|
| Cold plasma application | Chicken breast fillets | The authors evaluated the decontamination efficiency by cold plasma dielectric barrier discharge (DBD) at low voltage (20 kV) in chicken breast fillets. Chicken meat was exposed to oxygen and argon for 3 and 5 min in each gas. Samples treated with plasma revealed reduced bacterial count, shear force, and pH and showed no changes in thiobarbituric acid, volatile base nitrogen, and cooking loss. Sensorially, the attributes were better, except for the color and appearance of the meat treated with oxygen. Treatment with DBD plasma, especially with argon gas at 20 kV, for 3 min, proved to be an efficient decontamination method that can be safely applied in poultry processing plants without any adverse effects on the quality of chicken meat, with subsequent improvement in tenderness versus minimal destructive changes observed in oxygen-treated meat | Abdel-Naeem et al. (2022) |
| Ultraviolet light | Diced chicken breast fillet | Diced chicken breast fillets were exposed for 2, 4, 6, 8, and 10 min to UV light at 255, 280, 300, and 365 nm. Sample groups treated with wavelengths of 280, 300, and 365 nm had reductions ranging from 1.17 to 1.67 log CFU/g for total viable counts of total *Enterobacteriaceae,* mesophilic, and psychrophilic bacteria counts were determined. At the same time, *pseudomonas* and lactic acid bacteria showed reductions up to 2 log CFU/g. the 280 nm UV light treatment did not modify color, texture, and pH. Therefore, LED-based technologies can be applied to meat conservation technology without product changes | Soro et al. (2021) |
| Ozonation | Beef meat | This work evaluated the effect of treatment with ozonized water on psychrotolerant pathogenic bacteria on the surface of raw beef. The investigation was conducted at 1, 2, and 3 days and at 5, 15, and 20 °C temperatures. This study observed that an $O_3$ concentration of 0.5 ppm effectively reduced bacterial counts in meat blocks. However, increasing the time to 45 min allowed a 3 $\log_{10}$ CFU/mL decline. This way, the study shows that ozonated water can reduce pathogenic microorganisms in raw beef | Ali et al. (2022) |

revealed that frozen chicken breasts treated with 500 MPa for 1 min experienced fewer changes. This study demonstrates that higher pressures and shorter durations are more effective for chicken decontamination while preserving sensory aspects. Another study on chicken breasts was conducted by Evrendilek (2022), where HHP ranging from 200 to 600 MPa was applied for 5–15 min. According to the author, applying 600 MPa for 15 min effectively inactivated mesophilic aerobic bacteria, total mold, and yeast counts. Salmonella was undetectable after treatment with 400 MPa for 15 min, while 600 MPa for 5 min achieved the same effect. Sensory changes were observed in terms of texture springiness and color (lightness, L* value), but parameters such as aftertaste, aroma, and hardness were unaffected by HHP. Thus, HHP can be considered an alternative to traditional thermal processing for meats.

In addition to HHP, other nonthermal methods can be applied to meat preservation. Within this context, ozonation and cold plasma have been explored as promising alternatives (Roobab et al. 2022). Ozonation in food safety involves the application of ozone to surfaces and/or food products. Ozone, a triatomic molecule composed of oxygen, is generated artificially, possessing two unpaired electrons in its valence shell, making it highly reactive and unstable. Ozone is a potent oxidizing agent, surpassing chlorine and oxygen in terms of reactivity. Consequently, it exhibits antimicrobial properties suitable for food preservation (Adhikary et al. 2023). Ali et al. (2022) evaluated the effectiveness of ozone treatment in decontaminating raw beef blocks. They monitored psychrotolerant pathogenic bacteria in ozonized raw beef stored at temperatures of 5, 15, and 20 °C for 1, 2, and 3 days. The results showed that the addition of 0.5 ppm of $O_3$ in water significantly reduced bacterial counts in raw beef. A 45 min exposure time resulted in a 3 log10 CFU/mL reduction. Thus, this study demonstrates that ozone can effectively reduce pathogenic microorganisms in raw beef blocks.

Cold plasma represents an economical and environmentally friendly method for food preservation. In brief, plasma is considered the fourth state of matter, resulting from the introduction of additional energy into gases, leading to the disruption of all molecular interactions. Consequently, both negative and positive ions are released, and molecules and atoms become ionized, forming plasma. Plasma exhibits antimicrobial properties and can be used for food preservation (Laroque et al. 2022). Abdel-Naeem et al. (2022) applied cold plasma dielectric barrier discharge using oxygen and argon at 20 kV in chicken breast fillets. The treatment, especially with argon gas for 3 min, was an efficient decontamination method that could be safely applied to chicken meats without any adverse effects on meat quality, leading to improved tenderness. Chicken breasts exhibited minimal changes after oxygen treatment. In general, samples treated with plasma showed reduced bacterial counts, shear force, and pH, while no significant changes were observed in terms of thiobarbituric acid, volatile base nitrogen, or cooking loss.

Ultraviolet (UV) light is electromagnetic radiation with wavelengths between X-rays and the visible light spectrum, ranging from 100 nm to 400 nm. UV irradiation possesses strong antimicrobial properties and occurs naturally in sunlight. However, controlled UV light is generated using specific lamps in industrial and laboratory settings (Wang et al. 2023). Soro et al. (2021) exposed diced chicken breast fillets to UV light at wavelengths of 255, 280, 300, and 365 nm for 2, 4, 6, and 8 min. Chicken breasts treated with 280, 300, and 365 nm UV light exhibited reductions in psychrophilic bacteria, total *Enterobacteriaceae*, and mesophilic counts ranging from 1.17 to 1.67 log CFU/g. In particular, reductions of *Pseudomonas* and lactic acid bacteria were greater, reaching 2 log CFU/g. UV light at 280 nm did not induce changes in color, pH, or texture. Therefore, UV light can be considered a preservation technique suitable for chicken breast meats.

Pulsed light involves the application of very short pulses (1 μs to 0.1 s) covering a broad spectrum from ultraviolet to near-infrared (100–1100 nm) for a brief duration. These pulses act through photochemical and photothermal mechanisms, allowing for the inactivation of microorganisms in food matrices (Duma-Kocan et al. 2023). Baptista et al. (2022) evaluated 5 pulses for 1 min, delivered in 300 ms pulses at a rate of 1 Hz and a dose of 1 J/cm$^2$ on poultry. Pulsed light reduced *Enterobacteriaceae* counts but was not adequate for *Campylobacter* counts in the tested conditions, resulting in reductions of less than 1 log/g in poultry meat. Some sensory parameters, such as lightness, redness, and yellowness, were modified by pulsed light. Additionally, the proportion of aldehydes in volatile compounds decreased after pulsed light treatment. Therefore, pulsed light has the potential to be applied to poultry meat for microorganism reduction; however, the energy and doses need to be optimized for better results.

As demonstrated, various food preservation techniques are commonly used for traditional meats. Many of these methods can be adapted for the preservation of cultured meats, often in combination. For example, traditional preservation methods, such as salting or curing, followed by drying, can be combined with emerging technologies, followed by cooling. The choice of preservation methods depends on factors such as the type of meat, desired product characteristics, and the availability of technological resources. Preservation strategies have been applied to fresh, cured, and dried meats. It is conceivable that in the future, the processing of cultured meats may expand, leading to their incorporation into the production of other meat-based products. In conclusion, this topic has provided an overview of preserving fresh cultured meats and explored emerging trends and technologies that could be integrated into this process.

## 4 Challenges and Future Developments

Relying solely on the traditional method of meat production has become unfeasible and a hindrance in the global context of sustainability, environmental degradation, and natural resource utilization. The population projection for 2050 is approximately 10 billion, according to data from the United Nations (FAO) in partnership with the Organization for Economic Cooperation and Development (OECD) (OECD and FAO 2020). The increase in food demand is a consequence of the factors mentioned and can lead to gaps in the supply of animal-derived foods. Food production generates, on average, 26% of the greenhouse gases emitted into the atmosphere (Ritchie et al. 2020). Furthermore, many zoonotic viral diseases, such as the global outbreak of the SARS-CoV-2 virus that began in mid-2020, can be linked to the consumption and use of traditional livestock (Jelinek et al. 2021). This underscores the need to move away from traditional methods based on animal slaughter and reduce the number of animals used in food production to prevent future outbreaks of zoonotic diseases (Pajčin et al. 2022).

Therefore, combating what is consumed and how it is produced is crucial for slowing down climate change, reducing hybrid stress, aiding in the recovery of natural conservation areas, and protecting animal life. Thus, meats produced through cultivated cell production and reproduction methods, which can also be known as cultured meat, clean meat, synthetic meat, lab-grown meat, or in vitro meat, represent promising and viable alternatives to expand the supply of meat-based foods (Broucke et al. 2023).

Cultivated cell-based meat and cultured seafood and poultry are considered a trend in research and investment, both in the private sector and in government research institutions (Pajčin et al. 2022). Several companies, primarily startups, are allocating resources to research and develop clean meat-derived products, including companies like Aleph Farms, Mosa Meat, and Upside Foods (Broucke et al. 2023). The development of final products with lab-grown meat is still in its early stages. In this context, Eat Just has been selling cultivated chicken meat under the GOOD Meat brand since 2021, as a limited series in Singapore (Pajčin et al. 2022). However, no commercially manufactured and widely available product on the global market originates from cultivated meat.

Several aspects need to be meticulously observed and investigated before the effective implementation of in vitro meat. Nutritional value, food safety, sensory properties, industrial-scale production, process cost, and ethical safety are factors that require consideration until the development of a final product that resembles traditional meat (Chen et al. 2022). An incomplete understanding of health-related benefits or food safety risks is also a challenge and limitation in commercial implementation (Baum et al. 2023). The primary factor to consider for the commercialization of cultivated meat, and perhaps the main challenge to overcome, is consumer acceptance. Studies conducted in various countries and with various types of perceptions can attempt to elucidate the acceptability and rejection of in vitro animal-derived products, as Table 3 demonstrates (Baum et al. 2021; Mancini and Antonioli 2022; Fidder and Graça 2023).

As reported, consumer acceptability of clean meat is greatly influenced by factors extrinsic to the product, such as information about the subject, willingness to explore new food technology, neophobia, dietary habits, propensity to substitute conventional meat, familiarity with the concept of animal welfare, and the relationship between generations and acceptance of cultivated meat (Szejda et al. 2021). However, intrinsic factors related to the new concept of meat product consumption, such as taste, similarity to conventional products, and food safety, can influence consumer perception and subsequent acceptance of cultivated meat. Beyond these

**Table 3** Consumer perception of cultivated meat consumption

| Country | Number of individuals | Consumer perception | Considerations | References |
|---|---|---|---|---|
| Germany | 727 | Around 60% are willing to purchase cultivated meat | Neophobia plays an essential role in the level of acceptance | Baum et al. (2021) |
| United Kingdom | 1094 | 4.60 for purchase intention and 5.11 for willingness to try consumption (using a 7-point Likert scale) | Aligning the representation of cultivated meat with conventional consumption practices promotes more favorable evaluations, while emphasizing innovative characteristics has a negative effect. Consumers emphasize the importance of taste at the time of purchase | Fidder and Graça (2023) |
| Italy | 525 | 54% willing to try | Potential consumers are under 25 years old. A strong presence of Mediterranean cuisine contributes to lower acceptance. Informed populations are more likely to accept | Mancini and Antonioli (2022) |
| China | 4666 | 49.7% willing to try consumption; 3% willing to pay more for cultivated meat than conventional meat | Over half of the respondents believe cultivated meat will be less healthy and unsafe. A significant portion of those surveyed promotes replacing conventional meat in their regular diet, which may have contributed to the high level of acceptance | Liu et al. (2021) |
| EUA e UK | 4052 e 2034 | 40% very likely to try and consume regularly | A difference in acceptance regarding the generation of potential consumers has been observed, with generation Z being more inclined to accept, while earlier generations demonstrate more excellent resistance | Szejda et al. (2021) |
| Brazil | 304 | 41.1% stated that they had never heard of cultivated meat | Acceptance of cultivated meat was influenced by factors related to the product (food safety and naturalness), environmental/ ethical benefits (clean production and animal welfare), and psychological factors (neophobia) | Quevedo-Silva and Pereira (2022) |

factors, many challenging aspects can be evaluated in developing commercially cultivated meat products.

Due to the low popularity among potential consumers, there is an even more significant challenge when it comes to marketing related to the product. There is a need to emphasize the benefits associated with the marketing of cultured meat, such as the reduction of greenhouse gas emissions in the atmosphere and consumption of natural resources, water, and energy, which categorize aspects affected by conventional meat production (Siddiqui et al. 2022). Furthermore, cell-based cultured meat products support the United Nations Sustainable Development Program, part of the Agenda 2030, which enables, encourages, and mobilizes actions aimed at environmentally sustainable growth (United Nations 2015). Therefore, consuming cultured meat represents a significant contribution to achieving the goals established by the United Nations (UN).

## 5 Conclusion

The packaging of cultured meat is a multifaceted endeavor encompassing several crucial aspects. These include preserving freshness through airtight seals and modified atmospheres, safeguarding against contaminants with hygienic materials, and emphasizing transparency through clear labeling. Sustainability remains a cornerstone, with eco-friendly practices and waste reduction at the forefront. Packaging should also prioritize user-friendliness, enhancing convenience for consumers, and playing a pivotal role in branding and marketing efforts. Regulatory compliance is nonnegotiable, ensuring adherence to food safety and labeling regulations to guarantee legality and product safety. The ever-evolving cultured meat industry promises continuous innovation in packaging technology, with possibilities ranging from smart packaging to enhanced distribution strategies.

Moreover, the robustness of packaging materials is essential to withstand the rigors of distribution and transportation, ensuring products reach consumers in optimal condition. Lastly, minimizing food waste is an imperative goal, achieved through portion-controlled packaging and designs that encourage responsible consumption. In essence, the packaging of cultured meat not only assures the product's safety and quality but also aligns with the principles of sustainability and marketability. As the cultured meat industry continues to burgeon, ongoing innovations in packaging will play a pivotal role in delivering safe, sustainable, and delectable meat alternatives to consumers worldwide.

## References

Abdel-Naeem HHS, Ebaid EMSM, Khalel KHM, Imre K, Morar A, Herman V, El-Nawawi FAM (2022) Decontamination of chicken meat using dielectric barrier discharge cold plasma technology: the effect on microbial quality, physicochemical properties, topographical structure, and sensory attributes. LWT 165:113739. https://doi.org/10.1016/j.lwt.2022.113739

Adhikary M, Banerjee R, Chaudhuri S, Chanda A (2023) Food preservation by ozonation. In: Emerging technologies in food preservation, 1st edn. CRC Press, Boca Raton, pp 138–158

Aganovic K, Hertel C, RudiF V, Johne R, Schlüter O, Schwarzenbolz U, Jäger H, Holzhauser T, Bergmair J, Roth A, Sevenich R, Bandick N, Kulling SE, Knorr D, Engel K, Heinz V (2021)

Aspects of high hydrostatic pressure food processing: perspectives on technology and food safety. Compr Rev Food Sci Food Saf 20(4):3225–3266. https://doi.org/10.1111/1541-4337.12763
Ali MM, Al-Mousawi MR, Aljabawi RAA (2022) Studying the penetration ability of various pathogenic bacteria into raw beef meat surface and the antibacterial effect of ozonated water. J Pure Appl Microbiol 16(2):1252–1262. https://doi.org/10.22207/JPAM.16.2.54
Baptista E, Borges A, Aymerich T, Alves SP, Gama LTD, Fernandes H, Fernandes MJ, Fraqueza MJ (2022) Pulsed light application for campylobacter control on poultry meat and its effect on colour and volatile profile. Foods 11(18):2848. https://doi.org/10.3390/foods11182848
Baum CM, Bröring S, Lagerkvist C-J (2021) Information, attitudes, and consumer evaluations of cultivated meat. Food Qual Prefer 92:104226. https://doi.org/10.1016/j.foodqual.2021.104226
Baum CM, De Steur H, Lagerkvist C-J (2023) First impressions and food technology neophobia: examining the role of visual information for consumer evaluations of cultivated meat. Food Qual Prefer 110:104957. https://doi.org/10.1016/j.foodqual.2023.104957
Benvenutti L, Bortolini DG, Nogueira A, Zielinski AAF, Alberti A (2019) Effect of addition of phenolic compounds recovered from apple pomace on cider quality. LWT 100:348–354. https://doi.org/10.1016/j.lwt.2018.10.087
Borges A, Baptista E, Aymerich T, Alves SP, Gama LT, Fraqueza MJ (2023) Inactivation of listeria monocytogenes by pulsed light in packaged and sliced salpicão, a ready-to-eat traditional cured smoked meat sausage. LWT 179:114641. https://doi.org/10.1016/j.lwt.2023.114641
Broucke K, Van Pamel E, Van Coillie E, Herman L, Van Royen G (2023) Cultured meat and challenges ahead: a review on nutritional, technofunctional and sensorial properties, safety and legislation. Meat Sci 195:109006. https://doi.org/10.1016/j.meatsci.2022.109006
Bryant C, Barnett J (2018) Consumer acceptance of cultured meat: a systematic review. Meat Sci 143:8–17. https://doi.org/10.1016/j.meatsci.2018.04.008
Bryant C, Barnett J (2020) Consumer acceptance of cultured meat: an updated review (2018–2020). Appl Sci 10(15):5201. https://doi.org/10.3390/app10155201
Califano G, Furno M, Caracciolo F (2023) Beyond one-size-fits-all: consumers react differently to packaging colors and names of cultured meat in Italy. Appetite 182:106434. https://doi.org/10.1016/j.appet.2022.106434
Cap M, Paredes PF, Fernández D, Mozgovoj M, Vaudagna SR, Rodriguez A (2020) Effect of high hydrostatic pressure on *Salmonella* spp. inactivation and meat-quality of frozen chicken breast. LWT 118:108873. https://doi.org/10.1016/j.lwt.2019.108873
Cenci-Goga B, Iulietto M, Sechi P, Borgogni E, Karama M, Grispoldi L (2020) New trends in meat packaging. Microbiol Res 11(2):56–67. https://doi.org/10.3390/microbiolres11020010
Chen L, Guttieres D, Koenigsberg A, Barone PW, Sinskey AJ, Springs SL (2022) Large-scale cultured meat production: trends, challenges and promising biomanufacturing technologies. Biomaterials 280:121274. https://doi.org/10.1016/j.biomaterials.2021.121274
Cintoni M, Palombaro M, Egidi G, Leonardi E, Gardin F, Pulcini G, Rinninella E, Raoul P, Mele MC (2023) Impact of artificial meat as novel food: controversies and future perspectives. Nutrimentum et Curae
Dantec M, Allain H, Yvert N, Sigrist J, Bensafi M, Lafraire J (2022) Olfactory attributes and colors associated with naturalness. Food Qual Prefer 98:104495. https://doi.org/10.1016/j.foodqual.2021.104495
Datar I, Betti M (2010) Possibilities for an in vitro meat production system. Innov Food Sci Emerg Technol 11(1):13–22. https://doi.org/10.1016/j.ifset.2009.10.007
Dresen E, Pimiento JM, Patel JJ, Heyland DK, Rice TW, Stoppe C (2023) Overview of oxidative stress and the role of micronutrients in critical illness. J Parenter Enter Nutr 47(S1):S38. https://doi.org/10.1002/jpen.2421
Duma-Kocan P, Rudy M, Gil M, Stanisławczyk R, Żurek J, Zaguła G (2023) The impact of a pulsed light stream on the quality and durability of the cold-stored longissimus dorsal muscle of pigs. IJERPH 20(5):4063. https://doi.org/10.3390/ijerph20054063

Escribano AJ, Peña MB, Díaz-Caro C, Elghannam A, Crespo-Cebada E, Mesías FJ (2021) Stated preferences for plant-based and cultured meat: a choice experiment study of Spanish consumers. Sustainability 13(15):8235. https://doi.org/10.3390/su13158235
Evrendilek GA (2022) High hydrostatic processing of marinated ground chicken breast: exploring the effectiveness on physicochemical, textural and sensory properties and microbial inactivation. Food Control 142:109258. https://doi.org/10.1016/j.foodcont.2022.109258
Fidder L, Graça J (2023) Aligning cultivated meat with conventional meat consumption practices increases expected tastefulness, naturalness, and familiarity. Food Qual Prefer 109:104911. https://doi.org/10.1016/j.foodqual.2023.104911
Freixo APA, Silva ACRD, Castro MLRD, Quitério SL, Raices RSL (2022) Volatile organic compounds present in jerked beef, a traditional Brazilian meat product. Food Sci Technol 42:e99521. https://doi.org/10.1590/fst.99521
Fu H, Pan L, Wang J, Zhao J, Guo X, Chen J, Lu S, Dong J, Wang Q (2022) Sensory properties and main differential metabolites influencing the taste quality of dry-cured beef during processing. Foods 11(4):531. https://doi.org/10.3390/foods11040531
Furuhashi M, Morimoto Y, Shima A, Nakamura F, Ishikawa H, Takeuchi S (2021) Formation of contractile 3D bovine muscle tissue for construction of millimetre-thick cultured steak. npj Sci Food 5(1):6. https://doi.org/10.1038/s41538-021-00090-7
Gaspar MC, Braga MEM (2023) Edible films and coatings based on agrifood residues: a new trend in the food packaging research. Curr Opin Food Sci 50:101006. https://doi.org/10.1016/j.cofs.2023.101006
Goudarzi J, Moshtaghi H, Shahbazi Y (2023) Kappa-carrageenan-poly(vinyl alcohol) electrospun fiber mats encapsulated with Prunus domestica anthocyanins and epigallocatechin gallate to monitor the freshness and enhance the shelf-life quality of minced beef meat. Food Packag Shelf Life 35:101017. https://doi.org/10.1016/j.fpsl.2022.101017
Gupta RK, Dudeja P (2017) Food packaging. In: Food safety in the 21st century. Elsevier, New York, pp 547–553
Guzman-Puyol S, Benítez JJ, Heredia-Guerrero JA (2022) Transparency of polymeric food packaging materials. Food Res Int 161:111792. https://doi.org/10.1016/j.foodres.2022.111792
He J, Jia W, Lin Z, Zhang Y, Zhao Y, Fang Y (2023) Improving the quality and processing efficiency of beef jerky via drying in confined conditions of pre-stretching. Food Res Int 172:113171. https://doi.org/10.1016/j.foodres.2023.113171
Jelinek HF, Mousa M, Alefishat E, Osman W, Spence I, Bu D, Feng SF, Byrd J, Magni PA, Sahibzada S, Tay GK, Alsafar HS (2021) Evolution, ecology, and zoonotic transmission of betacoronaviruses: a review. Front Vet Sci 8:644414. https://doi.org/10.3389/fvets.2021.644414
Jiang C, Liu T, Wang S, Zou Y, Cao J, Wang C, Hang C, Jin L (2023) Antioxidant and ammonia-sensitive films based on starch, κ-carrageenan and Oxalis triangularis extract as visual indicator of beef meat spoilage. Int J Biol Macromol 235:123698. https://doi.org/10.1016/j.ijbiomac.2023.123698
Jiao X, Xie J, Du H, Bian X, Wang C, Zhou L, Wen Y (2023) Antibacterial smart absorbent pad with Janus structure for meat preservation. Food Packag Shelf Life 37:101066. https://doi.org/10.1016/j.fpsl.2023.101066
Khanashyam AC, Shanker MA, Kothakota A, Mahanti NK, Pandiselvam R (2022) Ozone applications in milk and meat industry. Ozone Sci Eng 44(1):50–65. https://doi.org/10.1080/01919512.2021.1947776
Lai J, Wang Y, Zhang X, Zhou L, Zhu Y (2023) Effects of active modified atmosphere on lipid and protein oxidation, moisture migration and quality properties of beef steaks. Packag Technol Sci 36(2):111–123. https://doi.org/10.1002/pts.2697
Laroque DA, Seó ST, Valencia GA, Laurindo JB, Carciofi BAM (2022) Cold plasma in food processing: design, mechanisms, and application. J Food Eng 312:110748. https://doi.org/10.1016/j.jfoodeng.2021.110748
Lei Y, Zhang Y, Cheng Y, Huang J, Huang M (2023) Monitoring and identification of spoilage-related microorganisms in braised chicken with modified atmosphere packaging during

refrigerated storage. Food Sci Human Wellness 12(1):28–34. https://doi.org/10.1016/j.fshw.2022.07.015
Liu S, Zhang Y, Zhou G, Bao Y, Ren X, Zhu Y, Peng Z (2019a) Protein degradation, color and textural properties of low sodium dry cured beef. Int J Food Prop 22(1):487–498. https://doi.org/10.1080/10942912.2019.1591444
Liu S, Zhang Y, Zhou G, Ren X, Bao Y, Zhu Y, Zeng X, Peng Z (2019b) Lipolytic degradation, water and flavor properties of low sodium dry cured beef. Int J Food Prop 22(1):1322–1339. https://doi.org/10.1080/10942912.2019.1642354
Liu S, Zhang Y, Harlina PW, Zhou G, Peng Z (2020) Sensory characteristics of low sodium dry-cured beef and their relation to odor intensity and electronic nose signals. Int J Food Prop 23(1):116–126. https://doi.org/10.1080/10942912.2019.1708927
Liu J, Hocquette É, Ellies-Oury M-P, Chriki S, Hocquette J-F (2021) Chinese consumers' attitudes and potential acceptance toward artificial meat. Foods 10(2):353. https://doi.org/10.3390/foods10020353
Mancini MC, Antonioli F (2022) Italian consumers standing at the crossroads of alternative protein sources: cultivated meat, insect-based and novel plant-based foods. Meat Sci 193:108942. https://doi.org/10.1016/j.meatsci.2022.108942
Manzoor A, Yousuf B, Pandith JA, Ahmad S (2023) Plant-derived active substances incorporated as antioxidant, antibacterial or antifungal components in coatings/films for food packaging applications. Food Biosci 53:102717. https://doi.org/10.1016/j.fbio.2023.102717
Mao Y, Yang S, Zhang Y, Luo X, Niu L, Holman BWB (2023) High-pressure processing and modified atmosphere packaging combinations for the improvement of dark, firm, and dry beef quality and shelf-life. Meat Sci 198:109113. https://doi.org/10.1016/j.meatsci.2023.109113
Maroldi WV, Maciel GM, Rossetto R, Bortolini DG, Fernandes IDAA, Haminiuk CWI (2022) Biosorption of phenolic compounds from *Plinia cauliflora* seeds in residual yeast: kinetic, equilibrium, and bioaccessibility studies. Food Process Preserv 46(12). https://doi.org/10.1111/jfpp.17156
Martín-Gómez A, Segura-Borrego MP, Ríos-Reina R, Cardador MJ, Callejón RM, Morales ML, Rodríguez-Estévez V, Arce L (2022) Discrimination of defective dry-cured Iberian ham determining volatile compounds by non-destructive sampling and gas chromatography. LWT 154:112785. https://doi.org/10.1016/j.lwt.2021.112785
Mirzazadeh M, Sadeghi E, Beigmohammadi F (2021) Comparison of the effects of microwave cooking by two conventional cooking methods on the concentrations of polycyclic aromatic hydrocarbons and volatile N-nitrosamines in beef cocktail smokies (smoked sausages). J Food Process Preserv 45(6). https://doi.org/10.1111/jfpp.15560
Muller L, Lacroix A, Ruffieux B (2019) Environmental labelling and consumption changes: a food choice experiment. Environ Resour Econ 73(3):871–897. https://doi.org/10.1007/s10640-019-00328-9
Nastasijević I, Lakićević B, Petrović Z (2017) Cold chain management in meat storage, distribution and retail: a review. IOP Conf Ser Earth Environ Sci 85:012022. https://doi.org/10.1088/1755-1315/85/1/012022
Nethra PV, Sunooj KV, Aaliya B, Navaf M, Akhila PP, Sudheesh C, Mir SA, Shijin A, George J (2023) Critical factors affecting the shelf life of packaged fresh red meat—a review. Meas Food 10:100086. https://doi.org/10.1016/j.meafoo.2023.100086
OECD, FAO (2020) OECD-FAO Agricultural Outlook 2020-2029
Osaili T, Savvaidis IN, Atoui A, Tsiraki MI, Karam L (2023) Microbiota and sensorial attributes of buffalo meat stored under vacuum packaging: combined impact of marination, citrox and oregano essential oil. Int J Food Prop 26(1):1953–1967. https://doi.org/10.1080/10942912.2023.2238919
Pajčin I, Knežić T, Savic Azoulay I, Vlajkov V, Djisalov M, Janjušević L, Grahovac J, Gadjanski I (2022) Bioengineering outlook on cultivated meat production. Micromachines 13(3):402. https://doi.org/10.3390/mi13030402

Qu P, Zhang M, Fan K, Guo Z (2022) Microporous modified atmosphere packaging to extend shelf life of fresh foods: a review. Crit Rev Food Sci Nutr 62(1):51–65. https://doi.org/10.1080/10408398.2020.1811635

Quevedo-Silva F, Pereira JB (2022) Factors affecting consumers' cultivated meat purchase intentions. Sustainability 14(19):12501. https://doi.org/10.3390/su141912501

Rathod NB, Elabed N, Punia S, Ozogul F, Kim S-K, Rocha JM (2023) Recent developments in polyphenol applications on human health: a review with current knowledge. Plants 12(6):1217. https://doi.org/10.3390/plants12061217

Reitznerová A, Semjon B, Bartkovský M, Šuleková M, Nagy J, Klempová T, Marcinčák S (2023) Comparison of lipid profile and oxidative stability of vacuum-packed and longtime-frozen fallow deer, wild boar, and pig meat. Appl Sci 13(6):4059. https://doi.org/10.3390/app13064059

Ren Q-S, Fang K, Yang X-T, Han J-W (2022) Ensuring the quality of meat in cold chain logistics: a comprehensive review. Trends Food Sci Technol 119:133–151. https://doi.org/10.1016/j.tifs.2021.12.006

Revilla I, Vivar-Quintana AM, González-Martín MI, Hernández-Jiménez M, Martínez-Martín I, Hernández-Ramos P (2020) NIR spectroscopy for discriminating and predicting the sensory profile of dry-cured beef "cecina". Sensors 20(23):6892. https://doi.org/10.3390/s20236892

Ritchie H, Rosado P, Max R (2020) Environmental impacts of food production. In: Our world in data. https://ourworldindata.org/environmental-impacts-of-food. Accessed 5 Sep 2023

Roobab U, Chacha JS, Abida A, Rashid S, Muhammad Madni G, Lorenzo JM, Zeng X-A, Aadil RM (2022) Emerging trends for nonthermal decontamination of raw and processed meat: ozonation, high-hydrostatic pressure and cold plasma. Foods 11(15):2173. https://doi.org/10.3390/foods11152173

Rossetto R, Maciel GM, Bortolini DG, Ribeiro VR, Haminiuk CWI (2020) Acai pulp and seeds as emerging sources of phenolic compounds for enrichment of residual yeasts (Saccharomyces cerevisiae) through biosorption process. LWT 128:109447. https://doi.org/10.1016/j.lwt.2020.109447

Rovira P, Brugnini G, Rodriguez J, Cabrera MC, Saadoun A, De Souza G, Luzardo S, Rufo C (2023) Microbiological changes during long-storage of beef meat under different temperature and vacuum-packaging conditions. Foods 12(4):694. https://doi.org/10.3390/foods12040694

Santo RE, Kim BF, Goldman SE, Dutkiewicz J, Biehl EMB, Bloem MW, Neff RA, Nachman KE (2020) Considering plant-based meat substitutes and cell-based meats: a public health and food systems perspective. Front Sustain Food Syst 4:134. https://doi.org/10.3389/fsufs.2020.00134

Shi Y, Wang Y, Hu X, Li Z, Huang X, Liang J, Zhang X, Zhang D, Zou X, Shi J (2023) Quantitative characterization of the diffusion behavior of sucrose in marinated beef by HSI and FEA. Meat Sci 195:109002. https://doi.org/10.1016/j.meatsci.2022.109002

Siddiqui SA, Bahmid NA, Karim I, Mehany T, Gvozdenko AA, Blinov AV, Nagdalian AA, Arsyad M, Lorenzo JM (2022) Cultured meat: processing, packaging, shelf life, and consumer acceptance. LWT Food Sci Technol 114192:114192

Siegrist M, Hartmann C (2020a) Consumer acceptance of novel food technologies. Nat Food 1(6):343–350. https://doi.org/10.1038/s43016-020-0094-x

Siegrist M, Hartmann C (2020b) Perceived naturalness, disgust, trust and food neophobia as predictors of cultured meat acceptance in ten countries. Appetite 155:104814. https://doi.org/10.1016/j.appet.2020.104814

Silva CPD, Semprebon E (2021) How about cultivated meat? The effect of sustainability appeal, environmental awareness and consumption context on consumers' intention to purchase. J Food Prod Mark 27(3):142–156. https://doi.org/10.1080/10454446.2021.1921090

Soro AB, Whyte P, Bolton DJ, Tiwari BK (2021) Application of a LED-UV based light technology for decontamination of chicken breast fillets: impact on microbiota and quality attributes. LWT 145:111297. https://doi.org/10.1016/j.lwt.2021.111297

Soro AB, Harrison SM, Whyte P, Bolton DJ, Tiwari BK (2022) Impact of ultraviolet light and cold plasma on fatty acid profile of raw chicken and pork meat. J Food Compos Anal 114:104872. https://doi.org/10.1016/j.jfca.2022.104872

Sucapane D, Roux C, Sobol K (2021) Exploring how product descriptors and packaging colors impact consumers' perceptions of plant-based meat alternative products. Appetite 167:105590. https://doi.org/10.1016/j.appet.2021.105590
Sutharsan J, Boyer CA, Zhao J (2023) Biological properties of chitosan edible films incorporated with different classes of flavonoids and their role in preserving the quality of chilled beef. Food Hydrocoll 139:108508. https://doi.org/10.1016/j.foodhyd.2023.108508
Szejda K, Bryant CJ, Urbanovich T (2021) US and UK consumer adoption of cultivated meat: a segmentation study. Foods 10(5):1050. https://doi.org/10.3390/foods10051050
Tian X, Li ZJ, Chao YZ, Wu ZQ, Zhou MX, Xiao ST, Zeng J, Zhe J (2020) Evaluation by electronic tongue and headspace-GC-IMS analyses of the flavor compounds in dry-cured pork with different salt content. Food Res Int 137:109456. https://doi.org/10.1016/j.foodres.2020.109456
Toldrá F, Hui YH (eds) (2007) Handbook of fermented meat and poultry, 1st edn. Blackwell, Ames, Iowa
Tuomisto HL, Teixeira De Mattos MJ (2011) Environmental impacts of cultured meat production. Environ Sci Technol 45(14):6117–6123. https://doi.org/10.1021/es200130u
United Nations (2015) Department of Economic and Social Affairs: Sustainable Development. https://sdgs.un.org/goals. Accessed 20 Sep 2023
Verbeke W, Sans P, Van Loo EJ (2015) Challenges and prospects for consumer acceptance of cultured meat. J Integr Agric 14(2):285–294. https://doi.org/10.1016/S2095-3119(14)60884-4
Wang Q, Chen Q, Xu J, Sun F, Liu H, Kong B (2022) Effects of modified atmosphere packaging with various CO2 concentrations on the bacterial community and shelf-life of smoked chicken legs. Foods 11(4):559. https://doi.org/10.3390/foods11040559
Wang J, Chen J, Sun Y, He J, Zhou C, Xia Q, Dang Y, Pan D, Du L (2023) Ultraviolet-radiation technology for preservation of meat and meat products: recent advances and future trends. Food Control 148:109684. https://doi.org/10.1016/j.foodcont.2023.109684
Wilks M, Phillips CJC, Fielding K, Hornsey MJ (2019) Testing potential psychological predictors of attitudes towards cultured meat. Appetite 136:137–145. https://doi.org/10.1016/j.appet.2019.01.027
Wójciak KM, Stasiak DM, Kęska P (2019) The influence of different levels of sodium nitrite on the safety, oxidative stability, and color of minced roasted beef. Sustainability 11(14):3795. https://doi.org/10.3390/su11143795
Xiang N, Yao Y, Yuen JSK, Stout AJ, Fennelly C, Sylvia R, Schnitzler A, Wong S, Kaplan DL (2022) Edible films for cultivated meat production. Biomaterials 287:121659. https://doi.org/10.1016/j.biomaterials.2022.121659
Xin X, Dell K, Udugama IA, Young BR, Baroutian S (2021) Transforming biomass pyrolysis technologies to produce liquid smoke food flavouring. J Clean Prod 294:125368. https://doi.org/10.1016/j.jclepro.2020.125368
Yan Z, Zhong Y, Duan Y, Chen Q, Li F (2020) Antioxidant mechanism of tea polyphenols and its impact on health benefits. Anim Nutr 6(2):115–123. https://doi.org/10.1016/j.aninu.2020.01.001
Yu Y, Wang G, Sun Y, Ge C, Liao G (2020) Changes in physicochemical parameters, free fatty acid profile and water-soluble compounds of Yunnan dry-cured beef during processing. J Food Process Preserv 44(4). https://doi.org/10.1111/jfpp.14380
Zainal Arifin M, Mohd Adzahan N, Zainal Abedin N, Lasik-Kurdyś M (2023) Utilization of food waste and by-products in the fabrication of active and intelligent packaging for seafood and meat products. Foods 12(3):456. https://doi.org/10.3390/foods12030456
Zawani CJ, Nor-Khaizura MAR, Mahyudin NA, Ismail-Fitry MR, Nirmal NP (2022) Microbiological and sensorial quality of beef meat (Longissimus dorsi) marinated with cinnamon extract and stored at various temperatures. Foods 11(24):3971. https://doi.org/10.3390/foods11243971
Zhang H, Tang D, Yang H, Liu X, Cheng J, Wang X, Zou J, Lin Y (2023) Effects of high hydrostatic pressure assisted enzymatic tenderization on goose meat texture and myofibril protein. LWT 184:114845. https://doi.org/10.1016/j.lwt.2023.114845

Zhao Y, Chen L, Bruce HL, Wang Z, Roy BC, Li X, Zhang D, Yang W, Hou C (2022) The influence of vacuum packaging of hot-boned lamb at early postmortem time on meat quality during postmortem chilled storage. Food Sci Anim Resour 42(5):816–832. https://doi.org/10.5851/kosfa.2022.e34

Zhu H, Li P, Wang L, Huang Q, Xu B (2023) Flavor profile of "Dao Ban Xiang" (a traditional dry-cured meat product in Chinese Huizhou cuisine) at different processing stages in winter and summer. Food Sci Nutr 11(6):2733–2750. https://doi.org/10.1002/fsn3.3225

# Quality and Risk Control in Cultivated Meat Production

**Renata Ernlund Freitas de Macedo, Guilherme Agostinis Ferreira, Luiza Poniewas, Fabiane Barchiki, Carmen Lúcia Kuniyoshi Rebelatto, Débora Regina Daga, Leandro Batista Costa, and Edvaldo Antonio Ribeiro Rosa**

**Abstract** Quality control and risk assessment are indispensable elements of the manufacturing process of cultivated meat, playing a pivotal role in ensuring reproducibility and mitigating variability stemming from the heterogeneity among cell donors and all the components involved in the entire process. The selection and characterization of cells for culture must prioritize safety, demanding meticulous consideration of donor selection, biopsy techniques, and transportation procedures. As cultivated meat is an innovative dietary product, global regulations and comprehensible supervisory frameworks are crucial for consistent quality and consumer trust. Qualifying and validating cell manufacturing processes are crucial for safety, including microbiological controls at different stages and chromosomal stability, genotoxicity, and tumorigenic potential evaluation. Environmental monitoring is essential to address risks such as hazardous materials, micro−/nanoplastics, heavy metals, toxins, and pathogens. Managing foreign objects and allergens introduced through cross-contamination is also important during upstream and downstream processes. Moreover, the safety and quality of the cultivated meat product, with a focus on key attributes such as pH, water-holding capacity, texture, flavour, and

R. E. F. de Macedo (✉) · G. A. Ferreira · L. Poniewas · L. B. Costa
Laboratory of Agrifood Research and Innovation, Graduate Program in Animal Science, School of Medicine and Life Sciences, Pontifícia Universidade Católica do Paraná, Curitiba, Paraná, Brazil
e-mail: renata.macedo@pucpr.br

F. Barchiki · C. L. K. Rebelatto · D. R. Daga
Core for Cell Technology, School of Medicine and Life Sciences, Pontifícia Universidade Católica do Paraná, Curitiba, Paraná, Brazil

E. A. R. Rosa
Xenobiotics Research Unit, Graduate Program in Animal Science, School of Medicine and Life Sciences, Pontifícia Universidade Católica do Paraná, Curitiba, Paraná, Brazil


C. R. Soccol et al. (eds.), *Cultivated Meat*,
https://doi.org/10.1007/978-3-031-55968-6_11

colour, are intricately shaped by decisions made at various stages of the production process. This encompasses the selection of diverse cell types, the application of various techniques for cell maturation, the utilization of scaffolds, and the incorporation of different ingredients.

**Keywords** Cell-based meat · Cultured meat · Good cell practices · Environmental monitoring · Safety assessment · Regulatory framework

## 1 Introduction

The production of cultivated meat is a multifaceted process that encompasses numerous stages, each entailing its unique array of challenges and intricacies. Starting from the selection and characterization of cells, the maintenance of cell banks, and the operation of bioreactors through to the delivery of a final product that is both safe and sensory accepted, every facet of this process is replete with potential hazards that necessitate meticulous attention and resolution.

In existing research, the intricate safety and regulatory aspects of cultivated meat have received limited attention. Furthermore, there is a noticeable dearth of comprehensive studies examining the potential risks and safety considerations associated with cultivated meat (Ketelings et al. 2021). Safety concerns should be given primary consideration, with subsequent attention directed towards factors such as taste, texture, and the cost of the product, as emphasized by Bryant and Barnett (2018). This prioritization is crucial, particularly because cultivated meat production is still in its nascent stages and is confronted with technological and production scale limitations.

This chapter provides an in-depth exploration of the significance of quality and hazard control in cultivated meat production, elucidating how the management of each stage impacts both the safety and the quality of the product. This aspect stands as one of the pivotal factors underpinning the reliability, safety, and consumer acceptance of the final product. When exploring the intricacies of this evolving field, four critical domains will be traversed. (1) Quality Control and Risk Assessment in Obtaining and Characterizing Cells: This section addresses the importance of implementing stringent controls during the collection of cells from donor animals, the characterization of cells, and the assessments required to ascertain the suitability of these cells for meat cultivation. (2) Quality Control and Risk Assessment in the Cell Bank: We will discuss potential risks associated with the maintenance of a cell bank and elucidate strategies aimed at their mitigation. (3) Quality Control and Risk Assessment in the Bioreactor: The operation of bioreactors necessitates robust quality control measures to uphold optimal conditions and prevent contamination. This section will shed light on the complexities involved in bioreactor operation, ensuring sterility, and emphasizing the pivotal role of process monitoring. (4) Quality Control and Safety Assessment of the Final Product: Lastly, we will explore various factors that influence the quality of the final product, factors that are profoundly shaped by the conditions and processes employed in the preceding stages of production.

## 2 Quality Control and Safety Assessment in Obtaining and Characterization of Cells

The selection and characterization of cells for culture represent a crucial stage that necessitates thorough consideration in order to optimize operational efficiency, enhance animal welfare, and ensure safety. This phase involves distinct aspects, including the choice of donor animals, techniques for conducting biopsies, the care and treatment of animals, and the vigilance in maintaining quality, all of which are integral to the process (Melzener et al. 2021).

In light of these considerations, certain questions come to the fore: Do potential safety risks inherent in commercial production techniques for cultivated meat differ from those associated with traditional meat production? Furthermore, what implications will these eventual commercial production methods have on the overall safety of the resulting product? These inquiries underscore the importance of endeavours to elucidate the risks and quality aspects linked to cultivated meat. Such efforts, in turn, serve as an added incentive for all stakeholders to invest in its production, with the aim of achieving validation and fostering increased acceptance.

As cultivated meat is considered an innovative dietary product, there is an ongoing need for the concurrent development of globally applicable, readily accessible, and easily comprehensible regulations and supervisory frameworks. This is essential to ensure consistency, quality, and safety in the production of the product. In any case, legislation pertaining to cultivated meat should encompass the entire production chain, ranging from the selection of cell donor animals to the packaged ready-to-eat product. According to Ye et al. (2022), key concerns related to the safety of cultivated meat encompass the following aspects: (1) the utilization of raw materials lacking a proven history of safe consumption; (2) the adoption of animal cell culture components and carriers that lack a track record of safe usage in the cultivated meat production process; (3) the novel procedure involved in cultivated meat production, which includes genetic modifications to alter the expression levels of certain nutrients; and (4) the consumption of transgenic cell lines. Building upon these concepts, the subsequent sections delve into the risks and quality control measures associated with various stages of cell collection and characterization.

### 2.1 *Donor Animals*

The selection of the stem cell donor animal has the potential to exert a significant impact on the overall process efficiency, thereby emphasizing the critical importance of identifying the essential attributes of an ideal donor. The choice of a suitable animal species, commonly raised for meat production, encompasses a multitude of factors, including taste, texture, and consumer preferences (Bhat and Fayaz 2011).

As this area garners increasing attention among consumers, it is imperative to recognize that the selection of the cell donor animal should have minimal implications on animal welfare. Typically, animals commonly raised for meat production fall within a category that already adheres to established animal welfare standards, including appropriate shelter, handling, and nutrition. Nonetheless, it is vital to ensure that the biopsy procedure is carried out in a manner that does not compromise the quality of life for these animals, thus minimizing stress and discomfort during the collection procedures. A myriad of characteristics associated with the donor animal can wield a significant influence on the yield and quality of the obtained cells:

Age: Typically, donor animals are selected from younger populations due to their higher cell proliferation rates. These cells possess the potential for genetic immortalization, enabling continuous proliferation and mitigating the need for additional animal donors (Rubio et al. 2020).

Sex: The sex of the donor animal may have an impact on the yield and quality of satellite cells. These stem cells are equipped with androgen receptors, and activating these receptors can potentially trigger cell proliferation. Moreover, androgens themselves have been shown to be positively associated with muscle strength and size (Chen et al. 2005).

Genetics: Genetic diversity plays a pivotal role in maintaining population stability and the capacity to adapt to changing circumstances. When making selections, it is imperative to consider genetic diversity to prevent adverse effects on animal populations (Bonny et al. 2017). Animal donors are likely to remain indispensable for preserving genetic diversity and providing non-genetically modified alternatives for cellular-based meat.

Nutrition: It is encouraged to provide animals with a well-balanced and appropriate diet that aligns with their nutritional requirements. Optimized nutrition not only enhances their overall well-being but also contributes to the quality of cells obtained for cultivated meat production. The relationship between nutrition, growth factors, and satellite cells in the context of skeletal muscle development can vary based on factors such as feed composition (nutrients in the diet) and the balance between roughage and concentrate in the diet (Vestergaard et al. 2000; Thornton 2019).

Animal welfare: Low levels of welfare can induce stress and atypical behaviours in animals. When used for cell sourcing, it is essential to minimize stressful conditions, ensuring that animals are accommodated in appropriate housing environments designed to meet their fundamental needs. This includes well-designed living spaces, effective ventilation, clean bedding, and protection against adverse weather conditions. For animals inclined towards social interactions, opportunities for conspecific engagement should be provided to prevent feelings of isolation and discomfort. Additionally, when handling, transporting, or conducting essential procedures, stress mitigation is crucial for maintaining animal well-being and preserving the quality of the obtained cells (Rushen and Passillé 1992; Carrasco-García et al. 2020; Lee et al. 2022).

However, amid all the aforementioned factors, the paramount consideration remains the consistent oversight of a veterinarian. Pre-biopsy inspections mirror the antemortem assessments conducted in traditional meat production. Following the globally recognized guidelines of the Food and Agriculture Organization (FAO) for meat inspection, certain prerequisites are mandatory before animal tissue collection. These include evaluating the animal's health condition; documenting any veterinary medications administered to the animal, such as antibiotics (including administration dates and the withdrawal period established for each medication); and identifying any illnesses that might potentially affect the safety of the resulting cultivated meat product (FAO 2023).

To meet these requirements, animals eligible for biopsy extraction should be in a veterinarian-certified healthy condition following a visual assessment, akin to the existing procedure. This holds paramount significance in monitoring animal health, addressing potential medical concerns, and ensuring their welfare and encompasses regular health assessments with immediate intervention if any health challenges arise. Veterinarians can primarily identify meat-related pathogens, particularly those with zoonotic implications like tuberculosis and brucellosis, as well as detect any potential faecal contamination.

In the context of cultivated meat production, pathogen identification can be achieved through the analysis of the donor animal's blood, while faecal contamination can be determined through laboratory analysis of the tissue sample itself. To ensure compliance with well-defined safety standards, veterinarians can also conduct specialized tests like DNA genotyping, which could be employed for cultivated meat products. This approach would ensure the adherence of these products to specified safety parameters (Melzener et al. 2021).

## 2.2 *Biopsy and Cell Collection*

The biopsy procedure should be minimally invasive and conducted by trained professionals. Typically, donor animals are anesthetized by a veterinarian, and a minimal tissue sample (<1 g) is excised (Rubio et al. 2020). The incorporation of cultivated meat production using cells obtained from non-sacrificed donor animals can seamlessly fit into the existing regulatory framework for meat inspections with only minor adjustments. Moreover, the application of advanced laboratory techniques for in-depth analysis of biopsied samples and harvested stem cells is expected to ensure an initial disease-free state in cultivated meat production. This level of accuracy is likely to be equivalent to that observed in traditional meat production. Prior to achieving widespread acceptance of cultivated meat on a commercial scale, it will be essential to establish effective channels of communication between cultivated meat producers and regulatory authorities (Treich 2021).

Nonetheless, advancements in cellular agriculture are exploring alternative methods of obtaining cells that do not require live animals, such as utilizing biopsy samples from deceased animals, cell lines, or genetic modification. However, there are drawbacks to this approach, including the limited size of the biopsy and the inherently uncertain nature of sample collection. This uncertainty can lead to instances where samples of the desired tissue type or quality may not be obtained, as highlighted by Stephens et al. (2018).

In evaluating the quality of the tissue extraction process, it is essential to consider distinct anatomical muscles or muscle groups. Different muscles exhibit varying muscle fibre type compositions, which may be associated with the proportion of satellite cells. Skeletal muscle tissue is the primary focus, as it constitutes the part of the animal used for meat consumption. Muscle cells, also known as myocytes or myoblasts, are responsible for the structure and function of muscle tissue and serve as a dependable source for acquiring cell lines for cultivated meat (Soice and Johnston 2021), ensuring both edibility and safety. Activated satellite cells, which play a crucial role in repairing damaged muscle tissue, are a type of adult stem cell that can be obtained from skeletal muscle tissue, as emphasized by Ben-Arye et al. (2020). These cells, serving as precursors to skeletal muscle cells, have the capability to undergo differentiation, giving rise to functional skeletal muscle cells. Through cell culture techniques, these activated satellite cells can effectively progress towards the formation of muscle fibres, taking on the characteristics of myocytes and myotubes (Yin et al. 2013).

To safeguard the health of the donor animal, which includes detecting infections (bacterial, viral, or parasitic), it is imperative to conduct a comprehensive examination. Health inspection is crucial in preventing the initial introduction of biohazards into the starter cells. Tissues and/or blood of the donor may contain pathogens. For instance, prions are typically located in specific tissues such as the brain, spinal cord, lymphoid tissue, tonsils, appendix, enteric nervous system, and blood of infected animals (Maddison et al. 2010). Consequently, it is advisable to avoid using these specific tissues to mitigate risks.

As for microbiological contamination, encompassing bacteria, fungi, and viruses, it remains a potential concern throughout the entirety of the cultivated meat production process, with the possibility of arising from various sources and individuals involved in the production cycle. Microbial contaminants can be found in the environment or on animal skin or hair and may contaminate the tissue during sampling. Therefore, performing trichotomy, cleaning, and antisepsis of the animal's region where the biopsy will be taken before collection, as well as the subsequent inspection of the sampled tissue for signs of contamination, are essential procedures to guarantee the safety of the collected cells. After evaluating the microbiological risk of the cells, it is necessary to check if it aligns with the guidance outlined by worldwide organizations such as the FAO and the World Health Organization (WHO) (FAO/WHO 2021). Hence, the procedure of gathering cells from animals can be meticulously managed using aseptic techniques, which may offer a means to mitigate or diminish the necessity for antibiotics and their associated residual risks (Coecke et al. 2005).

## 2.3 Cell Transportation and Storage

The shipping process should undergo validation to ensure no adverse impact on the cells and should be equipped with temperature monitors (Montgomery et al. 2021; Simsa et al. 2019). Shipments of materials subject to cold chain management ought to be dispatched with adequate refrigeration to sustain temperature during the shipping cycle, accommodating at least a 24 h delay in the anticipated arrival time (ISBER 2008). This implies that the critical risk factor for transporting cells is temperature. For short-term transportation of cells, a recommended approach involves transferring samples into sterile tubes containing Dulbecco's Modified Eagle Medium (DMEM) with 1% penicillin-streptomycin (PS). Subsequently, these tubes can be placed in a thermal box at 4 °C with a suitable cooling agent, such as ice. Dry ice can also be used without the need for media. This controlled environment helps maintain the desired temperature for preserving cell viability during transit. Additionally, the inclusion of an L-glutamine supplement in the medium is advisable to enhance cell health and stability throughout the transportation period.

For long-term transportation, the primary approved method is cryopreservation. The transportation of cryopreserved cells is typically carried out utilizing dry shippers containing absorbent material, which can be filled with liquid nitrogen to sustain vapour-phase temperatures for up to 2 weeks if appropriately charged. The United States Pharmacopeia specifies that the temperature for cryopreservation should be not warmer than −130 °C for nonclinical specimens or −150 °C for clinical materials. Therefore, factors such as container strength, pressure resistance, temperature controls, choice of refrigerants (whether liquid nitrogen or dry ice), and the heat-transfer capacity of the container material should be continually assessed and tested (USPC 2018).

It is advisable to conduct the freezing process in a controlled manner, employing insulation or specially designed boxes. This helps prevent additional challenges, such as the formation of extracellular ice, which can lead to osmotic stress and mechanical damage to the cells. It ensures an adequate cooling rate and consistency in freezing curves, typically resulting in better post-thaw recovery. When executed correctly to maintain the proper temperature, this process does not significantly impact the quality of the cells. As indicated in studies by Jang et al. (2017) and Montgomery et al. (2021), other precautions that should be considered include minimizing sample warming during the transfer of the cell product from the freezing device to storage, ensuring the strength and seal integrity of cryopreservation tubes, maintaining container cleanliness, and implementing continuous monitoring of freezer temperatures through a calibrated system. Importantly, access to cryopreserved products or samples should be restricted to individuals who have received proper training for this purpose. In certain instances, source traceability may necessitate verification by a second person.

Despite cells having already undergone anti-residual approval, it is essential to perform a new test upon their arrival at their new destination to detect any potential contamination that may have occurred during the transportation process. Following the US Food and Drug Administration (FDA) document on the production of cell lines for biological use, it is recommended to establish a protocol for cell bank stability, including a table for recording test results. The cell bank reports should be thoroughly documented and updated when necessary (FDA 1993).

Emerging alternatives to cryopreservation are in development, including a portable CO2 incubator designed for cell transportation via road, rail, or air. With this method, there is no requirement to defrost the biological material; it is ready for use upon arrival. This approach eliminates the use of toxic substances such as cryoprotective agents and prevents freeze-thaw loss, saving both cells and time. However, it is worth noting that this type of material is relatively new, and ongoing research is still assessing its associated risks (ATA Scientific Instruments 2020).

## 2.4 Risks and Quality Assessment of Collected Cells

The process of risk assessment commences with the evaluation of incoming raw materials. This assessment aims to scrutinize the properties of the raw materials and determine whether their attributes necessitate further analytical examination. It encompasses various factors, including the historical safety record of the raw material and its source. Additionally, regulatory directives also hold significant importance during this procedure, ensuring that the raw materials adhere to the benchmarks defined by the relevant authorities (Stephens et al. 2018; FAO and WHO 2021).

For instance, companies typically ensure that their supplier confirms the cell line was tested negative for known infectious agents that affect the donor animal. The cell line's product sheet should also include disclaimers, such as: 'This product is meant solely for laboratory research purposes. It should not be used for any animal or human therapeutic applications, consumed by humans or animals, or utilized for diagnostics'.

Food quality is a complex concept. Concerning meat, quality encompasses aspects of safety, nutrition, technology, and sensory perception. Adhering to the principles of Good Cell Culture Practice (GCCP) is crucial to ensure the consistent quality of materials, regardless of their nature or intended downstream use. GCCP aims to establish a set of guidelines for standardizing and implementing practices in cell and tissue culture, aligning with the principles of Good Laboratory Practice (GLP) (OECD 2022) and the Good Practices for Meat Industries (FAO 2004). These practices encompass the characterization and maintenance of crucial attributes, quality assurance, meticulous record-keeping and reporting, safety protocols, education and training, and ethical considerations (Coecke et al. 2005).

During the quality assessment of cultivated cell lines, rigorous microbiological measures ensure the complete absence of unintended agents. The detection of DNA plays a crucial role in most employed techniques. It is possible to utilize genetic indicators as quality controls for cell lines. These methods can encompass the entire genome or focus on specific segments, with each offering a trade-off among range, precision, sensitivity, and cost-effectiveness (Carrillo-Ávila et al. 2022).

# 3 Quality Control and Risk Assessment in the Cell Bank

Monitoring every step of the cultivated meat manufacturing process is essential to trace potential sources of danger and contamination. The phases where quality controls will be performed should be rigorously defined and described to ensure that the final product consistently meets its specifications across batches (Mebarki et al. 2021). Regulations that provide for Good Practices in cells for therapeutic use and clinical research suggest which quality controls must be performed during the manufacture of a product that is subjected to extensive manipulation. However, they do not define the phases of the manufacturing process and the methodologies that must be used to evaluate the different parameters.

Quality controls at the cell culture stage include cell characterization, cell viability, and cell differentiation, as well as safety tests such as microbiological, mycoplasma detection, endotoxin, genomic stability, and tumorigenicity. Regardless of the cell type, whether primary, modified, or immortalized cells, the quality controls are the same. However, when a primary culture is performed, it is suggested that the cells are cryopreserved in low passages to constitute the master cell bank. After thawing, the cells are expanded to create a working cell bank.

## 3.1 *Cell Characterization*

Flow cytometry serves as an immunophenotypic analysis method for characterizing cell types, helping to define and validate cell homogeneity in a tissue-specific manner and avoiding potential contamination with cells from other lineages (Santilli et al. 2022). This technique assesses cell expression profiling by incubating cells with conjugated monoclonal antibodies against different antigens, allowing observation of positivity and negativity for specific markers. Cell viability, crucial in primary cultures, can be determined using a flow cytometric assay based on 7-aminoactinomycin D (7-AAD).

In primary cultures, cells can proliferate and be subcultured but may change differentiated characteristics over time. Ensuring that cells isolated in the primary culture retain their characteristics in the final product is vital for cell culture quality control. Genetic sequencing, determining the sequence of nucleotide bases within an animal's or cell's genome (Dorado et al. 2021), is a helpful tool for identifying animal species for human consumption, such as pigs, cows, chickens, and sheep (Pareek et al. 2011). Comparing sequencing data to a reference database like Ensembl is a method to verify the animal origin of cells in the working cell bank and final product.

The cell differentiation assay, crucial for cell lineage characterization, depends on the cell type used, such as adult stem cells or induced pluripotent stem cells. Commercial media inducing cell differentiation can be utilized. In these assays, cells can be cultured in monolayers or 3D cultures, and cell differentiation can be evaluated through cytochemistry, immunocytochemistry, or immunohistochemistry.

## 3.2 Cell Manufacturing Process

Microbiological control to detect bacteria and fungi must be performed at different stages of the manufacturing process, from obtaining the biological material to the final product. The methodologies should adhere to compendial methods, and if alternative methods are used, they should be validated and approved by regulatory agencies. Automated methods are utilized for fungi and bacteria detection, while various methods, with PCR-based being the most common, can detect mycoplasma. For identifying endotoxins from Gram-negative bacteria, various methods exist, with the chromogenic kinetic method standing out as a fast and highly sensitive approach specifically designed for detecting lipopolysaccharides of Gram-negative bacteria.

Cells cultivated in vitro can accumulate genetic and epigenetic alterations, showcasing genetic instability that may explain their tumorigenic potential (Catalina et al. 2008; Sensebé and Bourin 2009). Evaluating chromosomal instability during cell culture is essential and is usually carried out through cytogenetic analyses that can be combined with other techniques (Borgonovo et al. 2015). Cytogenetics is performed using the G-banding technique, which can detect numerical and structural aberrations (Borgonovo et al. 2014).

The genotoxicity assessment verifies whether any reagents or materials used during the product manufacturing process could have a genotoxic effect, consequently affecting product quality (Sharma et al. 2015). The cytokinesis-block micronucleus (CBMN) assay can detect abnormalities that may be related to a risk of developing cancers, as they are indicators of genetic instability (Cornélio et al. 2014; Bonassi et al. 2011). Genomic instability could lead to tumours, ectopic tissue formation, and malignant transformation (Guadix et al. 2019). Therefore, evaluating the tumorigenic potential of expanded cells is recommended. Tumorigenicity can be assessed using in vitro assays, which evaluate anchorage-independent cell growth, and in vivo assays, conducted in immunocompromised animals. In vitro assays are a preferred option as there has been demonstrated correlation between in vivo and in vitro findings (Sharma and Bhonde 2020). A soft agar colony formation assay is indicated for detecting the malignant transformation of cells and allows for the assessment of the presence and quantification of colonies (Horibata et al. 2015). High-throughput screening methods for formed colonies enable the analysis of larger sample sizes simultaneously after a cell growth period of 6–7 days (Rebelatto et al. 2023).

During meat cultivation, quantities of impurities related to the product and the process can be introduced into the culture. The reagents used during the manufacturing process impact product quality and food safety; therefore, it is essential to determine the maximum amount of reagent that can be found in the product.

Impurities that must be tested in meat culture include various elements such as toxins and residues from products used in cell culture (medium, antibiotic, dimethyl sulfoxide, and animal proteins).

The assessment of impurities related to the cell manufacturing process must occur through risk management based on a guideline that defines the detection limits for the presence of impurities in the product and the best methods and analyses available for evaluating impurity presence. According to the European Medicines Agency, there must be qualitative and, whenever possible, quantitative information on impurities related to the process and the product, and the degree of determination of impurities must be substantiated (European Medicines Agency 2019). There is no consensus on which methodologies and parameters should be used; therefore, each institution should define and present sufficiently robust data that attest to the absence of impurities related to the process. In addition, the determination of impurities must be demonstrated at different processing stages and after specific interventions to purify the product of interest. If it is impossible to verify all components artificially added in the cell maintenance process, in that case, it is possible to choose an indicator to attest to the absence of impurities in the product.

Cell cryopreservation is another cell culture process that should be controlled to ensure safety in the production process. Cryoprotective agents may exhibit cell toxicity (Gao & Crister 2000); therefore, quality control evaluation after thawing is essential during manufacturing to demonstrate that cells maintain their characteristics, structural integrity, viability, and functionality potential after cryopreservation (Whaley et al. 2021). It is also essential to evaluate the likelihood of the presence of cryoprotectants in the final product. Cells should be stored in the vapour phase (−150 °C) to avoid contamination by microorganisms present in biological materials or cross-contamination (Grout and Morris 2009).

## 3.3 *Qualification and Validation of Cell Manufacturing Process*

Qualification and validation of critical processes are necessary to ensure product safety and effectiveness. Reference values are defined prior to validation and can be based on studies performed by the laboratory or data from published studies. Critical reagents used during cell manufacture should be qualified, as they may affect the quality or safety of the final product. All the reagents should have a qualification certificate, but some tests are remade for each batch. As part of the qualification process, microbiological control should be performed on reagents and equipment used during manufacturing (Coecke et al. 2005).

The choice of an adequate culture medium that provides pH conditions, osmolality, and substances essential for maintaining cell viability, proliferation, and functionality is essential because it can influence the product's final characteristics. Animal products, such as foetal bovine serum (FBS) and enzymes for cell dissociation used in cell culture, should be avoided; however, there are different cell expansion protocols where these products are present. Products of animal origin should have certification of the absence of infectious agents and contaminants following the recommendations of regulatory agencies (European Medicines Agency 2013). Due to their origin, both the FBS and the enzymes have a high content of xenogeneic proteins that may present risks in terms of contamination and transmission of pathogens, such as viruses, prions, bacteria, fungi, and endotoxins (Gottipamula et al. 2016). Strict quality control to avoid the transmission of unknown pathogens from animal-derived compounds is essential to avoid the risk of diseases such as bovine spongiform encephalopathies (Pinheiro de Oliveira et al. 2013).

Chemically defined xeno-free products, where all compounds are highly purified and of defined composition, are an alternative to animal-based products (O'Neill et al. 2021). These products facilitate process quality control, reducing production variability and increasing process reproducibility (Cimino et al. 2018; Moraes et al. 2008). Antibiotics and antifungals, which prevent microbial contamination of a primary cell culture, can only be used in the early stages of cell culture. The most used antibiotics are benzylpenicillin and streptomycin, and antimycotics are rarely used. The final product's residual antibiotics must meet the requirements established for conventional meat (WHO 2019).

All the equipment used in the production process, including cell cultivation and controls, must be suitable for the purpose for which they are intended and must not present any risk to the product. Equipment that has contact with the product must not have properties that could affect the quality of the product. The equipment should be qualified and have preventive maintenance (calibrations, inspections, or verifications) and controls for monitoring them at defined intervals to ensure adequate performance. Equipment cleaning is essential to avoid the risk of product contamination; for this reason, equipment cleaning and decontamination procedures must be validated, as well as their frequency.

## *3.4 Environmental Monitoring*

Cell cultivation must adhere to Good Manufacturing Practices through an aseptic process in classified environments to ensure the product's sterility. The room where cells will be cultivated must have an air filtration system that guarantees the area's classification according to ISO 14644-1 (Cleanrooms and Associated Controlled Environments, Classification of Air Cleanliness). Cell cultivation must be carried out in an environment with air quality with a particle count equivalent to ISO 5 classification. An environmental monitoring programme must be implemented to evaluate control measures and identify possible risks to the process and product. This programme should include the following parameters: non-viable and viable particle counts, air pressure differentials, temperature control, and, where appropriate, relative humidity. It is vital to achieve proper airflow from areas of higher air purity to adjacent areas of lower purity to maintain air quality. Rooms with higher air purity must have a substantial positive pressure differential over adjacent rooms with lower air purity. These pressure cascades should be clearly defined and continuously monitored with appropriate methods. Adjacent rooms of different grades should have a 10–15 Pa pressure differential (European Commission 2017).

The area should be certified by a specialized company at regular intervals to monitor non-viable particles as indicated by ISO 14644. Viable microorganisms must be monitored in classified environments intended for cell cultivation, environments where aseptic operations are performed, and equipment used in cell cultivation. Monitoring methods, such as sedimentation plates, volumetric air, and surface sampling (swabs and contact plates), should be frequent. Continuous monitoring is required during critical operations where the product is exposed to the environment. Surfaces and personnel must be monitored after critical operations.

## 4 Quality Control and Risk Assessment in Upstream and Downstream Steps

To conduct the risk assessment of cultivated meat and related cell cultures for food, the FAO recruited a panel of experts (FAO/WHO 2023). According to this panel, the risks associated with the production processes (upstream and downstream) include the presence of (1) potentially hazardous structural materials and related substances, (2) chemical contaminants, (3) micro−/nanoplastics, (4) heavy metals, (5) microbial toxins, (6) pathogens from cell culture media, (7) pathogens from unhygienic operators or equipment, (8) food additive residues, (9) food allergens, (10) antimicrobials, (11) novel toxins/allergens or an increase in endogenous toxicants/allergens, (12) foreign object contamination, and (13) allergens/pathogens/pathogenic agents (e.g. prions) introduced by cross-contamination, in the cell proliferation, cell differentiation, and bioreactor expansion.

### *4.1 Quality and Risk Control During Upstream Processes*

Producing cultivated meat requires special care, regardless of the production system (batch, fed-batch, continuous, or perfusion), and the bioreactors used, as it is susceptible to various interferences such as microbial contamination, oxidation, and autolysis. Lessons learned from cell culture and fermentation biotechnology used in the production of biotherapeutics can be applied. However, specific equipment and techniques must be developed for food industrial-scale production (Ben-Arye and Levenberg 2019). This includes cleanroom facilities for primary inoculum preparation with microbial contamination safeguards (McNamara 2023) and the implementation of a seed train process where cells are passed through a series of tanks of increasing volumes before reaching the main growth bioreactor. Quality control measures must be implemented, and specific sanitary conditions must be maintained throughout production. It is essential to strictly control the sterilization of culture media, tanks for seed training and cell growth, and the entire piping to prevent microbial and pyrogen/allergen contamination (Gouveia et al., 2015).

Fast-growing bacteria, yeast, and fungi can contaminate meat-growing bioreactors through contaminated broth components, failures in air filtering, unsatisfactory cleanliness, and incomplete sterilization of equipment, pipelines, and accessories or during handling operations when strict aseptic techniques are not followed. Even if the microbe is identified and its susceptibility to antimicrobials is known, using chemicals might not be the optimal solution, and it is advisable to dispose of the batch to prevent potential risks to consumers. Moreover, to prevent the spread of microbial contamination, it is essential to protect nearby bioreactors.

On the other hand, mycoplasmas are small (1–2 μm long and 0.1–0.2 μm wide), Gram-negative, slow-growing prokaryotes that are resistant to most antibiotics. As they are very thin and lack cell walls, they are very flexible and evade filtering systems, even those equipped with 0.22 μm membranes. Routine microscopic inspections do not detect them, so DNA-based methods must be implemented in the quality control protocols. Like contaminations from fast-growing microbes, the entire batch must be discarded. Contamination may occur from the same circumstances as other microbes, but they may come with the meat cell lines, demanding pre-upstream cell line decontamination.

Viral contaminations typically occur due to particles infecting the donor animal, which are then carried into cell lines. As their dimensions do not allow for microscopic observations, molecular techniques must be used for their detection during the acquisition of cell lines and to monitor their presence during upstream steps. They can also be introduced through animal-derived components such as FBS media and co-factors.

Due to their hydrophobic and thermostable nature, heat-stable lipopolysaccharides (LPS, endotoxin) may adhere to plastic components and resist autoclaving. LPS may arise from impure water, media components, and recombinant proteins produced using Gram-negative bacteria. Unlike active microbes that can coproliferate during cell growth, LPS concentrations do not vary, and there is no consensus on whether their mere presence can pose a threat to cell lines or consumers. However, a particular concern arises when disposable plastic bioreactors with a large internal surface area are employed.

Logical considerations prompt producers to adhere to Good Manufacturing Practices (GMPs) for sterile drug products, as advocated by the Food and Drug Administration (FDA) in the USA and the ECA Foundation in Europe. Both entities follow the guidelines published by the US Department of Health and Human Services (2004), specifying the necessity of working in such facilities.

Sterilization is an absolute term implying the complete inactivation or removal of living bacteria (including spores), fungi, archaea, algae, protozoa, microscopic metazoan, and infectious particles such as viruses, virus-like particles, prions, and aptamers from a solid surface, gas, or liquid bulk. In the cultivated meat industry, where intermediary and final products are highly sensitive to microbial contaminations, sterilization must adhere to rigorous protocols, and support laboratories must ensure sterility throughout the production pathway.

The process begins with the inter-batch cleansing of the bioreactor vessel, ensuring no retention of attached sediments (soil) from the previous fermentation. An effective and recommended method for this is the clean-in-place (CIP) system, employing high-velocity flow (~2.0 m/s), jet sprays, and agitation of hot solutions of surfactants, acids, alkalis, or sanitizers applied to internal surfaces of the vessel, sensors, valves, pumps, sanitary couplings, instrumentation (sensors, baffles, turbines, and impellers), and instrument ports (Chisti and Moo-Young 1994). Industrial bioreactors have intricate piping circuits for supplying sterile air/gases, acid/alkali, antifoam, and nutrients. Their cleansing is conducted with a flow rate dependent on the line diameter (McNulty 2016).

Cleaning chemicals employed for CIP must reduce the surface tension of water, disrupt bonding forces between soil and bioreactor/inner pipe surfaces, dissolve fats and soils to facilitate rinsing, and emulsify water-soluble soil for easy transport. Alkaline solutions of 0.5–2.0% NaOH are typically the initial agent added to the bioreactor, followed by the neutralizing 0.5% $HNO_3$ and subsequent rinsing with high-quality water.

After a thorough inspection, the cleaned system undergoes sterilization. The most suitable modality is sterilization-in-place (SIP), wherein the bioreactor vessel is filled with the liquid culture medium base containing thermostable components. Heat and pressure-tolerant sensors are immersed in the culture broth. In this manner, the bioreactor serves as an in situ autoclave, achieving microbial killing when correct temperature and pressure parameters are maintained (Junker et al. 1999). Interior ducts in the piping, sanitary couplings, and inlet/outlet gas filters are also sterilized (Junker et al. 2006). After cooling, heat-sensitive components are added following strict aseptic conditions, involving pre-solubilization and filtration, cells are introduced, and the growth phase begins.

Chemical sterilization with chlorine derivates and peroxides, as conducted in many standard fermentation processes, is contraindicated once internal crevices and other retentive areas can retain minimal residual soil that covers contaminants. Furthermore, after applying such sanitizers, it becomes necessary to rinse the surfaces with high volumes of sterile water, increasing the process expenses.

There is a growing trend towards replacing traditional stainless-steel bioreactors with single-use bioreactors, some of which have capacities as large as 6000 L (Schirmer et al. 2022). Different parts of these disposable bioreactors are constructed with food-safe plastics, such as polyethylene (PE), polypropylene (PP), polystyrene (PS), and polytetrafluoroethylene (PTFE) (Chen et al. 2022a). The advantages advocated are no sterilization needed, reduced microbial or chemical contamination risks, lower energy needs, and shorter downtime between batches. Even with the problem of the destination of disposed used vessels, they still represent an environmental impact smaller than the traditional bioreactors due to the lower energy and water footprints (Rawlings and Pora 2009). However, high costs and scaling-up issues persist as barriers to their adoption by the cultivated meat industry (Djisalov et al. 2021).

The stringent adherence to GMPs for sterile goods production ensures the desired elimination of the risk of microbial contamination. Consequently, there is no need for antibiotics and other inhibitory chemicals, such as sodium benzoate (Chodkowska et al. 2022). This is particularly beneficial for consumers, especially children, who may be allergic to these preservatives (Jacob et al. 2016). Similar considerations may be applied to pyrogens and toxins derived from bacterial or fungal origins (Wassenaar and Zimmermann 2018; Chodkowska et al. 2022).

The oxygen supply for muscle cells must guarantee a sterile condition, regardless of its source (compressed air or pure/mixture oxygen tank). Modern membrane filters retain microbes and particulate matter, maintaining their integrity even under positive pressures of 1 MPa (10.2 kg/cm$^{-2}$), which is more than sufficient to support cell growth (Pall Corporation, Gas Filter Makes Bioreactors Safer 2005).

## *4.2 Quality and Risk Control During Downstream Processes*

After upstream processing, the supply of oxygen and nutrients is discontinued, initiating cell death (Bohn et al. 2023). Biomass transfer, sedimentation, centrifugation, and filtration operations must be executed as soon as possible and under standardized sanitary conditions. These operations should take place in controlled rooms equipped with chillers to ensure low temperatures and air filters preventing the circulation of airborne particles carrying microbes (Del Valle 2004). Humbird (2021) proposed that upstream processes should be conducted in class ISO 8 cleanrooms (maximum 100,000 particles (≥0.5 μm) per cubic foot of air with a replacement rate of 20 exchanges per hour), and cell seedlings would require a class ISO 6 cleanroom (maximum 1000 particles (≥ 0.5 μm) per cubic foot of air with a replacement rate of 180 exchanges per hour). As the author did not specify the conditions for downstream, it is reasonable to suggest that a class ISO 8 cleanroom is suitable to maintain good cleanliness conditions. Minimizing contact between operators and downstream intermediary and final products following good hygiene practices (GHPs) must be implemented.

### 4.2.1 Supporting Laboratories and Facilities

The presence of well-equipped supporting laboratories and specialized personnel for controlling cell growth, monitoring contamination, and assessing performance parameters in batch-to-batch fermentations is a sine qua non condition for reliable cultivated meat production. The continuous monitoring of microbes, allergens, and toxins in all upstream and downstream steps ensures the high safety standards required by health authorities and helps prevent losses.

# 5 Quality Control and Safety Assessment of the Final Product

Quality and safety control are indispensable tools across all stages of cultivated meat production, including the final cultivated product. In this final stage, they are determined by assessing strict conformity of fresh or processed cultivated meat without any spoilage or damage to their sensory characteristics to meet consumers' safety and expectations. All these tools follow protocols that encompass a spectrum of measures linked to the chemical, physical, microbiological, and sensory attributes of cultivated meat, similar to conventional meat, but with the need for adaptations.

## *5.1 pH*

As a preventive analysis for the quality of the cultivated product that will reach the consumer's table, assessing the pH becomes essential because it often exhibits correlations with microbial and chemical reactions (Lu et al. 2022). After animal slaughter, the muscle pH is around 7.0 and decreases because of lactic acid accumulating in muscle tissue due to the anaerobic glycolysis process. In 24 h, this value is around 5.70, which is within the normal range, and most microorganisms can initiate their growth. When meat maintains a pH close to 7 (neutral) or slightly acidic, conditions are favourable for bacterial growth (Rahman and Rahman 2020). However, microorganisms are highly sensitive to pH variations, and a reduction in pH leads to a decline in the microbial growth rate, impacting bacteria. From a practical perspective, this implies that meat with a higher pH (equal to or exceeding 6.0) is more vulnerable to microbial activities and putrefaction when compared to meat with a pH below 6.0 (Feitosa 1999).

The pH can be measured in a simple way using a pH meter that must be calibrated with standardized buffers at pH 7.0 and pH 4.0 before meeting the sample surface (Sahar et al. 2019). To the extent of the current knowledge, no published report has assessed pH as a quality and safety indicator in cultivated meat. In a premarket study submitted to the FDA to obtain approval to market cultivated chicken meat, the pH of meat models cultivated in both serum-containing and serum-free medium was 6.2 and 5.91, respectively (Schulze 2022).

On the other hand, the importance of pH assessment is not limited only to the concern with bacterial growth and the risk of putrefaction in cultivated meat. The pH value exerts a noteworthy influence on protein stability. When the pH reaches the isoelectric point (5.5) of myofibrillar proteins in meat, the electrical charges in those proteins neutralize. This can result in the loss of proteins' ability to retain water, affecting the texture and water-holding capacity (WHC) of meat (Lu et al. 2022).

## 5.2 Water-Holding Capacity (WHC) and Other Technological Properties

Water-holding capacity (WHC), gelation, and emulsification are key technological characteristics of the quality, safety, and shelf life of meat. Their assessment allows the determination of quality indicators such as yield; exudation and loss of water-soluble proteins, vitamins, and minerals; and susceptibility to spoilage bacteria growth, as well as sensory properties in cooked meat (Garba et al. 2019; Karakaya et al. 2005; Oswell et al. 2021).

The formation of the actomyosin complex through myofibrillar proteins is an important property of conventional meat, and the WHC, gelation, and emulsification capacities depend on it. Embryonic and neonatal isoforms of actin and myosin, found in developing muscle fibres, need to undergo a maturation process to transform into adult isoforms (Fraeye et al. 2020; Broucke et al. 2023). To the extent of the current knowledge, cultivated meat production primarily relies on a small portion of embryonic and neonatal isoforms of actin and myosin, potentially exerting an influence on these properties (Zidarič et al. 2020).

WHC is characterized by the ability of fresh meat to retain water in different conditions, such as processing, storage, and cooking (Kamruzzaman et al. 2016). This property is commonly evaluated using methods that apply force to measure the extracted water. This force can be applied naturally, such as by gravimetric means (drip loss), weep or purge, and exudate scoring, or externally, such as by pressure, centrifugation, compression, and cooking (Warner 2023). Zheng et al. (2021) evaluated WHC in a cultivated meat model based on porcine smooth muscle cells, using the filter paper pressure method, and showed that when denatured, some collagen-type cells formed a block of macromolecules that prevented the water loss.

Emulsifying capacity (EC) and emulsion stability (ES) are important metrics for evaluating the emulsifying potential of proteins in meat, considering water and fat molecules. The widely used EC evaluation method is through emulsion interfacial shear rheology. Other methods to evaluate ES include sizing emulsion particles/droplets through light scattering (static or dynamic), electrical pulse counting, and ultrasonic spectrometry (Fu et al. 2021; Hu et al. 2017).

Striving for technological quality compliance in cultivated meat, scaffolds based on nanofibres and electrical and/or mechanical stimulation can be used during the cultivating stage to increase the maturation and diameter of myofibres, reaching the adult isoforms of actin and myosin and providing greater efficacy of sensory and technological properties (Arshad et al. 2017; Bomkamp et al. 2022; Khodabukus et al. 2019). Consumable scaffolds based on microcarriers, hydrogel, and collagen can be added to cultivated meat by mixing cellular biomass with the structural component provided by the scaffold to improve WHC, gelation, and emulsification of cell-based products (Bomkamp et al. 2022; Alexandretti et al. 2019).

Incorporating scaffolds into cultivated meat production requires the evaluation of these components' allergenic properties (FAO/WHO 2023). Additionally, any additives introduced in the formulation of cultivated meat products must undergo approval via analytical testing and quality control procedures. It is essential to scrutinize potential hazards and contaminants that could pose allergenic or residual risks, as this assessment is pivotal in approving the product for human consumption and ensuring consistent quality control.

## 5.3 Texture Properties

The intricate composition of meat, including muscle fibres, fats, organized connective tissues, and other muscle components, plays a significant role in determining its texture. Attributes like tenderness and juiciness are sensory qualities that play a pivotal role in determining meat product quality and overall consumer preference (Chen et al. 2022b).

Texture arises from an interplay of variables, including tissue composition, micro- and macrostructural tissue arrangement, and water and fat retention during cooking. Current cultivated meat methodologies confront obstacles such as inadequate actin and myosin levels, along with premature isoform expression (Thorrez and Vandenburgh 2019). This discovery raises the possibility of an altered protein response to potential post-mortem transformations. Should these transformations be absent, the conversion of muscle to meat, characterized by biochemical differences, would not take place (Datar and Betti 2010).

The conversion of muscle into meat through post-mortem metabolism, primarily driven by anaerobic glycolysis converting muscle glycogen to lactate, is a natural process occurring after slaughter. However, this essential biological phenomenon is not present in cultivated meat, and scientific evidence confirming this absence remains lacking (Balasubramanian et al. 2021). In the event of the non-occurrence of rigor mortis and the consequent non-formation of the actomyosin complex, the texture could be significantly affected, with the potential enhancement of tenderness and water-holding capacity compared to traditional meat, but these alterations might also have implications for the subsequent aging process (Fraeye et al. 2020).

The texture of cultivated meat presents varying technological challenges based on the type of meat being produced, with the task of achieving a satisfying texture resembling fresh meat being notably more demanding than that of crafting ground or finely minced meat products (Fraeye et al. 2020). Achieving consumer acceptance of cultivated meat products hinges on closely replicating the taste and texture of natural meat, yet this endeavour is complex due to the impact of several factors such as blood vessels, nerve tissues, intramuscular fats, and muscle cell maturation, all of which contribute to the overall texture and flavour of meat (Chen et al. 2022b).

Moreover, an inappropriate texture can be an indicator of issues in the production process, from cell cultivation to the formulation and processing of the final product. Problems in cell cultivation, such as an unsuitable growth environment, nutritional imbalance, or cellular stress, can adversely affect the formation of muscle fibres and, consequently, the final texture of cultured meat. Issues such as excessive hardness, disintegration, irregularities in tissue surface, or even excessive water loss can be indicators of texture-related problems, underscoring the importance of their monitoring and assessment for the quality and safety of the final product, whether it is marketed fresh or processed.

In general, the techniques used in the evaluation of conventional meat appear to apply to the assessment of cultivated meat, depending on its structure. Some techniques are more suitable for certain structures. Texture profile analysis (TPA), shear force, and sensory evaluation can be applied to both conventional and cultivated meat to assess texture and tenderness. On the other hand, ultrasonography, rheological analysis, magnetic resonance imaging (MRI), hyperspectral imaging (HSI), nuclear magnetic resonance (NMR), and digital image analysis can be useful for evaluating specific aspects of the cultivated meat structure, depending on the desired characteristics. For processed products, there is a greater preference for the extensive information provided by TPA as opposed to rheological analysis (Paredes et al. 2022). The use of this methodology can play a crucial role in optimizing processing strategies and elucidating unknown variables in the emerging field of cultivated meat.

Different techniques can be employed to enhance the texture of cultivated meat to meet the expectations associated with animal-derived products. One of them is 3D bioprinting, which enables the attainment of desired technological properties, including the incorporation of different tissues, such as muscle, adipose, blood vessels, and connective tissue (Kang et al. 2021). Edible microcarriers can also be an alternative to achieve satisfactory texture characteristics. The production of cultivated meat can be layered, using aggregates of microtissues that provide greater firmness and nutritional value, or homogenized in the form of a burger. This approach allows for the evaluation of the products using a conventional texture analyser to determine deformation, adhesiveness, cohesiveness, and chewiness, following the same methodology used for conventional meat evaluation (Yen et al. 2023).

Another alternative to enhance the texture of cultivated meat, which can be evaluated using established methodologies, involves the use of myocyte hydrogel modules to create a three-dimensional structure. This allows muscle bundles to contract in response to electrical stimulation, a technique also employed in conventional meat production for muscle tenderization (Furuhashi et al. 2021).

## *5.4 Flavour Properties*

Cultivated meat may be biologically equivalent to conventional meat, but the scientific evaluation of flavour characteristics is a challenging task at this stage due to the lack of sufficient experimental samples that could support a comprehensive analysis. Furthermore, the characteristic flavour found in conventional meat appears to be influenced by the intricate composition structure of the muscle, encompassing blood vessels, blood, nervous and connective tissues, and perhaps, most notably, intramuscular, intermuscular, and subcutaneous fats. While applying heat, complex thermally induced reactions occur among nonvolatile components such as amino acids, sugars, vitamins, lipids, and nucleotides, resulting in the production of numerous volatile compounds like thiophenes, thiazoles, oxazoles, pyrazines, furans, and others. Some of these constituents contribute to the distinctive flavour profile of meat (Chen et al. 2022b).

The intensity of meat flavour is also affected by the age of the animal at the time of slaughter, in addition to the animal rearing production system, which can impart a distinct taste to the final product (Arshad et al. 2018). These factors are mainly influenced by the fat composition, featuring unique fatty acid profiles and a diverse array of aromatic compounds, derived from the dietary sources, encompassed within the fats. Following post-mortem meat aging, umami-inducing amino acids and additional aromatic compounds are generated through the processes of proteolysis and lipolysis, respectively.

The Maillard reaction is important in forming meat flavour, elucidating the intricate interplay between carbonyl and amine reactions (Arshad et al. 2018). Typically observed in meat cooked at elevated temperatures, the browning phenomenon occurs due to the interaction between a free amino acid and a carbonyl group, resulting in the glycosylamine formation. The subsequent dehydration and rearrangement of glycosylamine lead to the formation of furanone derivatives, furfural, dicarbonyl compounds, and hydroxyketones, which enrich the meat's flavour profile (Fu et al. 2022).

Considering the significance of post-mortem ageing and cooking in the creation of flavour precursor molecules, it becomes essential to replicate both procedures when assessing the flavour of cultivated meat prototypes in comparison to animal-derived meat. Some cell types exhibit the ability for in vitro adipogenic differentiation, encompassing mesenchymal stem cells derived from bone and fat, de-differentiated fat cells, embryonic stem cells, induced pluripotent stem cells, and satellite cells.

The monitoring and evaluation of flavour in cultured meat can be conducted through various analytical approaches. One of these methods involves sensory evaluation conducted by a panel of tasters, who evaluate the flavour of the product through blind taste tests. They may be asked to evaluate attributes such as taste, flavour, texture, and overall acceptance. This approach is valuable for understanding how humans perceive the flavour of the product.

Another approach involves the analysis of volatile compounds through gas chromatography coupled with mass spectrometry (GC-MS). The aroma profile entails the analysis of volatile compounds contributing to the aroma of cultured meat. Mass spectrometry with olfactometry (GC-MS-O) is a technique that combines chemical analysis with sensory evaluation to identify odoriferous components. Some other instruments have been developed to objectively measure flavour characteristics as well, such as the electronic nose and tongue analyser, which uses sensors to detect and quantify flavours and aromas. These analyses have been employed in studies that conducted tests on conventional meat and seem promising for the evaluation of cultivated meat, as well.

The evaluation of cultured fat derived from fibro-adipogenic cells through sensory analysis showed that its texture can be creamy and meaty in taste after cooking while also resembling conventional fat in appearance, structure, and lipid accumulation. Conventional lipidomic analysis demonstrated an increase in lipid content during differentiation, with the cultured fat approaching subcutaneous fat and muscle composition. Notably, the cultured fat contained higher levels of monounsaturated fatty acids, associated with desirable flavour, resembling traditional beef fat in lipid content and taste (Dohmen et al. 2022). Moreover, cultured adipose cells from pigs exhibited a texture profile akin to subcutaneous fat. Furthermore, the analysis of volatile organic compounds revealed the presence of 28 compounds, with 17 of them being notably more abundant in the cultured fat. This finding suggests a promising avenue for achieving the anticipated flavour profiles in cultured meat products (Liu et al. 2023).

On the other hand, a preliminary hamburger prototype, primarily composed of muscle cells with a minimal presence of fibroblasts, lacked the essential aromatic compounds found in the fat portion of traditional meat. The inadequate fat content in the prototype resulted in a dry texture, negatively affecting the overall taste experience (Post and Hocquette 2017).

Using the electronic tongue to assess the sensory properties of cultivated beef and chicken meat, Joo et al. (2022) compared the amino acid composition between cultured and traditional muscle tissue and found significant differences in the amino acid profiles. Cultured muscle tissue showed higher levels of serine, proline, alanine, and arginine, whereas traditional meat exhibited a higher content of aspartic acid, isoleucine, leucine, and lysine. The presence of umami, associated with glutamic and aspartic acids, was lower in cultured meat compared to traditional meat, highlighting the importance of developing strategies to address this difference, such as adjusting the culture medium used for cell growth to enhance the umami intensity of cultured meat, aiming to attain a taste profile on par with that of traditional meat.

In a side-by-side sensory analysis comparison among a hybrid product consisting of cultured chicken and textured soy protein, a soy-based product, and conventionally farmed chicken, it was found that the cultured chicken meat exhibited fibre cross-sections similar to conventional chicken meat and the distinctive Maillard browning in the meat (Pasitka et al. 2023). The hybrid cultured chicken product had a closer resemblance in flavour to conventional chicken than the soy-based product and showed no visual differences compared to conventional chicken meat.

Understanding the complexities of the flavour of cultivated meat and implementing precise adjustments in cultivation and processing techniques is essential to provide a product that matches the taste of traditional meat. This requires a multidisciplinary approach, encompassing all the stages of production from cell cultivation techniques to sensory analyses, to ensure flavour excellence and consumer acceptance in cultivated meat.

## 5.5 Colour

The colour is one of the most important attributes that affect the quality and consumer acceptance of the meat (Tomasevic et al. 2021). The meat colour is attributed to the presence of myoglobin in the muscle cell, which behaves like a pigment

(Osório et al. 2009). Myoglobin, a conjugated protein, is generated through a bond between a prosthetic group (heme) and a molecule of a globular protein (globulin). The heme group has in its centrality an iron atom which can be found both in its ferrous form ($Fe^{2+}$) and in its ferric form ($Fe^{3+}$). When myoglobin is exposed to oxygen, it presents itself in its chemical form of oxymyoglobin, developing a bright cherry red colour, which is attractive to consumers (Passetti et al. 2016).

In cultivated meat, the presence of myoglobin is minimal, and therefore its appearance is almost colourless. Its expression is suppressed under environmental oxygen conditions (Post and Hocquette 2017). This colour failure can be mitigated by increasing myoglobin expression early in cell culture. In situations of hypoxia, this stimulation can be done by the addition of lipids and acetic acid and by the addition of stable haemoglobin or enough iron in the cell, although the impact on cell viability in vitro is still unknown (Broucke et al. 2023; Fraeye et al. 2020; Moritz et al. 2015; Siddiqui et al. 2022). A study performed by Simsa et al. (2019) demonstrated that the addition of metmyoglobin (oxidized form of myoglobin) directly in the culture medium increased the capacity of cell proliferation and, consequently, the amount of myoglobin present in the cultured cells. However, metmyoglobin has given a brown colour to the meat due to its inherent oxidation.

In addition to the incorporation of additives directly into the media, natural dyes like beet juice and turmeric can be used in the stage after the production of the cultivated meat, although taste may be affected (Broucke et al. 2023). The addition of binders is also an option for processed products from cell culture (GFI Brazil 2022). Plant-based binders, such as guar gum, xanthan gum, collagen, and psyllium, can be added not only to improve the texture of the food but also to fix the pigment particles in it (Pedroso and Demiate 2008).

The soy haemoglobin in *Pichia pastoris*, a genetically modified methylotrophic yeast approved by the FDA to be used in plant-based burger formulation to provide a similar colour and flavour as beef, could be an alternative colour enhancer for cultivated meat. However, it is not clear whether this soy haemoglobin can be used in cell culture since there is no evidence that *P. pastoris* is a food-grade host (Fraeye et al. 2020; Zhang et al. 2020).

Instrumental methods can be used for colour measurement using colorimeters or spectrophotometers, and both instruments measure tristimulus values such as Lab*, where L* indicates luminosity, a* indicates the red/green coordinate (+a values indicate redness and −a values indicate greenness), and b* indicates the yellow/blue coordinate (+b values indicate yellowness and −b values indicate blueness), providing combinations under illuminants and standardized observers (Ferreira and Spricigo 2017; King et al. 2023). Other devices can also be used to measure the colour of meat. Simsa et al. (2019) evaluated the colour of cultivated beef using a digital camera (Canon DS126311) with two fluorescent lamps. The images obtained were analysed in software for red, blue, and green (RBG) colours measuring L* a* b*.

Just as important as the colour of the meat is the colour of the fat. Conventional meat can have two different fat colours, white or yellow. Usually, white fat comes from animals raised in feedlot systems, which predominantly feed on grain. Yellow fat comes from animals raised in pasture systems, which consume more carotenoid-rich green vegetables (Meléndez-Martínez et al. 2022). Carotenoids are a large group of fat-soluble pigments responsible for the orange, yellow, and red coloration of various organisms, such as vegetables (Uenojo et al. 2007), and act as antioxidants, reducing lipid oxidation in meat. Stout et al. (2020) demonstrated that genetically modified primary bovine and immortalized murine skeletal muscle cells can produce endogenous carotenoids at levels capable of significantly reducing lipid oxidation, which is beneficial for the colour stability of meat. However, it is not yet known whether the use of these engineering cell approaches would be accepted by consumers.

In summary, comprehensive assessments play a crucial role in addressing technological and sensory defects in cultivated meat, managing the growth of foodborne and spoilage microorganisms, and mitigating oxidative reactions. These measures are essential for upholding both the safety and quality standards of cultivated meat products.

As illustrated in this chapter, there are distinct risks and potential hazards involved in each stage of cultivated meat production. Figure 1 provides an overview of the quality and safety controls that should be addressed during the different stages of cultivated meat production.

## 6 Final Considerations

The production of cultivated meat, from cell collection to the final product, demands a constant commitment to precision. Whether it involves fundamental controls over cell sourcing and characterization, the establishment of secure cell banks, the management of complex bioreactor systems, or the assessment of final product quality, quality control and risk analysis emerge as crucial pillars. In the landscape of cellular agriculture, these principles are not merely theoretical constructs but manifest as indispensable operational mechanisms. Their implementation is crucial to guarantee the production of high-quality cultivated meat products.

## References

Alexandretti C, Verlindo R, Hassemer GDS, Manzoli A, Roman SS, Fernandes IA, Backes GT, Cansian RL, Soares MBA, Schwert R, Valduga E (2019) Structural and techno-functional properties of bovine collagen and its application in hamburgers. Food Technol Biotechnol 57(3):369–377. https://doi.org/10.17113/FTB.57.03.19.5896

Arshad MS, Javed M, Sohaib M, Saeed F, Imran A, Amjad Z (2017) Tissue engineering approaches to develop cultured meat from cells: a mini review. Cogent Food Agric 3(1):1320814. https://doi.org/10.1080/23311932.2017.1320814

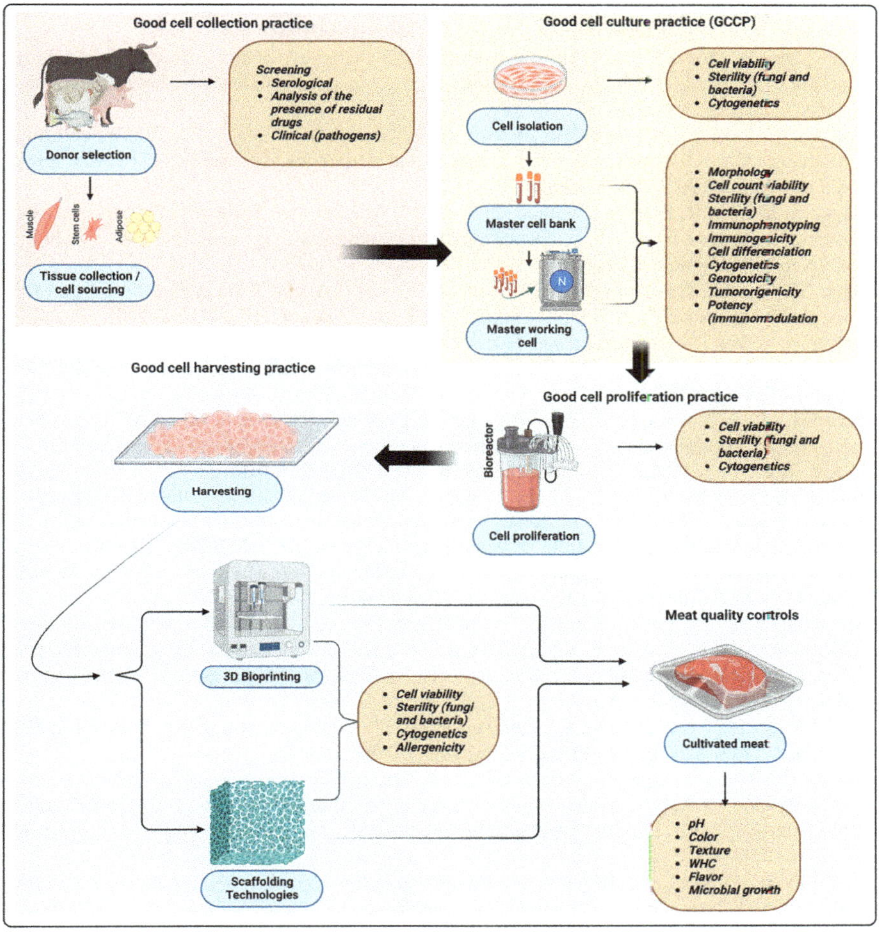


**Fig. 1** Hazards and quality controls in cultivated meat production stages. Created with BioRender.com

Arshad MS, Sohaib M, Ahmad RS, Nadeem MT, Imran A, Arshad MU, Kwon JH, Amjad Z (2018) Ruminant meat flavor influenced by different factors with special reference to fatty acids. Lipids Health Dis 17(1):223. https://doi.org/10.1186/S12944-018-0860-Z

ATA Scientific Instruments (2020) How to efficiently transport live cell cultures without freezing them. ATA Biomolecular Science—Guide. https://www.atascientific.com.au/how-to-efficiently-transport-live-cell-cultures-without-freezing-them/. Accessed 02 Sep 2023

Balasubramanian B, Liu W, Pushparaj K, Park S (2021) The epic of in vitro meat production—a fiction into reality. Foods 10(6):1395. https://doi.org/10.3390/FOODS10061395

Ben-Arye T, Levenberg S (2019) Tissue engineering for clean meat production. Front Sustain Food Syst 3:46. https://doi.org/10.3389/fsufs.2019.00046

Ben-Arye T, Shandalov Y, Ben-Shaul S, Landau S, Zagury Y, Ianovici I et al (2020) Textured soy protein scaffolds enable the generation of three-dimensional bovine skeletal muscle tissue for cell-based meat. Nat Food 1(4):210–220. https://doi.org/10.1038/s43016-020-0046-5

Bhat ZF, Fayaz H (2011) Prospectus of cultured meat—advancing meat alternatives. J Food Sci Technol 48:125–140. https://doi.org/10.1007/s13197-010-0198-7

Bohn S, Ung D, Moses T (2023) How to overcome the challenges of cost-effective cultured meat production at scale. CRB Group. https://www.crbgroup.com/insights/food-beverage/cultured-meat. Accessed 23 Aug 2023

Bomkamp C, Skaalure SC, Fernando GF, Ben-Arye T, Swartz EW, Specht EA (2022) Scaffolding biomaterials for 3D cultivated meat: prospects and challenges. Adv Sci 9(3):2102908. https://doi.org/10.1002/ADVS.202102908

Bonassi S, El-Zein R, Bolognesi C, Fenech M (2011) Micronuclei frequency in peripheral blood lymphocytes and cancer risk: evidence from human studies. Mutagenesis 26(1):93–100. https://doi.org/10.1093/mutage/geq075

Bonny SP, Gardner GE, Pethick DW, Hocquette JF (2017) Artificial meat and the future of the meat industry. Anim Prod Sci 57(11):2216–2223. https://doi.org/10.1071/an17307

Borgonovo T, Vaz IM, Senegaglia AC, Rebelatto CLK, Brofman PRS (2014) Genetic evaluation of mesenchymal stem cells by G-banded karyotyping in a cell technology center. Rev Bras Hematol Hemoter 36:202–207. https://doi.org/10.1016/j.bjhh.2014.03.006

Borgonovo T, Solarewicz MM, Vaz IM, Daga D, Rebelatto CLK, Senegaglia AC et al (2015) Emergence of clonal chromosomal alterations during the mesenchymal stromal cell cultivation. Mol Cytogenet 8:1–5. https://doi.org/10.1186/s13039-015-0197-5

Broucke K, Van Pamel E, Van Coillie E, Herman L, Van Royen G (2023) Cultured meat and challenges ahead: a review on nutritional, technofunctional and sensorial properties, safety and legislation. Meat Sci 195:109006. https://doi.org/10.1016/J.MEATSCI.2022.109006

Bryant C, Barnett J (2018) Consumer acceptance of cultured meat: a systematic review. Meat Sci 143:8–17. https://doi.org/10.1016/j.meatsci.2018.04.008

Carrasco-García AA, Pardío-Sedas VT, León-Banda GG, Ahuja-Aguirre C, Paredes-Ramos P, Hernández-Cruz BC, Murillo VV (2020) Effect of stress during slaughter on carcass characteristics and meat quality in tropical beef cattle. Asian Australas J Anim Sci 33(10):1656. https://doi.org/10.5713/ajas.19.0804

Carrillo-Ávila JA, Catalina P, Aguilar-Quesada R (2022) Quality control of cell lines using DNA as target. DNA 2(1):44–55. https://doi.org/10.3390/dna2010004

Catalina P, Montes R, Ligero G, Sanchez L, de la Cueva T, Bueno C et al (2008) Human ESCs predisposition to karyotypic instability: is a matter of culture adaptation or differential vulnerability among hESC lines due to inherent properties? Mol Cancer 7:1–9. https://doi.org/10.1186/1476-4598-7-76

Chen Y, Zajac JD, MacLean HE (2005) Androgen regulation of satellite cell function. J Endocrinol 186(1):21–31. https://doi.org/10.1677/joe.1.05976

Chen L, Guttieres D, Koenigsberg A, Barone PW, Sinskey AJ, Springs SL (2022a) Large-scale cultured meat production: trends, challenges and promising biomanufacturing technologies. Biomaterials 280:121274. https://doi.org/10.1016/j.biomaterials.2021.121274

Chen YP, Feng X, Blank I, Liu Y (2022b) Strategies to improve meat-like properties of meat analogs meeting consumers' expectations. Biomaterials 287:121648. https://doi.org/10.1016/J.BIOMATERIALS.2022.121648

Chisti Y, Moo-Young M (1994) Clean-in-place systems for industrial bioreactors: design, validation and operation. J Ind Microbiol Biotechnol 13(4):201–207. https://doi.org/10.1007/BF01569748

Chodkowska KA, Wódz K, Wojciechowski J (2022) Sustainable future protein foods: the challenges and the future of cultivated meat. Foods 11(24):4008. https://doi.org/10.3390/foods11244008

Cimino M, Gonçalves RM, Bauman E, Barroso-Vilares M, Logarinho E, Barrias CC, Martins MCL (2018) Optimization of the use of a pharmaceutical grade xeno-free medium for in vitro expansion of human mesenchymal stem/stromal cells. J Tissue Eng Regen Med 12(3):e1785–e1795. https://doi.org/10.1002/term.2588

Coecke S, Balls M, Bowe G, Davis J, Gstraunthaler G, Hartung T et al (2005) Guidance on good cell culture practice: a report of the second ECVAM task force on good cell culture practice. Altern Lab Anim 33(3):261–287. https://doi.org/10.1177/026119290503300311
Cornélio DA, Tavares JCM, Pimentel TVCDA, Cavalcanti GB Jr, Batistuzzo de Medeiros SR (2014) Cytokinesis-block micronucleus assay adapted for analyzing genomic instability of human mesenchymal stem cells. Stem Cells Dev 23(8):823–838. https://doi.org/10.1089/scd.2013.0383
Datar I, Betti M (2010) Possibilities for an in vitro meat production system. Innovative Food Sci Emerg Technol 11(1):13–22. https://doi.org/10.1016/J.IFSET.2009.10.007
de Oliveira TFP, Fonseca AA Jr, Camargos MF, de Oliveira AM, Cottorello ACP, dos Reis Souza A et al (2013) Detection of contaminants in cell cultures, sera and trypsin. Biologicals 41(6):407–414. https://doi.org/10.1016/j.biologicals.2013.08.005
Del Valle MA (2004) Keeping biopharmaceutical cleanrooms compliant. BioPharm Int 17(3):50–60
Djisalov M, Knežić T, Podunavac I, Živojević K, Radonic V, Knežević N, Bobrinetskiy I, Gadjanski I (2021) Cultivating multidisciplinarity: manufacturing and sensing challenges in cultured meat production. Biology 10(3):204. https://doi.org/10.3390/BIOLOGY10030204
Dohmen RGJ, Hubalek S, Melke J, Messmer T, Cantoni F, Mei A, Hueber R, Mitic R, Remmers D, Moutsatsou P, Post MJ, Jackisch L, Flack JE (2022) Muscle-derived fibro-adipogenic progenitor cells for production of cultured bovine adipose tissue. NPJ Sci Food 6(1):1–12. https://doi.org/10.1038/s41538-021-00122-2
Dorado G, Gálvez S, Rosales TE, Vásquez VF, Hernández P (2021) Analyzing modern biomolecules: the revolution of nucleic-acid sequencing – review. Biomolecules 11(8):1111. https://doi.org/10.3390/biom11081111
European Commission (2017) Guidelines on good manufacturing practice specific to advanced therapy medicinal products. Eur Commun J 4:1–32. https://ec.europa.eu/health/sites/health/files/files/eudralex/vol-4/2017_11_22_guidelines_gmp_for_atmps.pdf. Accessed 25 July 2023
European Medicines Agency (2013). Guideline on the use of bovine serum in the manufacture of human biological medicinal products. https://www.ema.europa.eu/en/documents/scientific-guideline/guideline-use-bovine-serum-manufacture-human-biological-medicinal-products_en.pdf. Accessed 25 July 2023
European Medicines Agency (2019) Guideline on quality, non-clinical and clinical requirements for investigational advanced therapy medicinal products in clinical trials. https://www.ema.europa.eu/en/guideline-quality-non-clinical-clinical-requirements-investigational-advanced-therapy-medicinal. Accessed 25 July 2023
FAO (2004) Good practices for the meat industry, vol 2. FAO Publishing Manager Service, Rome
FAO/WHO (2021) Microbiological risk assessment - guidance for food. Microbiological risk assessment series no. 36. Rome. https://doi.org/10.4060/cb5006en
FAO/WHO (2023) Food safety aspects of cell-based food. Food and Agriculture Organization of the United Nations-World Health Organization, Rome, p 146
FDA (1993) Characterization of cell lines used to produce biologicals. Congressional and Consumer Affairs Branch (HFM-12) - Food and Drug Administration. https://www.fda.gov/media/76255/download. Accessed on 02 Sep 2023
Feitosa TC (1999) Conservação e Alteração da Carne. EMBRAPA–CNTAP, Fortaleza, p 34
Ferreira MD, Spricigo PC (2017) Colorimetria-princípios e aplicações na agricultura
Fraeye I, Kratka M, Vandenburgh H, Thorrez L (2020) Sensorial and nutritional aspects of cultured meat in comparison to traditional meat: much to be inferred. Front Nutr 7:520512. https://doi.org/10.3389/FNUT.2020.00035/BIBTEX
Fu Y, Chen T, Chen SHY, Liu B, Sun P, Sun H, Chen F (2021) The potentials and challenges of using microalgae as an ingredient to produce meat analogues. Trends Food Sci Technol 112:188–200. https://doi.org/10.1016/J.TIFS.2021.03.050
Fu Y, Cao S, Yang L, Li Z (2022) Flavor formation based on lipid in meat and meat products: a review. J Food Biochem 46(12):e14439. https://doi.org/10.1111/JFBC.14439

Furuhashi M, Morimoto Y, Shima A, Nakamura F, Ishikawa H, Takeuchi S (2021) Formation of contractile 3D bovine muscle tissue for construction of millimetre-thick cultured steak. NPJ Sci Food 5(1):1–8. https://doi.org/10.1038/s41538-021-00090-7

Gao D, Critser JK (2000) Mechanisms of cryoinjury in living cells. ILAR J 41(4):187–196. https://doi.org/10.1093/ilar.41.4.187

Garba S, Sazili AQ, Mahadzir MF, Candyrine SCL, Jahromi MF, Ebrahimi M, Goh YM, Samsudin AA, Ronimus R, Meutzel S, Liang JB (2019) Effects of naturally-produced lovastatin on carcass characteristics, muscle physico-chemical properties and lipid oxidation and cholesterol content in goats. Meat Sci 154:61–68. https://doi.org/10.1016/J.MEATSCI.2019.04.008

GFI (2022) Estudo Regulatório sobre Proteínas Alternativas no Brasil—Carne Cultivada. https://gfi.org.br/wp-content/uploads/2022/11/Estudo-Regulatorio-Carne-Cultivada-GFI-Brasil.pdf. Accessed 25 July 2023

Gottipamula S, Muttigi MS, Chaansa S, Ashwin KM, Priya N, Kolkundkar U et al (2016) Large-scale expansion of pre-isolated bone marrow mesenchymal stromal cells in serum-free conditions. J Tissue Eng Regen Med 10(2):108–119. https://doi.org/10.1002/term.1713

Gouveia BG, Rijo P, Gonçalo TS, Reis CP (2015) Good manufacturing practices for medicinal products for human use. J Pharm Bioallied Sci 7(2):87. https://doi.org/10.4103/0975-7406.154424

Grout BWW, Morris GJ (2009) Contaminated liquid nitrogen vapour as a risk factor in pathogen transfer. Theriogenology 71(7):1079–1082. https://doi.org/10.1016/j.theriogenology.2008.12.011

Guadix JA, López-Beas J, Clares B, Soriano-Ruiz JL, Zugaza JL, Gálvez-Martín P (2019) Principal criteria for evaluating the quality, safety and efficacy of hMSC-based products in clinical practice: current approaches and challenges. Pharmaceutics 11(11):552. https://doi.org/10.3390/pharmaceutics11110552

Horibata S, Vo TV, Subramanian V, Thompson PR, Coonrod SA (2015) Utilization of the soft agar colony formation assay to identify inhibitors of tumorigenicity in breast cancer cells. J Vis Exp 99:e52727. https://doi.org/10.3791/52727

Hu YT, Ting Y, Hu JY, Hsieh SC (2017) Techniques and methods to study functional characteristics of emulsion systems. J Food Drug Anal 25(1):16–26. https://doi.org/10.1016/J.JFDA.2016.10.021

Humbird D (2021) Scale-up economics for cultured meat. Biotechnol Bioeng 118(8):3239–3250. https://doi.org/10.1002/bit.27848

International Society for Biological and Environmental Repositories (ISBER) (2008) Collection, storage, retrieval and distribution of biological materials for research. Cell Preserv Technol 6(1):3–58. https://doi.org/10.1089/cpt.2008.9997

ISO 14644-1: Cleanrooms and associated controlled environments, classification of air cleanliness

Jacob SE, Hill H, Lucero H, Nedorost S (2016) Benzoate allergy in children—from foods to personal hygiene products. Pediatr Dermatol 33(2):213–215. https://doi.org/10.1111/pde.12759

Jang TH, Park SC, Yang JH, Kim JY, Seok JH, Park US et al (2017) Cryopreservation and its clinical applications. Integr Med Res 6(1):12–18

Jankowiak H, Cebulska A, Bocian M (2021) A relação entre acidificação (pH) e características de qualidade da carne de porcos de raça branca polonesa. Eur Food Res Technol 247:2813–2820. https://doi.org/10.1007/s00217-021-03837-4

Joo ST, Choi JS, Hur SJ, Kim GD, Kim CJ, Lee EY, Bakhsh A, Hwang YH (2022) A comparative study on the taste characteristics of satellite cell cultured meat derived from chicken and cattle muscles. Food Sci Anim Resour 42(1):175. https://doi.org/10.5851/KOSFA.2021.E72

Junker BH, Beshty B, Wilson J (1999) Sterilization-in-place of concentrated nutrient solutions. Biotechnol Bioeng 62(5):501–508. https://doi.org/10.1002/(SICI)1097-0290(19990305)62:5<501::AID-BIT1>3.0.CO;2-Y

Junker B, Lester M, Leporati J, Schmitt J, Kovatch M, Borysewicz S et al (2006) Sustainable reduction of bioreactor contamination in an industrial fermentation pilot plant. J Biosci Bioeng 102(4):251–268. https://doi.org/10.1263/jbb.102.251

Kamruzzaman M, Makino Y, Oshita S (2016) Hyperspectral imaging for real-time monitoring of water holding capacity in red meat. LWT Food Sci Technol 66:685–691. https://doi.org/10.1016/J.LWT.2015.11.021

Kang DH, Louis F, Liu H, Shimoda H, Nishiyama Y, Nozawa H et al (2021) Engineered whole cut meat-like tissue by the assembly of cell fibers using tendon-gel integrated bioprinting. Nat Commun 12(1):5059. https://doi.org/10.1038/s41467-021-25236-9

Karakaya M, Saricoban C, Yilmaz MT (2005) The effect of various types of poultry pre- and post-rigor meats on emulsification capacity, water-holding capacity and cooking loss. Eur Food Res Technol 220(3–4):283–286. https://doi.org/10.1007/S00217-004-1068-1/TABLES/2

Ketelings L, Kremers S, de Boer A (2021) The barriers and drivers of a safe market introduction of cultured meat: a qualitative study. Food Control 130:108299. https://doi.org/10.1016/j.foodcont.2021.108299

Khodabukus A, Madden L, Prabhu NK, Koves TR, Jackman CP, Muoio DM, Bursac N (2019) Electrical stimulation increases hypertrophy and metabolic flux in tissue-engineered human skeletal muscle. Biomaterials 198:259. https://doi.org/10.1016/J.BIOMATERIALS.2018.08.058

King DA, Hunt MC, Barbut S, Claus JR, Cornforth DP, Joseph P et al (2023) American meat science association guidelines for meat color measurement. Meat Muscle Biol 6(4):10.22175/mmb.12473

Lee VE, Arnott G, Turner SP (2022) Social behavior in farm animals: applying fundamental theory to improve animal welfare. Front Vet Sci 9:932217. https://doi.org/10.3389/fvets.2022.932217

Liu P, Song W, Bassey AP, Tang C, Li H, Ding S, Zhou G (2023) Preparation and quality evaluation of cultured fat. J Agric Food Chem 71(9):4113–4122. https://doi.org/10.1021/ACS.JAFC.2C08004/ASSET/IMAGES/LARGE/JF2C08004_0008.JPEG

Lu N, Ma J, Sun DW (2022) Enhancing physical and chemical quality attributes of frozen meat and meat products: mechanisms, techniques and applications. Trends Food Sci Technol 124:63–85. https://doi.org/10.1016/J.TIFS.2022.04.004

Maddison BC, Baker CA, Terry LA, Bellworthy SJ, Thorne L, Rees HC, Gough KC (2010) Environmental sources of scrapie prions. J Virol 84(21):11560–11562. https://doi.org/10.1128/jvi.01133-10

McNamara E (2023) Cost-effective sterility controls for cultivated meat. Good Food Institute. https://gfi.org/solutions/cost-sterility-controls-cultivated-meat. Accessed 22 Aug 2023

McNulty C (2016) Cleaning bioreactors and fermenters with CIP systems. Pharm Technol 40:46–48. https://www.pharmtech.com/view/cleaning-bioreactors-and-fermenters-cip-systems. Accessed 27 July 2023

Mebarki M, Abadie C, Larghero J, Cras A (2021) Human umbilical cord-derived mesenchymal stem/stromal cells: a promising candidate for the development of advanced therapy medicinal products. Stem Cell Res Ther 12(1):1–10. https://doi.org/10.1186/s13287-021-02222-y

Meléndez-Martínez AJ, Mandić AI, Bantis F, Böhm V, Borge GIA, Brnčić M et al (2022) A comprehensive review on carotenoids in foods and feeds: status quo, applications, patents, and research needs. Crit Rev Food Sci Nutr 62(8):1999–2049. https://doi.org/10.1080/10408398.2020.1867959

Melzener L, Verzijden KE, Buijs AJ, Post MJ, Flack JE (2021) Cultured beef: from small biopsy to substantial quantity. J Sci Food Agric 101(1):7–14. https://doi.org/10.1002/jsfa.10663

Montgomery SA, Wheelwright S, Poon HF (2021) Contractor perspectives: best practices for transfer, handling, testing, and storage of cell banks. BioProcess International. https://bioprocessintl.com/analytical/cell-line-development/contractor-perspectives-best-practices-for-ensuring-cell-bank-stability-during-transfer-handling-testing-and-storage/. Accessed 02 Sep 2023

Moraes AM, Mendonça RZ, Suazo CAT (2008) Meios de cultura para células animais. Tecnologia do Cultivo de Células Animais: de Biofármacos a Terapia Gênica. Editora Roca, São Paulo, pp 105–119

Moritz MSM, Verbruggen SEL, Post MJ (2015) Alternatives for large-scale production of cultured beef: a review. J Integr Agric 14(2):208–216. https://doi.org/10.1016/S2095-3119(14)60889-3

OECD (2022) OECD series on principles of good laboratory practice and compliance monitoring. OECD Environment Directorate, Environment, Health and Safety Division

O'Neill EN, Cosenza ZA, Baar K, Block DE (2021) Considerations for the development of cost-effective cell culture media for cultivated meat production. Compr Rev Food Sci Food Saf 20(1):686–709. https://doi.org/10.1111/1541-4337.12678

Osório JCDS, Osório MTM, Sañudo C (2009) Características sensoriais da carne ovina. Rev Bras Zootec 38:292–300. https://doi.org/10.1590/S1516-35982009001300029

Oswell NJ, Gilstrap OP, Pegg RB (2021) Variation in the terminology and methodologies applied to the analysis of water holding capacity in meat research. Meat Sci 178:108510. https://doi.org/10.1016/J.MEATSCI.2021.108510

Pall Corporation (2005) Gas filter makes bioreactors safer. Membr Technol 6:2. https://doi.org/10.1016/S0958-2118(05)70422-5

Paredes J, Cortizo-Lacalle D, Imaz AM, Aldazabal J, Vila M (2022) Application of texture analysis methods for the characterization of cultured meat. Sci Rep 12(1):1–10. https://doi.org/10.1038/s41598-022-07785-1

Pareek CS, Smoczynski R, Tretyn A (2011) Sequencing technologies and genome sequencing. J Appl Genet 52:413–435. https://doi.org/10.1007/s13353-011-0057-x

Park J, Choi H, Shim K (2022) Inhibition of GSK3β promotes proliferation and suppresses apoptosis of porcine muscle satellite cells. Animals 12(23):3328. https://doi.org/10.3390/ani12233328

Pasitka L, Cohen M, Ehrlich A, Gildor B, Reuveni E, Ayyash M, Wissotsky G, Herscovici A, Kaminker R, Niv A, Bitcover R, Dadia O, Rudik A, Voloschin A, Shimoni M, Cinnamon Y, Nahmias Y (2023) Spontaneous immortalization of chicken fibroblasts generates stable high-yield cell lines for serum-free production of cultured meat Nature Food 4(1):35–50. https://doi.org/10.1038/s43016-022-00658-w

Passetti RC, Torrecilhas JA, Ornaghi MG, Mottin C, Guerrero A (2016) Determinação da coloração e a disposição de compra pelos consumidores da carne bovina. Pubvet 10(02):179–189. https://doi.org/10.22256/PUBVET.V10N2.179-189

Pedroso RA, Demiate IM (2008) Avaliação da influência de amido e carragena nas características físico-químicas e sensoriais de presunto cozido de peru. Food Sci Technol 28(1):24–31. https://doi.org/10.1590/S0101-20612008000100005

Post MJ, Hocquette JF (2017) New sources of animal proteins: cultured meat. In: New aspects of meat quality: from genes to ethics. Woodhead Publishing, pp 425–441. https://doi.org/10.1016/B978-0-08-100593-4.00017-5

Rahman MS, Rahman MRT (2020) pH in food preservation. In: Handbook of food preservation. CRC Press, pp 323–332

Rawlings B, Pora H (2009) Environmental impact of single-use and reusable bioprocess systems. BioProcess Int 7(2):18–26

Rebelatto CLK, Boldrini-Leite LM, Daga DR, Marsaro DB, Vaz IM, Jamur VR et al (2023) Quality control optimization for minimizing security risks associated with mesenchymal stromal cell-based product development. Int J Mol Sci 24(16):12955. https://doi.org/10.3390/ijms241612955

Rubio NR, Xiang N, Kaplan DL (2020) Plant-based and cell-based approaches to meat production. Nat Commun 11(1):6276. https://doi.org/10.1038/s41467-020-20061-y

Rushen J, Passillé AMBD (1992) The scientific assessment of the impact of housing on animal welfare: a critical review. Can J Anim Sci 72(4):721–743. https://doi.org/10.4141/cjas92-085

Sahar A, Allen P, Sweeney T, Cafferky J, Downey G, Cromie A, Hamill MR (2019) Online prediction of physico-chemical quality attributes of beef using visible—near-infrared spectroscopy and chemometrics. Foods 8(11):525. https://doi.org/10.3390/FOODS8110525

Santilli F, Fabrizi J, Pulcini F, Santacroce C, Sorice M, Delle-Monache S, Mattei V (2022) Gangliosides and their role in multilineage differentiation of mesenchymal stem cells. Biomedicines 10:3112. https://doi.org/10.3390/biomedicines10123112

Schirmer C, Eibl R, Maschke RW, Mozaffari F, Junne S, Daumke R et al (2022) Single-use technology for the production of cellular agricultural products: where are we today? Chemie Ingenieur Technik 94(12):2018–2025. https://doi.org/10.1002/cite.202200092

Schulze E (2022) Premarket notice for integral tissue cultured poultry meat. 2022D11D16)[2022D 11D16]. https://www.fda.gov/media/163262/download. Accessed 25 July 2023

Sensebé L, Bourin P (2009) Mesenchymal stem cells for therapeutic purposes. Transplantation 87(9S):S49–S53. https://doi.org/10.1097/TP.0b013e3181a28635

Sharma S, Bhonde R (2020) Genetic and epigenetic stability of stem cells: epigenetic modifiers modulate the fate of mesenchymal stem cells. Genomics 112(5):3615–3623. https://doi.org/10.1016/j.ygeno.2020.04.022

Sharma S, Venkatesan V, Prakhya BM, Bhonde R (2015) Human mesenchymal stem cells as a novel platform for simultaneous evaluation of cytotoxicity and genotoxicity of pharmaceuticals. Mutagenesis 30(3):391–399. https://doi.org/10.1093/mutage/geu086

Siddiqui SA, Bahmid NA, Karim I, Mehany T, Gvozdenko AA, Blinov AV, Nagdalian AA, Arsyad M, Lorenzo JM (2022) Cultured meat: processing, packaging, shelf life, and consumer acceptance. LWT 172:114192. https://doi.org/10.1016/J.LWT.2022.114192

Simsa R, Yuen J, Stout A, Rubio N, Fogelstrand P, Kaplan DL (2019) Extracellular Heme proteins influence bovine myosatellite cell proliferation and the color of cell-based meat. *Foods* 8(10):521. https://doi.org/10.3390/FOODS8100521

Soice E, Johnston J (2021) Immortalizing cells for human consumption. Int J Mol Sci 22(21):11660. https://doi.org/10.3390/ijms222111660

Song WJ, Liu PP, Meng ZQ, Ding SJ, Li HX (2023) N-acetylcysteine promotes the proliferation of porcine adipose-derived stem cells during in vitro long-term expansion for cultured meat production. Food Res Int 166:112606. https://doi.org/10.1016/j.foodres.2023.112606

Stephens N, Di Silvio L, Dunsford I, Ellis M, Glencross A, Sexton A (2018) Bringing cultured meat to market: technical, socio-political, and regulatory challenges in cellular agriculture. Trends Food Sci Technol 78:155–166. https://doi.org/10.1016/j.tifs.2018.04.010

Stout AJ, Mirliani AB, Soule-Albridge EL, Cohen JM, Kaplan DL (2020) Engineering carotenoid production in mammalian cells for nutritionally enhanced cell-cultured foods. Metab Eng 62:126–137. https://doi.org/10.1016/J.YMBEN.2020.07.011

Thornton KJ (2019) Impacts of nutrition on the proliferation and differentiation of satellite cells in livestock species. J Anim Sci 97(5):2258–2269. https://doi.org/10.1093/jas/skz081

Thorrez L, Vandenburgh H (2019) Challenges in the quest for clean meat. Nat Biotechnol 37(3):215–216. https://doi.org/10.1038/s41587-019-0043-0

Tomasevic I, Djekic I, Font-i-Furnols M, Terjung N, Lorenzo JM (2021) Recent advances in meat color research. Curr Opin Food Sci 41:81–87. https://doi.org/10.1016/J.COFS.2021.02.012

Treich N (2021) Cultured meat: promises and challenges. Environ Resour Econ 79(1):33–61. https://doi.org/10.1007/s10640-021-00551-3

U.S. Department of Health and Human Services (2004) Guidance for industry, sterile drug products produced by aseptic processing-current good manufacturing practice. http://www.fda.gov/CbER/gdlns/steraseptic.pdf. Accessed 22 Aug 2023

U.S. Pharmacopeia (2018) <1044> Cryopreservation of cells. USP General Chapter. https://www.usp.org/file/usp-general-chapter-cryopreservation-cells. Accessed 02 Sep 2023

Uenojo M, Maróstica MR, Pastore GM (2007) Carotenóides: Propriedades, aplicações e biotransformação para formação de compostos de aroma. Quim Nova 30(3):616–622. https://doi.org/10.1590/S0100-40422007000300022

Vestergaard M, Oksbjerg N, Henckel P (2000) Influence of feeding intensity, grazing and finishing feeding on muscle fibre characteristics and meat colour of semitendinosus, longissimus dorsi and supraspinatus muscles of young bulls. Meat Sci 54(2):177–185. https://doi org/10.1016/s0309-1740(99)00097-2

Warner RD (2023) The eating quality of meat: IV—water holding capacity and juiciness. In: Lawrie's meat science. Woodhead Publishing, pp 457–508. https://doi.org/10.1016/B978-0-08-100694-8.00014-5

Wassenaar TM, Zimmermann K (2018) Lipopolysaccharides in food, food supplements, and probiotics: should we be worried? Eur J Microbiol Immunol 8(3):63–69. https://doi.org/10.1556/1886.2018.00017

Whaley D, Damyar K, Witek RP, Mendoza A, Alexander M, Lakey JR (2021) Cryopreservation: an overview of principles and cell-specific considerations. Cell Transplant 30:0963689721999617. https://doi.org/10.1177/0963689721999617

World Health Organization (2019) Critically important antimicrobials for human medicine. https://Apps.Who.Int/Iris/Bitstream/Handle/10665/312266/9789241515528-Eng.Pdf?Ua=1. Accessed 27 July 2023

Yen FC, Glusac J, Levi S, Zernov A, Baruch L, Davidovich-Pinhas M, Fishman A, Machluf M (2023) Cultured meat platform developed through the structuring of edible microcarrier-derived microtissues with oleogel-based fat substitute. Nat Commun 14(1):1–11. https://doi.org/10.1038/s41467-023-38593-4

Yin H, Price F, Rudnicki MA (2013) Satellite cells and the muscle stem cell niche. Physiol Rev 93(1):23–67. https://doi.org/10.1152/physrev.00043.2011

Yurleni Y, Adriani A (2021) Antioxidant activity and bioactive components of curry leaves (Murraya koenigii) to lower red meat cholesterol: morphological, chemical content curry leaves and putrefaction test of meat

Zhang G, Zhao X, Li X, Du G, Zhou J, Chen J (2020) Challenges and possibilities for bio-manufacturing cultured meat. Trends Food Sci Technol 97:443–450. https://doi.org/10.1016/j.tifs.2020.01.026

Zheng YY, Zhu HZ, Wu ZY, Song WJ, Tang CB, Li CB, Ding SJ, Zhou GH (2021) Evaluation of the effect of smooth muscle cells on the quality of cultured meat in a model for cultured meat. Food Res Int 150:110786. https://doi.org/10.1016/J.FOODRES.2021.110786

Zidarič T, Milojević M, Vajda J, Vihar B, Maver U (2020) Cultured meat: meat industry hand in hand with biomedical production methods. Food Eng Rev 12(4):498–519. https://doi.org/10.1007/S12393-020-09253-W/FIGURES/6

# Scaling-Up of Cultivated Meat Production Process

**Maria Clara Manzoki, Ariane Fátima Murawski de Mello, Walter José Martinez-Burgos, Alexander da Silva Vale, Giuliana Biagini, Igor Negreiros Piazenski, Vanete Thomaz Soccol, and Carlos Ricardo Soccol**

**Abstract** Upscaling the production of cultured meat from laboratory flasks to an industrial scale is undeniably a great challenge, yet it is imperative for the process to achieve economic viability and commercial potential. The transition to a large-scale operation introduces a myriad of concerns spanning various themes, including the supply and storage of raw materials (cultivation media components, cells, high-quality water, microcarriers, and scaffolds) and the cell cultivation process itself (bioreactor design, mode of operation, sterilization, monitoring, and control tools), as well as downstream processes and final product development (formulation, stability, packaging, and reproducibility). The shift from a small-scale cell cultivation flask to large-scale bioreactors introduces several novel parameters that must be taken into account, such as gas exchange, shear stress, heat and mass transfers, mixing, and foaming. Expanding volumes while preserving cell productivity requires optimization across various parameters during the scale-up process, along with the identification of key parameters to be maintained constant throughout. This chapter also explores considerations on the sustainability of scaling-up cultured meat production, as well as economic aspects of the upscaled process and future developments crucial for achieving cost-effectiveness. These considerations are pivotal for extending cultured meat production beyond laboratory and pilot scales.

**Keywords** Process scale-up · Pilot scale · Industrial scale · Technoeconomic aspects

---

M. C. Manzoki · A. F. M. de Mello · W. J. Martinez-Burgos · A. da Silva Vale · G. Biagini · I. N. Piazenski · V. T. Soccol · C. R. Soccol (✉)
Bioprocess Engineering and Biotechnology Department, Federal University of Paraná, Curitiba, Paraná, Brazil
e-mail: soccol@ufpr.br


C. R. Soccol et al. (eds.), *Cultivated Meat*,
https://doi.org/10.1007/978-3-031-55968-6_12

# 1 Introduction

Cultivated meat presents several advantages when compared to livestock-based animal protein production, such as reduction of water consumption and greenhouse gas emissions, besides, of course, the decrease in animal cruelty and health concerns due to zoonotic diseases and overuse of antibiotics (Letti et al. 2021; Reiss et al. 2021). However, for these effects to really be put into practice, it is necessary to extrapolate the process of meat cultivation from laboratory flasks to industrial scale. Even further, the production process of cultivated meat needs to attain economic viability and commercial potential. In a survey conducted by CellRev (2023), which collected responses from scientists, executive managers, biologists, and bioprocessing engineers from companies of cultured meat and therapy development, it was identified that 82.8% of the surveyed cited scalability as their biggest barrier to commercial viability of cell manufacturing (CELLREV 2023).

The production of cultured meat is an emerging technology that does not present an established manufacturing protocol yet, but this process can be divided into four main steps: (i) isolation or obtention of cells and increasing volumes of cell cultivation, (ii) production of the cells on a large scale, (iii) differentiation of cells or muscle and formation of target tissue (maturation) on 3D biomaterial scaffolds, and (iv) processing into food products, either unstructured (e.g., hamburgers) or structured (e.g., "whole cut" such as steaks) (Kumar et al. 2023; Reiss et al. 2021). Growing cultivated meat at scale will require the use of bioreactors in volumes up to or beyond several thousands of liters, a scenario much different from usual cultivations in flasks at laboratory scale. According to McKinsey consulting company, cultivated meat could account for as much as 0.5% of the global meat supply by 2030, representing a $25 billion market. Those values will require the annual production of 1.5 million metric tons of cultivated meat; thus, the industry will depend on a cultivation capacity of 220–440 million liters (Brennan et al. 2021). Transitioning from a small scale to a large scale involves several new parameters that need to be considered, such as gas exchange, shear stress, heat and mass transfers, mixing, and foaming (GFI - Good Food Institute 2023). Furthermore, upscaling a process requires expanding volumes while maintaining cell productivity, which can be attained through optimization of several parameters during the scale-up and identification of the key parameters to be kept constant during this process.

To make the cultivated meat production process economically viable on a large scale, some other key aspects that require attention besides the upscaling process itself are the development of low-cost culture medium without animal components, the manufacture of growth factors, and the development and/or selection of new cell lines with high productivities (Ng and Kurisawa 2021; Chen et al. 2022).

In recent years, cultivated meat has attracted an influx of investment from both private organizations, which have funded around $320 million between 2015 and 2020, and public sectors from various countries in Asia, Latin America, North America, and Europe (Choudhury et al. 2020). A survey conducted by Choudhury et al. (2020) showed that of the 32 companies and startups that have shown interest

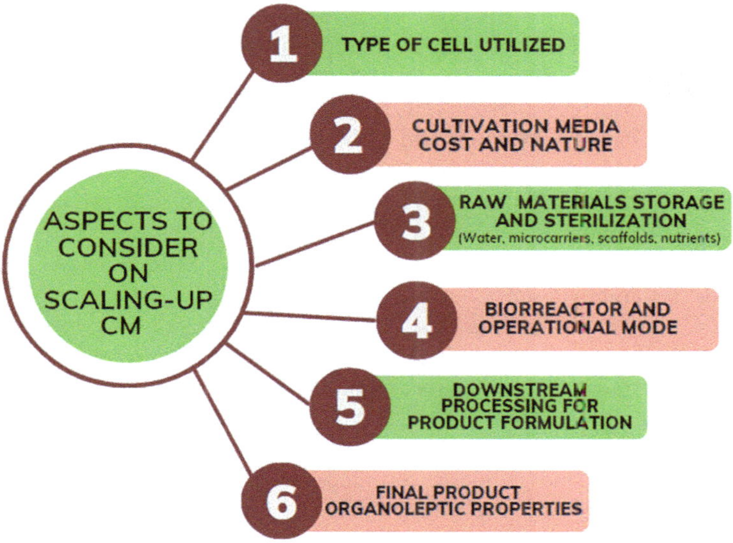


**Fig. 1** Principal aspects to be considered on cultivated meat scale-up process

in cultured meat, about 25% are focused on the development of beef, 22% on poultry, and 19% on pork and seafood, thus demonstrating the growth potential of this little-explored sector. Therefore, the objective of this chapter is to show the latest scientific and technological advances involved in the production and scaling-up of cultured meat, as well the challenges that need to be overcome on an industrial level. Different crucial aspects for guaranteeing a successful upscaling of meat cultivation and its further acceptance by consumers are addressed, which can be divided into the following principal thematics discussed in this chapter (Fig. 1).

## 2 Supply of Raw Materials and Nutrients for Industrial-Scale Production

When considering the production of cultivated meat (CM), the main raw materials generally consist of the cells that are going to constitute the product, the components of the culture medium, and the structures to support cell growth, such as scaffolds or microcarriers. On a large scale, however, the supply and storage of these ingredients have added difficulties.

A great variety of cells can be used to obtain CM, and even though the physical conditions for storing and culturing the cells can be the same (temperature, humidity, $CO_2$ and $O_2$ levels), each cell type can have different chemical requirements due to their own biology. The most commonly used cells are mesenchymal stromal cells (MSCs) and myosatellite cells, but some variations of these CM protocols include pluripotent (embryonic) stem cells or induced pluripotent stem cells (iPSCs) (Edelman et al. 2005; Reiss et al. 2021). In this case, because pluripotent stem cells and iPSCs differentiate naturally without the need for extra molecular signaling, it is crucial to not only prevent these cells from differentiating randomly but also guarantee their pluripotency (Romito and Cobellis 2016). This is achieved by using specific and expensive culture media and reagents, such as a ROCK inhibitor (Kurosawa 2012; Vernardis et al. 2017), which on a large scale would inevitably affect the CM's pricing.

On the other hand, MSCs and myosatellite cells are considered adult stem cells that do not differentiate unless specific molecular signaling is offered (Dominici et al. 2006; Caplan 2017; Pang et al. 2023), so no extra reagents are required to maintain the stemness of these cells. However, they can only divide a limited number of times (around 50 to 70 times) and eventually reach senescence and die (Izadpanah et al. 2008). Even before they die, their morphology and function are already altered (Izadpanah et al. 2008), so the same cell lineage cannot be used for long periods of time and should be replaced periodically. This not only demands highly controlled logistics but also puts the CM's quality at risk, since MSCs obtained from different individuals or different tissues from the same individual can behave differently in terms of differentiation capacity and proliferation rate. Such variability can complicate the reproducibility of the process at an industrial level (Jervis et al. 2019; Oyarzo et al. 2021). In this scenario, the advantage of using pluripotent stem cells or iPSCs could lie in the fact that they have a higher proliferation rate and present telomerase activity and, therefore, can divide indefinitely if no other external factor is involved (Li et al. 2020a; Azhar et al. 2023). Consequently, the same lineage of cells can be used for longer periods of time and guarantee the consistency of CM's quality in large scale.

An aspect shared by most of the cells used to produce CM is that they are adherent, which means that ideally they need a physical support to attach to. The support, called scaffold or microcarrier (MC), replicates the extracellular matrix, offering nutrients, such as differentiation factors, and facilitates the upscaling process (Bodiou et al. 2020) and can participate in the CM's sensory properties (Singh et al. 2023). In this case, the support's edibility is compulsory, though this is not a common characteristic of all scaffolds and MCs. In fact, there are a myriad of proposed compositions for them: a mix of polysaccharides (chitosan, cellulose) and proteins (Shouwei et al. 2021; Guembe Lapuente et al. 2022), mushroom mycelium (Bayer et al. 2019), and even decellularized plant of fungal cell wall (Pelling et al. 2017). Despite their variable composition, cell culture based on MCs is more straightforward and, therefore, easier to control compared to suspension cell culture (Bodiou et al. 2020). This allows for more consistency in the product's quality and can lower production costs. However, this is only achieved if the consistency and safety of the raw materials for MCs are ensured, which on a large scale can be a challenge in terms of material availability and sterilization.

It is also known that animal cells mutate at a higher rate when they are consistently cultured in the lab, and that could be a result of the manipulation techniques and the chemical substrates that are used (Kuijk et al. 2020). Certain mutations could affect the cell's behavior and the CM's characteristics. A recent study conducted in two of the largest iPSCs banks showed that more than 70% of a specific strain of preserved iPSCs carried genetic alterations attributed to ultraviolet light (UV) exposure that were acquired during their handling and manipulation (Rouhani et al. 2022). Therefore, there should be an evident effort in quality control of the cells being used for CM. Because these mutations can occur at any point during the cultivation process, including the phases of freezing and thawing, it is imperative that molecular quality control should be conducted not only at the start and conclusion of the process but also in real time during its progression (Stroncek et al. 2010). Considering the high cost of some of the genetic testing available, in large scale the added steps of cell and culture quality control could impact the product's final price.

DNA damage and alterations in the cell's metabolism can also be caused by mycoplasma contamination (Nikfarjam and Farzaneh 2012; Olarerin-George and Hogenesch 2015). This situation is known as "silent contamination" in cell culture because mycoplasma is a small bacterium (diameter < 1 μm) that can pass through the filtration processes used to purify primary cultures (Young et al. 2010; Olarerin-George and Hogenesch 2015) and is not easily detected by the usual means of daily quality control, such as visualization under an optic microscope (Nikfarjam and Farzaneh 2012; Olarerin-George and Hogenesch 2015). However, mutations and changes in cell behavior are not the only consequences. Mycoplasma contamination can be disastrous and hard to contain because it can go on for a long time without causing noticeable changes in cell behavior (Olarerin-George and Hogenesch 2015) and, hence, cause the loss of whole batches of cells and/or product. Also, the bacterium is resistant to the antibiotics commonly used in cell culture, and even though there are effective antibiotics against it, they are not indicated for cell culture due to their possible toxicity (Nikfarjam and Farzaneh 2012). The protocols for detecting mycoplasma contamination consist of several techniques (ELISA, agar culture, and fluorescent staining), but it is usually done through PCR tests (Young et al. 2010; Nikfarjam and Farzaneh 2012) in the early stages of cell culture, and in the case of CM it should also be performed periodically throughout the whole process. This scenario reinforces the need for a rigorous quality control of the cell cultures being supplied for CM.

Animal cells also require certain external factors to proliferate. Without them, they cannot enter the cell cycle and do not divide efficiently. In the lab, these factors are provided through fetal bovine serum (FBS) (Pilgrim et al. 2022). Even though the pharmaceutical industry uses FBS in cell culture to produce certain biological drugs, in the last couple of years there has been a growing concern over the fact that FBS can be contaminated by prions and viruses and risk human health (Piletz et al. 2018).

There are rigorous and specific guidelines that the pharmaceutical industry follows to ensure the safety of the FBS (European Medicines Agency 2013), and the CM industry could borrow those guidelines if necessary. However, one of the selling points of CM is the animal-free production process; thus, it would not be adequate to use an animal-derived substrate, such as FBS. The solution is to use nonanimal substitutes for FBS, and a lot of effort has been made toward finding the ideal candidate using several different sources ranging from algae (Ng et al. 2020) and plant extracts (Nahmias et al. 2021) to simply mixing isolated growth factors (Chieregato et al. 2011). The biggest issue is that even though some studies have been able to obtain adequate cell proliferation and survival through these substitutes (Price et al. 1997; Liu et al. 2016; Mullen et al. 2020), cultures containing FBS still perform better most of the time (Piletz et al. 2018). Adding to the challenge, the exact composition of FBS remains unknown, so it will still be a while before we can completely replace it and reach a consensus over which substitute formulation is more adequate.

Another issue is that most of these studies contemplate small-scale cell culture, and thereby their practical applicability in the industry still raises questions and remains to be thoroughly investigated (Lee et al. 2022). This uncertainty arises not only because of the ambiguous efficacy of the FBS replacements but also because it could be expensive and difficult to obtain the substrates in adequate quantities while also assuring the consistency of their composition (Hubalek et al. 2022; Pilgrim et al. 2022; Subbiahanadar Chelladurai et al. 2021). It is noteworthy that even the composition of FBS can vary from batch to batch, which leads to discernible distinctions in cell behavior in culture (Liu et al. 2023); thus, whether CM production incorporates FBS or not, the matter of consistent supplement composition should still be tackled to ensure the final product's quality and reproducibility.

Besides growth factors and supplements, water is another component of the culture medium that requires special attention. For producing cultivated meat on a large scale, considerable volumes of cultivation media are required, as are large volumes of appropriate water. Impurities and microorganisms present in the water, such as bacteria, viruses, endotoxins, and heavy metals, are related to impairments in cell growth and viability in vitro (Riche et al. 2023). Hence, the water used for this purpose should follow determined standards that define the ideal concentration of organic and inorganic matter. The water required for cell culture is known as highly purified water (HPW) or ultrapure water (UPW) (Stacey and Davis 2007). Several steps of purification are needed in the process of obtaining HPW, and they may vary according to cost and the source of the water, but on a large scale they usually consist of reverse osmosis, electrodeionization, and ultrafiltration (Stacey and Davis 2007). As with the quality control of cell mutations throughout the culture, it is also crucial to maintain the quality of the HPW during the process, and this can be achieved by recirculating the water at 80 °C (Stacey and Davis 2007). Nonetheless, because of the demand for specialized sets of equipment and the importance of HPW in CM production, these extra steps should be accounted for in the supply of raw materials for upscaling CM.

Another aspect that should be taken into account is that the physical structure for storing the cells (whether they are stem cells, progenitor cells, or any other type of animal cell) involves incubators, freezers, and liquid nitrogen gallons and can take up a lot of space, especially for a large-scale production system. In this case, perhaps it would be interesting to invest in specialized stem cells banks, as the pharmaceutical industry does for cell therapy purposes (Cherian et al. 2020). This would also provide a higher control over the product's quality and consistency throughout the batches (Cherian et al. 2020).

Ultimately, all these particularities highlight that the biggest challenge in supplying raw materials for the scalable expansion of CM is first and foremost the determination of the exact constituents of the culture medium that is going to be used. Only then will it be possible to reduce the price of CM substantially, given that culture medium stands as the most expensive component involved in the large-scale process (Soccol et al. 2022), and organize a reliable and effective supply chain. The unavailability of key materials needed for cell cultivation can result in delays and inefficiencies on the process, complicating an already intricate project even more (CELLREV 2023).

# 3 Upstream Processing

## *3.1 Process Overview*

In a survey conducted by CellRev (2023) with researchers, managers, and engineers of cell producer industries, it was identified that the biggest practical cell manufacturing challenge is upstream processing development (highlighted by more than 60% of the surveyed). Upstream processing usually follows three essential steps: collecting cells from an animal source, cell proliferation, and finally differentiation in bioreactors (Jairath et al. 2021). Usually, undifferentiated cells are collected from a healthy adult animal via biopsy. Adult stem cells are found in specific tissues, the most common being the mesenchymal stem cells (MSCs). Embryonic stem cells (ESC) and induced pluripotent stem cells (iPSCs) can also be applied; however, they carry the disadvantage of being highly proliferative and nonspecific, therefore inducing low yields of the specific cell types required for cultivated meat (Pajčin et al. 2022; Stephens et al. 2018). The selected cells are then cultured first in a proliferation bioreactor, in order to obtain the highest cell density possible (Allan et al. 2019). The cells can be cultured associated with a biomaterial, in aggregates, or in suspension in the liquid medium (Bomkamp et al. 2022; Reiss et al. 2021). Usually, dynamically agitated systems, such as stirred tank (STRs) and pneumatic bioreactors, are applied at this stage, making it easier to scale up as models for these kinds of equipment are already well established. Considering an average doubling time of 24–40 h for myogenic and adipogenic cells, it can be estimated that for reaching a final cell density of $10^7$ cells/mL, 4 days on the proliferation stage are necessary (Moutsatsou et al. 2023). After a high cell density is achieved, the cell culture can be transferred to another bioreactor for differentiation. In this step, the cells are generally immobilized in a solid matrix, the scaffold, which makes the scale-up of the applied reactors more complex (Letti et al. 2021). Differentiation requires more time than proliferation to happen and depends on the type of resulting cell (e.g., adipogenic cells require longer differentiation—around seven times the proliferation time) (Moutsatsou et al. 2023). In each step, the media will contain certain components that will enhance either cell growth or differentiation. Ideally, the cultivation media should not contain any animal-derived ingredient, as this would contradict the principle of the cultivated meat being slaughter- and animal-free (Jairath et al. 2021).

The mammalian cells applied for cultivated meat are anchorage-dependent; therefore, there is a need for a solid foundation for cell development (Allan et al. 2019). Polymeric structures (e.g., polysaccharides, polypeptides, bio- and synthetic polymers) are thus applied as scaffolds. These structures need to be either biocompatible and edible to be present in the final product or biodegradable for disassociation or degradation prior to cultured meat commercialization; however, they always need to be safe and in accordance with food regulations (Kong et al. 2022). Plant-derived or bio-based polymeric structures are preferred over animal-derived, and the production of these materials should be scalable and sustainable, as high quantities are required to meet the demand for cultivated meat (Bomkamp et al. 2022). If the scaffolds are present in the final product, they need to present good organoleptic characteristics and can be selected and/or engineered in order to provide certain properties or characteristics to the final product (e.g., enhanced protein content), enhancing therefore the overall nutritional value of the cultured meat (Bodiou et al. 2020).

Microcarriers (MCs) can be considered a type of scaffold (Bomkamp et al. 2022) and are commonly applied in the proliferation step, providing a high surface/volume ratio in the bioreactor. They present an advantage over MC-free culture, as cells naturally aggregate in these systems, which leads to difficulty on controlling the bioprocess and to the generation of nutrients' gradients inside the bioreactor (Bodiou et al. 2020). Usually, fresh MCs are added to the media during the bioprocess duration in order for the cells to keep reproducing in a bead-to-bead transfer manner (Seah et al. 2022). Separation of cells from MCs can be made by dissociation or degradation methods depending on the materials' properties; however, adding downstream steps will increase the overall cost of the bioprocess. If the MCs are present in the final product, they need to be necessarily edible, and cells can be further recovered from media for differentiation by simple mechanical methods (filtration, centrifugation, etc.) (Bomkamp et al. 2022). Examples of MC materials are cellulose, dextrans, polystyrene, silica, poly-glycolic acid (PGA) (animal-free), gelatin, and collagen (animal-derived) (Bodiou et al. 2020; Bomkamp et al. 2022; Kong et al. 2022). Other scaffolds are applied in the differentiation step, as they will provide a matrix for cell differentiation and final product formation (Kumar et al. 2023). The biomaterials can be porous or fibrous or in the form of hydrogels, each providing certain characteristics for cell development and, therefore, final product formation. These scaffolds must be able to provide the cells the passage of nutrients and oxygen for their proper development (Bomkamp et al. 2022). Examples of scaffolds are gelatin, collagen, keratin, myosin (animal-derived), cellulose, dextrans, PGA, poly-lactic acid (PLA), PGA and PLA copolymers (PGLA), hyaluronic acids, and plant-derived proteins (animal-free) (Bomkamp et al. 2022; Seah et al. 2022). It is important to acknowledge that numerous natural hydrogels can pose challenges for bioprocess scale-up since they may present batch variations (Moutsatsou et al. 2023).

## 3.2 Upscaling Process

The scaling-up of the cultivated meat bioprocess usually follows the steps of laboratory-, bench-, pilot-, and industrial-scale development, increasing the size of applied bioreactors in each step while maintaining some geometrical relations and biochemical parameters constant along scales. On laboratory and bench scales, tests are performed to optimize conditions for cell development in different kinds of flasks and bottles. Different cell lines are tested aiming the selection of the most productive ones in terms of specific growth rates ($\mu$ in $h^{-1}$). This selection is carried out taking into account several factors such as media composition requirements, presence or absence of animal-derived components on it, hormones, growth factors, carbon and nitrogen sources, vitamins, growth promoters, pH, temperature, and type of microcarriers/scaffolds and their concentration, among others (Hanga et al. 2020; Letti et al. 2021).

Usually on laboratory and bench scales, intermediate flasks are utilized, such as T-flasks, multilayered flasks, roller bottles, and spinner flasks. T-flasks are normally made of plastic, being essential for initial cell multiplication. Those flasks attain a maximum area of some hundreds of square centimeters, are labor-intensive, and become cost-inefficient when expanding cells beyond bench scale (Bellani et al. 2020). Multilayered flasks, otherwise, are large T-flasks composed of stacked flat surfaces, reaching up to 2.5 $m^2$ of surface. Each layer might have a different distribution of nutrients and gases, and it is also a labor-intensive dispositive due to its size and weight, which may complicate the usual daily operations (Bellani et al. 2020). To further increase the cell cultivation volume, roller bottles can be used. Those bottles are cylindrical vessels, which allow the cells to attach and cover the internal surface of them. Roller bottles present between 500 and 1700 squared centimeters for cell expansion, with volumes that range from 1 to 1.5 L (working volume of 0.1–0.3 L). Normally, those roller bottles are maintained under slow revolving, and there are certain constraints concerning the control of $O_2$ and $CO_2$ in both the gas and liquid phases of the culture (Bellani et al. 2020).

The next step in upscaling cell cultivation is using spinner flasks, which provide better aeration since they present side arms with loose screw caps to guarantee gas exchanges. Also, the system is mixed through a magnetic stir bar, normally at 50 rpm, which guarantees the suspension of the cells and the microcarriers, which at this point start to be introduced to the process (Bellani et al. 2020). At this step of upscaling meat cultivation, spinner flasks can be used to test different types of microcarriers and their concentrations, as well as to optimize the media composition, before taking the process to larger bioreactors. Higher cell numbers can be attained on spinner flasks, and the maximum work volume is usually 100–200 mL (Hanga et al. 2020).

Considering the biggest productivities reached in terms of meat biomass in g/L·h, the process is upscaled to bioreactors (up to 15 liters), also known as bench scale. Bioreactors introduce additional parameters to be considered, allowing finer control over culture conditions such as pH, $pCO_2$, $pO_2$, and temperature, and enable much simpler scalability. They also enable finer control over mechanical agitation and paddle design. When performing the upscaling of cell cultures from spinner flasks to bioreactors, some studies apply CFD (computational fluid dynamics) to evaluate and characterize flow patterns and shear stress that cells will undergo, especially considering the effects of microcarriers and impeller design (Cantarero Rivera and Chen 2022). The main critical hydrodynamic effects that act over the cells, especially when using microcarriers, are the Reynolds number, Kolmogorov eddy length, maximum mean aggregate size, and maximum shear stress (Merten 2015; Kahan et al. 2020). As an example, Jossen et al. (2018) used CFD and experimental data to evaluate how different rotational speeds of impellers would change the microcarrier dispersion in bioreactors of 125 and 500 mL. The results demonstrated that the suspension criteria $N_{s1}$—rpm is a key parameter to consider when scaling-up production and that in optimal cultivation conditions specific power inputs of 0.3–1.1 $W/m^3$ are ideal (Jossen et al. 2018; Cantarero Rivera and Chen 2022).

The scale-up is a challenge for cultured meat production, as there are large costs mainly from the applied materials (Stephens et al. 2018; Allan et al. 2019). Some strategies of scaling-out can also be applied: instead of only focusing on increasing reactor size, multiple smaller bioreactors can act in parallel. This can provide the opportunity to obtain thicker products, with more modulable characteristics (Pajčin et al. 2022; Stephens et al. 2018).

## 3.3 Industrial-Scale Bioreactors Design

There are several types and models of bioreactors that can be applied to cultivated meat production in the different steps (proliferation and cultivation). Distinct systems imply different cell densities and final product properties, along with the ease or difficulty to scale up the system. On a large-scale, the STRs are generally the most commonly used bioreactor in bioprocesses with well-defined scale-up strategies, being applicable for tissue engineering and meat cultivation, usually in the proliferation stage (Letti et al. 2021). This system consists of a vessel that can vary the height/diameter ratio from 1:1 up to 4:1. The agitation is provided by impellers, and usually the media homogenization is not a problem in the STRs. However, this kind of agitation can provoke deleterious effects on mammalian cells, as high shear forces are generated, compromising the cells' structure (Allan et al. 2019; Post et al. 2020). Foam formation and energy consumption are also factors that need to be considered when applying STRs. Cell densities up to $10^6$ cells/mL can be attained, and it is estimated that a 570 liter STR would be necessary to generate 1 kg of cultivated meat (Allan et al. 2019). STRs can be single-use bioreactors, which help prevent contamination; however, this implies higher costs and environmental problems as the bags cannot be reutilized (Pajčin et al. 2022).

On the other hand, pneumatic bioreactors are vertical vessels that use air in the form of bubbles (bubble column) or flow (airlift) for agitation and aeration of the system, which implies lower shear stress for the cells and therefore lower cell death rates. They can also be applied in the proliferation stage; however, scale-up of this kind of system is more complex when compared to the STRs (Letti et al. 2021; Post et al. 2020). The rocking bed or wave bioreactors can also be applied to proliferation and are systems that consist of bags disposed on a plate that moves in a wave/rocking manner, homogenizing the media. The bags are usually polymeric and disposable, which resembles the case of single-use STRs (Bellani et al. 2020). As the mixing is made in a rocking manner, the shear forces are lower, which implies lower cell death rates; however, homogenization can also be compromised. Besides, there is scale limitation in this kind of system when compared to the STRs models.

The hollow fiber bioreactors are constituted of semipermeable membranes in a vessel, in which the media pass through these membranes while also removing the broth with secondary metabolites. The cells are either immobilized or present in the outer space of membranes (Pajčin et al. 2022). These systems can be applied in either proliferation (reaching high cell densities up to $10^9$ cells/mL) or differentiation step (Letti et al. 2021). The main issues regarding these systems are the difficulty in controlling, homogenizing, and scaling-up. The fluidized bed bioreactors are vertical vessels in which solid particles are fluidized by air while the media flow upward, the particles being usually the MCs or scaffolds described previously. As there is a constant flow of media and air through the system, it can be considered well agitated with low shear forces (Pajčin et al. 2022; Post et al. 2020). Fixed or packed bed bioreactors are similar to the fluidized bed; however, the material is packed and condensed in an immobilization matrix that stays in the middle of the bioreactor so the media and air can pass through, but the cells are immobilized and not fluid, being mainly applied on the differentiation stage (Bellani et al. 2020; Pajčin et al. 2022; Post et al. 2020). Secondary metabolites contained in the broth are also removed constantly in this kind of system. There are serious mass and heat transfer difficulties in this case as nutrients and other parameters' (temperature, pH) gradients can be formed. Scaling-up can be tricky as several adjustments on media and gas flow need to be well made (Pajčin et al. 2022). Finally, rotating wall bioreactors are the ones in which the vessel rotates around its own axis, which creates a mixing with low shear stress for cells but good homogenization of the media (Allan et al. 2019; Pajčin et al. 2022). A survey recently conducted identified that around 50% of the industrial cell producers use stirred tank bioreactor technology, while 5% use airlift bioreactors, and 45% use "other options," which may include novel constructions to spinner flask techniques and indications of using multiple different methods (CELLREV 2023).

### 3.4 Industrial-Scale Process Monitoring

Especially in large-scale bioreactors, controlling the operation is essential to guarantee the successful growth of the cells and avoid losses of materials, time, and resources. Control systems are required to monitor and adjust important factors for cells (temperature, pH, dissolved oxygen, carbon source concentration, among others), since extreme conditions of these parameters can lead to cell death or poor development. Also, through cell concentration monitoring, dynamic feeding strategies can be put in place, which can be very important in large-scale CM systems. Some control strategies are already well defined for some industrial processes and can be replicated in the case of cultivated meat production. Each kind of bioreactor will require different strategies and sensors as distinct conditions are found (Pajčin et al. 2022).

For example, the monitoring of gases during cells cultivation is essential. Oxygen drops below cell metabolism requirement level can lead to slower respiration of the cells, impacting their growth and product quality. When $CO_2$ is above a certain level, cell growth can be inhibited due to a pH imbalance. This highlights the importance of sensors to monitor and feedback control systems on the operation (Bellani et al. 2020). Besides, measurement of substrate and reagents density at-line can be done through optical sensors, UV-Vis, ultrasound sensors, RAMAN, and fluorescence spectroscopy. Both RAMAN and NIR are noninvasive methods that can provide useful information about cell culture bioprocesses, providing a real-time measurement of several parameters (Bellani et al. 2020; GFI - Good Food Institute 2023).

The sensors applied to cultivation monitoring should have reliable and robust performance throughout the process. They must present accuracy in the desired control range, high sensibility, and selectivity, be easy to implement, and give a short-time response (Moutsatsou et al. 2023). The collection of data from the bioreactors can be performed in-line (in situ), at-line, online, or offline (Minnich et al. 2016). In the case of in-line measurements, sample analysis occurs within the bioreactor, providing real-time data suitable for monitoring dynamic processes. On at-line measurements, the sample is extracted from the bioreactor and analyzed nearby, introducing some time delay. In offline measurements, in turn, samples are taken to a different location (e.g., a control quality laboratory), causing significant time delays and making it unsuitable for real-time process control. Online measurements, at last, occur when a sample is diverted from the bioreactor using a bypass, and all steps, from sampling to data transmission, are automated, which allows real-time monitoring (Minnich et al. 2016; Moutsatsou et al. 2023). Investing in automation, digital transformation, and modern modeling methods is believed to be essential for the future development of the CM industry, as well as all other bioprocesses. These advancements are aligned with the concept of Industry 4.0 concept (integrating the Internet of Things, artificial intelligence, cloud computing and analytics, and machine learning into manufacturing processes), leading to better decisions and efficiencies (Moutsatsou et al. 2023).

Ensuring sterility and managing contamination present another significant practical hurdle in cultivated meat production (CELLREV 2023). Media and system sterilization are of extreme importance, as the media used for culturing mammalian cells are extremely nutrient-rich. Therefore, there is a higher risk of contamination in these systems. Some antibiotics can be used to prevent bacterial growth, but they need to be used with caution as resistance can develop and make the production medium more costly (Letti et al. 2021). Automation is necessary to guarantee the system's sterilization, and afterward, during cell cultivation, the careful evaluation of dissolved oxygen, pH, ammonia, and arginine over cultivation time may be used to early indicate contaminations on bioreactors, such as mycoplasma contamination (Fratz-Berilla et al. 2019).

### *3.5 Industrial-Scale Bioreactors' Operational Mode*

The bioreactors can be operated differently regarding feeding and/or withdrawal of media during the process duration. The choice of the mode of operation can directly affect bioreactor design and size, as different modes of operation will guarantee different productivities and have specific requirements (Allan et al. 2019). The batch operational mode is the simplest one when compared to the others, as there is no feeding or withdrawal of media over the process time, being an easy-to-handle and controlled operation. However, lower cell densities are achieved (Reiss et al. 2021). Differently from the batch, in the fed-batch there is feeding of fresh media along process time, but no withdrawal. In this case, the cell metabolites are still present in the media, which can cause a certain inhibition on cell development at the proliferation stage (Pajčin et al. 2022; Reiss et al. 2021). The continuous operation mode occurs when both feeding of fresh media to the bioreactor and withdrawal of broth take place. In suspension cultures, cells are also removed from the bioreactor. Therefore, the rate of feeding and withdrawal needs to be well established to avoid cell washout. Another solution is the cells' immobilization, which is the characteristics of perfusion mode of operation, in which the media pass through the immobilized cells while the metabolites are constantly removed, maintaining a constant volume throughout the process time. Perfusion is a great alternative for the differentiation step in cultivated meat production. The continuous mode offers the possibility of reduced costs, downtimes, and footprint when compared to the batch operational mode, along with greater quality, reproducibility, productivity, yield, and standardization (Allan et al. 2019; Pajčin et al. 2022).

## 4 Downstream Equipment for Industrial-Scale Production

The operations in the downstream process of cultivated meat are mainly divided into three stages: recovery of cells after differentiation, adaptation or mixture of biomass, and printing or molding (Fig. 2). One of the most critical stages is the recovery of the cells from the culture broth, since this process is affected by the following factors: type of cells (mesenchymal stem cells, fibro/adipogenic progenitors, and satellite cells) (Reiss et al. 2021), type of scaffolds, and culture technique. According to Brennan et al. (2021) and Pajčin et al. (2022), the separation of cells from the cultivation broth is mainly done by serial centrifugation processes, where

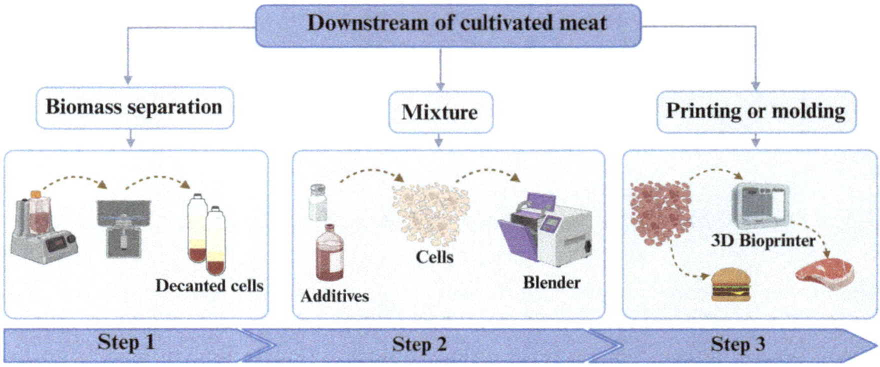


**Fig. 2** Downstream processing of cultivated meat (Created with BioRender.com)

the broth is exhausted until it manages to remove the maximum possible amount of cellular biomass.

According to Shekhawat et al. (2018), centrifugation is one of the most commonly used operations for efficient harvesting of mammalian cells. Other operations are not recommended because this type of biomass is highly sensitive. Therefore, the use of centrifuges to separate biomass requires a balance between productivity and shear. Among the centrifuges that have been used for separating mammalian cells, there are compact centrifuges with multiple separation zones and disk centrifuges, the latter being the most used, normally with speeds between 3000 rpm and 9000 rpm (Kempken et al. 1995; Takamatsu et al. 1996; Shekhawat et al. 2018).

The second downstream stage is mixing or adapting the cultivated biomass. After separating the biomass from the culture broth, it is mixed with different additives to provide the desired organoleptic characteristics, and sometimes preservatives are also added to increase the shelf life of cultivated meat (Brennan et al. 2021).

The rationale for cultured meat is the concept of traditional-natural meat. Consumers expect that synthetic meat, in addition to being easily accessible, is nutritionally similar or better than natural meat and that its organoleptic characteristics such as flavor, texture, and aroma are also similar to the traditionally known standard of meat (Ben-arye and Levenberg 2019; Zidarič et al. 2020). Among the additives that have been used, there are, for example, the soy protein leghemoglobin used to impart a reddish color to cultured meat (Brown et al. 2018). According to patent US10863761B2, this additive can be used in concentrations close to 50 g/L. Shouwei et al. (2020), in patent CN112515113A, present several additives to be used in cultivated meat for improving its sensory characteristics. For example, to adjust the flavor and color of the meat, they used a mixture of water, yeast extract, glutamic acid, nucleotide, glucose, thiamine, Monasco Red, soy sauce, and sodium tripolyphosphate in different proportions of up to 10%. Other compounds, which are not produced by myoblasts or adipocytes, such as vitamin B12 must also be supplemented (Allan et al. 2019).

The biomass is further molded depending on the type of final product aimed at. For example, when the target product is a hamburger-type meat, molding can be done manually and is easier due to the state of biomass. In the case of forming specific pieces of meat, similar to conventional ones, such as steak, pork shoulder, etc., sophisticated technologies should be used, such as 3D bioprinting (3DB) (Dick et al. 2019; Balasubramanian et al. 2021). Those kinds of tissues contain vasculature and intramuscular fat, which affects their flavor and texture. The three major bioprinting techniques are inkjet printing, extrusion printing, and light-assisted printing (Kačarević et al. 2018; Handral et al. 2022). Extrusion printing is, among those three, the most widely used for bioprinting applications due to its versatility, low cost, and simplicity (Levi et al. 2022). Besides, this technique is recommended due to its final semi-solid consistency and the presence of muscle fibers (MacQueen et al. 2019). This technique is based on either pneumatic pressure or mechanical screw plunger to dispense bioink in the form of continuous filaments (fibers) instead of droplets (Handral et al. 2022). Multiple bioinks can be mixed since a lot of printer heads can be present, and it is possible to attain customized and well-regulated compositions, shapes, porosities, cell distributions, etc. The post-processing of 3D bioprinting is also an important step: parameters such as cooking loss, fat retention, moisture retention, shrinkage, and textural changes need to be evaluated before meat commercialization (Handral et al. 2022).

For example, in 2021 the company Aleph Farms produced for the first time a sirloin steak uniting 3DB technology, natural meat building blocks, and cow cells (Newswire 2021). According to the manufacturers, Aleph Farms technology is based on printing incubated live cells. The system presents vascularity similar to that naturally occurs in tissues, allowing the perfusion of nutrients through the thicker tissue and contributing to the steak's shape and structure being similar to those found in natural meat.

## 5 Economics of Scale-Up

The economic aspects of upscaling cultured meat production are crucial in determining the feasibility and commercial viability of this industry. As the demand for sustainable and ethical alternatives to conventional meat grows, it is crucial to understand the economic implications of scaling-up cultured meat production.

Cultured meat production comes with a significant cost, which is intricately tied to multiple factors throughout the entire process. The complex and sensitive nature of mammalian animal cells requires advanced technology and significant investment to be cultivated efficiently (Risner et al. 2020; Humbird 2021). This impacts the design of bioreactors, specialized equipment, culture medium, process duration, and the quality of utilities like water, air, and energy, among other crucial aspects.

Researchers have conducted several analyses to evaluate the production cost of cultured meat, offering valuable insights by considering various factors, limitations, and assumptions. It is essential to consider multiple reports to gain a comprehensive understanding of the cost dynamics and challenges involved. Besides that, each study operates under its own set of premises and methodologies, leading to variations in the findings and conclusions. Understanding these variations provides valuable information about the uncertainties and potential risks associated with cultured meat production.

## 5.1 Culture Media and Growth Factors

Among the factors influencing the economics of scaling-up CM production, the culture medium holds the most prominent position, as animal cells need to be fed with high-quality components and require multiple changes or perfusion of the media (Risner et al. 2020; Humbird 2021). It plays a key role in the cultivation of animal cells, as it must provide all essential nutrients necessary to support cell growth and metabolic requirements, enabling proliferation and subsequent differentiation (Soccol et al. 2022).

The major contributors to the cost of the media are the complex set of amino acids (e.g., glutamine, arginine, methionine) and growth factors (e.g., TGF-β, FGF, transferrin, insulin) needed for the production (Risner et al. 2020; Humbird 2021; Negulescu et al. 2023). Since these are components with high purity, high sensibility, and low market offer, the base scenario considering bench-scale component pricing is around \$380/L of media (Specht 2020). This may account for 40–95% of the total marginal costs of the final product, even in scenarios with cheaper or lower use, e.g., \$1/L (Humbird 2021; Negulescu et al. 2023; Soccol et al. 2022).

Therefore, minimizing the utilization of medium components while expanding their production scale worldwide is a great strategy for cost reduction of cultured meat cultivation, thereby enhancing the economic viability of the process (Risner et al. 2020; Vergeer et al. 2021). Promising alternatives such as plant protein hydrolysates (source for amino acids) and recombinant plants (source for growth factors) have been identified. These options hold the potential to facilitate the attainment of both cost-effective and sustainable production of cultured meat (Liu et al. 2015; Humbird 2021).

## 5.2 Mammalian Cells

Mammalian cells have a low specific growth rate. While some microorganism cells can double in hours, for these animal cells to double, it takes at least a day. Furthermore, mammalian cells are more sensitive to catabolite inhibition, nutrient deprivation, and mechanical stresses, which affect their viability and control during the growth (Humbird 2021).

Risner et al. (2020) conducted a sensitivity analysis test that examined 67 variables in the established process and identified key factors that significantly influenced capital and annual operating expenses in the production of cultured meat. The findings revealed that among the most influential factors, cell characteristics had a substantial impact on the economic feasibility of the production. Such factors include average single-cell volume, average single-cell density, fibroblast growth factor-2 (FGF-2) concentration, glucose concentration, glucose consumption rate per cell, maturation time, transforming growth factor beta (TGF-β) concentration, and oxygen consumption per cell (Risner et al. 2020). Given this scenario, improvements in mammalian cells through genetic engineering are necessary to produce cells with positive economic traits, such as absence of cell anchorage dependence; metabolic alterations aimed at lowering media consumption or activating pathways without the requirement for high concentrations of growth factors; capacity of increased maximum cell density; shorter production time; and more (O'Neill et al. 2021; Humbird 2021; Negulescu et al. 2023).

## *5.3 Facility Costs*

After the culture medium, the facility cost is the most significant factor in the overall expenses of the production process due to several factors (Negulescu et al. 2023). Cleanliness is the most critical factor in the entire process and limits its scalability since the equipment, environment, and every input have to be totally free of any contamination, preventing batch losses due to bacterial growth. In this manner, the bioreactors and equipment end up being constrained by the sterility conditions and, furthermore, by the cells' inability to handle mechanical stress and spatial heterogeneity (Humbird 2021).

Bioreactors for this purpose generally work at smaller volumes (approximately 20m$^3$) resulting in a larger number of vessels, elevating the maintenance and capital expenses (Risner et al. 2020; Humbird 2021). However, computational fluid dynamics (CFD) modeling has already described bioreactors that may be suitable for animal cell culture that reach volumes of 300 m$^3$, which offer the potential for increased productivity and scalability in large-scale production with great perspectives in conceptual large-scale bioreactor designs by private companies (Li et al. 2020b; Negulescu et al. 2023).

Finally, as in all bioprocesses, energetical demands should be considered in scaling-up CM production. The process is energy-intensive, since energy is required for maintaining the ideal temperature for cell cultivation, as well as to produce the culture media or the ingredients that compose it (Sinke et al. 2023). According to Sinke et al. (2023), the use of renewable energy, especially in the upstream supply chain, can ensure that CM is still a sustainable alternative to conventional meat in terms of energy (Sinke et al. 2023).

# 6 Sustainability, Challenges, and Future Developments

Among the different stages of cultivated meat production, aiming for large-scale production in a cost-effective way, most challenges are related to engineering, process costs, and sustainability aspects. Firstly, since cultivation media clearly have a major impact on the final product price, as stated before, optimizing them is an essential step in making scaled-up cultivated meat accessible and widespread. A lot of research has been done on different chemical compositions of the media with low value and sustainable components, which may include hydrolysates or vegetal, microalgal, bacterial, or fungal extracts (Hocquette 2016). The replacement of lab-scale cultivation media components can also be done through fermentation-derived alternatives. For example, abundant proteins such as albumin, transferrin, and insulin are being studied to be produced via fermentation through modified microorganisms at large scales and focusing on lowering production costs (Hubalek et al. 2022).

Especially at a large scale, media recycling strategies can be employed aiming at the recovery of valuable nutrients not consumed during cultivation, such as amino acids, albumin, growth factors, etc. In turn, products of cell metabolism (such as ammonia and lactate) can be recovered and commercialized, or they can even be applied in other processes within the CM industry, including on the cultivation of other microorganisms capable of generating nutrients from those by-products (Moutsatsou et al. 2023; Yang et al. 2023).

Imperial focus is also necessary concerning the upscaling of bioreactors for cell cultivation. Lots of parallels between cell cultivation bioreactors and fermentation bioreactors or the ones in the pharmaceutical industry have been drawn, but since CM technology is an innovative and disruptive technology, it should not be limited to those parallels but bring its own contributions to bioprocess scale-up technologies. In designing bigger bioreactors for upscaled cell cultivation, CFD (computational fluid dynamics) is a useful tool to predict constraints in bioreactor transfer capacity, mixing, dissolved oxygen concentration/distribution, and shear stress that can damage animal cells (Bellani et al. 2020). In the study by Li et al. (2020a), the potential of CFD is exemplified. The authors proposed a $300m^3$ air-lift bioreactor for cell cultivation, using CFD to optimize the reactor geometry and internal apparatus (such as perforated plate flow distributors and concentric conical flow diffusers) in order to meet the mass transfer, mixing, and energy dissipation rate criteria typical for animal cells (Li et al. 2020b). The results indicated that the maximum achievable cell density would be of $2 \times 10^8$ cells per milliliter, assuming oxygen mass transfer is the limiting factor. Those sorts of studies can serve as an encouragement for accelerating the upscaling and commercialization of cultured meat technology, given the bioreactor's current design limitations (Li et al. 2020b).

In 2022, the company ABEC has partnered with Good Meat—the only company that commercializes cultivated meat in the world nowadays—to construct ten 250,000 liter bioreactors for both chicken and beef meat cultivation (ABEC and Good Meat 2022). In September 2023, the European industry Extracellular has opened Europe's largest contract pilot facility dedicated to cultivated meat and seafood production, in Bristol. To date, the company operates with 200 L of bioreactor capacity, with prospects to attain 2000 L of capacity by December 2023 and, in the future, 8000 L of capacity. The aim of the company is to eliminate the need for cultivated meat companies to develop pilot-scale facilities altogether, thus helping them to de-risking the scale-up process (Extracellular 2023). In addition, batch cultivations may not be enough to improve cell proliferation beyond the values currently achieved, and continuous cultivation is expected to be more explored as a strategy (Bellani et al. 2020).

Finally, as scaling-up is related to the need for a larger production volume to meet consumer demands, public acceptance of the final product characteristics is essential for a CM industry to achieve payback. Consumer behavior toward artificial meat varies across regions of the world and cultures (Zhang et al. 2021). Texture, color, odor, and taste of cultivated media are expected to be similar to those of common meat for great acceptance by consumers; thus, a lot of efforts should be destined to render CM the ideal organoleptic properties and guarantee its market. It is difficult to reproduce those characteristics since they result, mostly, from complex interactions between lipids, carbohydrates, and proteins of the common meat (Hocquette 2016; Zhang et al. 2021). Luckily, 3D bioprinting is a field in rapid development (Levi et al. 2022). Bioprinted cultivated meat can be customized and optimized to meet consumer demands by varying the cocktail of components used in the product formulation (cellular and acellular ones), printing patterns, printing speed, nozzle diameter and height, extrusion rate, addition of porous materials, and infill percentage (Handral et al. 2022).

## 7 Conclusion

Upscaling a process is inherently challenging since new variables emerge, such as the technology applied in the storage of raw materials, cells, packaging, stability, and reproducibility of the final product (Zhang et al. 2021). It is even more challenging in the case of a bioprocess involving the cultivation of complex and sensitive cells, as it is the case of cultivated meat production. The development of new pluripotent cell lines or the improvement of the proliferation abilities of existent cell lines to attain greater productivity, although difficult to execute, is key and can have a significant impact on process economics and viability, making its scaling-up much simpler (Chodkowska et al. 2022).

Furthermore, the bioprocess engineering constraints found in upscaling cultivated meat production, related especially to a bioreactor design that guarantees great and equal cell growth throughout, are pivotal. A full-scale cultured meat bioprocess is still in development, and attaining cost-effectiveness is essential to the feasibility and commercialization of CM in a broad range of regions. Nevertheless, being the cultivation medium responsible for the main part of the process costs on a large scale, it is expected that intense research in this area can lead to more cost-effective and sustainable CM production. All these challenges can be explored and overcome with intense research, both from industries and research establishments, at laboratory and bench scales, but mostly searching to demonstrate and optimize the feasibility of the process at pilot scale, focusing on future large-scale productions.

## References


ABEC, Good Meat (2022) GOOD meat partners with industry leader to build the world's first large-scale cultivated meat facility. Web page. https://www.goodmeat.co/all-news/good-meat-partners-with-industry-leader-to-build-first-large-scale-cultivated-meat-facility. Accessed 25 Sep 2023

Allan SJ, De Bank PA, Ellis MJ (2019) Bioprocess design considerations for cultured meat production with a focus on the expansion bioreactor. Front Sustain Food Syst 3. https://doi.org/10.3389/fsufs.2019.00044

Azhar A, Zeyaullah M, Bhunia S et al (2023) Cell-based meat: the molecular aspect. Front Food Sci Technol 3. https://doi.org/10.3389/frfst.2023.1126455

Balasubramanian B, Liu W, Pushparaj K, Park S (2021) The epic of in vitro meat production—a fiction into reality. Foods 10:1–21. https://doi.org/10.3390/foods10061395

Bayer E, McIntyre G, Mueller P, et al (2019) Methods of generating mycelial scaffolds and applications thereof. WO Patent 2020106743A1

Bellani CF, Ajeian J, Duffy L et al (2020) Scale-up Technologies for the manufacture of adherent cells. Front Nutr 7. https://doi.org/10.3389/fnut.2020.575146

Ben-arye T, Levenberg S (2019) Tissue engineering for clean meat production. Front Sustain Food Syst 3:1–19. https://doi.org/10.3389/fsufs.2019.00046

Bodiou V, Moutsatsou P, Post MJ (2020) Microcarriers for upscaling cultured meat production. Front Nutr 7. https://doi.org/10.3389/fnut.2020.00010

Bomkamp C, Skaalure SC, Fernando GF et al (2022) Scaffolding biomaterials for 3D cultivated meat: prospects and challenges. Adv Sci 9:1–40. https://doi.org/10.1002/advs.202102908

Brennan T, Katz J, Quint Y, Spencer B (2021) Cultivated meat: out of the lab, into the frying pan. Web page. https://www.mckinsey.com/industries/agriculture/our-insights/cultivated-meat-out-of-the-lab-into-the-frying-pan. Accessed 25 Sep 2023

Brown PO, Vrljic M, Varadan R, et al (2018) Methods and compositions for consumables. US patent 10863761B2

Cantarero Rivera FJ, Chen J (2022) Computational fluid dynamics modeling of cell cultures in bioreactors and its potential for cultivated meat production—a mini-review. Future Foods 6:100195. https://doi.org/10.1016/j.fufo.2022.100195

Caplan AI (2017) Mesenchymal stem cells: time to change the name! Stem Cells Transl Med 6:1445–1451. https://doi.org/10.1002/sctm.17-0051

CELLREV (2023) The future of commercial-scale cell manufacturing 2023 Report

Chen L, Guttieres D, Koenigsberg A et al (2022) Large-scale cultured meat production: trends, challenges and promising biomanufacturing technologies. Biomaterials 280:121274. https://doi.org/10.1016/j.biomaterials.2021.121274

Cherian DS, Bhuvan T, Meagher L, Heng TSP (2020) Biological considerations in scaling up therapeutic cell manufacturing. Front Pharmacol 11. https://doi.org/10.3389/fphar.2020.00654

Chieregato K, Castegnaro S, Madeo D et al (2011) Epidermal growth factor, basic fibroblast growth factor and platelet-derived growth factor-bb can substitute for fetal bovine serum and compete with human platelet-rich plasma in the ex vivo expansion of mesenchymal stromal cells derived from adipose tissue. Cytotherapy 13:933–943. https://doi.org/10.3109/14653249.2011.583232

Chodkowska KA, Wódz K, Wojciechowski J (2022) Sustainable future protein foods: the challenges and the future of cultivated meat. Foods 11. https://doi.org/10.3390/foods11244008

Choudhury D, Tseng TW, Swartz E (2020) The business of cultured meat. Trends Biotechnol 38:573–577. https://doi.org/10.1016/j.tibtech.2020.02.012

Dick A, Bhandari B, Prakash S (2019) 3D printing of meat. Meat Sci 153:35–44. https://doi.org/10.1016/j.meatsci.2019.03.005

Dominici M, Le Blanc K, Mueller I et al (2006) Minimal criteria for defining multipotent mesenchymal stromal cells. The International Society for Cellular Therapy position statement. Cytotherapy 8:315–317. https://doi.org/10.1080/14653240600855905

Edelman PD, McFarland DC, Mironov VA, Matheny JG (2005) Commentary: In vitro-cultured meat production. Tissue Eng 11(5–6):659–662. https://doi.org/10.1089/ten.2005.11.659

European Medicines Agency (2013) Committee for Medicinal Products for Human Use (CHMP) guideline on the use of bovine serum in the manufacture of human biological medicinal products

Extracellular (2023) Meet the team scaling up cultivated meat. Web page. https://www.extracellular.com/insights/meet-the-team-scaling-up-cultivated-meat/. Accessed 25 Sep 2023

Fratz-Berilla EJ, Faison T, Kohnhorst CL et al (2019) Impacts of intentional mycoplasma contamination on CHO cell bioreactor cultures. Biotechnol Bioeng 116:3242–3252. https://doi.org/10.1002/bit.27161

GFI - Good Food Institute (2023) Deep dive: cultivated meat bioprocess design. Web page. https://gfi.org/science/the-science-of-cultivated-meat/deep-dive-cultivated-meat-bioprocess-design/. Accessed 2 Oct 2023

Guembe Lapuente A, Zúñiga Arrarás T, Recalde Irurzun JI, Izco Zaratiegui JM (2022) Collagen hydrogels useful as cell carriers

Handral HK, Hua Tay S, Wan Chan W, Choudhury D (2022) 3D printing of cultured meat products. Crit Rev Food Sci Nutr 62:272–281. https://doi.org/10.1080/10408398.2020.1815172

Hanga MP, Ali J, Moutsatsou P et al (2020) Bioprocess development for scalable production of cultivated meat. Biotechnol Bioeng 117:3029–3039. https://doi.org/10.1002/bit.27469

Hocquette JF (2016) Is in vitro meat the solution for the future? Meat Sci 120:167–176. https://doi.org/10.1016/j.meatsci.2016.04.036

Hubalek S, Post MJ, Moutsatsou P (2022) Towards resource-efficient and cost-efficient cultured meat. Curr Opin Food Sci 47. https://doi.org/10.1016/j.cofs.2022.100885

Humbird D (2021) Scale-up economics for cultured meat. Biotechnol Bioeng 118:3239–3250. https://doi.org/10.1002/bit.27848

Izadpanah R, Kaushal D, Kriedt C et al (2008) Long-term in vitro expansion alters the biology of adult mesenchymal stem cells. Cancer Res 68:4229–4238. https://doi.org/10.1158/0008-5472.CAN-07-5272

Jairath G, Mal G, Gopinath D, Singh B (2021) An holistic approach to access the viability of cultured meat: a review. Trends Food Sci Technol 110:700–710. https://doi.org/10.1016/j.tifs.2021.02.024

Jervis M, Huaman O, Cahuascanco B et al (2019) Comparative analysis of in vitro proliferative, migratory and pro-angiogenic potentials of bovine fetal mesenchymal stem cells derived from bone marrow and adipose tissue. Vet Res Commun 43:165–178. https://doi.org/10.1007/s11259-019-09757-9

Jossen V, Eibl R, Kraume M, Eibl D (2018) Growth behavior of human adipose tissue-derived stromal/stem cells at small scale: numerical and experimental investigations. Bioengineering 5. https://doi.org/10.3390/BIOENGINEERING5040106

Kačarević ŽP, Rider PM, Alkildani S et al (2018) An introduction to 3D bioprinting: possibilities, challenges and future aspects. Materials 11. https://doi.org/10.3390/ma11112199

Kahan S, Camphuijsen J, Cannistra C, et al (2020) Cultivated meat modeling consortium: inaugural meeting whitepaper. https://doi.org/10.22541/au.158057683.31004563

Kempken R, Preissmann A, Berthold W (1995) Assessment of a disc stack centrifuge for use in mammalian cell separation. Biotechnol Bioeng 46:132–138. https://doi.org/10.1002/bit.260460206

Kong Y, Jing L, Huang D (2022) Plant proteins as the functional building block of edible microcarriers for cell-based meat culture application. Crit Rev Food Sci Nutr 0:1–11. https://doi.org/10.1080/10408398.2022.2147144

Kuijk E, Jager M, van der Roest B et al (2020) The mutational impact of culturing human pluripotent and adult stem cells. Nat Commun 11. https://doi.org/10.1038/s41467-020-16323-4

Kumar A, Sood A, Han SS (2023) Technological and structural aspects of scaffold manufacturing for cultured meat: recent advances, challenges, and opportunities. Crit Rev Food Sci Nutr 63:585–612. https://doi.org/10.1080/10408398.2022.2132206

Kurosawa H (2012) Application of rho-associated protein kinase (ROCK) inhibitor to human pluripotent stem cells. J Biosci Bioeng 114:577–581. https://doi.org/10.1016/j.jbiosc.2012.07.013

Lee DY, Lee SY, Yun SH et al (2022) Review of the current research on fetal bovine serum and the development of cultured meat. Food Sci Anim Resour 42:775–799. https://doi.org/10.5851/kosfa.2022.e46

Letti LAJ, Karp SG, Molento CFM et al (2021) Cultivated meat: recent technological developments, current market and future challenges. Biotechnol Res Innov 5:e2021001. https://doi.org/10.4322/biori.202101

Levi S, Yen FC, Baruch L, Machluf M (2022) Scaffolding technologies for the engineering of cultured meat: towards a safe, sustainable, and scalable production. Trends Food Sci Technol 126:13–25. https://doi.org/10.1016/j.tifs.2022.05.011

Li F, Ge Y, Liu D, Songyang Z (2020a) The role of telomere-binding modulators in pluripotent stem cells. Protein Cell 11:60–70. https://doi.org/10.1007/s13238-019-0651-y

Li X, Zhang G, Zhao X et al (2020b) A conceptual air-lift reactor design for large scale animal cell cultivation in the context of in vitro meat production. Chem Eng Sci 211. https://doi.org/10.1016/j.ces.2019.115269

Liu Y-K, Li Y-T, Lu C-F, Huang L-F (2015) Enhancement of recombinant human serum albumin in transgenic rice cell culture system by cultivation strategy. N Biotechnol 32:328–334. https://doi.org/10.1016/j.nbt.2015.03.001

Liu L, Wang J, Duan S et al (2016) Systematic evaluation of sericin protein as a substitute for fetal bovine serum in cell culture. Sci Rep 6. https://doi.org/10.1038/srep31516

Liu S, Yang W, Li Y, Sun C (2023) Fetal bovine serum, an important factor affecting the reproducibility of cell experiments. Sci Rep 13. https://doi.org/10.1038/s41598-023-29060-7

MacQueen LA, Alver CG, Chantre CO et al (2019) Muscle tissue engineering in fibrous gelatin: implications for meat analogs. NPJ Sci Food 3:1–12. https://doi.org/10.1038/s41538-019-0054-8

Merten OW (2015) Advances in cell culture: anchorage dependence. Philos Trans R Soc B Biol Sci 370. https://doi.org/10.1098/rstb.2014.0040

Minnich C, Hardy S, Krämer S (2016) Stopping the Babylonian confusion: an updated nomenclature for process analyzers in PAT applications. Chem Ing Tech 88:694–697. https://doi.org/10.1002/cite.201500188

Moutsatsou P, Cameron P, Dages B et al (2023) Scaling cell production sustainably in cultured meat product development. In: Post M, Connon C, Bryant C (eds) Advances in cultured meat technology, 1st edn. Burleigh Dodds Science Publishing Limited

Mullen N, Park N, Jones C, et al (2020) In vitro avian food product. US patent 20200392461A1

Nahmias Y, Cohen M, Caspi J (2021) Utilization of plant protein homologues in culture media. WO patent 2021148955A1

Negulescu PG, Risner D, Spang ES et al (2023) Techno-economic modeling and assessment of cultivated meat: impact of production bioreactor scale. Biotechnol Bioeng 120:1055–1067. https://doi.org/10.1002/bit.28324

Ng S, Kurisawa M (2021) Integrating biomaterials and food biopolymers for cultured meat production. Acta Biomater 124:108–129. https://doi.org/10.1016/j.actbio.2021.01.017

Ng JY, Chua ML, Zhang C et al (2020) Chlorella vulgaris extract as a serum replacement that enhances mammalian cell growth and protein expression. Front Bioeng Biotechnol 8. https://doi.org/10.3389/fbioe.2020.564667

Nikfarjam L, Farzaneh P (2012) Prevention and detection of mycoplasma contamination in cell culture. Cell J 13:203–212

O'Neill EN, Cosenza ZA, Baar K, Block DE (2021) Considerations for the development of cost-effective cell culture media for cultivated meat production. Compr Rev Food Sci Food Saf 20:686–709. https://doi.org/10.1111/1541-4337.12678

Olarerin-George AO, Hogenesch JB (2015) Assessing the prevalence of mycoplasma contamination in cell culture via a survey of NCBI's RNA-seq archive. Nucleic Acids Res 43:2535–2542. https://doi.org/10.1093/nar/gkv136

Oyarzo R, Valderrama X, Valenzuela F, Bahamonde J (2021) Bovine fetal mesenchymal stem cells obtained from omental adipose tissue and placenta are more resistant to cryoprotectant exposure than those from bone marrow. Front Vet Sci 8. https://doi.org/10.3389/fvets.2021.708972

Pajčin I, Knežić T, Azoulay IS et al (2022) Bioengineering outlook on cultivated meat production. Micromachines (Basel) 13. https://doi.org/10.3390/mi13030402

Pang KT, Loo LSW, Chia S et al (2023) Insight into muscle stem cell regeneration and mechanobiology. Stem Cell Res Ther 14. https://doi.org/10.1186/s13287-023-03363-y

Pelling AE, Cuerrier CM, Modulevsky DJ, Hickey RJ (2017) Decellularised cell wall structures from plants and fungus and use thereof as scaffold materials. WO patent 2017136950A1

Piletz JE, Drivon J, Eisenga J et al (2018) Human cells grown with or without substitutes for fetal bovine serum. Cell Med 10:215517901875514. https://doi.org/10.1177/2155179018755140

Pilgrim CR, McCahill KA, Rops JG et al (2022) A review of fetal bovine serum in the culture of mesenchymal stromal cells and potential alternatives for veterinary medicine. Front Vet Sci 9. https://doi.org/10.3389/fvets.2022.859025

Post MJ, Levenberg S, Kaplan DL et al (2020) Scientific, sustainability and regulatory challenges of cultured meat. Nat Food 1:403–415. https://doi.org/10.1038/s43016-020-0112-z

PR Newswire (2021) Aleph Farms and The Technion Reveal World's First Cultivated Ribeye Steak. Web page. https://www.prnewswire.com/il/news-releases/aleph-farms-and-the-technion-reveal-worlds-first-cultivated-ribeye-steak-301224800.html. Accessed 15 August 2023

Price PJ, Gorfien S, Danner D (1997) Animal cell culture media comprising peptides derived from rice abstract. US patent 6103529A

Reiss J, Robertson S, Suzuki M (2021) Cell sources for cultivated meat: applications and considerations throughout the production workflow. Int J Mol Sci 22. https://doi.org/10.3390/ijms22147513

Riche E, Bôle J, Mabic S (2023) Endotoxin-free water for tissue culture. In: Merck, Milli-Q® Lab Water Solutions, Application note, Guyancourt, France

Risner D, Li F, Fell JS et al (2020) Preliminary techno-economic assessment of animal cell-based meat. Foods 10:3. https://doi.org/10.3390/foods10010003

Romito A, Cobellis G (2016) Pluripotent stem cells: current understanding and future directions. Stem Cells Int 2016. https://doi.org/10.1155/2016/9451492

Rouhani FJ, Zou X, Danecek P et al (2022) Substantial somatic genomic variation and selection for BCOR mutations in human induced pluripotent stem cells. Nat Genet 54:1406–1416. https://doi.org/10.1038/s41588-022-01147-3

Seah JSH, Singh S, Tan LP, Choudhury D (2022) Scaffolds for the manufacture of cultured meat. Crit Rev Biotechnol 42:311–323. https://doi.org/10.1080/07388551.2021.1931803

Shekhawat LK, Sarkar J, Gupta R et al (2018) Application of CFD in bioprocessing: separation of mammalian cells using disc stack centrifuge during production of biotherapeutics. J Biotechnol 267:1–11. https://doi.org/10.1016/j.jbiotec.2017.12.016

Shouwei W, Meng L, Shunliang Z, et al (2020) Cell culture meat, preparation method and application. CN patent 112515113A

Shouwei W, Yang F, Li Y, et al (2021) Composite microcarrier for manufacturing cell culture meat and preparation method of composite microcarrier. WO patent 2023104768A1

Singh A, Kumar V, Singh SK et al (2023) Recent advances in bioengineered scaffold for in vitro meat production. Cell Tissue Res 391:235–247. https://doi.org/10.1007/s00441-022-03718-6

Sinke P, Swartz E, Sanctorum H et al (2023) Ex-ante life cycle assessment of commercial-scale cultivated meat production in 2030. Int J Life Cycle Assess 28:234–254. https://doi.org/10.1007/s11367-022-02128-8

Soccol CR, Karp SG, Letti LAJ et al (2022) Is the development of low-cost media one of the greatest challenges to produce cultivated meat on an industrial scale? Biotechnol Res Innov 6:e2022005. https://doi.org/10.4322/biori.20226201

Specht L (2020) An analysis of culture medium costs and production volumes for cultivated meat. The Good Food Institute (GFI) Report

Stacey G, Davis J (2007) Medicines from animal cell culture. Wiley

Stephens N, Di Silvio L, Dunsford I et al (2018) Bringing cultured meat to market: technical, socio-political, and regulatory challenges in cellular agriculture. Trends Food Sci Technol 78:155–166. https://doi.org/10.1016/j.tifs.2018.04.010

Stroncek DF, Jin P, Ren J et al (2010) Quality assessment of cellular therapies: the emerging role of molecular assays. Korean J Hematol 45:14. https://doi.org/10.5045/kjh.2010.45.1.14

Subbiahanadar Chelladurai K, Selvan Christyraj JD, Rajagopalan K et al (2021) Alternative to FBS in animal cell culture - an overview and future perspective. Heliyon 7. https://doi.org/10.1016/j.heliyon.2021.e07686

Takamatsu H, Hamamoto K, Ishimaru K et al (1996) Large-scale perfusion culture process for suspended mammalian cells that uses a centrifuge with multiple settling zones. Appl Microbiol Biotechnol 45:454–457. https://doi.org/10.1007/BF00578455

Vergeer R, Sinke P, Odegard I (2021) TEA of cultivated meat. Future projections of different scenarios - corrigendum. Web page. https://cedelft.eu/publications/tea-of-cultivated-meat/. Accessed in 26 Sep 2023

Vernardis SI, Terzoudis K, Panoskaltsis N, Mantalaris A (2017) Human embryonic and induced pluripotent stem cells maintain phenotype but alter their metabolism after exposure to ROCK inhibitor. Sci Rep 7. https://doi.org/10.1038/srep42138

Yang M, Wang Q, Zhu Y et al (2023) Cell culture medium cycling in cultured meat: key factors and potential strategies. Trends Food Sci Technol 138:564–576. https://doi.org/10.1016/j.tifs.2023.06.031

Young L, Sung J, Masters JR (2010) Detection of mycoplasma in cell cultures. Nat Protoc 5:929–934. https://doi.org/10.1038/nprot.2010.43

Zhang L, Hu Y, Badar IH et al (2021) Prospects of artificial meat: opportunities and challenges around consumer acceptance. Trends Food Sci Technol 116:434–444. https://doi.org/10.1016/j.tifs.2021.07.010

Zidarič T, Milojević M, Vajda J et al (2020) Cultured meat: meat industry hand in hand with biomedical production methods. Food Eng Rev 12:498–519. https://doi.org/10.1007/s12393-020-09253-w

# Waste Management in Cultivated Meat Production

**Adenise Lorenci Woiciechowski, Thamarys Scapini, Maria Clara Manzoki, Walter José Martínez-Burgos, Denisse Tatiana Molina-Aulestia, Susan Grace Karp, and Carlos Ricardo Soccol**

**Abstract** Waste management and environmental impacts remain fields that require assessment alongside the advances in cultivated meat production. Water and energy consumption, the demand for natural resources, and the generation of residues throughout the production chain must all be considered. These aspects can be more effectively adjusted in the early stages of development. By practicing "reduce, reuse, repurpose, refuse, and recycle," we can optimize processes and explore innovative alternatives in the cultured meat market while prioritizing food and environmental safety. Despite the exponential increase in studies and patents dedicated to advancing cultivated meat production in recent years, there remains a gap in identifying and analyzing the environmental impacts associated with this process, particularly in understanding the generation of liquid, solid, and gaseous effluents with the objective of minimizing their production or repurposing them into other process streams. Considering the importance of developing and solidifying this bioprocess within the field of biotechnology, this chapter aims to address the latest advances in waste management within the cultivated meat production process. To achieve this goal, recent research focusing on characterizing the generated effluents has been conducted, exploring conventional treatments for these streams and integration strategies aligned with circular economy.

**Keywords** Wastewater · Effluent · Artificial meat · Medium recycling · Biological treatment

A. L. Woiciechowski (✉) · T. Scapini · M. C. Manzoki · W. J. Martínez-Burgos · D. T. Molina-Aulestia · S. G. Karp · C. R. Soccol
Bioprocess Engineering and Biotechnology Department, Federal University of Paraná, Curitiba, Paraná, Brazil
e-mail: adenise@ufpr.br


C. R. Soccol et al. (eds.), *Cultivated Meat*,
https://doi.org/10.1007/978-3-031-55968-6_13

# 1 Introduction

The development of cultivated meat (CM) is an emerging field of biotechnological research and development, bridging aspects of cell biology and food technology. It is emerging as an alternative to the traditional animal production industry, with the insight of minimizing environmental damage, food safety, human health, and animal welfare concerns, while also alleviating the production overload of the global livestock sector (Scharf et al. 2020). Concerning to environmental impacts, there is a suggestion that CM has the potential to significantly reduce greenhouse gases (GHGs), land usage, and water consumption (Tuomisto and Teixeira de Mattos 2011). However, environmental aspects of CM production remain a controversial discussion (Chriki and Hocquette 2020).

Waste management is a critical aspect of industrial bioprocesses that must be considered in technoeconomic and bioeconomic analyses. Projecting the adequate waste treatment to follow environmental legislation and avoid harmful impacts on environment and health is a prerequisite for a successful downstream process.

The production of cultivated meat is carried out in nutrient-rich media containing amino acids, carbohydrates, growth factors, and sometimes antibiotics. Residual amino acids and carbohydrates can be treated as "regular" effluents together with cell metabolism products such as ammonia and lactate. Antibiotics, however, can be persistent in effluent treatment systems and, thus, be released to the environment. Growth factors, even though expected to have a short half-life, can release degradation products that should be evaluated for toxicity and other environmental impacts.

Despite the significant attention given to CM process development in recent years, the processes are still in the maturation stage for pilot and industrial scales. Additionally, there are limited studies assessing the environmental impacts of CM production or identifying effluent streams with the aim of minimizing their generation or reusing them for other purposes (Tuomisto et al. 2022; Kim et al. 2022; Sinke et al. 2023; Trinidad et al. 2023). Bridging this knowledge gap is essential as the development progresses. By incorporating environmental impact information into the developmental phase, adjustments can be made proactively. This approach facilitates the integration of industrial processes into reuse and recycling streams, to minimize the generation of effluents or even to ensure that the treatments are effective and reduce the polluting potential.

Given the continuously growing scenario of research, patents, and pilot-scale implementations in cultivated meat production systems, the aim of this chapter is to address into the most recent advancements in waste management within these processes. The pursuit of an environmentally sustainable process demands that concerns about reducing environmental impacts through the reduction and proper management of waste be conducted in parallel. Consequently, this chapter reviews recent research on CM production, identifying the effluents generated in these processes, exploring potential treatments for the streams, and integrating strategies aligned with the principles of circular economy concept.

## 2 Effluents Generated by Cultivated Meat Production

The production of cultivated meat in scale, within a facility where laboratory, bench, and production scales are operating, entails the generation of different types of wastes, which can be categorized as liquid, solid, or gaseous wastes. Substantial volumes of culture media are required for producing cultivated meat. It is estimated that to produce 1 tonne of meat, considering both the proliferation and differentiation phases, 140,000 liters of media will be required, with a full medium exchange every 2 days, or 80,000 liters if a recycling approach that alternates 80% recycling with fresh medium every 3 days is used (Specht 2020; Hubalek et al. 2022). Within the culture media, only some of the components are degraded or metabolized by the cells, while others act principally on maintaining pH balance or osmotic balance. The pH buffers, for example, are responsible for a substantial fraction of the basal medium cost, accounting for at a minimum 38% of the total medium cost (Specht 2020). The research from Myers et al. (2023) estimated that for each kilogram of cultivated meat produced, 1.57 kg of chemical oxygen demand (COD) is generated, considering carbonaceous wastes present on the wastewaters such as residual glucose and amino acids (Myers et al. 2023). Those high values imply on high wastewater treatment costs. Thus, at a large scale, medium recycling is expected as a solution to reduce the overall process wastes and costs.

To understand the transformations that occur within the cultivation media along cell cultivation, Haraguchi and Shimizu (2021) cultivated myoblast cells for 3 days and evaluated the media composition at the end and at the beginning of the process. Practically all of glucose was consumed along the cell growth and 69% of all glutamines. However, phosphorus was consumed by 16%, and the amount of 14 amino acids from the basal medium (DMEM), except for glutamine, decreased only by 26% on the residual media. Components such as pyruvate; vitamins B1, B2, and B6; folic acid; sodium; potassium; calcium; and magnesium in the medium were kept almost constant. In turn, as expected, ammonium ion concentration increased during the cultivation (Haraguchi and Shimizu 2021).

As well as ammonia, lactate is also normally identified on residual cultivation media. It is generated through two main pathways: glycolysis and amino acid catabolism, being a product of the primary metabolic processes of the cell. It also can disturb cell growth when above certain concentrations (normally 40 mM) and alters the pH of the culture environment (Yang et al. 2023). By appropriate methods, metabolic wastes produced during cultivation can be identified and removed. Great amounts of lactate and ammonium ions will be produced in a large-scale facility of cultivated meat, with the possibility of recovery and recycling these co-products (Nahmias and Wissotsky 2021; Sinke et al. 2023). Non-metabolized components like salts, buffers, and growth factors can be retained and recycled, while nutrients significantly consumed by cells can be replenished (Kumar et al. 2021). Those strategies can substantially help with economics and environmental aspects of the cultivated meat production process.

Besides, in cell cultivation antibiotics are usually applied to prevent contaminations. The control of antibiotics presence and concentration in the media after cultivation is imperative (Cai et al. 2024). Fortunately, there is a tendency among CM producers to eliminate the use of antibiotics in their process or to utilize extremely low quantities of those (Saied et al. 2023).

Concerning to solid wastes from cultivated meat production, especially on laboratory and bench scales, plastic residues are emphasized. Those plastic residues are the discarded flasks and bottles utilized for the maintenance of the cell lineages and for their expansion into large volumes to generate the inoculum necessary to the larger production bioreactors. Normally those flasks, such as T-flasks, roller bottles, and spinner flasks, are composed of polyethylene terephthalate (PET), high- and low-density polyethylene (PE), polyvinyl chloride (PVC), and polypropylene (PP), but the most frequently used plastic in labs today is polystyrene (PS). The affordability of PS, combined with its nonreactive chemistry, renders it an excellent option for a disposable culture surface (Thermo Fisher Scientific 2023).

Another source of solid wastes on cultivated meat production can be related to the solid residues generated for culture media production. Since the cultivation media accounts for most of the process cost, raw materials have been explored to generate nutrients for culture media, being submitted to a pre-processing, like hydrolysis. Depending on the chosen feedstock for nutrient recovery, solid wastes can be generated, e.g., from the hydrolysis of cyanobacteria, soy protein, wheat, etc. (Haraguchi et al. 2022; Yang et al. 2023; Ghosh et al. 2023). The solid waste produced during the hydrolysis of alternative nutrients for cell cultivation must be managed in accordance with its final properties. This waste can be either utilized for re-extraction of biocompounds of interest, energy production, and agricultural fertilizer production or disposed of in a landfill.

Gaseous wastes are also intrinsic to the production of cultivated meat but are generated in much lower amounts when compared to conventional meat production according to some studies (Mattick et al. 2015; Tuomisto et al. 2022). Carbon dioxide ($CO_2$) is emitted from different steps of the production chain of cultivated meat. One of the $CO_2$ sources is the cell cultivation step, since $CO_2$ is fed into the bioreactors for pH control of the media, and an amount of the gas is lost during this process. Other sources may include the $CO_2$ emitted on the generation of the energy utilized all along the cultivated meat production chain and to produce the cultivation media components, which can be assessed more deeply through *life-cycle assessment* (LCA) studies.

## 3 Waste Treatment Technologies in Cultivated Meat Production

One of the main environmental impacts of CM production arises from the liquid effluents produced, which can be attributed to both its physical-chemical properties and the sheer volumes generated (Tuomisto et al. 2022). As discussed in the previous topic, these effluents contain significant concentrations of lactate, nitrogen, and phosphorus (Haraguchi and Shimizu 2021). Nitrogen and phosphorous play a major role in promoting eutrophication processes in water bodies (Dodds and Smith 2016; Wang et al. 2019). Moreover, certain forms of nitrogen, including ammonia, nitrites, and nitrates, exhibit toxicity towards aquatic organisms and even terrestrial life forms (Lee et al. 2003). Consequently, it is imperative to ensure efficient treatment of this wastewater before its release into the environment. The main challenge lies in the fact that conventional biologic wastewater treatment systems are not efficiently equipped to remove phosphorus and nitrogen, meaning that complementary (secondary) treatments are necessary (Yang et al. 2021).

Considering this scenario, the possibility of using conventional technologies will be presented, extrapolating through the characteristics recently presented for the treatment of cultivated meat wastewater. One approach is the biological processes for phosphorus removal and wastewater treatment, including the utilization of various microorganisms (bacteria, fungi, and microalgae), macroalgae, higher plants, and occasionally enzymes or nanoenzymes.

Microalgae have gained considerable attention in effluent treatment, mainly for nitrogen and phosphorus removal. According to Delgadillo-Mirquez et al. (2016), microalgae can convert phosphorus and nitrogen from wastewater into biomass and other bioproducts (Shi et al. 2007). In a study conducted by Shi et al. (2007), phosphorus and nitrogen were removed from wastewater employing two microalgae (*Chlorella vulgaris* and *Scenedesmus rubescens*) immobilized in a double-layer system, where microalgae adhered to an ultrafine and microporous substrate. Results indicated that microalgae could effectively eliminate nitrate, and they also exhibited a minimum removal rate of 10% of nitrite and ammonia.

Delgadillo-Mirquez et al. (2016) utilized high-rate algal ponds, employing a microbial consortium of microalgae and bacteria, for phosphorus and nitrogen removal. Through this process, 83% of nitrogen and 100% of phosphorus were removed. In another study, Haraguchi and Shimizu (2021) employed the microalgae *Chlorococcum littorale* and *Chlorella vulgaris* to remove nitrogen and phosphorus from cultivated meat effluents. The *C. littorale* strain removed approximately 80% of the ammonia content, while the *C. vulgaris* strain removed 26%. Regarding phosphorus, both strains removed around 15%.

Another interesting conventional effluent treatment based on biological approach is the wastewater treatment pond, which can operate in anaerobic or aerobic mode. Anaerobic systems are mainly employed for pathogens and organic matter removal, along with biogas production, while aerobic ponds are generally used to remove nitrogen and phosphorus compounds, among others (Craggs et al. 2004).

Biological treatment processes in ponds are carried out by microorganisms, including bacteria, that can also be used individually to remove nutrients like nitrogen and phosphorus from wastewater. As reported by Wan et al. (2017), the bacteria *Enterobacter cloacae* is capable of denitrification and phosphorous accumulation. The authors reported that this strain can effectively remove between 60% and 99% of nitrogen compounds in wastewater and approximately 73% of phosphorus. The *Enterobacter cloacae* HW-15 strain was also effective in denitrifying and accumulating phosphorus, with reported removal percentages of 97% for $NH_4^+$ and 74% of $PO_4^{3-}$ (initial concentrations of 100 mg/L and 20 mg/L, respectively) under aerobic conditions (Li et al. 2021). Nitrogen removal occurred via nitrification-denitrification (regulated by the dissolved oxygen concentration in the liquid phase), while phosphorus removal involved phosphate assimilation and poly-P accumulation. Zhou et al. (2022) developed a packed biofilter consisting of zerovalent iron (ZVI)/poly-3-hydroxybutyrate-cohyroxyvelate (PHBV)/sawdust for secondary treatment of effluents, with nitrogen and phosphorus removal efficiencies ranging from 95 to 99%.

Floating aquatic plants are also utilized for secondary treatment to remove phosphorus and nitrogen since these elements are crucial for the photosynthesis of these organisms. For example, Sudiarto et al. (2019) demonstrated that *Pistia stratiotes* could remove 63% of nitrogen (initial: 151 mg/L), while *Lemna* sp. removed around 35% of phosphorus (initial: 82 mg/L). Su et al. (2019) evaluated the removal of phosphorus and nitrogen over a 46-day period using *Ipomoea aquatica* and *Salvinia natans*, with nitrogen removal rates of 91% and 67%, respectively, (initial: 25 mg/L) and a maximum phosphorous removal rate of 14% (initial: 2 mg/L). These biological technologies can shed light on the possibility of treatments for cultivated meat effluents.

Another approach involves chemical processes which are also being utilized for nitrogen and phosphorous removal from wastewater. Through these processes, phosphate and nitrogen react with other compounds to form products that subsequently precipitate and can be removed. Lee et al. (2003) achieved nitrogen and phosphorous removal through struvite crystallization ($MgNH_4PO_4{\cdot}6H_2O$). The precipitation of nitrogen and phosphorus mainly relies on ionic reactions (Sengupta et al. 2015). Huang et al. (2017) removed 98% of phosphorus and 58% of nitrogen using magnesium salts. In this method, maintaining the pH at a correct level is crucial for effective struvite production, typically occurring at alkaline pH ranging from 8.0 to 9.5.

Currently, electrochemical processes have been under investigation as they can enhance the efficiency of pollutant recovery. In the case of nitrogen, efficiencies of up to 92% can be achieved, with the possibility of recovery in the form of $NH_3$ (gas) or $(NH_4)_2SO_4$ (Sengupta et al. 2015). Regarding phosphorus, removal efficiencies of up to 99% can be attained.

In this way, it has been possible to provide an overview of the conventional secondary treatments that can be evaluated for the liquid effluents generated in the production of cultivated meat. Gaseous effluents that may arise throughout the supply chain demand logistical and management approaches to minimize their impacts.

# 4 Waste Management and Circular Economy Strategies in Cultivated Meat Production

Beyond conventional treatments that can be used to treat effluents generated in the production of cultivated meat, new integrative technologies can be explored. In pursuit of bioprocesses innovation, there are already studies evaluating the potential for integrating or reusing the medium for alternative purposes or for reintroducing it into cell cultivation. An overview of these studies is presented in Table 1.

Cultivation media consist of essential components for cell proliferation, such as organic growth factors like hormones, enzymes, and coenzymes, along with amino acids, animal sera, and salts to maintain the medium's osmolarity (Nir 2023). Although the carbon source in the culture medium is reported to be depleted, other compounds often persist in bioreactor wastewater, including amino acids, vitamins, pyruvate, and organic salts (Haraguchi and Shimizu 2021). Considering the complexity of the cell culture system, it is essential to keep the composition of the culture medium constant to ensure optimal growth (Nir 2023).

In standard processes, the residual medium is discarded and replaced, as some nutrients become depleted and waste products accumulate. For instance, the amount of ammonia and lactate increases post-cultivation, which can negatively affect cell

**Table 1** Overview of recent studies that have either developed technologies for reusing cultured meat effluent or proposed treatment technologies

| Cell culture | Strategies | References |
|---|---|---|
| C2C12 mouse myoblasts (muscle cell) | Cultivation of *Chlorella vulgaris* and *Chlorococcum littorale* to remove ammonia and phosphorus from cell culture media | Haraguchi and Shimizu (2021) |
| C2C12 mouse myoblasts (muscle cell) | Circular cell culture system with resource recycling with microalgae cultivation (*Chlorococcum littorale*) and nutrient extraction | Haraguchi et al. (2022) |
| C2C12 mouse myoblasts (muscle cell) | Circular cell culture system using lactate assimilation with cyanobacterium *Synechococcus* sp. PCC 7002 | Haraguchi et al. (2023) |
| C2C12 mouse myoblasts | Production of mammalian cells using extracts produced from microalgae (*Chlorococcum littorale* NBRC 102761; *Stichococcus* sp. NBRC 102709; *Chlorella vulgaris* NIES-2170; *Euglena gracilis*; *Spirulina subsalsa* NIES-27; and *Arthrospira platensis* NIES-39) | Okamoto et al. (2020) |
| Tissues (plant, animal, and microorganism) | Culture medium recycling system by removing inhibitory substances such as lactates and ammonia by electrodialysis | Nir (2023) |
| Mammalian, chicken, crustacean, fish, and other cells | System of filtration and recycling of culture medium through receptacles and absorbent materials | Nahmias and Wissotsky (2021) |
| Skin, muscle, adipose, and bone cells | A circular cell culture system with two units using dialysis systems to reduce contaminants or inhibitors after several production stages | Chin et al. (2021) |

growth even at low concentrations (in millimolar) (Schneider 1996; Mattick et al. 2015; Nir 2023; Yang et al. 2023). Moreover, if not properly treated, the effluent can lead to watercourse eutrophication resulting in significant environmental consequences (Haraguchi and Shimizu 2021).

The development of selective separation processes presents an alternative for recycling water from the culture medium, thereby reducing the water footprint of the CM production (Chin et al. 2021; Nahmias and Wissotsky 2021; Nahmias et al. 2022; Nir 2023). Electrodialysis membranes appear promising for this purpose, as they can incorporate materials with high affinity for specific compounds, demonstrating results of over 90% removal of ammonium ions and 99% removal for lactate (Nir 2023). The concentrate containing separated compounds can find utility in other applications, while the water could be reintroduced into the system (Nir 2023). Nevertheless, more comprehensive studies on the implications of recirculation are required.

In this context, the use of biological reactors emerges as a promising approach for harnessing the electrodialysis concentrate, potentially yielding a nutrient-rich medium that can promote cell growth. Recently, processes involving nitrifying bacteria and enzymes have been patented for this purpose (Chin et al. 2021).

Beyond the treatment and selective removal of the water from the culture medium, strategies involving the integration of secondary bioprocesses could be an alternative solution. Moreover, the reuse of culture media has already been the subject of recent studies, with a focus on cultivating microalgae in the residual medium of cell cultures (Okamoto et al. 2020; Haraguchi and Shimizu 2021; Haraguchi et al. 2022, 2023). The presence of compounds such as phosphorus, amino acids, and salts can facilitate the growth of microalgal cells, yielding a new product line rich in protein and lipids that can be exploited for food or energy development or as a culture medium for reintroduction into cell cultivation (Okamoto et al. 2020; Haraguchi et al. 2022, 2023; Yang et al. 2023).

The versatility of microalgal cells presents an opportunity for food companies to expand their offerings and serves as a viable approach for recycling culture medium. Microalgae cultivation can effectively reduce the ammonia content in the cultivation medium. Subsequently, this enriched medium can be reintroduced into the cell culture, potentially eliminating the need for additional nutrient supplementation (Shimizu and Haraguchi 2022; Haraguchi et al. 2022; Ghosh et al. 2023).

Moreover, there is potential for further development in the field of bioenergy production, including the generation of biogas or biohydrogen. This can be a strategic approach for utilizing solid waste generated from the hydrolysis of alternative biomass sources (such as microalgae and wheat) (Yao et al. 2018; Zhu et al. 2022). Additionally, the cultivation of other microorganisms, such as bacteria and fungi with industrial relevance, opens pathways for process integration in various industrial segments.

It should be noted that a large amount of plastic waste is generated from cell culture at bench scale, and with the exponential growth of studies in the area, it is essential to highlight this problem. Single-use plastic consumables are convenient and efficient; still, they generate large amounts of waste. Currently, research is aimed at finding alternatives to reuse plastic, such as cleaning methods. The possibility of reusing the plastic flasks could save up to US$ 12,000 per year in a facility that consumes around 125 kg of plastic per year (Clancy et al. 2023). Another type of research is oriented towards producing bioplastics that are compatible with cell cultures to reduce the mass and volume of waste generated (Réu et al. 2019).

## 5 Challenges and Future Developments

The main challenges of waste management within the CM production process are linked to the scarcity of data. The unit operations of the treatment process need accurate evaluation to enhance the efficiency of CM production, both within the bioprocess scenario and in the economic perspective.

Environmental impacts generated from these processes have been identified in LCA studies, which extrapolate data generated through scientific research, revealing significant potential for reducing GHG emissions (Tuomisto and Teixeira de Mattos 2011; Mattick et al. 2015; Scharf et al. 2020; Tuomisto et al. 2022; Kim et al. 2022; Risner et al. 2023; Sinke et al. 2023; Trinidad et al. 2023). However, there has been limited research on fully harnessing the solid and liquid waste generated in large volumes. As a result, the key challenges for research into the environmental impacts of waste generated from cultivated meat include:

- Identifying the flows and characteristics of the waste generated during the cultivated meat production chain.
- Developing production processes integrated with waste management strategies aimed at minimizing effluents generation.
- Integrating effluents into processes that offer the possibility of recovering other high value-added products, thereby expanding the company's product portfolio.

In future research, it is crucial to comprehensively characterize the effluents generated. This characterization will provide the necessary data to explore innovative waste treatment and management technologies. Additionally, as biotechnology advances to produce CM, waste management needs to advance in tandem. This will ensure that future industrial processes are established with truly green and sustainable processes.

## 6 Conclusion

The exponential advances in CM development have significantly reduced production costs and supported a consolidation of processes that are progressing towards making this biotechnology economically feasible. Simultaneously, environmental impacts remain a field that requires assessment alongside these advances in CM production. Throughout this chapter, we have highlighted characteristics provided in some studies to date, mainly focusing on wastewater volumes and composition. This has allowed us to create an overview of the application of conventional wastewater treatment technologies and expand the discussion to integrated processes for recovering other bioproducts or recycling for cell culture. While there is still a considerable way ahead before cultivated meat achieves full economic viability, it is essential that economic development aligns with efforts to mitigate environmental impacts. This approach is crucial for establishing a cultivated meat production process that is both economically viable and environmentally safe.

**Acknowledgments** The authors want to thank the National Council for Scientific and Technological Development (CNPq), the Coordination for the Improvement of Higher Education Personnel (CAPES), the Araucária Foundation (Fundação Araucária, NAPI Proteínas Alternativas), and the Bioprocess Engineering and Biotechnology Department of the Federal University of Paraná (UFPR).

## References

Cai J, Wang S, Li Y et al (2024) Industrialization progress and challenges of cultivated meat. J Future Foods 4:119–127

Chin PSM, Li CW, Chan KYC, Spitters T (2021) System for producing cultivated meats, tissues and associated products from cells

Chriki S, Hocquette J-F (2020) The myth of cultured meat: a review. Front Nutr 7. https://doi.org/10.3389/fnut.2020.00007

Clancy M, Wade IS, Young JJ (2023) Facile methods for reusing laboratory plastic in developmental biology experiments. Differentiation 130:1–6. https://doi.org/10.1016/j.diff.2022.11.001

Craggs RJ, Sukias JP, Tanner CT, Davies-Colley RJ (2004) Advanced pond system for dairy-farm effluent treatment. N Z J Agric Res 47:449–460. https://doi.org/10.1080/00288233.2004.9513613

Delgadillo-Mirquez L, Lopes F, Taidi B, Pareau D (2016) Nitrogen and phosphate removal from wastewater with a mixed microalgae and bacteria culture. Biotechnol Rep 11:18–26. https://doi.org/10.1016/j.btre.2016.04.003

Dodds WK, Smith VH (2016) Nitrogen, phosphorus, and eutrophication in streams. Inland Waters 6:155–164. https://doi.org/10.5268/IW-6.2.909

Ghosh J, Haraguchi Y, Asahi T et al (2023) Muscle cell proliferation using water-soluble extract from nitrogen-fixing cyanobacteria Anabaena sp. PCC 7120 for sustainable cultured meat production. Biochem Biophys Res Commun 682:316–324. https://doi.org/10.1016/j.bbrc.2023.10.018

Haraguchi Y, Shimizu T (2021) Microalgal culture in animal cell waste medium for sustainable 'cultured food' production. Arch Microbiol 203:5525–5532. https://doi.org/10.1007/s00203-021-02509-x

Haraguchi Y, Okamoto Y, Shimizu T (2022) A circular cell culture system using microalgae and mammalian myoblasts for the production of sustainable cultured meat. Arch Microbiol 204:615. https://doi.org/10.1007/s00203-022-03234-9

Haraguchi Y, Kato Y, Inabe K et al (2023) Circular cell culture for sustainable food production using recombinant lactate-assimilating cyanobacteria that supplies pyruvate and amino acids. Arch Microbiol 205:266. https://doi.org/10.1007/s00203-023-03607-8

Huang H, Liu J, Zhang P et al (2017) Investigation on the simultaneous removal of fluoride, ammonia nitrogen and phosphate from semiconductor wastewater using chemical precipitation. Chem Eng J 307:696–706. https://doi.org/10.1016/j.cej.2016.08.134

Hubalek S, Post MJ, Moutsatsou P (2022) Towards resource-efficient and cost-efficient cultured meat. Curr Opin Food Sci 47:100885

Kim S, Beier A, Schreyer HB, Bakshi BR (2022) Environmental life cycle assessment of a novel cultivated meat burger Patty in the United States. Sustainability 14:16133. https://doi.org/10.3390/su142316133

Kumar P, Sharma N, Sharma S et al (2021) In-vitro meat: a promising solution for sustainability of meat sector. J Anim Sci Technol 63:693–724

Lee SI, Weon SY, Lee CW, Koopman B (2003) Removal of nitrogen and phosphate from wastewater by addition of bittern. Chemosphere 51:265–271. https://doi.org/10.1016/S0045-6535(02)00807-X

Li B, Jing F, Wu D et al (2021) Simultaneous removal of nitrogen and phosphorus by a novel aerobic denitrifying phosphorus-accumulating bacterium, Pseudomonas stutzeri ADP-19. Bioresour Technol 321:124445. https://doi.org/10.1016/j.biortech.2020.124445

Mattick CS, Landis AE, Allenby BR, Genovese NJ (2015) Anticipatory life cycle analysis of in vitro biomass cultivation for cultured meat production in the United States. Environ Sci Technol 49:11941–11949. https://doi.org/10.1021/acs.est.5b01614

Myers GM, Jaros KA, Andersen DS, Raman DR (2023) Nutrient recovery in cultured meat systems: impacts on cost and sustainability metrics. Front Nutr 10. https://doi.org/10.3389/fnut.2023.1151801

Nahmias Y, Wissotsky G (2021) Methods and devices for filtering cell culture media

Nahmias Y, Halaf G, Shidlovsky Z, Wissotsky G (2022) Systems and methods for recycling cell culture medium

Nir O (2023) Selective separation of ammonium and lactate from cell culture media, pp 1–74

Okamoto Y, Haraguchi Y, Sawamura N et al (2020) Mammalian cell cultivation using nutrients extracted from microalgae. Biotechnol Prog 36:e2941. https://doi.org/10.1002/btpr.2941

Réu P, Svedberg G, Hässler L et al (2019) A 61% lighter cell culture dish to reduce plastic waste. PLoS One 14:e0216251. https://doi.org/10.1371/journal.pone.0216251

Risner D, Negulescu P, Kim Y et al (2023) Cradle to production gate life cycle assessment of cultured meat growth media: a comparison of Essential 8™ and Beefy-9. BioRxiv

Saied ARA, Chandran D, Chopra H et al (2023) Cultivated meat could aid in reducing global antimicrobial resistance burden - producing meat without antibiotics as a safer food system for the future. Int J Surg 109:189–190. https://doi.org/10.1097/JS9.0000000000000199

Scharf A, Breitmayer E, Carus M (2020) Review and gap-analysis of LCA-studies of cultured meat

Schneider M (1996) The importance of ammonia in mammalian cell culture. J Biotechnol 46:161–185. https://doi.org/10.1016/0168-1656(95)00196-4

Sengupta S, Nawaz T, Beaudry J (2015) Nitrogen and phosphorus recovery from wastewater. Curr Pollut Rep 1:155–166. https://doi.org/10.1007/s40726-015-0013-1

Shi J, Podola B, Melkonian M (2007) Removal of nitrogen and phosphorus from wastewater using microalgae immobilized on twin layers: an experimental study. J Appl Phycol 19:417–423. https://doi.org/10.1007/s10811-006-9148-1

Shimizu T, Haraguchi Y (2022) Compositions and methods for culturing algae

Sinke P, Swartz E, Sanctorum H et al (2023) Ex-ante life cycle assessment of commercial-scale cultivated meat production in 2030. Int J Life Cycle Assess 28:234–254. https://doi.org/10.1007/s11367-022-02128-8

Specht L (2020) An analysis of culture medium costs and production volumes for cultivated meat

Su F, Li Z, Li Y et al (2019) Removal of total nitrogen and phosphorus using single or combinations of aquatic plants. Int J Environ Res Public Health 16:1–12. https://doi.org/10.3390/ijerph16234663

Sudiarto SIA, Renggaman A, Choi HL (2019) Floating aquatic plants for total nitrogen and phosphorus removal from treated swine wastewater and their biomass characteristics. J Environ Manag 231:763–769. https://doi.org/10.1016/j.jenvman.2018.10.070

Thermo Fisher Scientific (2023) Culture plastics. https://www.thermofisher.com/br/en/home/references/gibco-cell-culture-basics/cell-culture-environment/culture-plastics.

html#:~:text=Cell%20culture%20plastic%20surface,-Many%20different%20plastic&text=These%20are%20polyethylene%20terephthalate%20(PET,today%20is%20polystyrene%20(PS). Accessed 4 Oct 2023
Trinidad KR, Ashizawa R, Nikkhah A et al (2023) Environmental life cycle assessment of recombinant growth factor production for cultivated meat applications. J Clean Prod 419:138153. https://doi.org/10.1016/j.jclepro.2023.138153
Tuomisto HL, Teixeira de Mattos MJ (2011) Environmental impacts of cultured meat production. Environ Sci Technol 45:6117–6123. https://doi.org/10.1021/es200130u
Tuomisto HL, Allan SJ, Ellis MJ (2022) Prospective life cycle assessment of a bioprocess design for cultured meat production in hollow fiber bioreactors. Sci Total Environ 851:158051. https://doi.org/10.1016/j.scitotenv.2022.158051
Wan W, He D, Xue Z (2017) Removal of nitrogen and phosphorus by heterotrophic nitrification-aerobic denitrification of a denitrifying phosphorus-accumulating bacterium Enterobacter cloacae HW-15. Ecol Eng 99:199–208. https://doi.org/10.1016/j.ecoleng.2016.11.030
Wang Y, Zhu Y, Sun P et al (2019) Augmenting nitrogen removal by periphytic biofilm strengthened via upconversion phosphors (UCPs). Bioresour Technol 274:105–112. https://doi.org/10.1016/j.biortech.2018.11.079
Yang Z, Zhou Q, Sun H et al (2021) Metagenomic analyses of microbial structure and metabolic pathway in solid-phase denitrification systems for advanced nitrogen removal of wastewater treatment plant effluent: a pilot-scale study. Water Res 196:117067. https://doi.org/10.1016/j.watres.2021.117067
Yang M, Wang Q, Zhu Y et al (2023) Cell culture medium cycling in cultured meat: key factors and potential strategies. Trends Food Sci Technol 138:564–576. https://doi.org/10.1016/j.tifs.2023.06.031
Yao Y, Bergeron AD, Davaritouchaee M (2018) Methane recovery from anaerobic digestion of urea-pretreated wheat straw. Renew Energy 115:139–148. https://doi.org/10.1016/j.renene.2017.08.038
Zhou Q, Sun H, Jia L, Wu W (2022) Simultaneously advanced removal of nitrogen and phosphorus in a biofilter packed with ZVI/PHBV/sawdust composite: deciphering the succession of dominant bacteria and keystone species. Bioresour Technol 347:126724. https://doi.org/10.1016/j.biortech.2022.126724
Zhu J, Song W, Chen X, Sun S (2022) Integrated process to produce biohydrogen from wheat straw by enzymatic saccharification and dark fermentation. Int J Hydrog Energy 48:11153. https://doi.org/10.1016/J.IJHYDENE.2022.05.056

# Environmental Impacts of Cultivated Meat

**Hanna L. Tuomisto and Toni Ryynänen**

**Abstract** Cultivated meat is developed as an alternative to livestock meat in reducing the negative externalities of the animal-based food systems. Prospective life cycle assessment studies have been used to estimate the potential environmental impacts of cultivated meat production. Results indicate that cultivated meat production has relatively high industrial energy demand, but lower land use requirements compared to livestock meat production. The climate impacts of cultivated meat depend on the source of energy used. Due to the low land use requirements, cultivated meat could have benefits to biodiversity by reducing the pressure to convert forests and natural habitats to agricultural land. Cultivated meat production could also reduce water use and emissions to waterbodies. Although the development of cultivated meat technologies offers credible potential to reduce the environmental impacts of the meat products in the long term, they may not be feasible for reaching the sustainable development goals by 2030.

**Keywords** Life cycle assessment · Biodiversity · Land use · Energy use · Water use · Climate change · Carbon footprint

H. L. Tuomisto (✉)
Department of Agricultural Sciences, University of Helsinki, Finland & Natural Resources Institute Finland, Helsinki, Finland
e-mail: hanna.tuomisto@helsinki.fi

T. Ryynänen
Ruralia Institute, University of Helsinki, Mikkeli, Finland


C. R. Soccol et al. (eds.), *Cultivated Meat*,
https://doi.org/10.1007/978-3-031-55968-6_14

# 1 Introduction

Agriculture and food production systems are fundamental in provision of human life. However, these systems correspond to over 80% of the global biodiversity loss, major land use changes and nutrient losses (Campbell et al. 2017). Food systems produce approximately a third of anthropogenic greenhouse gas (GHG) emissions (Crippa et al. 2021). Livestock production is causing 60% of these food system-related emissions (Xu et al. 2021). The consumption of animal-based products has been estimated to increase simultaneously with the expanding global population and cumulating wealth (WRI 2019). Meat consumption is also justified with the four N's: meat eating is claimed to be necessary, natural, normal and nice (Piazza et al. 2015). These trends will increase supply and demand of food products from livestock food systems and therefore directly cause more negative environmental impacts from the food systems.

Major socio-technical and system-level transitions or even transformations (Moritz et al. 2023) are needed in food production and diets to reduce the current environmental impacts. Switching animal-based diets to plant-based diets is shown to reduce the environmental impacts: transition to vegan diets globally could lead to 80% reduction of GHG emissions from food production systems compared to status quo scenario in 2050 (Willett et al. 2019). Similarly, achieving the set climate targets and limiting the increase in atmospheric temperature to 1.5° above preindustrial levels need a major restructuring of the food supply and demand systems (Clark et al. 2020). Consumer acceptance of vegan diet is currently rather low: approximately 1–4% of population in the European countries follow a vegan diet (Statista 2023a). Plant-based meat alternatives are widely available on the market globally, and consumer interest in the products is steadily increasing (Statista 2023b). However, the market share of the alternative protein sources is small compared to preferred livestock products. Cultivated meat has the potential to improve the sustainability of food production systems: it offers appealing, ethical and environmentally sustainable analogue—not just an alternative—to livestock meat without the need to change culturally and socially meaningful eating habits and practices.

This chapter addresses the environmental impacts of livestock meat and cultivated meat production by reviewing recent environmental life cycle assessment (LCA) studies. An overview of the LCA methods and environmental challenges associated with livestock production is followed by a review of the LCA studies considering environmental impacts of cultivated meat. The chapter concludes by presenting a summary of the current scientific knowledge on the environmental impacts of cultivated meat production.

# 2 How Environmental Impacts of Cultivated Meat Are Studied?

## 2.1 *Background Information about the Life Cycle Assessment*

Environmental impacts of food products are estimated with LCA methods (Clark et al. 2020; Poore and Nemecek 2018) standardised by the International Organisation for Standardisation (ISO 2006). LCA studies are typically attributional LCAs (ALCAs) that calculate the emissions and resource use along the life cycle of a product and use allocation methods to share the impacts between the co-products (Poore and Nemecek 2018). Consequential LCA (CLCA) is less used method that assesses the marginal impacts of changes in a product system or production of an additional product (Smetana et al. 2015). CLCA quantifies the impacts of changes in all product systems affected rather than quantifying the impacts of the production of inputs needed for the product system. CLCA does not use allocation methods, but the system considers all co-products. Most of the available LCA studies of cultivated meat use the ALCA approach.

LCA studies follow four steps: (1) goal and scope definition, (2) inventory analysis, (3) impact assessment and (4) interpretation. The goal and scope definition details the purpose of the study and sets the functional unit (FU), system boundaries, data requirements and allocation processes. The data for a study is collected during an inventory analysis step, whereas an impact assessment step is used to modify the data to the environmental impact results. The assessment of the robustness of the analysis, conclusions and recommendations are presented in the interpretation step.

## 2.2 *Setting Goals, Functional Units and System Boundaries*

The first step is to define the goal of the LCA study and the intended application, such as policy design, hotspot analysis, product development, eco-design, marketing or research. Studies can target a specific group such as consumers, policymakers, researchers or business stakeholders. The aim of the study outlines the quality of data required from each step of the life cycle. The cultivated meat LCAs published are prospective studies aiming to estimate the potential impacts of future food technologies and therefore are based on a set of assumption rather than factual data needed for studies assessing existing processes.

The environmental impacts are allocated according to a functional unit (FU). The FU answers to what, how much, how well and how long: the FU of a cultivated meat product can be 1 kg unpackaged cultivated meat (with X% protein content) at the factory gate. Complex indexes considering the nutritional content of the foods such as the health-promoting and health-reducing nutrients have been developed in addition to using simple FU measures as weight (McAuliffe et al. 2020). Although nutritional indexes are useful in comparing foods with differing nutritional profiles, their use as FU is not simple as diets consist of several food items: the nutritional content of the whole diet is more valuable information than the nutrients in single-food items. The environmental impacts of protein-rich foods are often compared using the protein content as an FU (Poore and Nemecek 2018).

Setting the system boundaries for an LCA study determinates the included and excluded processes. It is not necessary to include packaging, retail and consumption steps if the study aims to compare different food production methods. Often used assessment types are a cradle-to-gate assessment and a cradle-to-grave assessment. The extant cultivated meat LCA system boundaries include processes from extraction of natural resources up to the harvesting of the cultured cell biomass from the bioreactors. These studies exclude the product processing or packaging. The production of bioreactors is included to some of the cultivated meat LCA studies, but other industrial infrastructures such as buildings are excluded. The limitation of the studied cultivated meat system boundaries to the cradle-to-gate approach is justified when compared to livestock meat: most of the environmental impacts of livestock production originate from the agricultural activities before the products leave the farm (Poore and Nemecek 2018).

## 2.3 Inventory Analysis and Co-Product Allocation

The inventory analysis step includes the collection of the material and energy input and emission flow data within the defined system boundaries. The data quality requirements are lower for a screening analysis aiming to understand the hotspots of impacts in the product system and to estimate the general impacts than for a more detailed LCA study. A higher quality of data is often required for analysing core processes such as cell culturing than for upstream processes such as production of inputs. LCA databases have available life cycle inventory (LCI) data for a wide range of products. Activity data for the core processes considering the material and energy input requirements and outputs are collected first, and the existing databases are used as a data source for the material and emission flows of the input production processes.

Several methods are used in ALCAs for allocating the impacts of the studied production process between the co-products. Creating sub-processes should be avoided, but not often possible in the food product assessments. A system expansion can also be used, where the impacts of alternative ways of producing the co-products are calculated. In the exemplary case of dairy production, meat from dairy cows is assigned similar impacts than producing meat in beef herd systems. System expansion is associated with the CLCA approach and only rarely used in ALCA studies, whereas allocation methods based on the dry or wet mass, energy content or economic value of the co-products are used (Kyttä et al. 2022). Different allocation methods have a major impact on the results and can change the conclusions. Kyttä (2022) showed that when system expansion was used for comparing the climate impacts of cultivated meat production with livestock and plant-based systems, the relative differences were different compared to using standard allocation methods. When the system expansion was used, cultivated meat had lower impacts compared to dairy, beef and pork systems. When economic allocation was used, cultivated meat had a higher impact than pork and beef from dairy system, but a lower impact compared to beef from beef cattle.

Lactate and ammonia are produced as cultivated meat co-products. Some of the impacts of the whole cultured meat production process can be allocated to these co-products if they are used as inputs in other product systems. A study compared alternative allocation methods that included product substitution that assumed lactate to replace lactic acid production and dry mass- and wet mass-based allocations (Tuomisto et al. 2022). The substitution-based allocation resulted in 5–48% of the impacts of the whole production process being allocated to lactate depending on the impact category. Dry mass allocation and wet mass allocation resulted in 85% and 63%, respectively, of the impacts being allocated to lactate. Economic allocation is often the most reasonable choice but needs reliable price information.

## 2.4 *Assessing the Life Cycle Impacts and Interpreting the Results*

The life cycle impact assessment (LCIA) phase converts the LCI data to selected environmental impacts. The results of different LCA studies are challenging to be directly compared due to several LCIA method options. Some LCIA methods consist of up to 20 different environmental impact categories ranging from climate change impacts to eutrophication potential. Often only the most relevant environmental impact categories are selected and analysed using a LCA software.

The first step of LCIA is characterised by grouping and characterisation of the resource and emissions flows to form the environmental impact results for each impact category. In addition, normalisation and weighting steps may be used for aggregating the results to a single score in cases when a study involves several impact categories. The weighting factors aim to describe the relative importance of different environmental impact categories, for instance, based on the opinions of stakeholders and experts or based on the distance to targets (Tuomisto et al. 2012).

The last interpretation step evaluates the robustness of the results based on the data quality. Monte Carlo method is commonly used for the uncertainty assessment in LCA studies, whereas sensitivity analyses are used for testing how set assumptions or uncertainties in the data impact the results. Contribution assessments are used to identify hotspots showing the impacts of each life cycle stage. The interpretation step includes limitations of the study analysis, whereas conclusions and recommendations are drawn from reflecting the results with the uncertainty assessments and limitations.

# 3 Baseline: Environmental Impacts of Conventional Livestock Production

## 3.1 Climate Impact

Approximately 20% of the global anthropogenic GHG emissions originate from livestock production, when direct and indirect emissions are considered (Xu et al. 2021; Crippa et al. 2021). The direct atmospheric emissions consist of methane emissions from the ruminants' digestion and manure management processes and nitrous oxide emission from manure and use of nitrogen fertilisers. Indirect emissions consist of the production of agricultural inputs such as fertilisers, pesticides and fuels, transportation of livestock and fuel used at the farm. In addition, indirect emissions include land use change such as carbon emissions from deforestation caused by expansion of agricultural land area. The direct global GHG emissions from crop and livestock production account for approximately 11%, and the indirect emissions are approximately 14% of anthropogenic GHG emissions (WRI 2019). Over 80% of the GHG emissions of livestock products are allocated to land use change and agricultural practices. The processing, transporting, packaging and retailing have only a small overall contribution (Fig. 1).

A majority of all GHG emissions of global livestock production are related to bovine and sheep cattle production (Xu et al. 2021). Beef has approximately six times higher emissions than pork meat and eight times higher than poultry meat per

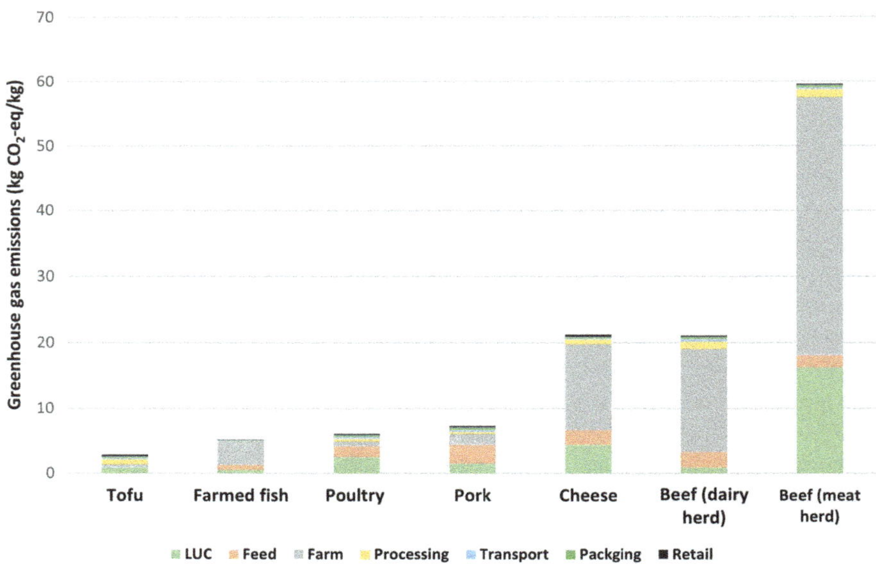


**Fig. 1** Greenhouse gas emissions of livestock products, fish and tofu by life cycle stages (LUC = land use change, source of data: Poore and Nemecek 2018)

unit of product (Poore and Nemecek 2018). This is explained by the methane emissions from ruminants and their greater feed requirements. The production of 1 kg of poultry needs approximately 3 kg, pork 6–7 kg and beef 25 kg of feed (Alexander et al. 2016).

## 3.2 Land Use and Biodiversity Impacts

Livestock feed production covers a third of the world's total land area and 80% of all agricultural land area (WRI 2019). Livestock production is one of the main causes of global land use change accounting for 67% of agriculture-derived deforestation (Poore and Nemecek 2018). The deforestation is mainly driven by increasing production of soybean feed and beef for export. Global demand for feed and consumption of livestock meat have direct and indirect consequences to the deforestation of tropical forests, which support two-thirds of the world's biodiversity although covering only 10% of the Earth's land area (Giam 2017). Approximately 10 million hectares is globally deforested per year, which may lead to a major loss of biodiversity (FAO and UNEP 2020).

Food production contributes to farmland biodiversity and agrobiodiversity losses via changes in agricultural practices (Dudley and Alexander 2017). Decreasing livestock grazing practices have been associated with decline of wild species such as butterflies and beetles dependent on extensive grazing (Filazzola et al. 2020). Increased use of fertilisers and pesticides, reduced number of crop varieties and general intensification of agriculture have all contributed to the decline of farmland biodiversity. In addition, reduced mosaic of farmland landscapes due to favouring of larger fields and shrinking non-crop habitats such as field boundaries, set-aside lands, wildflower meadows and agroforestry has contributed to increased number of endangered wild plant and animal species.

The land use changes such as deforestation and intensification of agriculture are often caused by the demand to produce more and cheaper food for the growing and more affluent global population in increasingly competitive and global agricultural product markets. The gross domestic product per capita tends to correlate with the consumption of livestock products per capita (Tilman and Clark 2014). It is anticipated that livestock product demand will potentially increase in currently low meat-consuming and the developing economies (WRI 2019). Global agricultural land area would need to expand on an alarming rate by 2050 to an area that equals approximately the size of all current forests and savannah on the Earth if anticipated consumption trends will hold and the increasing demand of livestock products are met with current agricultural practices (WRI 2019).

### 3.3 Water Use and Emissions to Waterbodies

The environmental assessment methods produce differing results in estimating the contribution of livestock production to global water use. Simpler methods calculate only blue water usage including freshwater extraction from rivers, reservoirs, lakes and aquifers. In addition to blue water, the more complex water footprint method also includes the usage of green water or rainwater and grey water describing the amount of water needed to dilute wastewater to meet the regionally specific water quality standards. Livestock production contributes approximately 30% of all anthropogenic freshwater use globally (Mekonnen and Hoekstra 2012). Similarly, blue water use is approximately eight percent (FAO 2006).

Food products from ruminants have relatively high blue, green and grey water usage or water footprint due to the larger agricultural land area requirements for feed production. The green water footprint is therefore significantly higher compared to monogastric animal meat production such as pork of poultry. The differences between livestock products tend to be low when only blue water usage is assessed. Studies reporting only blue water footprint results may include caveats: some water footprint methods use country-specific water scarcity factors in assessing the potential impact of livestock water use to the availability of water resources in the specific locations (Boulay et al. 2019). Water use in water scare areas has more severe consequences than using the same amount in a water-rich area.

Livestock production contributes to the pollution of waterbodies due to the leaching of nitrogen and phosphorus from fields and manure handling systems (Willett et al. 2019). Regionally centralised livestock production units and expanded farm sizes create additional challenges. Traditionally smaller and versatile farm units could use produced manure fully to fertilise the fields and produce the needed feed on-site. The concentration of individual agricultural production sectors to certain regions has created a mismatch: excessive manure cannot be efficiently used as fertiliser in nearby fields, whereas crop production in different regions depends on industrially produced fertilisers. Manure is therefore spread to the fields excessively or managed in unprotected systems outdoors. The industrial livestock farms import feed from global supply chains. Excessive nutrients leach to waterbodies causing eutrophication, which increases the growth of water plants and algae disturbing the aquatic ecosystems and causing negative impacts to aquatic species. The eutrophication impact of producing meat from ruminants is higher compared to pork, poultry and plant-based proteins (Poore and Nemecek 2018).

# 4 Environmental Impacts of Cultivated Meat Production

## *4.1 LCA Studies on Cultivated Meat*

The environmental impacts of cultivated meat production have been studied only with prospective LCA (Table 1). The studies have assumed several factors in modelling the impacts of cultivated meat production as large-scale industrial production units do not exist and data availability from real-world applications are largely unavailable or missing. The extant LCA studies involve uncertainties due to these assumptions. It is uncertain whether the assumed production amounts of cultivated meat are realistic or achievable with modelled production system designs and whether assumptions considering the needed ingredients of the culture medium, design or types of the bioreactors, the duration of the cultivation process and the nutritional content of cultivated meat align with the actual future production processes.

The first study that assessed the potential environmental impacts of cultivated meat (Tuomisto and Teixeira de Mattos 2011) is a screening analysis providing estimates of the environmental impacts at the level of magnitudes. The LCA study is based on a hypothetical production scenario using cyanobacterial hydrolysate as the main source of amino acids and sugars in culture medium. The cyanobacteria were assumed to be produced in large-scale commercial production units using open ponds. Other ingredients of the culture medium such as vitamins, minerals and growth factors were excluded from the analysis. The modelling used a stir-tank

**Table 1** Comparison of the climate impact and land use of cultivated meat based on published life cycle assessments

| | Carbon footprint (kg $CO_2$-eq/kg cultivated meat) | Land use ($m^2$/kg cultivated meat) | Details |
|---|---|---|---|
| Tuomisto and Teixeira de Mattos (2011) | 1.9–2.2 | 0.2 | Cyanobacteria as a main source of nutrients in the culture medium, stir-tank bioreactor |
| Smetana et al. (2015) | 23.9–24.6 | 0.4–0.8 | Tuomisto and Teixeira de Mattos (2011), but cyanobacteria produced in bioreactors instead of open ponds |
| Mattick et al. (2015) | 3.0–25.5 | 1.5–9.5 | Chinese hamster ovarian cells, standard serum-free culture medium, stir-tank bioreactors |
| Tuomisto et al. (2022) | 4.9–25.2 | 1.8–6.9 | C2C12 cells, standard culture medium with and without FBS, hollow fibre bioreactors |
| Kim et al. (2022) | 15.4 | 0.08 | Primary bovine cells, standard culture medium, data from a company |
| Sinke et al. (2023) | 2.9–14.3 | 2.5 | Aggregated data from companies (several cell types and culture medium ingredients) |

bioreactor, the bioreactor energy use requirements were simplified and assumed production locations were selected from three continents: Spain, Thailand and California in the USA. The seminal study had the lowest land use estimates compared to the other LCA studies conducted later due to the use of cyanobacteria as the main source of nutrients (Table 1). Also, low climate impacts resulted partly due to the usage of cyanobacteria-based culture medium and potentially underestimated energy requirements for bioreactor processes. Bioreactor cleaning processes were not included in the study.

Smetana et al. (2015) drew from the seminal study (Tuomisto and Teixeira de Mattos 2011) by modelling the production of cyanobacteria in bioreactors, which increased the energy usage and climate impacts. However, the land use requirements increased only marginally (Table 1).

Another study assessed the environmental impacts of cultivated meat production in the USA (Mattick et al. (2015) using Chinese hamster ovarian (CHO) cells cultivated in a stir-tank bioreactor and serum-free culture medium. The medium contained standard basal medium (synthetic amino acids, vitamins and minerals) and soybean hydrolysate. The study extended Tuomisto's and Teixeira de Mattos' (2011) study by including the cleaning of the bioreactors and energy requirements of the production facilities. The results reported higher environmental impacts for cultivated meat production compared to the previous study (Tuomisto and Teixeira de Mattos 2011). However, the uncertainty ranges of the results were extensive (Table 1).

Yet another study presents an LCA based on data collected from start-up companies producing cultivated meat supplemented with modelling and data from literature (Sinke et al. 2023). The company data included various cultivated meat cell types and culture media options. The study estimated the potential ranges of the environmental impacts rather than presented the impacts of specific cultivated meat product type. The study included the future scenarios of improved cultivated meat and livestock production systems reaching to 2050. The study reported lower climate and land use impacts from the previous study (Mattick et al. 2015).

A recent study assessed the environmental impacts of cultivated meat production in the UK using hollow-fibre bioreactors (Tuomisto et al. 2022). The study focused on different cultivated meat process designs and their potential environmental impacts. The baseline scenario used immortalised murine myoblast cell line C2C12 grown in standard culture medium containing FBS and using standard average UK electricity mix for the production process. The alternative scenarios included serum-free culture medium, CHO cells, renewable energy use and various cell metabolism assumptions. The climate and land use impacts aligned with the uncertainty ranges presented in Mattick et al.'s (2015) study.

Kim et al. (2022) estimated the environmental impacts of cultivated meat beef patty produced in the USA. The data was obtained from a start-up company and, therefore, was not published in detail. The carbon footprint result of cultivated meat cells was in the upper end of the finding of Sinke et al. (2023) and in the middle of the range found in Tuomisto et al. (2022), whereas the land use result was in the lowest of all studies.

## *4.2 Issues Affecting the Environmental Impacts of Cultivated Meat*

### 4.2.1 Culture Medium

The published LCA studies show that the production of the culture medium has a major contribution to the environmental impacts of cultivated meat production. The volume of culture medium needed per kg cultured meat in the LCA studies had a wide variation ranging from 21 litres to 200 litres. The medium requirements depend on the cell type, medium type, cultivation time and possible reuse of the nutrients of the spent medium. The published LCA studies assumed relatively short differentiation and maturation periods in cultivated meat production (Mattick et al. 2015; Tuomisto et al. 2022; Sinke et al. 2023). Muscle tissue may need more cell-culturing time especially when whole meat cuts are cultivated, which also increase the medium consumption. However, also cells with short differentiation time or even undifferentiated cells can be used as ingredients in processed products (e.g. sausages). For instance, Mattick et al. (2015) and Tuomisto et al. (2022) assumed seven days differentiation period in the baseline cases, which means that undifferentiated cells would require less medium. A shorter cultivation time would reduce the environmental impacts of cultivated meat production due to lower impacts related to the medium production and lower energy use requirements for the bioreactor processes. The data published in the supplementary information in Tuomisto et al. (2022) shows that in the baseline scenario, 58% of the total medium and 61% of the bioreactor energy inputs were spent during the differentiation phase. This would mean that the environmental impacts of cultivated meat production would be only around 40% of those reported in Tuomisto et al. (2022). However, currently it is not known what the protein content and quality would be in the undifferentiated cells.

The environmental impacts also depend on how well the nutritional content of the medium is optimised to meet the requirements of the cells. Tuomisto et al. (2022) showed that adjusting the amino acid content of the medium to the cell requirements reduced the environmental impacts of cultivated meat over 20% compared to using a standard culture medium. The total nutrient use conditions depend also on factors such as recycling potential of unused nutrients in the cultivation process or utilisation of the leftover nutrients as ingredients for producing fertilisers or other suitable products. Advanced filtration solutions can be used in removing the surplus nutrients from the outflow medium and circulate them in cultivated meat production process or collect, store and sell unused nutrients to other industrial sectors. Part of the environmental impacts of used and recycled nutrition ingredients can be allocated to the co-products reducing the environmental impact of cultivated meat production. Also, environmental impacts of waste management processes decrease when recycling of nutrients is possible.

The sources for the nutrients used in the culture medium have a major effect on the environmental impacts. The culture medium often used in mammalian cell-cultures consists of two main components: foetal bovine serum (FBS) and basal medium, which includes glucose, synthetic amino acids, vitamins and minerals. This culture medium is expensive and contains animal-based ingredients challenging its usability in large-scale cultivated meat production. Alternative culture media are being developed using non-animal-based growth factors derived from microalgae, yeast, fungi or plant extracts that would perform equally well or better in cultivated meat production (e.g. Skrivergaard et al. 2023). Trinidad et al. (2023) estimated environmental impacts of common growth factors used in serum-free media and found that the differences in the impacts of different growth factors were high. They estimated the contribution of the growth factor production to the carbon footprint of cultivated meat production would vary between 0.4% and 23% depending on the type of growth factor used.

The synthetic production processes of amino acids, vitamins and minerals needed for cultivated meat production are energy intensive and require a relatively high amount of glucose sourced from agricultural crops (Mattick et al. 2015). Therefore, alternative sources for these ingredients are actively searched and developed within the industry. Recycled nutrients from agricultural and food-processing side-streams are considered promising in composing culture medium for cultivated meat production (Upcraft et al. 2021).

### 4.2.2 Bioprocess Design

The bioprocess design has a major impact on the environmental implications of cultivated meat production. The energy use requirements for the functions of the bioreactor (e.g. temperature control, pumping, mixing, aeration), cleaning of the bioreactor, type of scaffold materials used, downstream processing of the cells and wastewater treatment are key factors affecting the level of environmental impacts. As cultured meat production is still at the development stage and data from larger-scale bioreactors are not available, the published LCA studies are based on models assuming either stir-tank or hollow-fibre bioreactors. The process designs included in the studies are not necessarily fully optimised to minimise the energy use.

The environmental impacts of cultivated meat production are also highly dependent on the type of energy sources used for the production processes (Tuomisto et al. 2022). As most of the climate impact of cultivated meat production is related to energy use, the carbon footprint can be lowered by using renewable or low-emission energy sources.

## *4.3 Impacts to Land Use and Biodiversity*

Land use in cultivated meat production is mainly associated to production of ingredients to culture medium. Most of the land use is related to agricultural production of maize or sugar cane for source of glucose that is directly used in the culture medium and indirectly in the production of synthetic amino acids, vitamins and growth factors. Cultivated meat is estimated to have lower land area requirements compared to conventionally produced beef but not necessarily to poultry production (Fig. 2).

The land use requirements of cultivated meat production potentially decrease in cases when novel and sustainable sources of feedstock ingredients are discovered. A potential source of nutrients for cultivated meat production is hydrogen-oxidising bacteria (HOB), which sources carbon from carbon dioxide and hydrogen from water molecule. HOB have minimal land area requirements: agricultural land is not needed in the production, and only indirect land use is allocated for the energy and other input material production (Sillman et al. 2020; Järviö et al. 2021).

The potential biodiversity benefits of cultivated meat could be realised via reduced land use requirements. Especially pressure on land use changes and deforestation could decline (Machovina et al. 2015). Alternative use for lands released from agriculture would define the potential biodiversity advantages. Biodiversity impacts would be only neutral in cases where the freed land area would be converted to intensive and industrial usage such as forestry, bioenergy crop production or material crop production. The potential biodiversity benefits of cultivated meat production are not self-evident: expanding protected and nature conservation areas and use of wildlife-promoting land management practices could be advanced with appropriate policy measures.

In addition, grazing livestock provides ecosystem services and maintains farmland biodiversity as some of the endangered wild plants and insects depend on extensive grazing systems (Alkemade et al. 2013). Grazing livestock provide landscape maintenance by keeping up high nature value farmland areas such as

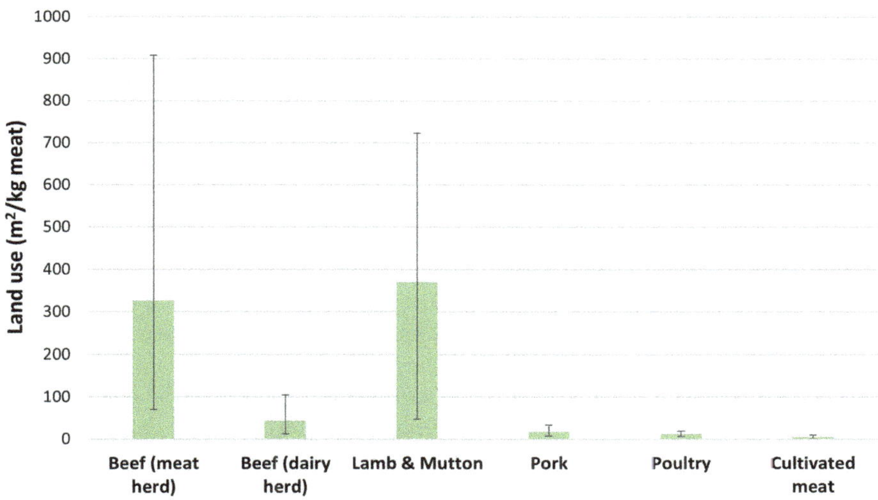


**Fig. 2** Land use of cultivated meat (as reported in the LCA studies in Table 1) and livestock meat (based on global means, error bars represent the 5th and 95th percentiles, source of data for livestock products: Poore and Nemecek (2018))

ecological and cultural benefits provided by semi-natural grasslands (Bengtsson et al. 2019). However, positive farmland biodiversity impacts provided by extensive grazing livestock production systems can be reached with a fraction of current livestock and farm animal numbers.

The biodiversity impacts depend also on what type of sources are used for the nutrients used in culture medium and how they are produced. Monocropping-based intensive production of crops for the sources of glucose and amino acids would have negative impacts on biodiversity. Cropping systems based on agroecological farming practices would provide higher biodiversity benefits. The type of crop has also biodiversity impacts: grass-clover leys and legume crops (e.g. beans and peas) are important part of the crop rotations used in agroecological farming systems. Grass and legume crops can also be used as a source of glucose and amino acids for culture medium used in cultivated meat production. Therefore, agroecological farming systems could be optimal for providing nutrients for cultivated meat production.

## 4.4 *Impacts of Energy Use*

Direct energy use of cultivated meat production comprises pumping, mixing, aeration, temperature control, cleaning and drying processes. Indirect energy use consists of the production of ingredients used in culture medium and industrial infrastructure materials. The total energy requirements depend on the culture medium ingredients and bioreactor design: the production processes of synthetic culture medium ingredients are energy intensive (Tuomisto et al. 2022), and the energy requirements of the bioreactor processes consist of the level of heat integration or the use of heat exchangers, insulation and the possibility to utilise cold air or fresh- or seawater for cooling.

Cultivated meat production is estimated to require higher amounts of industrial energy such as electricity, heat or gas than even beef production (Mattick et al. 2015; Tuomisto et al. 2022). Livestock produces energy from feed and uses it for maintaining the bodily functions. The energy in feed is replaced with industrial energy in cultivated meat production processes. However, cultivated meat required only a third of the total energy (industrial and feed combined) needed for beef production (Fig. 3).

Food production with cellular agriculture technologies will increase industrial energy consumption and would therefore require investments to sustainable energy production. However, the need for producing sustainable energy is a concern not only in food production but also in other sectors participating in the green energy transition. Conventional meat production uses efficient and natural energy production process: livestock feed with plants that convert direct sunlight into energy without the need to build complex industrial infrastructure and processes for energy conversion and supply. Livestock grazing systems especially benefit from taking advantage of these natural energy-generating processes eliminating energy usage for fertilising, harvesting, processing and transporting feed. Future food

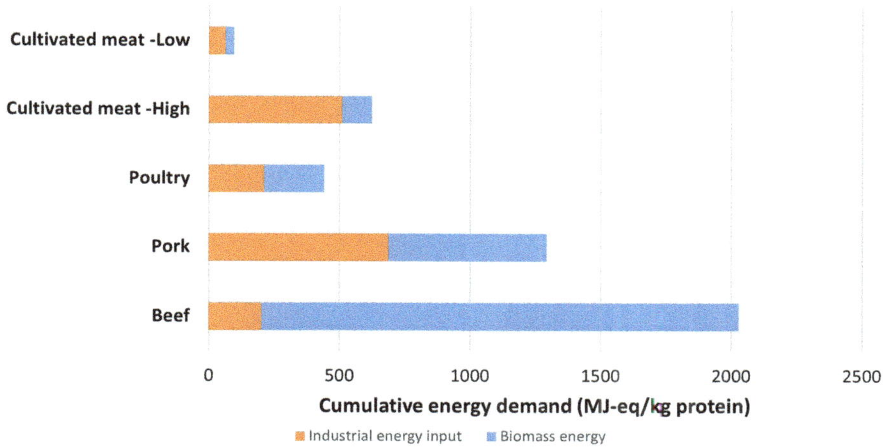


**Fig. 3** Cumulative energy demand of cultivated meat (low and high indicate the lowest and highest impacts reported in Tuomisto et al. (2022)) compared to livestock meat (data from Agribalyse (2022) database). Industrial energy input includes electricity, oil and gas, whilst biomass energy includes energy content of feed

manufacturing in bioreactors such as cultivated meat production needs to be optimised for energy efficiency as sustainable industrial energy production is limited and increasingly competitive.

## *4.5 Climate Impact*

Most GHG emissions of cultivated meat production originate from industrial energy consumption. Therefore, the energy source dictates most of the climate impact of the cultivated meat products. Fossil fuel usage in cultivated meat production leads often to higher climate impact in comparison with pork and poultry. However, GHG emissions from conventional beef production are nevertheless higher (Fig. 4, Mattick et al. 2015; Tuomisto et al. 2022). Cultivated meat production processes that use clean or renewable energy sources reduce substantially GHG emissions and consequently climate impacts (Sinke and Odegaard 2021; Tuomisto et al. 2022).

In addition, the production of culture medium ingredients used in cultivated meat production causes climate impacts. The ingredients sourced from agriculture or require agricultural inputs in the processes such as the production of synthetic amino acids cause GHG emissions from the agriculture. These include nitrous oxide emissions from the use of nitrogen fertilisers and fuel usage in planting, weeding, harvesting, drying, transporting and storing culture medium input materials from agriculture.

As cultivated meat production requires often less land area compared to conventional livestock production, it provides indirect climate benefits through reduced

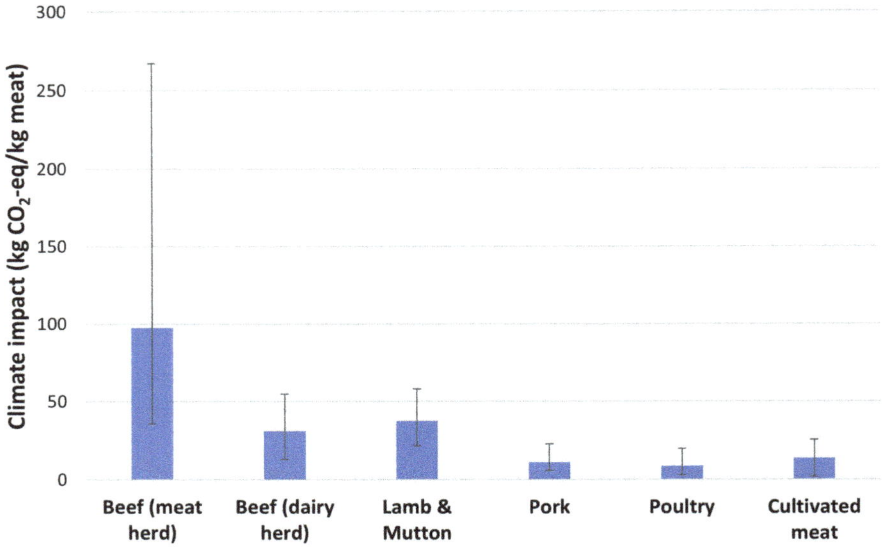


**Fig. 4** Climate impact of cultivated meat (as reported in the LCA studies in Table 1) and livestock meat (based on global means, error bars represent the fifth and 95th percentiles, source of data for livestock products: Poore and Nemecek (2018))

need to convert forests to agricultural land. In some areas, agricultural land previously used for feed production could potentially be afforested increasing the carbon storages in soil and vegetation. However, it is still uncertain how much large-scale cultivated meat production will need land area as such systems do not exist yet.

Climate impacts are reported by converting all GHG emissions to carbon dioxide equivalents ($CO_2$-eq). The resulting figure represents the climate impacts of the emitted gases comparable to carbon dioxide during a 100-year time frame. Nitrous oxide is 265 and methane 28 times stronger GHGs than carbon dioxide. It is criticised that the commonly used methods for reporting climate impacts unwarrantedly emphasise methane emissions, although methane has an atmospheric lifetime of only 12 years, whilst carbon dioxide remains for longer time periods in the atmosphere. Lynch and Pierrehumbert (2019) conducted a study using a 1000-year time frame for comparing the climate impacts of cultivated meat with beef. They concluded that in the longer-term analysis, some of the cultivated meat systems produced higher climate impacts than selected beef production systems.

Methane has a higher climate impact during shorter periods because it traps more heat per mass unit than carbon dioxide: 84 times in a 20-year period and 28 times during a 100-year period. Therefore, focusing on rapid reduction of especially methane emissions has faster impacts on climate change mitigation. In addition, methane cannot be removed from the atmosphere during its lifetime by currently known methods, whilst carbon dioxide is consumed by growing vegetation and absorbed in waterbodies. The anthropogenic GHG emissions generated by 2030 will determine the seriousness of climate change (IPCC 2018). In this context, switching from the 100-year time frame to the 20-year time frame would emphasise the short-term impacts of methane emissions.

### *4.6 Water Use and Discharge to Waterbodies*

Cultivated meat production uses direct and indirect water inputs. Water is used directly in preparing culture medium, controlling temperature and cleaning the bioreactors. Indirect water use considers production of inputs: culture medium ingredients, materials and energy. The water use of cultivated meat production depends on several factors such as the type of the culture medium ingredients and their production processes, used energy sources along the cultivated meat value chain, options for recycling water used in culture medium and whether fresh- or seawater is used in the cooling processes.

Water use between cultivated meat and livestock meat production depends on the methods used in calculating the relative differences. The water use of livestock meat production is relatively higher (Tuomisto and Teixeira de Mattos 2011) when calculated using water footprint approach that considers green, blue and grey water than when using methods that include only blue water usage (Tuomisto et al. 2022). The green water or rainwater use is significantly higher as livestock production requires agricultural land for feed production.

Cultivated meat production facilities are assumed to discharge only minor amounts of wastewater to waterbodies due to closed-loop production design and recycling of valuable ingredients. Ingredients used for culture medium production determine potential emissions to waterbodies. Culture medium ingredients originating from agricultural production such as nitrogen and phosphorus tend to leach to waterbodies as agricultural fields are open systems in natural environments where plants use only a part of the applied nutrients. Cultivated meat production is anticipated to require substantially less agricultural inputs than livestock meat production leading to decreasing eutrophication impact. A study estimated that the eutrophication impact of cultivated meat could be in parity with poultry production, significantly lower (−70%) than pork production and just a fraction (−97%) compared to beef production in the USA (Mattick et al. 2015).

Agricultural production and livestock farms produce other water emissions such as pesticide residues, antibiotics and pathogens from manure that impact on the quality of surrounding waterbodies. Cultivated meat production systems will completely avoid or drastically reduce these emissions to waterbodies due to closed-loop production system design and assuming that antibiotics are not needed in cultivated meat production. However, total avoidance of livestock-based emissions to waterbodies is not anticipated due to the need to maintain cell-donating farm animals. Regardless, livestock-based emissions in cultivated meat production would be rather negligible: the number of livestock needed for cultivated meat production would be only an infinitesimal compared to the current amount of animal production globally. Cultivated meat production could be independent from rearing livestock or cell-donating animals if the development of immortalised cell lines proves to be successful in the future.

## 5 Conclusions

The LCA studies conducted on a product level indicate that cultivated meat production has the potential to reduce environmental impacts compared to pork and beef production systems. Environmental impacts of cultivated meat production depend mainly on bioreactor energy use and the ingredients needed for the culture medium production. Large-scale cultivated meat production units are not yet in operation. Therefore, environmental impact assessments are anticipatory and based on a set of assumptions. Technological development and innovations in bioreactor and production process design, energy efficiency and culture media ingredients have the potential to reduce environmental impacts.

Anticipatory product-level assessments provide baseline information about environmental impacts of a cultivated meat product. However, real-world data from state-of-the-art cultivated meat production processes are needed to verify the results from anticipatory studies that are mainly based on sets of assumptions. These data are currently hard to obtain and only available in start-up companies developing the novel products of cellular agriculture. In addition, global-level assessments are needed to better understand the total impacts of cultivated meat production and its implications to the food systems. Major changes in land use dynamics caused by cultivated meat production call for global earth system impact modelling (Humpenoder et al. 2022). Several studies have assessed the potential climate benefits of switching to more sustainable vegan and vegetarian diets that originate from reforesting land previously used for livestock production (Hayek et al. 2021; Sun et al. 2022). Similar global assessments could be used in modelling various rates of the cultivated meat product usage, in different parts of the globe and at several time scales. The environmental assessments should also model the potential development of energy technologies: cultivated meat production needs clean and sustainably produced energy in achieving the climate benefits.

The pressing environmental challenges and global targets set to 2030 (e.g. United Nations Sustainable Development Goals and climate targets) call for quick and impactful environmental measures in the food production systems. However, the rather slow upscaling of cultivated meat production systems, the associated challenges with technologies and the culture medium composition indicate that products of cellular agriculture may not provide feasible solutions in time. The development of novel food products such as cultivated meat is needed as a part of the grand food system transformation (Moritz et al. 2023). However, environmentally beneficial changes in all parts of the current food systems need to be implemented: urgent reduction of food loss and waste, switching to plant-based diets, developing the sustainability of current agricultural practices, stopping fossil fuel usage and deforestation and tackling the farmland biodiversity loss. Alternative protein sources including algae, fungi and insects, novel food innovations such as cell-cultured microbial cells and recombinant proteins need to be also considered in improving the sustainability of global food systems before 2030.

## References


Agribalyse (2022). AGRIBALYSE LCA database. https://doc.agribalyse.fr/documentation-en/

Alexander P, Brown C, Arneth A, Finnigan J, Rounsevell MDJGEC (2016) Human appropriation of land for food: the role of diet. Global Environ Change 41:88–98

Alkemade R, Reid RS, van den Berg M, de Leeuw J, Jeuken M (2013) Assessing the impacts of livestock production on biodiversity in rangeland ecosystems 110(52):20900–20905

Bengtsson J, Bullock J, Egoh B, Everson C, Everson T, O'Connor T, O'Farrell P, Smith H, Lindborg R (2019) Grasslands—more important for ecosystem services than you might think. Ecosphere 10(2):e02582

Boulay A-M, Benini L, Sala S (2019) Marginal and non-marginal approaches in characterization: how context and scale affect the selection of an adequate characterization model. The AWARE model example. Int J Life Cycle Assess 25:1–13

Campbell BM, Beare DJ, Bennett EM, Hall-Spencer JM, Ingram JSI, Jaramillo F, Ortiz R, Ramankutty N, Sayer JA, Shindell D (2017) Agriculture production as a major driver of the Earth system exceeding planetary boundaries. Ecol Soc 22(4):8

Clark MA, Domingo NGG, Colgan K, Thakrar SK, Tilman D, Lynch J, Azevedo IL, Hill JD (2020) Global food system emissions could preclude achieving the 1.5° and 2°C climate change targets. Science 370(6517):705–708. https://doi.org/10.1126/science.aba7357. PMID: 33154139.

Crippa M, Solazzo E, Guizzardi D, Monforti-Ferrario F, Tubiello FN, Leip A (2021) Food systems are responsible for a third of global anthropogenic GHG emissions. Nat Food 2:198–209

Dudley N, Alexander SJB (2017) Agriculture and biodiversity: a review. Biodiversity 18:45–49

FAO (2006) Livestock's long shadow. FAO, Rome

FAO AND UNEP (2020) The state of the world's forests 2020: forests, biodiversity and people. FAO AND UNEP, Rome

Filazzola A, Brown C, Dettlaff MA, Batbaatar A, Grenke J, Bao T, Peetoom Heida I, Cahill JF Jr (2020) The effects of livestock grazing on biodiversity are multi-trophic: a meta-analysis. Ecol Lett 23:1298

Giam X (2017) Global biodiversity loss from tropical deforestation. Biol Sci 114:5775–5777

Hayek MN, Harwatt H, Ripple WJ, Mueller ND (2021) The carbon opportunity cost of animal-sourced food production on land. Nature Sustainability 4(1):21–24

Humpenoder F, Bodirsky BL, Weindl I, Lotze-Campen H, Linder T, Popp A (2022) Projected environmental benefits of replacing beef with microbial protein. Nature 605:90–96

IPCC, (2018) Global Warming of 1.5 °C - an IPCC special report on the impacts of global warming of 1.5 °C above pre-industrial levels and related global greenhouse gas emission pathways, in the context of strengthening the global response to the threat of climate change, sustainable development, and efforts to eradicate poverty. Intergovernmental Panel on Climate Change, Incheon.

ISO (2006) ISO14044; Environmental management - life cycle assessment - requirements and guidelines. International Organization for Standardization, Geneva

Järviö N, Maljanen N-L, Kobayashi Y, Ryynänen T, Tuomisto HL (2021) An attributional life cycle assessment of microbial protein production: a case study on using hydrogen-oxidizing bacteria. Sci Total Environ 776:145764

Kim S, Beier A, Schreyer HB, Bakshi BR (2022) Environmental life cycle assessment of a novel cultivated meat burger Patty in the United States. Sustainability 14(23):16133

Kyttä V (2022) Circular bioeconomy in life cycle assessment - addressing multifunctionality of agriculture. Faculty of Agriculture and Forestry, University of Helsinki, Helsinki, p 60

Kyttä V, Roitto M, Astaptsev A, Saarinen M, Tuomisto HL (2022) Review and expert survey of allocation methods used in life cycle assessment of milk and beef. Int J Life Cycle Assess 27:191–204

Lynch J, Pierrehumbert R (2019) Climate Impacts of Cultured Meat and Beef Cattle. Frontiers of Sustainable Food Systems 3:5

Machovina B, Feeley KJ, Ripple WJ (2015) Biodiversity conservation: The key is reducing meat consumption. Sci Total Environ 536:419–431

Mattick CS, Landis AE, Allenby BR, Genovese NJ (2015) Anticipatory life cycle analysis of in vitro biomass cultivation for cultured meat production in the United States. Environ Sci Technol 49:11941–11949

McAuliffe GA, Takahashi T, Lee MR (2020) Applications of nutritional functional units in commodity-level life cycle assessment (LCA) of agri-food systems. Int J Life Cycle Assess 25:208–221

Mekonnen MM, Hoekstra AYJE (2012) A global assessment of the water footprint of farm animal products. Ecosystems 15:401–415

Moritz J, McPartlin M, Tuomisto HL, Ryynänen T (2023) A multi-level perspective of potential transition pathways towards cultured meat: Finnish and German political stakeholder perceptions. Res Policy 52(9):104866

Piazza J, Ruby MB, Loughnan S, Luong M, Kulik J, Watkins HM, Seigerman M (2015) Rationalizing meat consumption. The 4Ns. Appetite 91:114–128

Poore J, Nemecek TJS (2018) Reducing food's environmental impacts through producers and consumers. Science 360:987–992

Sillman J, Uusitalo V, Ruuskanen V, Ojala L, Kahiluoto H, Soukka R, Ahola J (2020) A life cycle environmental sustainability analysis of microbial protein production via power-to-food approaches. Int J Life Cycle Assess 25:2190–2203

Sinke P, Odegaard I (2021) LCA of cultivated meat. Future projections for different scenarios, CE Delft

Sinke P, Swartz E, Sanctorum H, van der Giesen C, Odegard I (2023) Ex-ante life cycle assessment of commercial-scale cultivated meat production in 2030. Int J Life Cycle Assess 28:234–254

Skrivergaard S, Young JF, Sahebekhtiari N, Semper C, Venkatesan M, Savchenko A, Stogios PJ, Therkildsen M, Rasmussen MK (2023) A simple and robust serum-free media for the proliferation of muscle cells. Food Res Int 172:113194

Smetana S, Mathys A, Knoch A, Heinz V (2015) Meat alternatives: life cycle assessment of most known meat substitutes. Int J Life Cycle Assess 20(9):1254–1267

Statista (2023a) Share of vegans in European countries. https://www.statista.com/forecasts/1256518/share-of-vegans-in-european-countries. Retrieved 17 Oct 2023

Statista (2023b) Meat substitutes—worldwide. https://www.statista.com/outlook/cmo/food/meat/meat-substitutes/worldwide#volume. Retrieved 17 Oct 2023

Sun Z, Scherer L, Tukker A, Spawn-Lee SA, Bruckner M, Gibbs HK, Behrens P (2022) Dietary change in high-income nations alone can lead to substantial double climate dividend. Nature Food 3:29–37

Tilman D, Clark M (2014) Global diets link environmental sustainability and human health. Nature 515:518–522

Trinidad KR, Ashizawa R, Nikkhah A, Semper C, Casolaro C, Kaplan DL, Savchenko A, Blackstone NT (2023) Environmental life cycle assessment of recombinant growth factor production for cultivated meat applications. J Clean Prod 419:138153

Tuomisto HL, Teixeira de Mattos MJ (2011) Environmental impacts of cultured meat production. Environ Sci Technol 45(14):6117–6123

Tuomisto HL, Hodge ID, Riordan P, MacDonald DW (2012) Exploring a safe operating approach to weighting in life cycle impact assessment—a case study of organic, conventional and integrated farming systems. J Clean Prod 37:147–153

Tuomisto HL, Allan SJ, Ellis MJ (2022) Prospective life cycle assessment of a bioprocess design for cultured meat production in hollow fiber bioreactors. Sci Total Environ 851:158051
Upcraft T, Tu W-C, Johnson R, Finnigan T, Van Hung N, Hallett J, Guo M (2021) Protein from renewable resources: Mycoprotein production from agricultural residues. Green Chem 23(14):5150–5165
Willett W, Rockström J, Loken B, Springmann M, Lang T, Vermeulen S, Garnett T, Tilman D, Declerck F, Wood AJTL (2019) Food in the anthropocene: the EAT–Lancet Commission on healthy diets from sustainable food systems. Lancet 393:447–492
WRI (2019) Creating a sustainable food future. A menu of solutions to feed nearly 10 billion people by 2050. World Resources Report. World Resources Institute, Washington, DC
Xu X, Sharma P, Shu S, Lin T-S, Ciais P, Tubiello FN, Smith P, Campbell N, Jain AK (2021) Global greenhouse gas emissions from animal-based foods are twice those of plant-based foods. Nat Food 2:724–732

# Regulatory Aspects of Cultivated Meat

**Alexandre Cabral, Isabela de Oliveira Pereira, and Amanda Leitolis**

**Abstract** Cultivated meat is among the most disruptive, innovative alternatives to meet the growing global demand for food. Historically, regulators have always come after innovators and are challenged to build regulatory frameworks without excess and lack of risks. However, cultivated meat regulation has additional complications, such as the novelty of the production process and the need to create a new baseline. For now, very few regulatory bodies have received applications for cultivated meat, and regulatory frameworks are diverse across different countries. This chapter discusses the complex interrelationship between science and public policy in the global mobilization to regulate cultivated meat and tries to anticipate potential impacts in the geopolitics, labeling and food safety regulations.

**Keywords** Cultivated meat · Regulation · Geopolitics · Food safety and labeling

## 1 Introduction

Over recent years, scientists have developed innovative alternatives to meet the growing global demand for food. They presented innovations as part of the solution to build a more efficient, resilient and sustainable food system (Frezal et al. 2022). Among them, we can point out cultivated meat as one of the most disruptive. Using improved versions of cell culture techniques established decades ago, it became possible to take a small number of cells from a live animal and place it in an environment that provides the conditions for the cells to grow and multiply, developing muscle and other tissues to create meat. Meat without slaughtering and its

A. Cabral
Policy Department, The Good Food Institute Brazil, São Paulo, Brazil
e-mail: acabral@gfi.org

I. de Oliveira Pereira · A. Leitolis (✉)
Science and Technology Department, The Good Food Institute Brazil, São Paulo, Brazil
e-mail: isabelap@gfi.org; amandal@gfi.org


C. R. Soccol et al. (eds.), *Cultivated Meat*,
https://doi.org/10.1007/978-3-031-55968-6_15

byproducts. Meat without livestock and its costs. Meat without adverse roles in zoonotic diseases and the antibiotics used to prevent them. Meat without using an inefficient biofactory (the live animal) can consume several times the amount of protein in the form of feed that it delivers in the form of meat.

Obtaining human-edible animal tissues in a different way from slaughtering a live animal is as profound a technological disruption as starting to breed animals for slaughtering has been in the distant past. We are facing a new revolution in obtaining animal protein for human consumption. This time, we can open up this tissue to the countless possibilities of cultivation, such as growing meat from cell lines of animals that are already extinct or even mixing tissues from different species that could never occur in nature as a viable form of living animal.

Historically, regulators have always come after innovators, often blocking the way of the following steps by trying to regulate the steps already taken. When building a regulatory framework, there are both excess and lack risks. If we control too much, we can hinder the next leaps. If we regulate too little, we could jeopardize the safety aspects of the food we produce to the detriment of the consumer. In cultivated meat, we have additional complications to this optimum level of regulation, which is the need to create a new baseline. Suppose we adopt the strict standards of health-grade cell therapies, where a hospital must provide a set of cells with high purity because it will be injected directly into your bloodstream. In that case, we may miss that food tissue will be submitted to a gastroenteric tract accustomed to higher degrees of impurity. From another perspective, if we copy the rules of animal production, we could incorporate food safety rules that are stricter than necessary for an environment where there is no blood or viscera, for example, in the production process.

This chapter discusses the complex interrelationship between science and public policy in the global mobilization to regulate cultivated meat. It's been a long time, perhaps from the time of the regulatory debate on genetically modified organisms (GMOs), since these two very different entities have often had to talk about so many issues.

## 2 Regulatory Agencies for Cultivated Meat Products

The ruler that measures the pioneering countries in regulating cultivated meat is hard to explain. It is not the ruler of economic size nor the richness and complexity of the innovation system, much less the size of the incumbent animal protein market. But it is nonetheless a ruler that combines parts of each component.

Suppose today we can explain ex-post why Dr. Mark Post (excuse the pun) presented his first experiment to the world in 2013 in the Netherlands, based on public funding to tackle a food sovereignty issue. In that case, we can hardly do so ex-ante based on his career or the consistency of the Dutch innovation ecosystem. We would have the same difficulty explaining ex-ante that the pioneer startup Modern Meadow, hosted in the American innovation ecosystem of Singularity University and also announcing at the end of 2013 its search for cultivated meat through bioprinting techniques, would pivot its business plan to cultivated leather. It would be even more challenging to explain why, even though both have their names on the proceedings of the early days of cultivated meat, the companies they lead are not today the leading companies in the sector.

A relevant aspect of trying to convene an explanatory apparatus for the nature of this ruler is that it needs to incorporate diverse elements involving science, business, public policy, and society. Morais-da-Silva et al. (2022) point out that these elements only form the basis of an explanatory apparatus if we consider that they are associated with a sociotechnical transition in agriculture. According to the authors, "Sociotechnical transition theory states that when a more efficient technological proposal is discovered, there is support from the landscape environment and actors are interested in developing and consuming the new technology, opportunities open up to integrate niche initiatives into the regime level."

Thus, the complexity of the ruler derives from the complexity and multifaceted nature of the sociotechnical transition. The speed with which certain countries define their regulatory frameworks for cultivated meat is part of this complexity. Considering that a cultivated meat production ecosystem is linked mainly to the existence of a supply chain (which can be attracted to a given location) rather than to a stock of natural resources (which need to be present there), we can conclude that what drives a country to develop a regulatory framework more or less fast is the perspective of the position it wants to occupy in the short and medium term scenario of the technology. Countries that want to be protagonists in the supply of cultivated meat or urgently need to incorporate it into their food and nutrition security programs tend to regulate more quickly than others tied to the animal protein paradigm or that do not have food sovereignty issues among their priorities. This helps explain the list of countries listed below and some notable absences.

The Singapore Food Agency (SFA) was the first national regulator to green-light the sale of a cultivated meat product, approving GOOD Meat's cultivated chicken for use as an ingredient in the company's chicken bites in November 2020. SFA has since approved additional cultivated meat products from GOOD Meat. In early 2023, SFA approved using serum-free media in producing GOOD Meat's products, allowing the company to scale up production further and reduce costs. Singapore treats cultivated meat as a novel food requiring premarket authorization. However this explanatory picture becomes more comprehensive if we connect these facts to the "30by30" Program (Singapore Food Agency 2023). This government action aims to ensure the country produces at least 30% of the food consumed in Singapore by 2030. Rather than seeking to play a leading international role in cultivated meat, Singapore is trying to solve its domestic problems because, as the program says, "With the majority of our food imported, local production is important to reduce our reliance on food imports. It can also mitigate the impact of supply disruptions." For a country with such a small territory and still having to use part of it to host one of the largest ports in the region, producing meat without having to use the land for it is a great opportunity.

In 2023, The United States became the second country after Singapore to approve cultivated meat. In a significant milestone, Upside Foods and Eat Just went through the US government's process to obtain full approval, which included endorsements from both the FDA and USDA. FDA completed their first premarket consultation for Upside Foods' cultivated chicken and the second for Eat Just's cultivated chicken, announcing they had "no questions" about the safety of the production process and formulation (FDA 2022, 2023). After that, both companies get the approval process of the USDA, which has jurisdiction over product labeling and clearing for the safety of the meat-making facilities. In the United States, all cultivated meat producers must register their facilities with the FDA, and companies producing cultivated terrestrial meat or catfish must also obtain a grant of inspection from the USDA. USDA inspectors will oversee these products' processing, packaging, and labeling and verify that cultivated meat products are safe, wholesome, and unadulterated (free from contamination). FDA will retain jurisdiction over the processing, packaging, and labeling of cultivated seafood. According to the FAO (2022b), North America is the third largest meat producer in the world (after Asia and Europe) and the second largest exporter (behind South America and ahead of Europe). In addition, two of the four largest meatpacking companies in the world and two of the four largest cultivated meat companies in the world are American. If the animal meat market is big, so is the cultivated meat. Furthermore, the cultivated meat market is still "up for grabs," which is one of the reasons for the speed with which the United States has been implementing its regulatory framework.

Oceania became the third region to face a regulatory issue on cultivated meat. In this case, a startup's demand drove the regulatory framework's building. The local scenario contains all the classic elements of the alternative protein development, repeated in several other countries: an agricultural commercial agenda based on commodities and the government's perception that the path towards a higher value-added export agenda is the best choice, an NGO dedicated to the subject lobbying in favor (Food Frontier, active in the country since 2017), a mobilization of traditional agribusiness lobbying against the development of "non-meat meats" and a growing number of companies, mainly startups, going to market dedicated to alternative proteins. One element that may have spurred government action in favor of alternative proteins is the geographical proximity to Asia, which is shaping up as a substantial importer of food of all kinds (according to FAO 2022b). Asia is both the largest producer and the largest importer of animal protein. Oceania has an inherent competitive advantage in not facing the similar high logistical costs as North and South America in its exports to Asia. One element that seems to have been decisive in advancing the issue (but which is also not exclusive to the region) was the engagement of the National Innovation Agency (CSIRO), whose 2022 study "Australia's Protein Roadmap" involved proteins of animal and plant origin in a single plan.

The Food Standards Australia New Zealand (FSANZ) received an application for cultivated meat in March 2023. FSANZ has commenced assessing an application from Vow Group Pty Ltd seeking approval for cultivated quail meat. The assessment is expected to take around 12 months and will include two rounds of public consultation. If FSANZ approves the product, Australian and New Zealand food ministries will have 60 days to review FSANZ's decision. Cultivated meat can be regulated under their novel food frameworks in these countries (FSANZ 2023).

In Europe, cultivated meat is subject to the Novel Food Regulation (EU 2015/2283). According to the guidance established, the European Food Safety Authority (EFSA) procedure begins with submitting a safety dossier (EFSA NDA Panel 2016). The analysis can take at least 9 months, assuming EFSA has no questions. After that, the European Commission's Standing Committee on Plants, Animals, Food and Feed (PAFF committee) must approve the application. Europe's first-ever cultivated meat application was filed in 2023 by Aleph Farms, an Israeli startup, asking the Swiss food safety regulators to approve selling its cultivated beef steaks in its first European market.

As Singapore and the United States are leading regulatory advances, many countries still have a long way to go to produce robust and efficient regulatory frameworks. Still, early signs in many regions are promising. Like the European Union and Australia, Canada and Israel have indicated that cultivated meat can be regulated under their novel food frameworks. Health Canada's Food Directorate has jurisdiction over cultivated meat in the country. The National Food Control Service (FCS) within the Ministry of Health has regulatory authority over cultivated meat produced and marketed in Israel.

Brazil has yet to set forth a regulatory framework. The General Food Office at the National Health Agency (ANVISA) and the Animal Products Inspection Department within the Ministry of Agriculture have jurisdiction over cultivated meat. Like in the other countries cited previously, companies will need to apply for authorization, and the agencies will conduct a safety evaluation under the Novel Foods and Novel Ingredients regulation, which is currently being updated and now explicitly includes cultivated meat under its scope.

Besides countries' initiatives, multilateral organizations are leading global studies about food safety, regulation and trade of cultivated meat products and have expressed increasing support for continued field innovation. World Health Organization (WHO), Food and Agriculture Organization of the United Nations (FAO) and Codex Alimentarius are working to provide technical support and leading subject matter experts' roundtables on cell-based food (a neutral terminology proposed by the FAO to avoid conflicts with the different stages of the debates on the use of names commonly used by animal products by cultivated meat in different countries). The aspects of cultivated meat production drive these efforts, aiming to propose global patterns and address the mitigation of international hurdles such as mitigation of climate change effects, public health risks, and food insecurity. On April 5, 2023, FAO and WHO officially launched a publication entitled "Food safety aspects of cell-based food" (FAO and WHO 2023), convening the results of an FAO-led expert consultation held in Singapore in November 2022, where a comprehensive food safety hazard identification was conducted. After the Expert Consultation, FAO prepared a short brief, presented as a Conference Room Document (CRD) at the 45th Codex Alimentarius Commission (CAC) session (CAC 2022).

## 3 Geopolitics, Cultivated Meat, and Regulation

We've been consuming animal meat for millennia. Anthropological dietary studies (like Teaford and Ungar 2000) argue that even the early hominids were not dentally preadapted to eat meat, evolutions gradually allowed them to do so, and meat may have provided critical nutrients for both young and old hominids. In the last 40 years, when official statistics have formed reliable historical series (OECD-FAO 2023), we've eaten less beef and more poultry, but we've always eaten more and more meat.

The existence of a piece of edible tissue on a human's plate obtained from the slaughter of a live animal requires several factors. Let's start with a live and healthy animal interested in remaining alive at the entry door of the slaughterhouse. This live animal results from modified genetics to develop faster or more of the most desired edible parts. This healthy animal is the result of the use of medication throughout its life to prevent the effects of pathogens, which ultimately reduce the quantity and quality of the tissues obtained and can even lead to their total loss in the case of sanitary culling. This live and healthy animal results from a growth process carried out on a piece of land dedicated to this purpose, consuming feed and water throughout this period. This growth process begins with the birth of the offspring from a live and healthy female parent, fertilized by genetic material obtained from a live and healthy male, involving zootechnical technology from the embryo. Inside the slaughterhouse, we have an industrial process surrounded by productivity technologies, a workforce trained in its activities, an extensive set of process controls aimed at eliminating risks that could mean product contamination, the refrigerated storage of the parts of the slaughtered animal and the sending of byproducts to secondary downstream processing chains. From the exit door, an equally refrigerated logistics chain distributes these products worldwide, sold in wholesale, retail and food service.

Up to the slaughterhouse entrance, cattle raisers and animals. From there, meatpackers. Around them all, the regulators. At the end of the line is the consumer. All these actors and scenarios make up meat's geopolitics (Jakobsen and Hansen 2020; Flexor 2023), which radically changes the way we obtain cultivated meat.

Firstly, the cattle raiser is entirely out of the picture in its current role. The meatpacker doesn't need the many thousands (or even millions) of animals that are used today to produce cultivated meat, but maybe only dozens of cell donors, which its own units can maintain. This represents a real tectonic shock because these gigantic clusters operate together for a long time and are umbilically connected in animal meat production, which is now becoming detached due to disruptive innovation. The coexistence between the two clusters has become increasingly tense since the demands for sustainability in the entire chain have fallen more heavily on meatpackers than cattle raisers. Cultivated meat is not just another point of tension but also has disruptive colors.

Secondly, the role of the cell grower appears on the scene, responsible for managing the bioreactors and the relationship with a complex and ultra-specialized supply chain, which, in principle, seems to be made up of companies migrated from the health supply sector. This place in the chain requires knowledge that neither cattle raisers nor meatpackers have mastered in their current profiles. But what is happening is that while cattle raisers focus on promoting legislative actions in various countries around the world to ban the use of terminologies or the manufacture of the product itself, meatpackers have focused on acquiring positions in startups or developed internal departments focused on cultivated meat. One of the central environmental liabilities of meatpackers today is the traceability of the animals they slaughter, so it makes sense that it would be advantageous for them to verticalize production and easily trace the entire meat process, from cell to steak. On the other hand, some cattle and broiler chicken producers seem less organized around resistance to disruptive meats and more interested in diversifying their activities, provided they have access to knowledge on how to engage and an assured demand (Bogueva et al. 2023). Such a stance allows for new possibilities, where the need for crops tailored for cultivated meat and replacing chickens with on-farm bioreactors may become part of the new landscape.

An additional element is brought into play by technology. In the case of livestock farming, the extensiveness of the activity generates cost reduction, which fits perfectly in the economic concept of scale gaining. In cultivated meat, the size of a bioreactor for industrial production is a function of the growth cell efficiency achieved. The bioreactor becomes more compact when kinetics is favored. In the case of cell growth kinetics, that means achieving higher replication rates and an increase in the number of passages before critical senescence thresholds are reached. The challenge scientists haven't yet defined is how to increase the cell density to its maximum point without losing the cells' properties and allowing them to form tissues. This uncertainty leads to whether the ideal model will be a scale-up (a small set of large bioreactors in a large factory) or a scale-out (a large group of small bioreactors spread over medium-sized factories). Apparently restricted to engineering, this decision has substantial implications for geopolitics.

On some industrial lines, the assembly process does not significantly affect the volume of things between the factory's entrance and exit doors. On an electronics assembly line, for example, a volume of "x" cubic meters of raw materials comes in, assembled in processes of medium and high complexity, but at the end, generates a volume very close to "x" cubic meters of finished products. In cultivated meat biofactories, the cell proliferation process can multiply the number of cells that enter the process and what comes out of it by trillions. In addition, recycling culture media means reusing the volumetric quantity supplied at the starting point of the process as many times as possible before being replenished. As a result, in a characteristic of bioprocesses in general, the volumetric content leaving the production line is greater than the volumetric content entering it.

This leads us to believe that if the option of scale-out (stacking medium-sized bioreactors) proves both technically viable and economically advantageous, we will probably have a global protein production chain model different from the current one. The new pattern would be the concentration of suppliers of manufactured inputs in large plants and the scattering of industrial cell multiplication plants in smaller plants, placed closer to their consumer markets. This model has positive environmental appeal because it can reduce the carbon footprint of outbound logistics, as the distance between the factory and the point of sale is shorter than in the current model, based on production concentrated in large meatpacking structures. In a nascent industry, other models are also being considered. RESPECTfarms (2023), for example, defends a business model in which cultivated meat is produced on a farm scale, establishing a chain in which farmers diversify meat production by growing cultivated meat and using their land to produce plants that are useful for nourishing cells and functioning as inputs for growing media.

But what does this have to do with geopolitics? If deglobalization (Balsa-Barreiro et al. 2020; Witt et al. 2021; Chase-Dunn et al. 2023) gains traction in global geopolitical design in the near future, food sovereignty issues could find numerous advantages in this new design of the protein production chain for human consumption provided by cultivated meat. A country with little land available for livestock farming or with unfavorable climatic conditions for this activity needs to import the animal proteins it offers to its population. Many issues beyond the country's control can weaken this supply relationship and threaten the country's food sovereignty, especially in high-volume commodity trade flows. These could include bilateral issues generated by suppliers, such as the intensification of United States-China or United States-Russia relations, and multilateral issues beyond the control of the country and its suppliers that affect the logistics routes they use, such as the war in Ukraine or conflicts that affect maritime routes.

A possible advance in the decentralized model of cultivated meat production could make it viable to have local units large enough to serve just one country in a configuration, meaning the country could carry out most of the process inside its borders. This kind of self-sufficiency mitigates a good part of this threat. Still, it does not eliminate it because it does not seem feasible on the visible horizon that the production of inputs and bioreactors will also migrate to a decentralized model. With this, raw materials and equipment producers and the licensors of cell multiplication technologies become relevant geopolitical actors.

This kind of concern is the basis of Singapore's 30by30 Program. It also prompted Chinese President Xi Jinping to call food security "guozhi dazhe" (a national priority) and to bring the issue up for debate at the Central Committee of the Communist Party of China in 2023. In the same direction, Aleph Farms, the Israeli startup best known for producing the world's first lab-grown steak, announced in July 2023 that DisruptAD, the venture capital arm of Abu Dhabi's Sovereign Wealth Fund, led its last $100 million investment round.

Regulators are also relevant players in this geopolitics, whether they are technical regulators of issues related to food safety, labeling and identity and quality standards, or legislative regulators (deputies, senators and city councilors) who define terminology and can even ban products from being produced or marketed in a country.

Constant change is the rule in frontier technology areas, and state-of-the-art changes frequently. In food regulations, these variations must never jeopardize food safety aspects. Two primary models in the world deal with foods that do not yet have a history (usually between 10 and 30 years) of safe consumption in a country. The first is the GRAS model (generally recognized as safe), where the manufacturer is responsible for submitting the food safety studies, and the regulatory agent replies with a "letter of no objection," in which it certifies that the authors of the study are reliable and that the results contained therein do not point to an imminent risk to the consumer, but where the regulator also states that its agreement immediately ceases to exist if supervening facts point to elements of risk to the consumer, in line with the "generally" prefix of the concession. The second model defines a "Regulation for Novel Foods and Novel Ingredients," which functions as an "Office for Further Analysis" within the agencies and where the volume of information required from the applicant for approval is much higher than in the usual applications, but where the agency's position is more forceful concerning the safety of that food for the consumer. The GRAS model is expensive for companies, who have to buy dossiers, but it tends to be quicker for the agency to issue its position. The "Novel Food" model is less expensive for those who apply to register, although the companies need to present toxicity studies and risk prevention models in the production process. Still, it tends to take longer for the regulatory agent to analyze because there is no prior knowledge about the consumption of that product. Expensive and fast vs. cheap and slow are not just rhetorical figures but scenarios that can influence a company's decision on where to manufacture and market its products, affecting the geopolitics of meat.

The requirements influence the analysis process, especially in segments where the dynamics of innovation are still undergoing intense change. Over-regulation restricts and discourages innovation. Under-regulation increases regulatory uncertainty, which also deters innovation. The optimum point always combines food safety science with the interests and objectives of the country's political project.

Bills proposing banning cultivated meat from the country or admitting its existence but proposing that the term "meat" cannot be used for it have appeared in Legislative Branches in different countries, presented by legislators of different representativeness, but usually with a mandate linked to livestock farming. Politics is the locus where different interests meet, and it is legitimate and natural for the interests of cattle farmers to be represented as much as those who defend cultivated meat. But this kind of decision doesn't seem to have the same power to affect geopolitics as, for example, a hostile regulatory environment defined by the health authorities because it's easier to circumvent or avoid their consequences than restrictions incorporated into health surveillance protocols. If it's label censorship, you change the label. If it's a ban, you can introduce another bill to allow it again.

At the end of this line are the consumers, without whom any reflection on the geopolitics of meat and any other category of human food is just rhetoric. In scenarios of free and informed consumption decisions (where the consumer has access to the product and can afford to buy it), various studies screened the factors determining consumer perception and acceptance of Cultivated Meat. Pakseresht et al. (2022) compiled some studies and listed the different dimensions of positive and negative perceptions. The authors detail the following main dimensions: product awareness and familiarity, knowledge of the production process, knowledge of the technology advantages, risk-benefit trade-offs, perception of healthiness, labeling, animal welfare and environmental concerns, unnatural production process (tampering with nature), fear of unfamiliar technology and unknown effects (neophobia).

However, surveys measured these issues with respondents who, for the most part, have never tried the product. To date, customers can only order the product in a single Singapore restaurant or attend trial sessions sponsored by the manufacturers. Perceiving a food without the definitive judgment of the tongue's taste buds may seem strange. Still, the act of eating is so culturalized that it has anthropological dimensions (Pollan 2006; Crowther 2013) that allow us to make consistent and well-founded considerations even without the researcher being able to answer the interviewee's crucial question: "But what does it taste like?" More than just a question of flavor and taste, political dimensions permeate the perception of this product. Raised under the paradigm of meat obtained from a live slaughtered animal, perceptions of "what's behind this new product" (politics, sustainability, animal welfare, science) make consumers feel like they are scratching the surface that separates them from the future of food and having the opportunity to have a say in the food of the future. This multifaceted set of elements of such a diverse nature makes cultivated meat unique in the food world.

## 4 Food Safety Regulations for Cultivated Meat

Making cultivated meat a viable commercial option goes beyond overcoming all the technical challenges for production on an industrial scale and in a cost-effective way; it also involves some social challenges and their regulation. The diversity of regulatory frameworks across different countries reported in previous publications (FAO and WHO 2023; Miyake et al. 2023; Guan et al. 2021) highlights the opportunity (and advantages) of having harmonized regulatory frameworks in the future to ensure safe consumption.

To date, food safety authorization and the regulatory approval processes for cultivated meat products have chiefly occurred on a case-by-case basis (GFI IN 2022; FAO and WHO 2023). Most governments that have already established regulatory frameworks for cultivated meat have stated that such products would fall under current novel food regulations, as is the case of the United Kingdom, the United States, Singapore and Israel. However, some governments, such as the Japanese, have created specialized regulations on cultivated food products (Miyake et al. 2023). For the time being, only four regulatory bodies, Singapore, the United States, Swiss and Australia, have received applications for cultivated meat, and very few national competent authorities are fully prepared to assess the safety of cell culture food products and approve a market entry (GFI 2022; GFI IN 2022; GFI Europe 2023; Ong et al. 2023).

To guarantee the innocuousness of the food, the food safety regulations are made up of a set of rules (from laws to operational normative instructions) aiming at protecting each country's consumer populations and ensuring that its food industry will be able to export its products to other countries. The regulatory pieces may vary among countries, influenced by their technological and economic development as well as religious beliefs, but are mainly based on consumer concerns, nutrition and food safety risks (FAO and WHO 2023). Due to the novelty of the production process and limited safe consumption data, establishing regulations for cultivated meat is a unique challenge common to all countries. This topic focuses on key technological aspects and safety issues the countries' regulatory approaches must address.

Meat is a nutritious product that contains high biological value protein and various other nutrients such as vitamin B12, iron, zinc, selenium, phosphorus and many bioactive compounds such as creatine and taurine (Watanabe and Bito 2018; FAO 2023b). In this sense, cultivated meat products are expected to deliver consumers a nutritional profile, including proteins, fats, minerals, and vitamins, equivalent to conventional meat. Due to the low number of registered products and the industrial secrecy surrounding the ongoing research and development processes, independent scientific analyses to verify whether this equivalence is the reality of the products being developed by the industry are absent. To date, only a limited number of studies addressing the topic is available (Broucke et al. 2023; Fraeye et al. 2020); some companies report similarities between conventional and their product's (Steakholder Foods 2023) amino acid profiles, and data from the first pre-marketing notices were made public, as of FDA-approved Upside Foods cultivated chicken which exhibits nutrient composition within safe ranges (Upside Foods 2021).

Many of the strategies to improve the nutritional properties of cultivated meat focus on methods to directly induce protein synthesis through electrical or mechanical stimulation, genetic engineering focusing on expressing certain proteins and delivering signals by adjusting culture media composition to orientate cell fate (Broucke et al. 2023; Fraeye et al. 2020). In addition to these means, culture media supplementation with some minerals and vitamins is a required action in cases where, for example, those elements are not produced by animal cells but accumulated from the diet (e.g., vitamin B12) (Fraeye et al. 2020; Pereira and Vicente 2013) or whether the objective is to create fortified meat. Formulating and developing cell-based products, combining cultivated meat with plant-based ingredients or other alternative protein sources, such as fruits, mushrooms, spices, and ingredients derived from fermentation, is a strategy companies (FoodNavigator USA 2021) can use to overcome some technical challenges and simultaneously improve the nutritional characteristics of the product. Hybrids blending cultured cells with plant-based elements have been the first cell-based products placed on the market, representing the first generation of cultivated meat available to consumers (Green Queen 2020).

Hybrid meat, as a concept, can be differentiated from meat products with plant-based functional ingredients with a basis in the purpose of that mixture. In general, functional ingredients are used because of their properties (e.g., emulsification, gelling, water binding) and for cost savings. This concept is changed for hybrids to include positive connotations such as healthiness, lower environmental impact and a way of decreasing meat consumption (Grasso 2020; Grasso and Jaworska 2020). From a cultivated meat perspective, plant-based inputs can be added to any step of production mostly to achieve economic and technological purposes. They can be used as culture medium supplements or microcarriers, enabling cell growth and adherence, working as structural elements, like scaffolds and bio-inks contributing to texture and thickness. They can also be applied as ingredients during food processing. These days, plant-based ingredients have been used in cultivated meat products to contribute to a balanced production volume and cost, impacting the final product's nutritional and sensory attributes. Equally important, formulating cultivated meat products with plant-based ingredients is an alternative allowing for further reduction of the environmental impact of cell-based products. The countless ways to formulate animal cell cultured foods with plant-based ingredients could undoubtedly result in a range of new products with different nutritional values and percentages of plant derivatives, leading to potential compositional differences compared to their conventional equivalents. Developing products that use these elements to enable the characteristics defined when the product was designed seems worth intentionally pursued by companies shortly. Such differences may require a differentiated regulatory approach and impact how the product will be named and labeled and which health and nutrition claims can be made.

Potential food safety concerns regarding hybrids would be related mainly to the plant source and their integrated or residual presence in the final product. Plant sources commonly used as ingredients or media supplements for cultivated meat include fungus or algae derivatives such as cellulose, alginate, and protein hydrolysates from soy and pea (Bomkamp et al. 2022). Thus, the potential allergic responses elicited by known or new allergenic sources are one of the main issues once animal meat cells do not commonly express allergenic epitopes (FAO and WHO 2023). Furthermore, some food business operators may wish to use genetically modified (GM) plants as ingredients or processing aids. In that case, those hybrids would likely be subject to existing GM regulations depending on the jurisdiction.

Regulating cultivated meat under GM may also occur based on the approach used by regulatory authorities regarding the industry's application of genetic technologies in foods. Since the development of modern agriculture, the food industry has had an extensive list of methods for genetic engineering of plants, microorganisms and animals, including those based on recombinant DNA technology widely used in plant and animal breeding and generally addressed by countries under GM regulations. In the last decade, the emergence of gene-editing techniques, such as ZFN, TALEN, and CRISPR, has allowed the precise introduction of targeted changes in genome organisms. Those techniques do not usually involve transgenics, resulting in organism improvement without safety issues surrounding foreign genes and their effects on food safety quality and trade are not expected to differ from foods derived from pre-existing breeding techniques (FAO 2022a, 2023b; Giraldo et al. 2019). Differences between genetic engineering and gene editing motivated discussions in some countries where gene-edited organisms and the food derived from them are treated as novel foods or even conventional products. In other words, they are generally subject to much less prohibitive legislation than GMOs, and requirements result in a less costly and time-consuming approval process (FAO 2023b; Rozas et al. 2022).

However, there is a lack of consensus among countries, and no one has created a new and separate category for regulating gene-edited food products (FAO 2023b). Some jurisdictions require a case-by-case consideration, and in some cases, every gene-edited product is considered to be a GMO. Brazil, for example, applies a case-by-case evaluation according to the methods used to generate the organisms (a process-oriented regulation), and products derived from gene-editing can be excluded from the scope of the legislation on GMOs (FAO 2023b; Vieira et al. 2021). On the other hand, the European Union has restrictive regulations stating organisms obtained by genome editing and related techniques are subject to the requirements of the GMO legislation (FAO 2023b).

Methods for genetic improvement are not a requisite for the production of cultivated meat to be feasible. Depending on the target species, cells with features such as shorting doubling times and viability after many passages can be selected as initial cells for the process. Furthermore, methods not involving direct genetic manipulation to establish immortal cell lines, which means cells capable of proliferating indefinitely, such as spontaneous immortalization protocols, can be applied to allow cell infinite divisions to reach a commercial production scale (Soice and Johnston 2021). However, optimizing cell lines using genetic tools is a powerful way to produce highly proliferative and bioprocess-resistant cells and obtain better taste and nutritional value. That combination would probably result in the most appropriate cell types to achieve commercial production scale at competitive costs and a final product reaching the characteristics it was designed to have. Based on recent advances in the cultivated meat industry and the efficiency of modern biotechnology methods, companies that would choose to perform cell optimizations are more likely to do so through gene editing instead of genetic engineering, generating products that will most likely not be regulated as GM foods in many countries.

Potential safety concerns related to GMOs are allergenicity, gene transfer, and outcrossing (WHO 2014). In the context of cultivated meat, allergenicity can be generated after any intentional genetic modification of the cell line or cell donor animal, potentially leading to increased expression of known allergens or the expression of new allergens. That hazard can be controlled by validating that the modification did not cause further changes in the genome beyond those desired and through analysis of the expression and allergenicity testing of new proteins (FAO and WHO 2023). Regarding gene transfer, there is limited scientific evidence demonstrating that food-derived DNA can integrate into the genome of somatic cells and gut bacteria. In addition, because of the absence of evidence indicating that this integration has gene expression/function and because clinical effects were never reported, recent publications and experts' opinion is that it is impossible to describe a consistent sequence of events supporting that gene transfer from food could lead to harm to consumers (FAO and WHO 2023; Nawaz et al. 2019). Last, outcrossing refers to migrating genes from modified plants into conventional crops or related species in the wild. Concerns about the spread of modified organisms in nature also apply to other organisms besides plants, such as microorganisms (bacteria, fungi) and animals. The negative results of the spread of GMOs in the environment vary according to the type of organism. In the case of microorganisms, for example, a concern is spreading bacteria or fungi with pathogenic potential and possible implications for public health (WHO and FAO 2001). In the case of the culture of cells isolated from multicellular organisms, as occurs for cultivated meat, the cell survival outside the controlled environment and consequent environmental spread is not plausible (FAO and WHO 2023). Thus, as anticipated by Ong et al. (2023), the environmental safety issues related to releasing animal cells into the environment need to be evaluated, but significant effects would not be expected.

Another technological feature subjected to different regulatory approaches is the inputs used for cultured media. Cultivated meat still relies on supply chains that exist for the pharmaceutical industry and, because of it, is subject to pharmaceutical industry regulation. Still, these ingredients need to be treated as food ingredients or processing aids, driven to a safety downgrade to a food-grade ingredient. Moreover, the application of pharma-grade inputs in cultivated meat production has prohibitive economic constraints; therefore that needed downgrade should potentially result in cost reduction associated with producing cultivated meat and stimulate innovation, especially in culture media formulations, for which researchers have been focusing on developing elements as recombinant proteins and growth factors (Swartz 2023; Vergeer et al. 2021).

Using ingredients derived from recombinant microorganisms in food production is not a novelty, and this step of the value chain of cultivated meat production offers more established regulatory frameworks globally. For example, Kallscheuer (2018) and Deckers et al. (2020) listed several genetically modified microorganism (GMM) strains approved for food additives and processing aids (enzymes) production in the EU, respectively. Codex Alimentarius established a Guideline that supports the Principles for the Risk Analysis of Foods Derived from Modern Biotechnology and addresses the safety and nutritional aspects of foods produced through the actions of recombinant-DNA microorganisms (FAO/WHO 2003). According to the guideline, the recombinant-DNA microorganisms that are used to produce these foods are typically derived using modern biotechnology techniques from strains that have a history of safe, purposeful use in food production. However, in instances where the recipient strains do not have a history of safe use, their safety must be established. The safety of foods produced using recombinant-DNA microorganisms is assessed relative to the conventional counterparts with a history of safe use, not only for the food produced using a recombinant-DNA microorganism but also for the microorganism itself. Therefore, choosing host microorganisms and target molecules with a history of safe use in food production or safe consumption in foods will avoid complexities in the production approval by regulators. This is an assured first step of the process.

Finally, from a food safety standpoint, scientists generally agree that, based on the controlled conditions in which cultivated meat is produced, it should be less prone to food-safety risks than conventional meat. Thereby, core differences between cultivated and traditional production, such as production environment and the absence of slaughter, can make traditional meat policy considerations, laws and regulations not applicable to cultivated meat (GFI IN 2022). In order to formulate international recommendations, FAO/WHO captured the key food safety issues related to cell-based foods to equip competent authorities worldwide with scientific information to support regulatory actions worldwide (FAO and WHO 2023). From an exhaustive list of potential food safety issues gathered in one of the document sections, the results have indicated that many hazards identified in a production based on cell culture are also present in conventional foods as well as existing control measures and traditional food safety systems are applicable to address those hazards and ensure safe production. Identified hazards not common in the food industry are already reported and managed in industries that apply production techniques and bioprocess similar to cultivated meat, like those found in fermented products, GM foods, and therapeutics and biosimilars industries.

## 5 Product Labeling

The use of the term "meat" for these products, as well as the use of the traditional names for presentations of meat products of animal origin (sausage, hamburger, etc.), are less controversial from the consumer's point of view than plant-based products but will be decided in a more heated political environment, due to the nature of the incumbents.

While for plant-based food analogous to animal products, there is a large consensus on the rules for claims and nutritional information (they must be the same as for animal products), and there is intense controversy over terminology (equal names may or may not be used, followed by suffixes such as plant-based or vegetable), in terms of cultivated meat the opposite is perceived. There seems to be remarkable consensus on the name referring to animal products, differentiated by an identifying suffix such as cultivated or cell-based. The debate on the terminology of analogous products is usually based on avoiding disinformation for the consumer, allowing them to safely make an informed purchasing decision. However, the nature of the incumbents involved is significant in the difference between the points of controversy in the two product categories.

In plant-based products, startups began the terminology debate by launching their products and facing resistance from incumbent companies, arguing that animal and plant-based products were different in their structures. Under a microscope, their component structures are other; therefore, vegetable mimics meat. For a while, until the big animal product companies had plant-based products in their portfolios, it was a David versus Goliath struggle. Cultivated meat, a much more capital-intensive business that drastically reduces the number of active entrants and attracts the big meatpackers right at the start, does not mimic the animal product. On the contrary, it is a product of animal origin, and the difference lies in how it is obtained. Under the microscope, all the structures of the animal product are there, so it is meat. And if there are other structures, the manufacturing process intentionally added them. From the beginning, it has been ultra-capitalized startups or meatpacking giants facing off against cattle raisers, as we explored above when commenting on the geopolitics of Cultivated Meat. In other words, from the start, it's a fight between two different Goliaths.

Regarding terminology, researcher Cristopher Bryant has conducted studies with co-authors covering different countries and regions. Studies covered the United States, India and China, the United Kingdom (Bryant et al. 2019), Germany and France (Bryant et al. 2020), Brazil (Bryant and Krelling 2020), Belgium (Bryant and Sanctorum 2021), the United States and the United Kingdom (Szejda et al. 2021), Brazil, Germany, India, the United Kingdom and the United States (Zollman Thomas and Bryant 2021). A broad study, including seafood, is in Malerich and Bryant (2022). We can summarize that cell-based food should use the terms usually used for animal products in conjunction with a differentiating word, which can vary among cultivated, cultured or cell-based.

However, the United States leads the first consistent survey the regulator itself takes. The US Department of Agriculture (USDA), which since 2019 announced a joint regulation of cultivated foods with the US Food and Drug Administration (FDA), published in September 2021 an advance notice of proposed rulemaking (ANPR) on the labeling issue (USDA 2021). They asked 14 questions about describing these products on packaging labels, especially compared to animal-derived products. Which words work best for this type of product? Which terms would be misleading consumers? A summary of the more than a thousand contributions can be found in previous publications (FoodDive 2022a, b). Overall, the dichotomy between incumbents and newcomers is reflected in their contributions. While some argue in the name of science and the future that products should have the same terminology not to misinform consumers, others go in the opposite direction, arguing that it is necessary to prohibit cultivated products from using the terms because of established standards and the possibility of misinforming consumers. In that case, it is up to the American regulator to decide who is right. According to WHO and FAO (2023), there is no universally accepted term; the most essential thing should be a term understood at the local level.

## 6 Challenges and Future Developments

Cultivated meat is initially thought to obtain the same products traditionally earned through conventional livestock farming, potentially more efficiently and sustainably. However, as the technological tools relevant to its production are explored and improved and the market advances, new opportunities for creating innovative products emerge and could change how this industry will operate. Characteristics inherent to the production method, such as the targeted choice of which and how key elements (cells, scaffolds, and culture media) will be part of the product composition, open new routes to developing entirely new animal-derived food products. As a result, more than mimicking meat pieces of conventional production animals such as cows, chickens, fish and others, there is the chance of producing animal tissues not usually used in human food besides already extinct or exotic animals, for instance (Abrahao et al. 2023; Vow 2023; Independent.co 2022). The mixture of cells from different animals, even in a combination of species that would not generate a viable living being in nature, can create meats with unprecedentedly varied textures, flavors and nutrition, but for which there may be no conventional comparator. Conventional meat also presents a wide range of variations in terms of composition and nutrition. Still, those examples would be new foods for which safety assessments and the regulatory process may be more challenging and will require the construction of collective knowledge involving all stakeholders.

## 7 Conclusions

Science advises policy decisions and is influenced by policymaking. In this chapter, relevant themes discussed this interrelationship in the context of cultivated meat regulation. The conclusions obtained from this chapter are the following:

- The costs associated with the regulatory risk of an underregulated environment are as significant as the extra costs caused by excessive regulation. Therefore, it makes sense to think that regulatory parameters for cultivated meat are defined first in multilateral bodies (like Codex Alimentarius) and then disseminated worldwide.
- From a systemic perspective, the production of cultivated meat can cause disruptive changes mainly in the upstream chain of the current meatpacker. They can verticalize their business and carry out the cell growing stage, dispensing with the cattle raiser for this production, with undeniable gains in traceability and accountability and focusing positive marketing by showing the steps of your process.
- Besides being appropriate to cover the safety assurance of food products, safety regulations could impact the path to market of those products, speeding up or hindering their commercialization. Despite recent advances, food safety data remains essential to assist regulators in establishing regulatory frameworks, reviewing applications, and adjusting their food regulations in response to new scientific evidence.
- The use of the term "meat" for these products and the use of the traditional names for presentations of meat products of animal origin will be decided in a more heated political environment due to the nature of the incumbents.

**Compliance with Ethical Standards** A. C. is the Policy Vice-President, and I.P. and A.L. are employees of The Good Food Institute Brazil, an affiliate of The Good Food Institute (GFI). GFI is a nonprofit think tank funded by philanthropic resources that promotes and accelerates alternative protein innovation. GFI funders had no role in the design of this chapter, in the writing of the manuscript, and/or in the analysis and interpretation of data. The authors have no conflicts of interest to declare.

## References

Abrahao AFS, Rufino JPF, Reis GG, Cabral A (2023) Cultivated manatee meat aiding Amazon biodiversity conservation: discussing a proposed model. Conservation 3(2):303–318
Balsa-Barreiro J, Vié A, Morales AJ, Cebrián M (2020) Deglobalization in a hyper-connected world. Palgrave Commun 6(1):1–4. https://doi.org/10.1057/s41599-020-0403-x
Bogueva D, Marques M, Molento CFM, Marinova D, Phillips CJC (2023) Will the cows and chickens come home? Perspectives of Australian and Brazilian beef and poultry farmers towards diversification. Sustainability 15:12380. https://doi.org/10.3390/su151612380
Bomkamp C et al (2022) Scaffolding biomaterials for 3D cultivated meat: prospects and challenges. Adv Sci 9(3):1–40
Broucke K et al (2023) Cultured meat and challenges ahead: a review on nutritional, technofunctional and sensorial properties, safety and legislation. Meat Sci 195(September 2022):109006. https://doi.org/10.1016/j.meatsci.2022.109006
Bryant C, Krelling F (2020) Alternative proteins in Brazil: nomenclature for plant based & cultured meat. The Good Food Institute
Bryant C, Sanctorum H (2021) Alternative proteins, evolving attitudes: comparing consumer attitudes to plant-based and cultured meat in Belgium in two consecutive years. Appetite 161(February):105161. https://doi.org/10.1016/j.appet.2021.105161
Bryant C et al (2019) A survey of consumer perceptions of plant-based and clean meat in the USA, India, and China. Front Sustain Food Syst 3(February)
Bryant C, van Nek L, Rolland NCM (2020) European markets for cultured meat: a comparison of Germany and France. Foods 9(9):1–15
CAC, Codex Alimentarius Commission. Joint FAO/WHO Food Standards Programme (2022) Food safety aspects of cell-based food. Forty-fifth Session FAO Headquarters, Rome, Italy. 21–25 November and 12–13 December 2022
Chase-Dunn C, Álvarez A, Liao Y (2023) Waves of structural deglobalization: a world-systems perspective. Soc Sci 12(5)
Crowther G (2013) Eating culture: an anthropological guide to food. University of Toronto Press
Deckers M, Deforce D, Fraiture MA, Roosens NHC (2020) Genetically modified microorganisms for industrial food enzyme production: an overview. Foods 9(3)
EFSA NDA Panel (2016) Guidance on the preparation and presentation of an application for authorisation of a novel food in the context of Regulation (EU) 2015/2283. EFSA J 14(11)
FAO (2022a) Gene editing and agrifood systems. Rome
FAO (2022b) Meat market review emerging trends and outlook 2022. Rome. https://www.fao.org/3/cc3164en/cc3164en.pdf
FAO (2023a) Contribution of terrestrial animal source food to healthy diets for improved nutrition and health outcomes—an evidence and policy overview on the state of knowledge and gaps. Rome
FAO (2023b) Gene editing and food safety—technical considerations and potential relevance to the work of codex Alimentarius. Rome
FAO/WHO (2003) Guideline for the conduct of food safety assessment of foods derived from the recombinant-DNA plants. Risk Anal 1–17. http://www.fao.org/docrep/007/y5819e/y5819e04.htm
FAO & WHO (2023) Food safety aspects of cell-based food safety aspects of cell-based food. Rome. https://doi.org/10.4060/cc4855en
FDA (2022) Cell culture consultation notification file: CCC 000002. https://www.fda.gov/media/163260/download
FDA (2023) Cell culture consultation notification file: CCC 000001. https://www.fda.gov/media/166347/download
Flexor (2023) Food challenges, technological changes, and food geopolitics. Rev Econ Sociol Rural 62(3). https://doi.org/10.1590/1806-9479.2022.269655

FoodDive (2022a) How should cell-based meat be labeled? What 1,179 comments to USDA say about stakeholders' priorities. https://www.fooddive.com/news/cell-based-cultivated-meat-comments-usda/623028/
FoodDive (2022b) Tracking the comments on cell-based meat labeling. https://www.fooddive.com/news/cell-based-cultivated-meat-comments-tracker-usda/623608/. 30 Aug 2023
FoodNavigator USA (2021) Mission Barns, and Silva Sausage Test-drive hybrid sausages blending cell-cultured animal fat and plant-based protein. https://www.foodnavigator-usa.com/Article/2021/10/20/Mission-Barns-and-Silva-Sausage-test-drive-hybrid-sausages-blending-cell-cultured-animal-fat-and-plant-based-protein#. 30 Aug 2023
Fraeye I, Kratka M, Vandenburgh H, Thorrez L (2020) Sensorial and nutritional aspects of cultured meat in comparison to traditional meat: much to be inferred. Front Nutr 7(March)
Frezal C, Nenert C, Gay H (2022) OECD Trade and Agriculture Directorate meat protein alternatives opportunities and challenges for food systems' transformation
FSANZ (2023) Statement on cultured quail as a novel food application. https://www.foodstandards.gov.au/media/Pages/Statement-on-Cultured-Quail-as-a-Novel-Food-application%2D%2D.aspx. 20 Aug 2023
GFI (2022) State of global policy report—public investment in alternative proteins to feed a growing world
GFI Europe (2023) Aleph farms make Europe's first cultivated meat approval submission to Swiss regulators. https://gfieurope.org/blog/aleph-farms-make-europes-first-cultivated-meat-approval-submission-to-swiss-regulators/. 30 Aug 2023
GFI IN (2022) The regulatory status of cultivated meat: overview of regulatory best practices with recommendations for India. https://gfi-india.org/wp-content/uploads/2022/03/The-Regulatory-Status-of-Cultivated-Meat.pdf
Giraldo PA et al (2019) Safety assessment of genetically modified feed: is there any difference from food? Front Plant Sci 10(December):1–17
Grasso S (2020) Hybrid meat. Food Sci Technol 34(3):48–51
Grasso S, Jaworska S (2020) Part meat and part plant: are hybrid meat products fad or future? Foods 9(12):1–13
Green Queen (2020) Eat just serves cultured meat to diners at 1880 restaurant Singapore in a world first. https://www.greenqueen.com.hk/eat-just-serves-cultured-meat-to-diners-at-1880-restaurant-singapore-in-a-world-first/. 30 Aug 2023
Guan X et al (2021) Trends and ideas in technology, regulation and public acceptance of cultured meat. Future Foods 3(December 2020):100032. https://doi.org/10.1016/j.fufo.2021.100032
Independent.co (2022) World's first lab-grown lion meat as climate-friendly 'cultivated' food arrives in UK. https://www.independent.co.uk/climate-change/lion-meat-lab-grown-world-first-b2047237.html. 30 Aug 2023
Jakobsen J, Hansen A (2020) Geographies of meatification: an emerging Asian meat complex. Globalizations 17(1):93–109. https://doi.org/10.1080/14747731.2019.1614723
Kallscheuer N (2018) Engineered microorganisms for the production of food additives approved by the European Union—a systematic analysis. Front Microbiol 9(AUG):1–20
Malerich M, Bryant C (2022) Nomenclature of cell-cultivated meat & seafood products. npj Sci Food 6(1)
Miyake Y, Tachikawa M, Kohsaka R (2023) Policy frameworks and regulations for the research and development of cell-based meats: systematic literature review. Food Res Int 167(July 2022):112599. https://doi.org/10.1016/j.foodres.2023.112599
Morais-da-Silva RL et al (2022) The expected impact of cultivated and plant-based meats on jobs: the views of experts from Brazil, the United States and Europe. Humanit Soc Sci Commun 9(1):1–14
Nawaz MA et al (2019) Addressing concerns over the fate of DNA derived from genetically modified food in the human body: a review. Food Chem Toxicol 124:423–430. https://doi.org/10.1016/j.fct.2018.12.030
OECD-FAO (2023) Meat. In agricultural outlook 2021–2030, pp 184–201

Ong KJ et al (2023) Cultured meat safety research priorities: regulatory and governmental perspectives. Foods 12(14):2645
Pakseresht A, Kaliji SA, Canavari M (2022) Review of factors affecting consumer acceptance of cultured meat. Appetite 170:105829. https://doi.org/10.1016/j.appet.2021.105829
Pereira PMDCC, Vicente AFDRB (2013) Meat nutritional composition and nutritive role in the human diet. Meat Sci 93(3):586–592. https://doi.org/10.1016/j.meatsci.2012.09.018
Pollan M (2006) The omnivore's dilemma: a natural history of four meals. Penguin, New York
RESPECTfarms (2023) RESPECTfarms: The first cultivated meat farm. https://respectfarmers.wordpress.com/participate/. 30 Aug 2023
Rozas P, Kessi-Pérez EI, Martínez C (2022) Genetically modified organisms: adapting regulatory frameworks for evolving genome editing technologies. Biol Res 55(1):1–14. https://doi.org/10.1186/s40659-022-00399-x
Singapore Food Agency (2023) 30 by 30—strengthening our food security. https://www.ourfoodfuture.gov.sg/30by30/. 30 Aug 2023
Soice E, Johnston J (2021) Immortalizing cells for human consumption. Int J Mol Sci 22(21):1–14
Steakholder Foods (2023) Steakholder Foods® demonstrates the same nutritional value as beef in its cultivated meat. https://steakholderfoods.com/steakholder-foods-demonstrates-the-same-nutritional-value-as-beef-in-its-cultivated-meat/. 30 Aug 2023
Swartz E (2023) Anticipated growth factor and recombinant protein costs and volumes necessary for cost-competitive cultivated meat. Gfi, January
Szejda K, Bryant CJ, Urbanovich T (2021) US and UK consumer adoption of cultivated meat: a segmentation study. Foods 10(5)
Teaford MF, Ungar PS (2000) Diet and the evolution of the earliest human ancestors. Proc Natl Acad Sci USA 97(25):13506–13511
Upside Foods (2021) Premarket notice for integral tissue cultured poultry meat (C). https://www.fda.gov/media/163262/download
USDA (2021) 86 labeling of meat or poultry products comprised of or containing cultured animal cells. Food Safety and Inspection Service, USDA. https://www.regulations.gov/document/FSIS-2020-0036-0001
Vergeer R, Pelle S, Odegard I (2021) TEA of cultivated meat TEA of cultivated meat. 44. https://cedelft.eu/publications/tea-of-cultivated-meat/
Vieira LR et al (2021) Chapter 5—Regulatory framework of genome editing in Brazil and worldwide. In: CRISPR technology in plant genome editing: biotechnology applied to agriculture, vol 1, pp 169–194
Vow (2023) The mammoth meatball. 2023. https://www.mammothmeatball.com/. 30 Aug 2023
Watanabe F, Bito T (2018) Vitamin B12 sources and microbial interaction. Exp Biol Med 243(2):148–158
WHO (2014) Food, genetically modified—questions and answers. https://www.who.int/news-room/questions-and-answers/item/food-genetically-modified. 30 Aug 2023
WHO & FAO (2001) Report of a Joint FAO/WHO expert consultation on foods derived from biotechnology WHO safety assessment of foods derived from genetically modified microorganisms. Geneva, Switzerland. http://www.embase.com/search/results?subaction=viewrecord&from=export&id=L34113277
WHO & FAO (2023) Nine things to know about: food safety aspects of cell-based food. https://www.fao.org/3/cc6419en/cc6419en.pdf
Witt MA, Li PP, Välikangas L, Lewin AY (2021) Deglobalization and decoupling: game changing consequences? Manag Organ Rev 17(1):6–15
Zollman Thomas O, Bryant C (2021) Don't have a cow, man: consumer acceptance of animal-free dairy products in five countries. Front Sustain Food Syst 5(June):1–14

# Forecasting the Market Potential and Consumer Acceptance of Cultivated Meat

**Marina Sucha Heidemann, Stéphanie Massaki, and Germano Glufke Reis**

**Abstract** The cultivated meat market is expected to grow significantly in the coming years, driven by increased population and demand for sustainable meat, animal welfare concerns, and advances in biotechnology, aligning with evolving consumer values. This chapter provides a comprehensive overview of commercial products utilizing this technology, the cultivated meat market, and consumer acceptance. The market addresses several aspects such as market segmentation, key players, growth projections, and the challenges faced by the industry. Market size projections vary based on various factors, including technological advancements, investments, cost competitiveness, regulatory support, consumer acceptance, and supply. However, the industry faces several challenges, including consumer acceptance, cost, and taste. Consumer acceptance is influenced by several factors, including health and safety concerns, cultural differences, neophobia, sensory characteristics, and price. To overcome these challenges, a comprehensive and holistic approach is required, which should consider the social, economic, regulatory, and environmental impacts of alternative protein systems, as well as the needs and concerns of the workforce and the communities they support. Companies and policymakers should develop strategies to promote the acceptance of cultivated meat and contribute to a more sustainable food system. Sensory attributes, terminology, education, and marketing are crucial to shaping consumer perceptions. Emphasizing the environmental and health benefits, ensuring safety and transparency, and addressing neophobia is essential for the widespread adoption of cultivated meat, paving the way for a more sustainable and ethical food industry. In addition, it is essential to simultaneously provide consumers with information about cultivated meat to increase familiarity by reinforcing perceived benefits, addressing social norms, and ensuring cultural relevance. Overall, the potential benefits of cultivated meat make it an exciting and

M. S. Heidemann (✉)
SFoodLab - Sustainable Food Chains Laboratory, Federal University of Parana, Curitiba, Parana, Brazil

S. Massaki · G. G. Reis
SFoodLab – Sustainable Food Value Chains Laboratory, Curitiba, Parana, Brazil

School of Management, The Federal University of Parana, Curitiba, Parana, Brazil



C. R. Soccol et al. (eds.), *Cultivated Meat*,
https://doi.org/10.1007/978-3-031-55968-6_16

promising area of innovation in the food industry. However, addressing the challenges will require continued research and development, collaboration between industry and regulatory bodies, and consumer education.

**Keywords** Cultivated meat market · Consumer acceptance · Cultivated meat products · Food innovation · Sustainable food systems

## 1 Introduction

The success of cultivated meat depends on several factors, such as consumer acceptance, market potential, and the ability to achieve economies of scale. Understanding consumer perceptions and preferences is crucial for improving products and developing effective communication and positioning strategies (Bryant and Barnett 2020). According to Swartz and Bomkamp (2023), consumer research shows that most consumers are willing to try cultivated meat, especially when they receive information about its production process and the benefits it offers compared to conventional meats, such as sustainability and animal ethics (Bryant and Barnett 2020).

Cultivated meat has the potential not only to match but also to surpass the taste and nutritional value of conventional meat due to the selection of cell lines from specific animals with good flavor and nutritional value, such as salmon (Roth et al. 2023). Additionally, cultivated meat is viewed as a viable option by familiarized consumers who want to adopt more sustainable eating habits without eliminating meat from their diet (Shapiro 2018; Pakseresht et al. 2022). However, as a new product, it is still not well-known among consumers, and companies entering the sector have limited knowledge of how consumers will receive the products and their willingness to pay for them (Roth et al. 2023). For instance, individuals educated regarding the ethical and environmental advantages of cultivated meat are likely to pay a premium price (Pakseresht et al. 2022). Therefore, knowledge concerning cultivated meat technology, from its processes to its benefits, has an important role in consumer acceptance (Mancini and Antonioli 2020; Pakseresht et al. 2022).

Another question regarding the success of cultivated meat is its market potential, especially when compared to animal-based products. If consumers embrace these products, the market for cultivated meat could reach $25 billion by 2030 (Brennan et al. 2021). However, to fully realize its potential, the industry must address health and safety concerns, job displacement, and economic ripple effects (Morais da Silva et al. 2022; Roth et al. 2023).

Companies need to sell a substantial amount and secure a significant market share to fulfill the predictions regarding the popularity of cultivated meat, which will become as common as regular meat by the early 2030s (Stephens et al. 2018). Consultancy firms like the Boston Consulting Group (BCG), McKinsey, and AT Kearney have addressed those aspects, striving to forecast the future of the cultivated meat market. Currently, the industry faces challenges related to culture media cost, bioreactor availability, and tailoring production processes to meet specific requirements (Stephens et al. 2018; Roth et al. 2023). To achieve economies of scale, the cultivated meat industry must prioritize demonstrating commercial viability and scaling up production, while achieving efficient and cost-effective production as well as optimizing supply chain management (Brennan et al. 2021). Maximizing sales and obtaining a significant market share can benefit the industry from economies of scale, thus further benefiting the companies involved (Roth et al. 2023). Also, from a microeconomic perspective, how consumers categorize and interpret this new offer is an important topic to be explored because the ultimate success of cultivated meat depends on consumer acceptance (Pakseresht et al. 2022).

This chapter explores commercial products based on cultivated meat, market projections, and consumer acceptance of cultivated meat products. It initially delineates the diverse product categories that could be introduced to the market and then examines market potential projections developed by AT Kearney, BCG, McKinsey, and Rethink X and their implications. Additionally, the chapter offers a comprehensive overview of consumer research and its evolving trends.

## 2 Commercial Products Based on Cultivated Meat

In 2013, the first cultivated meat burger, made from cow muscle cells grown in the lab, was introduced by a team led by Dr. Mark Post at Maastricht University in the Netherlands (BBC News 2013). This groundbreaking event marked the beginning of the cultivated meat industry. Since then, several companies have emerged as key players in the cultivation of meat products. For instance, Upside Foods (formerly Memphis Meats) focuses on producing beef, chicken, and seafood from cells (Upside Foods 2023). The company was founded in 2015, launched its first cultivated meatball in 2016 (FOX News 2016), and today has one of the greatest cultivated meat factories worldwide (Food Business 2021), and develops meat, chicken, and seafood from cells. Another notable company is GOOD Meat (GOOD Meat 2023), an EAT JUST company, which started in 2016 and produces cultivated chicken nuggets. EAT JUST was already in the market with their plant-based egg (JUST Egg 2023) when they started their cultivated meat history.

In terms of cost, cultivated meat initially had a high price tag due to small-scale production and technical challenges (Stephens et al. 2018). However, advancements and scaling up production processes are expected to significantly reduce costs. Brennan et al. (2021) suggested that by enhancing production processes and investing in research and development, it is feasible to reduce the cost from thousands of dollars per pound to less than $5 per pound. Also, the regulation of cultivated meat has been a topic of discussion in various countries. Singapore was the first to grant regulatory approval for selling GOOD Meat's cultivated chicken nuggets to a premium restaurant in 2020 (CNBC 2020). In 2023, the United States Food and Drug Administration (FDA) allowed two cultivated meat startups, Upside Foods being one of them, to commercialize their products, including GOOD Meat (Wiener-Bronner 2023).

Regarding cultivated meat technology improvement, in recent years, there have been advancements and new products in the cultivated meat space. In 2021, Wildtype unveiled its sushi-grade cultivated salmon (Vegeconomist 2021). In addition, Aleph Farms announced the production of the world's first slaughter-free ribeye steak (Bandoim 2021). This achievement showcased the potential of cultivated meat to replicate complex meat structures.

Looking ahead, there are promising technologies and prospective products on the horizon. The development of 3D bioprinting for cultivated meat holds the potential to create intricate tissues and textures. The use of plant-based scaffolds and edible structures could further enhance the fibrous qualities of cultivated meat (Bomkamp et al. 2022). Furthermore, there is a potential emergence of hybrid products or meat blends, which combine plant-based components or mushrooms, different types of fermentation-based ingredients (e.g., mycoproteins), and cultivated meat (Tarrega et al. 2020; Barone et al. 2021; Sogari et al. 2021; Grasso et al. 2022; Caputo et al. 2023).

Also, several companies, such as BioCraft, Wild Earth, Because Animals, CULT Food, and Science Corp, are actively pursuing the creation of cultivated meat for inclusion in dog and cat food. However, there is a long way to go before this product is accepted. Furthermore, while some individuals who reduce their meat consumption choose to provide plant-based diets to their pets, others have expressed concerns (Oven et al. 2022). In addition, Attwood et al. (2023) explore Halal-certified cultivated meat as a crucial factor that could impact the consumption of cultivated meat among Muslims, who make up a substantial part of the global population. Halal certification is essential for the acceptance of cultivated meat by the Muslim population, serving as a key determinant for the widespread adoption of cultivated meat in the mass market.

In the recent GFI study by O'Donnell and Murray (2023), a notable trend emerges, as the world's most prominent protein producers are showing increasing interest and active involvement in the alternative protein industry, as well as great enthusiasm from investors. The examination extends to the FAIRR (2021), a comprehensive tracker of the 60 largest global meat, dairy, and seafood companies. As of 2022, it is clear that half of these industry giants have started investing in alternative proteins. This trajectory underlines a significant increase in participation in contrast to the figures observed in 2019. In summary, cultivated meat has seen significant progress since its inception in 2013. Companies like Upside Foods and GOOD Meat have played pivotal roles in advancing the field. As costs continue to decrease and regulatory frameworks are established, cultivated meat has the potential to revolutionize the meat industry and provide a sustainable and ethical alternative to traditional animal agriculture. Hence, Fig. 1 exemplifies the diverse range of products derived from cultivated meat.

Products may stem from various cell sources, including cattle, poultry, fish, and pork meat. Additionally, the industry may witness the emergence of hybrid products, blending cultivated meat with plant-based or fermentation-based ingredients. These products can be processed into items such as patties, hamburgers, and meatballs, or they can be complex cuts resembling traditional beef. Moreover, there is

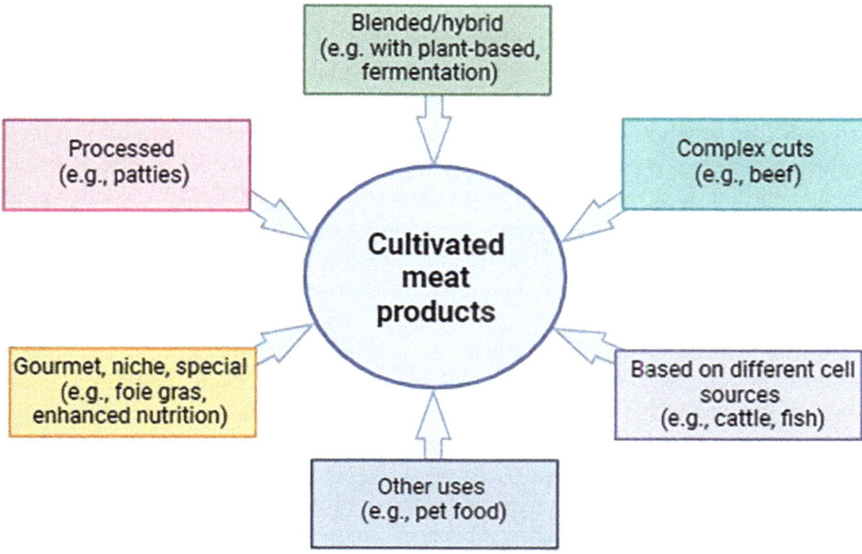


**Fig. 1** Diverse prospects and applications of cultivated meat products. Note: Created in Biorender.com

potential for gourmet, niche, and specialty products like *foie gras*, those with enhanced nutrition, and those adapted to local culinary traditions. Furthermore, cultivated meat can find utility in other sectors, including the pet food industry.

## 3 Market Analysis and Projections for Cultivated Meat Products

The cultivated meat market is expected to experience significant growth in the coming years. Based on AT Kearney (Gerhardt et al. 2020), McKinsey (Brennan et al. 2021), and BCG (Von Koeller et al. 2023) reports, it is projected that cultivated meat and seafood will approach price parity with conventional animal proteins in the early 2030s. Projections for the global market need to consider that developing countries must approach this issue differently from poor countries, mostly because they have different economic priorities. While both may face challenges in meeting the demand for protein-rich diets, developing countries may have more resources to invest in alternative protein production methods. Factors such as economic development, cultural differences, infrastructure, regulatory environment, and consumer preferences may all play a role in shaping the market for alternative protein sources in different regions of the world. In addition, for the main suppliers of conventional meat, such as Brazil and the United States, the transition from one protein source to another may require a longer period.

AT Kearney Consulting forecasts an optimistic scenario in which alternative proteins, including cultured meat, will make up 45% of the global protein market by 2035 and are expected to disrupt the billion-dollar global meat sector and make 60% of the market by 2040 (Gerhardt et al. 2020). Also, it predicts that these new alternatives will gain ground over the next 20 years, potentially rendering traditional meat products a marginal presence, with a market share potentially declining to as low as 40%. According to a more conservative estimate by the BCG (Von Koeller et al. 2023), the share of alternative proteins, including cultivated meat, will be 11% of the global protein market by 2035. Taking a middle path between several projections, FAIRR (2021) reports that the growth of alternative proteins will rise to about 28% of the market by 2035.

Cultivated meat market growth is related to the increasing demand for sustainable protein sources, animal ethics, and advancements in biotechnology (Global Cultivated Meat Market 2022). This novel technology addresses these concerns by eliminating the need to raise and slaughter animals, offering a solution that aligns with changing consumer values (Letti et al. 2021; Heidemann et al. 2020). The forecasts indicate a significant potential for the cultivated meat market to disrupt the traditional meat industry and gain a substantial share of the global protein market in the coming years (Gerhardt et al. 2020; Brennan et al. 2021; Von Koeller et al. 2023). Public and private funding will be crucial for the industry's growth (GFI 2022).

According to McKinsey & Company (Brennan et al. 2021), the cultivated meat industry is projected to contribute to less than 1% of the global meat supply by 2030, with plant-based alternatives making up a significant portion. Also, the authors project that if cellular agri-foods replace processed and other meat products in some Western (United States of America, Canada, Australia, New Zealand, and countries in Europe except for the former Eastern Europe and Turkey) and Asian countries, the market size will grow to \$20 billion by 2030, and to \$25 billion if they spread to other countries with high meat consumption rates. This scenario indicates that the actual growth and market penetration will depend on various factors, including technological advancements, substantial investments, cost competitiveness, regulatory support, consumer acceptance, and supply (Stephens et al. 2018; GFI 2022). The industry will require investments and continued technological innovation to achieve its growth potential (Letti et al. 2021).

As explained by Treich (2021), a pessimistic scenario, as mentioned by McKinsey, for the cultivated meat market is due to the many challenges faced by the industry, including how to lower costs, raise funds to achieve economies of scale and win consumers' trust. Furthermore, a report by Rethink X provides insights into the probability distribution for the cultivated meat market, revealing a prevalent trend toward lower production volumes. The aggregated probabilities from their panel include a 54% probability that less than 100,000 metric tons of cultivated meat (where >51% of the "meat" is produced directly from animal cells) will be produced and sold at any price in 12 months before the end of 2051. In summary, the substantial difference in projections for the cultivated meat industry can be attributed to a combination of differences in methodologies, geographical focus, underlying assumptions, and the dynamic nature of the industry. It is crucial to take these factors into account when assessing and contrasting diverse predictions for nascent markets such as cultivated meat.

The challenges cultivated meat has to circumvent to discourage pessimistic forecasts can be categorized into scientific, technological, market-related, and socio-political and economic factors. Figure 2 exemplifies these challenges for the cultivated meat market and possible solutions. According to Post et al. (2020) and Treich (2021) one of the key scientific challenges is selecting and accessing the right and high-quality cell lines and optimizing the growth conditions. Companies need to develop and maintain cell lines that can be used to produce meat products at scale. Improving the efficiency of cell culture and differentiation processes is crucial for reducing production costs and increasing the scalability of cultivated meat production. This includes optimizing cell growth media, developing better cell lines, and enhancing the differentiation of stem cells into muscle cells. Cell culture media are a sustainability challenge because they often contain animal-based components like fetal bovine serum (FBS), which can introduce contamination risks and unknown substances, and are highly inadequate in terms of animal ethics concerns.

According to Post et al. (2020), the cultivated meat industry faces technological challenges in scaling up production. One of the key challenges is the current reliance on small-scale bioreactors, which may not be conducive to large-scale manufacturing (Post et al. 2020; Hong et al. 2021; Treich 2021; Roy et al. 2023; Santos

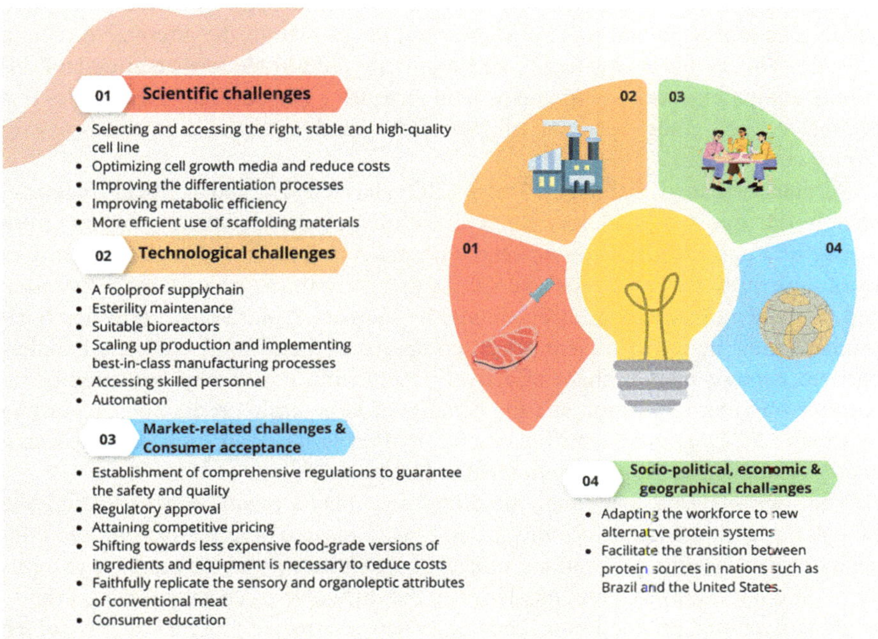


**Fig. 2** Challenges and possible solutions. Source: Elaborated by the authors

et al. 2023). According to several authors, including Post et al. (2020), Lanzoni et al. (2022), and Treich (2021), substantial technological advancements are imperative to render cultivated meat economically competitive with traditional meat. These advancements include enhancing cell media utilization efficiency, reducing costs of media components, growth factors, and scaffolding materials, as well as developing efficient bioreactor systems and cell culture techniques capable of facilitating large-scale production (Stephens et al. 2018). Additional necessities involve minimizing capital expenditure, augmenting maximum cell density and volume, and decreasing production cycle durations. This necessitates research in immortalizing cell lines, media, and scaffolding recycling, and prioritizing the utilization of locally sourced ingredients as inputs (Letti et al. 2021). Moreover, the pursuit of environmentally sustainable cultivated meat production must be a priority, emphasizing minimal resource consumption and waste generation.

By doing so, companies can reduce the cost of facilities, labor, utilities, and input materials per kilogram of finished product. Additionally, using less expensive food-grade ingredients and efficiently recycling the media can reduce costs, according to BCG (Von Koeller et al. 2023). Another important aspect mentioned in this report is replicating the fibrous quality of conventional meats in cultivated meat. This involves adding non-meat ingredients to induce the growing cells to form fibers and meat-like fat. Estimates by von Koeller et al. (2023) suggest that almost 30 million tons of bioreactor capacity will be needed by 2035 to achieve an 11% adoption rate. This would require substantial investment capital.

In the study conducted by Morach et al. (2021), technology development and supply chain play crucial roles in supporting the growth of the alternative protein industry. This includes developing and supplying food-grade fermentation and cell culture equipment, such as bioreactors and measuring devices. Shifting towards less expensive food-grade versions of ingredients and equipment is necessary to reduce costs.

As market-related challenges, Treich (2021) has underscored a critical challenge, namely, the aspect of regulatory approval within the cultivated meat industry. Given the novelty and innovation associated with cultivated meat, the regulatory framework governing its production and sale is presently in the process of formulation and refinement. In addition, as elucidated by Lanzoni et al. (2022), cultivated meat products must be competitively priced compared to their traditional meat counterparts to capture market share effectively. Achieving cost parity with traditional requires ongoing innovation and the benefits of large-scale production. According to Treich (2021), governments are crucial in offering financial support for research and development in the cultivated meat industry.

The socio-political and economic dimensions play a pivotal role in shaping the acceptance and economic viability of cultivated meat. These factors are interconnected with consumer perceptions, ethical considerations, and the products (Morais da Silva et al. 2022). A great challenge is the difficulty faced by people currently working in animal production in adapting to new alternative protein systems, which demand the acquisition of fresh skills and capabilities, as noted by Morais da Silva et al. (2022).

The implementation of alternative protein systems, such as indicated by Treich (2021), can also have significant ethical and environmental implications. Nevertheless, it is essential to acknowledge that advancements within the food industry frequently give rise to unforeseen adverse consequences. This occurs, for instance, when efficiency enhancements in production inadvertently compromise facets like animal welfare, global health, the interests of small-scale producers, or the stability of ecosystems.

According to Lanzoni et al. (2022), an additional market challenge and socio-political aspects relate to the acceptance of cultivated meat by consumers, with certain individuals perceiving it as unconventional or atypical. Morais da Silva et al. (2022) emphasize the need to faithfully replicate the sensory and organoleptic attributes of conventional meat. Nonetheless, insights from consumer studies, as exemplified by Post (2012), demonstrate that a substantial majority of consumers are open to trying cultivated meat, especially when they receive detailed information about its production techniques and associated benefits.

The industry must overcome several obstacles to achieve its full potential, including how to lower costs, raise funds to achieve economies of scale and win consumers' trust (Von Koeller et al. 2023). The main challenge is to bring the cost of production down to a level that is competitive with traditional meat. The industry also needs to win consumers' trust, who may be skeptical about the safety and quality of cultivated meat products. On the other hand, Gerhardt et al. (2020) suggest a reduction in the cost of cultivated meat from \$80 per 100 g in 2018 to below \$4 per 100 g within the next 12 years. Educating society about the benefits of cultivated meat is crucial to ensure consumer acceptance. The funding of cultivated meat companies has exceeded \$50 million, with early-stage investments in these companies.

As outlined by Tubb and Seba (2021), the prospective trajectory of the alternative protein market anticipates the emergence of alternative meat products between 2021 and 2023, with cost-competitive products entering the market by 2025–26. By 2030, there is projected to be a significant reduction in the demand for animal-derived ground beef and tissue beef in the United States, driven by factors such as the direct replacement of traditional products with alternatives and a shift in consumption patterns. Addressing these challenges requires continued research and development, collaboration between industry and regulatory bodies, and consumer education. Despite the obstacles, the potential benefits of cultivated meat make it an exciting and promising area of innovation in the food industry.

## 4 Consumer Acceptance

Consumer acceptance of cultivated meat is a crucial barrier to its mass-market adoption. It is influenced by various factors, including health and safety concerns, cultural differences, neophobia, sensory characteristics, and price (Pakseresht et al. 2022). Understanding these drivers (Fig. 3) can help companies and policymakers

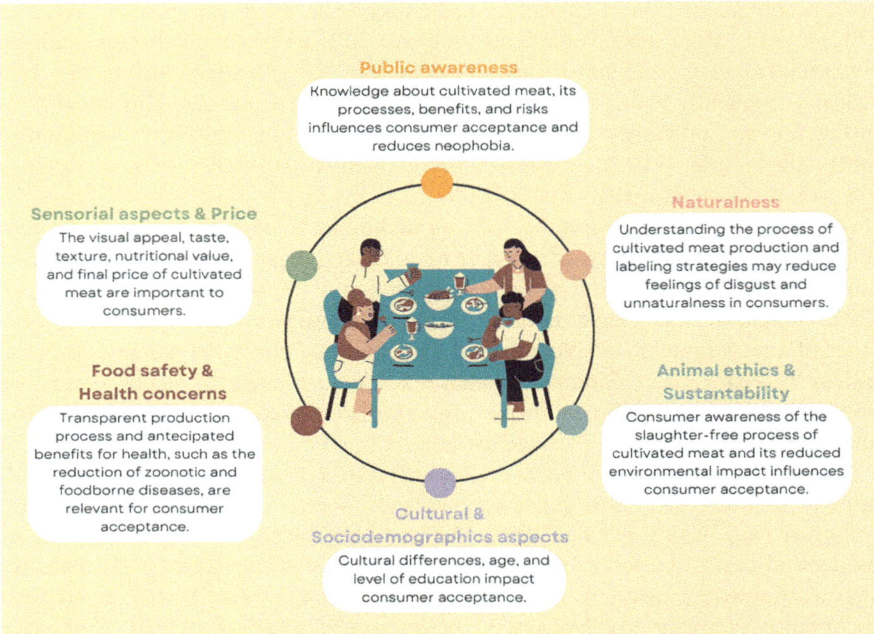


**Fig. 3** Consumer acceptance of cultivated meat. Source: Elaborated by the authors

develop strategies to promote the acceptance of cultivated meat and contribute to a more sustainable food system.

The pivotal moment of the first cultivated meat burger prototype marked the emergence of a new industry, with numerous companies striving not only to develop their own cultivated meat products but also to enhance their ability to mimic the sensory qualities of conventional meat. The sensory attributes of meat hold great importance for consumers, as the acceptance of cultivated meat hinges on its ability to exhibit positive sensory characteristics, including visual appeal, flavor, texture, and nutritional value (Pakseresht et al. 2022; Lin-Hi et al. 2023). Surveys of consumer preferences (Profeta et al. 2021; Caputo et al. 2023) indicate that conventional meat products are perceived to be tastier than alternative meats, including blended products (McNamara 2023).

In the case of Dr. Post's hamburger, the issue was the lack of fat. Meat consists not only of muscle tissue but also fat, incorporating juiciness and flavor (Duhaime-Ross 2013), collagen, and other components (Fraeye et al. 2020). Therefore, cultivated meat production aims to replicate all parts of meat. Currently, several companies are working on producing not only muscle but also fat and other cells together (Aleph Farms) or separately (Cellva).

Another important characteristic to consider is texture. Initially, cultivated meat products were mostly processed. However, in 2015, Aleph Farms made a breakthrough by creating structured meat that resembled small beef parts. They used the scaffolding technique to stimulate fiber layer development and enable the production of meat cuts. More recently, companies have been using 3D techniques to print muscle and other tissues together, creating a complete piece of meat. Each print needle contains one type of cell, and the layers are formed with these tinted cells, resulting in improved marbling, structure, and texture of the final product (Bomkamp et al. 2022).

However, another important aspect for consumers is the sense of naturalness. Consumers' spontaneous reactions to the idea of laboratory-produced meat are the perception of unnaturalness and disgust (Siegrist and Hartmann 2020). Factors such as the perceived naturalness, familiarity, and taste of the product are just as important as carefully framing and communicating farmed meat to potential consumers (Fidder and Graça 2023). Therefore, how cultivated meat is described and labeled plays a crucial role in shaping consumer perceptions. Initially, cultivated meat was referred to as artificial, lab-grown, or synthetic. However, it is important to highlight the similarities between cultivated meat and conventional meat rather than focusing on the technical production process (Siegrist and Hartmann 2020). The choice of name is also a marketing strategy. In the study conducted by Szejda et al. (2021), it was determined that terms such as "cultivated" or "cultured" exhibit a higher degree of acceptability compared to expressions like "cell-based" or "cell-cultured." This conclusion was reached through a comprehensive assessment encompassing various criteria, including neutrality, attractiveness, descriptiveness, and distinctiveness from other categories of meat (Szejda et al. 2019, 2021).

Another significant issue is the lack of consumer familiarity with cultivated meat, which requires education and exposure. As reported by GFI (2022) a large percentage of consumers (ranging from 38 to 64%) admit to being unfamiliar with cultivated meat. To combat neophobia, which is the apprehension toward novel concepts and fear of novel things (Baum et al. 2023), it is crucial to provide education and opportunities for individuals to become acquainted with this food technology, as suggested by Palmieri et al. (2020). One effective approach may involve comparing cultivated meat production to conventional animal production (Siegrist and Hartmann 2020), highlighting artificial techniques such as insemination and feeding. In addition, other marketing benefits of cultivated meat include environmental and health implications compared to traditional meat production (Siddiqui et al. 2022). Additionally, Baum et al. (2021) noted that implicit and explicit attitudes serially mediated the relationship between information provision and purchasing evaluations, thereby affecting the overall effectiveness of information. Similarly, Rolland et al. (2020) report a surprising 58% acceptance of cultivated meat in the Netherlands when previous information is provided.

Moreover, consumer willingness to try, purchase, and pay a premium for cultivated meat is positively influenced by factors such as food curiosity, perceived meat importance, and the view of cultivated meat as a realistic alternative to conventional meat (Rolland et al. 2020; Rombach et al. 2022). For instance, in 2022, GFI-APAC published the report "What do Asian consumers crave in alternative seafood?", compiling unique data on consumer sentiments regarding alternative seafood products in key Asian markets: Singapore, Thailand, South Korea, and Japan. The results revealed that local consumers are curious about cultivated seafood but expect these products to meet or exceed the taste, texture, freshness, and health benefits of conventional seafood. Studies report that implicit and explicit attitudes serially mediated the relationship between information provision and purchasing evaluations (Baum et al. 2021), and acceptance may change with a lack of prior knowledge (Baum et al. 2022).

As a sustainable and slaughter-free process, cultivated meat significantly reduces environmental impact and the need for animals in meat production, as it starts with cells (Heidemann et al. 2020). The pre-knowledge of the cultivated meat process, in this sense, is reported to be a stronger positive driver in consumer acceptance (Weinrich et al. 2020; Palmieri et al. 2020). Additionally, consumers are increasingly drawn to the environmental benefits and ethical aspects of alternative proteins. Consumers who prioritize ethics tend to be more open to trying and regularly consuming cultivated meat (Siegrist and Hartmann 2020) and even pay more for it (Pakseresht et al. 2022). Highlighting the environmental and health implications of cultivated meat compared to traditional meat production is advantageous (Palmieri et al. 2020; Siddiqui et al. 2022). In addition, the sustainability aspect of cultivated meat positively influences consumers' purchase intentions (Silva and Semprebon 2021).

Also, factors such as the anticipated risk of zoonotic diseases, perceived healthiness, and food safety conditions have been identified as important attributes influencing consumer acceptance (de Oliveira et al. 2021). Transparent production processes, high safety, health standards, and government assurances are important requirements for cultivated meat to gain consumer trust (Szejda et al. 2021). There are uncertainties regarding safety and health that may create resistance towards cultivated meat (Pakseresht et al. 2022), and this can be addressed through education and institutional efforts, as well as effective marketing and labeling strategies.

As cultivated meat is seen as unlikely to possess a price advantage, competitiveness can only be realized through non-price-related determinants, such as health (Tiberius et al. 2019). Consumer market price is a major barrier to the consumption of cultivated meat (Verbeke et al. 2015; Mancini and Antonioli 2020). However, arguing that cultivated meat will form part of a varied landscape of future protein sources and that cultivated meat is a solution to existing food safety problems may be an effective approach to increasing acceptance (Bryant and Barnett 2020).

Demographic factors such as gender, age, income, and education level have been found to influence consumer acceptance of cultivated meat. Younger consumers (Szejda et al. 2021), men, and those who consume meat more frequently tend to be more accepting of cultivated meat (Chriki et al. 2021). But the perception of whether cultivated meat is (un)natural does not depend on a person's age (Chong et al. 2023). Furthermore, being informed about meat alternatives, having prior experience with them, possessing a higher level of education, having a higher income, and having children all increase individuals' willingness to pay to try or consume cultivated meat (Siddiqui et al. 2022; Chen et al. 2023). Nevertheless, with increasing age, there is a decrease in willingness to pay for these alternatives (Chen et al. 2023). On the other hand, meat professionals and consumers with higher incomes may have a lower willingness to consume cultivated meat regularly (Chriki et al. 2021). Meat professionals, in particular, may be resistant to cultivated meat due to potential threats to animal production and a lack of perceived market opportunities (Heidemann et al. 2020).

Additionally, cultural differences also have a significant impact on consumer acceptance. Research conducted in China, England, France, Germany, Mexico, South Africa, Spain, Sweden, and the United States has revealed differences in acceptance levels, with French consumers being less accepting of cultivated meat compared to consumers in other countries (Siegrist and Hartmann 2020). Factors such as the perceived naturalness of cultivated meat, trust in the food industry, food neophobia, and food disgust sensitivity have been identified as influencers of acceptance across different cultural contexts (Siegrist and Hartmann 2020). Food technology neophobia is a determining factor in consumer rejection and acceptance of cultivated meat (Baum et al. 2023). Fu et al. (2023) conclude that care must be taken when using technical terminology to avoid triggering resistance related to food technology neophobia. The results of the study by Shen and Chen (2020) indicated that clean production practices and food safety increase consumers' willingness to buy cultivated meat, as opposed to the perception of naturalness.

The increased willingness can be driven by the perception that cultivated meat is on par with real meat in terms of health, safety, sensory quality, and societal benefits (Leung et al. 2023). Effective communication regarding the production and advantages of cultivated meat can enhance consumer acceptance (Rosenfeld and Tomiyama 2023).

Therefore, cultivated meat offers a sustainable and ethical alternative to traditional meat production. However, consumer acceptance is a pivotal factor in its success. Sensory attributes, terminology, education, and marketing play key roles in shaping consumer perceptions. Demographic factors and cultural differences also impact acceptance. Emphasizing the environmental and health benefits, ensuring safety and transparency, and addressing neophobia are essential for the widespread adoption of cultivated meat, paving the way for a more sustainable and ethical food industry. In addition, it is essential to simultaneously provide consumers with information about cultivated meat to increase familiarity by reinforcing perceived benefits, addressing social norms, and ensuring cultural relevance (Siddiqui et al. 2022; Rosenfeld and Tomiyama 2023). When individuals are provided with information about the nutritional content of the product and the environmental advantages it offers, their willingness to pay for it increases (Chen et al. 2023). These efforts are likely to be most effective when concerns about the price, taste, and the satiating capacity of cultivated meat are also addressed (Rosenfeld and Tomiyama 2023).

## 5 Concluding Remarks

In summary, the cultivated meat market is expected to grow significantly in the coming years, driven by increased demand for sustainable meat coupled with the presence of key players in the region. The benefits of alternative proteins are clear, and the industry has the potential to become a major player in the global food market. Cultivated meat represents a promising and innovative solution to some environmental, ethical, and resource-related challenges associated with traditional meat production. As technology and investment continue to advance, cultivated meat will likely play an increasingly significant role in meeting the protein needs of a growing global population. However, overcoming cost, taste, and consumer acceptance challenges will be critical for its continued growth and success.

To address the challenges, a comprehensive and holistic approach is required. This approach should consider the social, economic, regulatory, and environmental impacts of alternative protein systems, as well as the needs and concerns of the workforce and the communities they support. By working together, stakeholders can navigate the transition to a more sustainable and resilient food system while ensuring the well-being of those affected by these disruptive innovations.

However, consumer acceptance of cultivated meat may be a crucial barrier to its mass-market adoption. The review of factors affecting consumer acceptance emphasizes that consumer acceptance is influenced by various factors, including health and safety concerns, cultural differences, neophobia, sensory characteristics, and price. Understanding these drivers can help companies and policymakers develop strategies to promote the acceptance of cultivated meat and contribute to a more sustainable food system.

Sensory attributes, terminology, education, and marketing play key roles in shaping consumer perceptions. Demographic factors and cultural differences also impact acceptance. Emphasizing the environmental and health benefits, ensuring safety and transparency, and addressing neophobia are essential for the widespread adoption of cultivated meat, paving the way for a more sustainable and ethical food industry. In addition, it is essential to simultaneously provide consumers with information about cultivated meat to increase familiarity by reinforcing perceived benefits, addressing social norms, and ensuring cultural relevance.

## References

Attwood S, Jameel S, Fuseini A, AlKhalawi E, Hajat C (2023) Halal cultivated meat: an untapped opportunity. Front Nutr 10:1196475

Bandoim L (2021) World's first 3D bioprinted and cultivated Ribeye steak is revealed. Forbes https://www.forbes.com/sites/lanabandoim/2021/02/12/worlds-first-3d-bioprinted-and-cultivated-ribeye-steak-is-revealed/?sh=3bfd919f4781 Accessed 2 Nov 2023

Barone AM, Banovic M, Asioli D, Wallace E, Ruiz-Capillas C, Grasso S (2021) The usual suspect: How to co-create healthier meat products. Food Res Int 143:110304. ISSN 0963-9969. https://doi.org/10.1016/j.foodres.2021.110304
Baum CM, Bröring S, Lagerkvist CJ (2021) Information, attitudes, and consumer evaluations of cultivated meat. Food Qual Prefer 92:104226
Baum CM, Verbeke W, De Steur H (2022) Turning your weakness into my strength: How counter-messaging on conventional meat influences acceptance of cultured meat. Food Qual Prefer 97:104485
Baum CM, De Steur H, Lagerkvist CJ (2023) First impressions and food technology neophobia: Examining the role of visual information for consumer evaluations of cultivated meat. Food Qual Prefer 110:104957
BBC News (2013) World's first lab-grown burger is eaten in London. https://www.bbc.com/news/science-environment-23576143. Accessed 2 Nov 2023
Bomkamp C, Skaalure SC, Fernando GF, Ben-Arye T, Swartz EW, Specht EA (2022) Scaffolding biomaterials for 3D cultivated meat: prospects and challenges. Adv Sci 9(3):2102908
Brennan T, Katz J, Spencer B (2021) Cultivated meat: Out of the lab, into the frying pan. McKinsey & Company, 16
Bryant C, Barnett J (2020) Consumer acceptance of cultured meat: an updated review (2018–2020). Appl Sci 10:5201. https://doi.org/10.3390/app10155201
Caputo V, Sogari G, Van Loo EJ (2023) Do plant-based and blend meat alternatives taste like meat? A combined sensory and choice experiment study. Appl Econ Perspect Policy 45(1). https://doi.org/10.1002/aepp.13247
Chen B, Zhou G, Hu Y (2023) Estimating consumers' willingness to pay for plant-based meat and cultured meat in China. Food Qual Prefer 111:104962
Chong M, Fernandez TM, Ng ST (2023) Effects of framing, nomenclature, and aversion to tampering with nature on consumer acceptance of cultivated meat in Singapore. J Environ Psychol 91:1–16
Chriki S, Payet V, Pflanzer SB, Ellies-Oury MP, Liu J, Hocquette É et al (2021) Brazilian consumers' attitudes towards so-called "cell-based meat". Foods 10(11):2588
CNBC (2020) This restaurant will be the first ever to serve lab-grown chicken (for $23). https://www.cnbc.com/2020/12/18/singapore-restaurant-first-ever-to-serve-eat-just-lab-grown-chicken.html. Accessed 2 Nov 2023
de Oliveira GA, Domingues CHDF, Borges JAR (2021) Analyzing the importance of attributes for Brazilian consumers to replace conventional beef with cultured meat. PLoS One 16(5):e0251432
Duhaime-Ross A (2013) Test-tube burger: lab-cultured meat passes taste test (sort of). Sci Am. https://www.scientificamerican.com/article/test-tube-burger-lab-culture/
FAIRR (2021) Appetite for disruption: the last serving. 10 Sept 2021. FAIRR
Fidder L, Graça J (2023) Aligning cultivated meat with conventional meat consumption practices increases expected tastefulness, naturalness, and familiarity. Food Qual Prefer 109. https://doi.org/10.1016/j.foodqual.2023.104911
Food Business (2021) Upside Foods opens cultivated meat production facility. https://www.foodbusinessnews.net/articles/19980-upside-foods-opens-cultivated-meat-production-facility. Accessed 2 Nov 2023
FOX News (2016) 'World's first' lab-grown meatball revealed. https://www.foxnews.com/food-drink/worlds-first-lab-grown-meatball-revealed. Accessed 2 Nov 2023
Fraeye I, Kratka M, Vandenburgh H, Thorrez L (2020) Sensorial and nutritional aspects of cultured meat in comparison to traditional meat: much to be inferred. Front Nutr 7:35
Fu W, Zhang H, Whaley JE, Kim Y-K (2023) Do consumers perceive cultivated meat as a sustainable substitute to conventional meat? Assessing the facilitators and inhibitors of cultivated meat acceptance. Sustainability 15(15):11722. https://doi.org/10.3390/su151511722

Gerhardt C, Suhlmann G, Ziemßen F, Donnan D, Warschun M, Kühnle HJ (2020) How will cultured meat and meat alternatives disrupt the agricultural and food industry? Ind Biotechnol 16(5):262–270
GFI (2022) State of the industry report cultivated meat and seafood. https://gfi.org/wp-content/uploads/2023/01/2022-Cultivated-Meat-State-of-the-Industry-Report.pdf
GFI-APAC (2022) What do Asian consumers crave in alternative seafood? https://gfi-apac.org/industry/what-do-asian-consumers-crave-in-alternative-seafood/
Global Cultivated Meat Market (2022) Verified market research, October 2022. https://www.verifiedmarketresearch.com/product/cultivated-meat-market/
GOOD Meat (2023) How we make meat. https://www.goodmeat.co/process. Accessed 2 Nov 2023
Grasso S, Asioli D, Smith R (2022) Consumer co-creation of hybrid meat products: A cross-country European survey. Food Quality Prefer 100:104586. ISSN 0950-3293. https://doi.org/10.1016/j.foodqual.2022.104586
Heidemann MS, Molento CFM, Reis GG, Phillips CJC (2020) Uncoupling meat from animal slaughter and its impacts on human-animal relationships. Front Psychol 11:1824
Hong TK, Shin DM, Choi J, Do JT, Han SG (2021) Current issues and technical advances in cultured meat production: a review. Food Sci Anim Resour 41(3):355–372. https://doi.org/10.5851/kosfa.2021.e14
JUST Egg (2023) A new way to egg. https://www.ju.st/eat/just-egg. Accessed 2 Nov 2023
Lanzoni D, Bracco F, Cheli F, Colosimo BM, Moscatelli D, Baldi A, Rebucci R, Giromini C (2022) Biotechnological and technical challenges related to cultured meat production. Appl Sci 12:6771. https://doi.org/10.3390/app12136771
Letti LAJ, Karp SG, Molento CFM, Colonia BSO, Boschero RA, Soccol VT et al (2021) Cultivated meat: Recent technological developments, current market and future challenges. Biotechnol Res Innov J 5(1)
Leung AKY, Chong M, Fernandez TM, Ng ST (2023) Higher well-being individuals are more receptive to cultivated meat: An investigation of their reasoning for consuming cultivated meat. Appetite 184:106496
Lin-Hi N, Reimer M, Schäfer K et al (2023) Consumer acceptance of cultured meat: an empirical analysis of the role of organizational factors. J Bus Econ 93:707–746. https://doi.org/10.1007/s11573-022-01127-3
Mancini MC, Antonioli F (2020) To what extent are consumers' perception and acceptance of alternative meat production systems affected by information? The case of cultured meat. Animals 10:656. https://doi.org/10.3390/ani10040656
McNamara E (2023) Hybrid products to optimize nutrition, taste, cost, and sustainability. GFI. https://gfi.org/solutions/hybrids-blends-nutrition-taste-cost-sustainability/
Morach B, Witte B, Walker D, von Koeller E, Grosse-Holz F, Rogg J et al (2021) Food for thought: the protein transformation. Ind Biotechnol 17(3):125–133
Morais da Silva RL et al (2022) The social impacts of a transition from conventional to cultivated and plant-based meats: Evidence from Brazil. Food Policy 111:102337
O'Donnell M, Murray S (2023) A deeper dive into alternative protein investments in 2022: The case for optimism. GFI, 16 Feb 2023. https://gfi.org/blog/alternative-protein-investments-update-and-outlook/
Oven A, Yoxon B, Milburn J (2022) Investigating the market for cultivated meat as petfood: A survey analysis. PLoS One 17(12):e0275009. https://doi.org/10.1371/journal.pone.0275009
Pakseresht S, Bogueva D, Marinova D (2022) Consumer acceptance of cultivated meat: A systematic review. Trends Food Sci Technol 118:1–14
Palmieri N, Perito MA, Lupi C (2020) Consumer acceptance of cultured meat: Some hints from Italy. Br Food J 123(1):109–123. https://doi.org/10.1108/BFJ-02-2020-0092
Profeta A, Baune MC, Smetana S, Broucke K, Van Royen G, Weiss J et al (2021) Consumer preferences for meat blended with plant proteins–Empirical findings from Belgium. Future Foods 4:100088. https://doi.org/10.1016/j.fufo.2021.100088

Post MJ (2012) Cultured meat from stem cells: challenges and prospects Meat Sci 92(3):297–301. https://doi.org/10.1016/j.meatsci.2012.04.008
Post MJ, Levenberg S, Kaplan DL et al (2020) Scientific, sustainability and regulatory challenges of cultured meat. Nat Food 1:403–415. https://doi.org/10.1038/s43016-020-0112-z
Rolland NC, Markus CR, Post MJ (2020) The effect of information content on acceptance of cultured meat in a tasting context. PLoS One 15(4):e0231176
Rombach M, Dean D, Vriesekoop F, de Koning W, Aguiar LK, Anderson M et al (2022) Is cultured meat a promising consumer alternative? Exploring key factors determining consumer's willingness to try, buy and pay a premium for cultured meat. Appetite 179:106307. https://doi.org/10.1016/j.appet.2022.106307
Rosenfeld DL, Tomiyama AJ (2023) Toward consumer acceptance of cultured meat. Trends Cogn Sci 27(8):689–691. https://doi.org/10.1016/j.tics.2023.05.002. ISSN 1364-6613. https://www.sciencedirect.com/science/article/pii/S1364661323001195?via%3Dihub
Roth A, Katz J, Brennan T (2023) What is cultivated meat? McKinsey & Company, 13 Sept 2023. https://www.mckinsey.com/featured-insights/mckinsey-explainers/what-is-cultivated-meat
Roy NK, Panda S, Dey G (2023) Engineering a sustainable protein revolution: Recent advances in cultured meat production. Food Bioeng 1–16. https://doi.org/10.1002/fbe2.12066
Santos ACA, Camarena DEM, Roncoli Reigado G, Chambergo FS, Nunes VA, Trindade MA, Stuchi Maria-Engler S (2023) Tissue engineering challenges for cultivated meat to meet the real demand of a global market. Int J Mol Sci 24(7):6033. https://doi.org/10.3390/ijms24076033
Shapiro P (2018) Clean meat: how growing meat without animals will revolutionize dinner and the world. Simon and Schuster
Shen Y-C, Chen H-S (2020) Exploring consumers' purchase intention of an innovation of the agri-food industry: a case of artificial meat. Foods 9(6):745. https://doi.org/10.3390/foods9060745
Siddiqui SA, Alvi T, Sameen A, Khan S, Blinov AV, Nagdalian AA, Mehdizadeh M, Adli DN, Onwezen M (2022) Consumer acceptance of alternative proteins: a systematic review of current alternative protein sources and interventions adapted to increase their acceptability. Sustainability 14:15370. https://doi.org/10.3390/su142215370
Siegrist M, Hartmann C (2020) Perceived naturalness, disgust, trust and food neophobia as predictors of cultured meat acceptance in ten countries. Appetite 155:104814. https://doi.org/10.1016/j.appet.2020.104814
Silva CPD, Semprebon E (2021) How about Cultivated Meat? the effect of sustainability appeal, environmental awareness and consumption context on consumers' intention to purchase. J Food Prod Mark 27(3):142–156
Sogari G, Li J, Wang Q, Lefebvre M, Gómez MI, Mora C (2021) Factors influencing the intention to purchase meat-mushroom blended burgers among college students. Food Quality Prefer 90:104169. ISSN 0950-3293. https://doi.org/10.1016/j.foodqual.2020.104169
Stephens N, Di Silvio L, Dunsford I, Ellis M, Glencross A, Sexton A (2018) Bringing cultured meat to market: Technical, socio-political, and regulatory challenges in cellular agriculture. Trends Food Sci Technol 78:155–166. https://doi.org/10.1016/j.tifs.2018.04.010
Swartz E, Bomkamp C (2023) The science of cultivated meat. GFI. https://gfi.org/science/the-science-of-cultivated-meat/
Szejda K, Allen M, Cull A, Banisch A, Stuckey B, Dillard C, Urbanovich T (2019) Meat cultivation: embracing the science of nature. The Good Food Institute, Washington, DC
Szejda K, Bryant CJ, Urbanovich T (2021) US and UK consumer adoption of cultivated meat: a segmentation study. Foods 10(5):1050
Tarrega A, Rizo A, Murciano A, Laguna L, Fiszman S (2020) Are mixed meat and vegetable protein products good alternatives for reducing meat consumption? A case study with burgers. Curr Res Food Sci 3:30–40. ISSN 2665-9271. https://doi.org/10.1016/j.crfs.2020.02.003
Tiberius V, Borning J, Seeler S (2019) Setting the table for meat consumers: an international Delphi study on in vitro meat. npj Sci Food 3(1):10
Treich N (2021) Cultured meat: promises and challenges. Environ Resour Econ 79:33–61. https://doi.org/10.1007/s10640-021-00551-3

Tubb C, Seba T (2021) Rethinking food and agriculture 2020-2030: the second domestication of plants and animals, the disruption of the cow, and the collapse of industrial livestock farming. Rethink X. Ind Biotechnol 17(2):57–72

Upside Foods. (2023). Upside Foods. Available at: https://upsidefoods.com/. Accessed 8 Mar. 2024.

Vegeconomist (2021) Wildtype unveils "world's first" cell-cultured salmon pilot plant and sushi bar. https://vegconomist.com/cultivated-cell-cultured-biotechnology/wildtype-unveils-worlds-first-cell-cultured-salmon-pilot-plant-and-sushi-bar/. Accessed 2 Nov 2023

Verbeke W, Sans P, Van Loo EJ (2015) Challenges and prospects for consumer acceptance of cultured meat. J Integr Agric 14(2):285–294. ISSN 2095-3119. https://doi.org/10.1016/S2095-3119(14)60884-4

Von Koeller E, Ravi N, Tanovic E, Taylor L, Clausen M (2023) Taking alternative proteins mainstream. Alternative proteins report. Boston Consulting Group, 22 Feb 2023. https://www.bcg.com/publications/2023/taking-alternative-protein-trends-mainstream

Weinrich R, Strack M, Neugebauer F (2020) Consumer acceptance of cultured meat in Germany. Meat Sci 162:107924

Wiener-Bronner D (2023) Lab-grown meat is cleared for sale in the United States. CNN Business. Available at: https://edition.cnn.com/2023/06/21/business/cultivated-meat-us-approval/index.html. Accessed 8 Mar. 2024

# Cultivated Meat and Bioeconomy in Tandem: Unlocking the Sustainability Implications of Transformative Food Supply Chains

**Vinicius Picanço Rodrigues and Mariel Alem Fonseca**

**Abstract** Food production and consumption systems are at the center of the sustainability debate. The alternative meat sector is emerging as a promising response to food production and consumption's overwhelming environmental and social impacts, such as high carbon emissions, biodiversity destruction, pollution, food insecurity, and animal suffering. This chapter provides an in-depth exploration of the transformative view of cultivated meat supply chains when intertwined with the concepts and principles of the bioeconomy. By analyzing their fundamental connections, we explore the broad sustainability implications that arise from this relationship. As the field of cultivated meat production evolves, understanding its impacts and its profound implications becomes critical. In addition to contributing to the ongoing dialogue on these emerging issues, this chapter offers a potential blueprint for framing the strategic interplay between the development of cultivated supply chains, the bioeconomy, and the ensuing sustainability implications. This conceptual incursion into many emerging areas underscores the importance of these interactions, particularly in the context of future food production, consumption systems, and overarching global sustainability goals.

**Keywords** Alternative meat · Sustainable protein · Sustainable development · Transformative supply chains

---

V. Picanço Rodrigues (✉)
Department of Design, Manufacturing and Engineering Management, University of Strathclyde, Glasgow, UK

Insper Institute of Education and Research, São Paulo, Brazil

School of Arts, Sciences and Humanities, University of São Paulo, São Paulo, Brazil
e-mail: vinicius.picanco-rodrigues@strath.ac.uk

M. A. Fonseca
Department of Engineering, School of Technology, Institute for Manufacturing (IfM), University of Cambridge, Cambridge, UK


C. R. Soccol et al. (eds.), *Cultivated Meat*,
https://doi.org/10.1007/978-3-031-55968-6_17

## 1 Introduction

Cultivated meat production embodies the potential of a major disruptive innovation in food systems, with far-reaching implications for animals, society and the environment (Lewis 2022; Negulescu et al. 2023). Still in its early days, cell-based production technologies have been projected to grow steadily from a niche, expensive market of limited products in the next few years to a more affordable and available product range over time, which are *poised to boom* (Reis et al. 2020). Cultivated meat uses biotechnology and biological processes to transform unsustainable conventional meat production, which has major environmental impacts. Widespread adoption of cultivated meat could greatly reduce the land, water, feed and other resources required for animal agriculture (Bryant 2020; O'Neill et al. 2021). This would potentially restrain traditional meat production's massive contribution to climate change, biodiversity loss and land degradation.

Recent developments have lit up enthusiasm in the sector. In December 2020, GOOD Meat in Singapore became the first company to be granted regulatory approval for commercializing cultivated meat. With that, Singapore witnessed the debut of GOOD Meat's cultivated chicken in several of its renowned hawker stalls. This move underscores the adaptability of cultivated meat in culinary contexts (GOOD Meat 2022; Shu 2023). In the US, the Food and Drug Administration (FDA) finalized its inaugural pre-market review for a cultivated meat product, specifically UPSIDE Foods' cultivated chicken meat (UPSIDE Foods 2023; US Food and Drug Administration 2023). In July 2023, Aleph Farms became the first company to apply for permission to commercialize cultivated meat in Europe after submitting a safety dossier to regulators in Switzerland (The Good Food Institute 2023a). This development is significant, as several other reviews are still underway; the successful completion of this review not only pushes cultivated meat closer to being a staple on consumer tables but also acts as a crucial signal for investors, indicating a reduced risk for investment in the cultivated meat sector (The Good Food Institute 2023b).

Turning to the investment landscape, 2022 was a milestone year. Both UPSIDE Foods and Wildtype clinched the largest investment deals ever recorded in the cultivated meat and seafood sectors, respectively. Moreover, the pool of unique investors in this arena expanded by 19%, reaching 679 investors. Notably, when comparing 2022 to the previous year, which had already set funding records, the median deal sizes in 2022 were generally more substantial (The Good Food Institute 2023b).

Metaphors come in handy to explain the rationale behind the buzz. Drawing from a mixed set of qualitative materials, Broad (2020) thoroughly documented two main metaphors used by advocates of plant-based and cultivated meats. The *meat is made* metaphor redefines meat as a manufactured product rather than something requiring animal slaughter, fundamentally decoupling the production of meat from traditional animal agriculture. This metaphor suggests meat can be reproduced molecularly and sensorially using the state-of-the-art in food science, big data and biotechnology. The *market for alternative proteins* metaphor centralizes capitalist economic activity, enabling additional supporting metaphors about innovation, investment, industry collaboration, behavioral economics, and marketing as key drivers of cultivated meat adoption. As these metaphors have roots in (1) the biology of meat and (2) the traditional mechanisms of the market economy, the concept of bioeconomy emerges as a fundamental structure underpinning the development of cultivated meat.

These metaphors align with a broader bioeconomy's political and technological rationale, which seeks to widen the concept of the bioeconomy beyond a technoscientific standpoint (Goven and Pavone 2015; Broad 2020). Overall, these metaphors documented by Broad (2020) conceive of meat production shifting away from animal agriculture toward alternative protein meats, facilitated by technology, science and market-based drivers. With that, a *post-animal bioeconomy* is portrayed as the primary solution to global food system issues. The metaphors emphasize industrial manufacturing and capitalist economics while potentially downplaying a wide array of public concerns around these disruptive technologies, such as the impacts on public health, the economy, the cultural elements and the environment as a whole (Jönsson et al. 2019; Broad 2020).

In general, the bioeconomy typically encompasses global economic activity derived from renewable, raw biological resources such as plants, animals and microorganisms; it entails the large-scale use of biotechnology (Scarlat et al. 2015). Major sectors include agriculture, forestry, fisheries, pulp and paper, food production, biotechnology, and parts of the chemical and pharmaceutical industries. The bioeconomy aims to sustainably produce food, materials, chemicals, fuels and energy using biomass as feedstocks instead of finite fossil resources, thus reducing critical carbon emissions, waste, and environmental degradation (Tan and Lamers 2021) The integration of cultivated meat into mainstream food systems creates a symbiotic relationship with the bioeconomy. Not only would it foster the growth of a sustainable bioeconomy, but it would also rely on the continuous development of the bioeconomy. It would meet nutritional needs while realizing positive economic, social, and environmental impacts throughout food supply chains.

In this context, cultivated meat development faces challenges in reducing production costs, scaling up manufacturing, overcoming ethical and technological issues, and gaining consumer acceptance (Mouat and Prince 2018; Chriki and Hocquette 2020; Bryant 2020; Benny et al. 2022). Overcoming these hurdles is essential for cultivated meat to provide an alternative to replace traditional animal agriculture and live up to the expectations around its transformative potential. The success of cultivated meat exemplifies how biotechnology can transform sectors like meat production to align with circular, ecologically conscious bioeconomy principles, which has profound sustainability implications. This is the rationale of this chapter, whose storyline is depicted in Fig. 1.

In this chapter, our objective is to thoroughly explore the transformative view of cultivated meat supply chains. We will examine their foundational ties to the concepts and principles of the bioeconomy and delve into the extensive sustainability implications stemming from this relationship. As the production of cultivated meat evolves, mechanisms to track its impacts and profound implications become

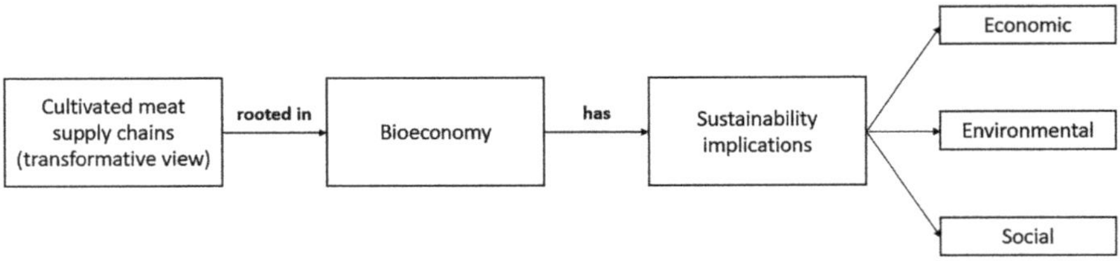


**Fig. 1** Rationale underpinning this chapter: the transformative view of cultivated meat supply chains is rooted in the bioeconomy's concepts and principles. This, in turn, has profound implications for sustainability in economic, environmental, and social dimensions

imperative. This chapter contributes to that dialogue, presenting a potential blueprint for framing the strategic interplay between cultivated meat supply chains, the bioeconomy, and their sustainability implications. Grasping these implications is vital, as they carry significant consequences for the future of food production, consumption systems, and global sustainability objectives, such as the Sustainable Development Goals (Rosati et al. 2023). In particular, cultivated meat directly addresses SDG 2 (No Hunger), SDG 3 (Good Health and Well-Being), SDG 8 (Decent Work and Economic Growth), SDG 9 (Industry, Innovation and Infrastructure), SDG 12 (Responsible Consumption and Production), SDG 13 (Climate Action) and SDG 15 (Life on Land) (UNDP 2023). Moreover, researchers have advocated for an "18th SDG" centered on animal health, welfare, and rights, given its deep interconnection with many other goals (Visseren-Hamakers 2020).

This chapter is organized as follows. Section 2 offers an overview of transformative supply chains applied to cultivated meat. Section 3 brings the opportunities and challenges related to the bioeconomy in the context of cultivated meat supply chains. Then, we open up the sustainability implications of the development of cultivated meat supply chains, exploring their speculative nature. The economic dimension is explored in Sect. 4, followed by the environmental dimension (Sect. 5) and the social dimension (Sect. 6). We finish the chapter by providing concluding remarks and potential avenues forward (Sect. 7).

## 2 Cultivated Meat Supply Chains: A Transformative View

When we broaden our perspective beyond the individual production processes unique to each technological platform for cultivated meat, we are met with a supply chain that is both detailed and layered. This isn't just a simple sequence of steps; it's a complex network of interconnected activities and participants that collaborate to bring alternative protein products from conception to consumers (Hearnshaw and Wilson 2013). These supply chains play a foundational role in rapidly expanding not only the cultivated meat industry but also the alternative protein industry as a whole (Miranda and Picanco Rodrigues 2023). One of the characteristics that will likely define the success of these emerging supply chains is their resilience—as the supply chains demonstrate a resistance to unexpected external challenges, whether economic, environmental, social or otherwise. Resilience ensures a more stable supply and can potentially lead to more consistent pricing and availability for consumers (Wood et al. 2023; Wieland et al. 2023).

What makes the cultivated meat supply chain particularly complex is the heterogeneity of products and how they relate to the broader alternative protein landscape. Product mix has a profound impact on supply chain's configuration and performance. The development of cultivated meat supply chains will not be limited to producing and bringing to market solely fully cultivated meat products. Indeed, there's already a noticeable trend where several companies begin their exploration of cultivated meat technologies by offering hybrid products, combining both plant-based and cultivated meats (The Good Food Institute 2023b).

The importance of fully understanding these complex supply chains cannot be overstated. By closely examining their structure and operations, not only can we identify the primary stakeholders involved but also uncover specific areas that might benefit from targeted interventions. This knowledge could pave the way for overarching strategies that enhance the efficiency and performance of the industry. What's particularly interesting is that despite the differences in sources—whether they are produced from plant-based sources, through cell-based methods, or via fermentation—there's a discernible commonality in how the value is created and distributed across the supply chains. This shared structure underscores the universality of the principles guiding the movement of value in this sector, extending from the initial production stages all the way to the final consumption by end-users.

Within the emergent streams of literature on supply chain management, transformative supply chains have been gaining traction as new interpretations and worldviews of how to analyze and intervene in supply chain considering the contemporary and interconnected challenges we face (Wieland 2021; Mollenkopf et al. 2021; Wieland et al. 2023) Transformative supply chains fundamentally challenges the traditional perspective, which often portrays supply chains as insular, self-contained entities (machines). Instead, the transformative view invites us to reimagine supply chains through the lens of social-ecological systems. This fresh interpretation emphasizes the intricate and ever-evolving relationships between human communities and the broader natural ecosystems in which supply chains are embedded. By acknowledging and integrating these dynamics, we can cultivate a more comprehensive understanding of supply chains, especially in food production and distribution.

The move towards cultivated meat supply chains exemplifies this new wave of thinking as the emergent supply chain is certainly transformative. This transition embodies a deep-seated transformation, characterized by rigorous exploration and iterative learning, to find sustainability leverages across the supply chain (Coenen et al. 2012; Abson et al. 2017; Silvestre et al. 2020). The endgame isn't just efficiency or profitability but the creation of a supply chain that is fundamentally free from the infliction of suffering and has reduced environmental footprint and social harms. The implications of such a radical change are vast and multifaceted.

Eradicating non-human suffering and drastically reducing environmental and social impact from the traditional meat production and distribution processes reverberates far beyond the immediate confines of the supply chain (Gunderson 2013; Mouat and Prince 2018; Cockshaw 2021). Our analytical lens must broaden to grasp the magnitude and nuances of these effects. Thus, we've opted to delve deeper into two additional analytical dimensions: the macroscopic political-economic landscape and the overarching planetary context. Though often overlooked, these dimensions are integral to a holistic understanding of transformative supply chain dynamics (Holling 2001; Carter et al. 2015; Wieland 2021). These insights guide transformative action, enabling a more layered and interconnected approach to supply chain management in our era of grand challenges.

The supply chain for cultivated meat can be broken down into several stages, each contributing to the final product. The following description of each one of the stages is based on specialized literature depicting and describing the manufacturing processes and the supply chain in more detail (Reis et al. 2020; Santo et al. 2020; Humbird 2021; Benny et al. 2022; The Good Food Institute 2023b).

Below we provide an overall description of each stage in the cultivated meat supply chain:

1. **Cell collection:**
    (a) **Source animals:** A small biopsy, often no larger than the size of a sesame seed, is taken from a live animal. This procedure is performed under anesthesia to ensure the animal experiences no discomfort. Interestingly, this process means that potentially thousands of kilograms of meat can originate from a single biopsy.
    (b) **Cell lines**: Scientists isolate the primary cells once the biopsy is taken. These cells can be turned into cell lines that have the ability to proliferate indefinitely under the right conditions. This ensures a consistent and ongoing cell source without repeatedly returning to the animal.
2. **Cell culturing:**
    (a) **Growth medium:** For the cells to thrive and multiply, they need a nutrient-rich medium. Historically, this medium contained animal-derived serums like fetal bovine serum. However, due to ethical and scalability concerns, many companies are innovating to develop plant-based or fully synthetic serums that are more sustainable and cruelty-free.
    (b) **Bioreactors:** These are controlled environments where cells are cultured. Bioreactors ensure the right conditions for cell growth, including temperature, pH, and nutrient supply. They can range in size from small lab-scale flasks to massive industrial tanks capable of producing large quantities of meat.
    (c) **Scaffold (for structured products):** To create products with complex structures (like steaks), cells are often cultured on a scaffold. This 3D matrix, which can be edible and made of various materials, provides a framework

for muscle and fat cells to grow in desired patterns, mimicking the texture and mouthfeel of traditional meat cuts.

3. **Harvesting and processing:** After a certain growth period, the cells reach a density that prepares them for harvest. Advanced techniques ensure cells are separated from the growth medium and scaffold without damaging the cellular structure. The harvested cell mass is then processed like traditional meat, undergoing techniques to enhance flavor, texture, and appearance. Depending on the product, natural fats, flavors, or plant-based ingredients might be added to fine-tune taste, texture, appearance and mouthfeel.
4. **Product formulation:** Once processed, the meat can be formulated into various consumer-ready products with intense activities around product design and development. This includes common products like burgers and sausages, but the possibilities are vast. As the technology matures, we might see intricate products that resemble gourmet meat cuts. Cultivated meat behaves much like its conventional counterpart when cooked.
5. **Packaging and distribution:** The formulated meat is packaged using contemporary food preservation techniques to ensure freshness and longevity. This could include vacuum sealing, flash freezing, or even modified atmosphere packaging. The products are then channeled through distribution networks to reach various points of sale. This includes retail stores, specialty food stores, food services, restaurants, or even direct-to-consumer online platforms (last-mile operations).
6. **End consumer:** At the end of this supply chain stands the consumer, equipped with the choice to purchase cultivated meat products. As awareness grows about conventional meat production's environmental and ethical concerns, many consumers will potentially turn to cultivated meat as a compelling alternative. They seek assurance regarding its ethical sourcing, sustainable origins, and processing methods. This underscores the importance of supply chain visibility and traceability in the cultivated meat sector. Furthermore, consumer feedback can refine specific stages of the supply chain, aiming for enhanced performance.

## 3 An Overview of the Bioeconomy: Opportunities and Challenges for Cultivated Meat

The bioeconomy encompasses the innovative exploration and utilization of biological resources, processes, and products, especially in fields like agriculture, health, and biotechnology (Rampasso et al. 2021; Palmer et al. 2022). Specifically, when referring to the food industry and its associated supply chains, the bioeconomy addresses the foundational phases of primary production. This includes the enhancement and cultivation of plants through breeding and crossbreeding techniques. Furthermore, it encapsulates the advancement of groundbreaking applications on an industrial scale, such as bioproducts, biofuels, and bioprocesses. These terms refer to inputs, products and fuels derived from biological sources and the processes to produce them, respectively (OECD 2019; Rampasso et al. 2021).

There is documented evidence of the rising significance of the bioeconomy on the global stage. As documented by Bracco et al. (2018) and Dietz et al. (2018), and underscored by Rampasso et al. (2021), the bioeconomy is at the forefront of public policy formulation in over 40 countries worldwide. Notably, European nations have been pioneers in this movement. The root of these policies is to harness economic benefits from emerging opportunities in the bioeconomy, all while maintaining and enhancing both environmental and societal value. Moreover, these policy and regulatory frameworks provide explicit directives regarding the trade dynamics—including both import and export—of products linked to the bioeconomy.

As the concept of bioeconomy quickly emerges, a critical perspective must be considered. It is dangerous and misleading to frame *sociobiodiverse economies* (e.g., Amazon rainforest) and *plantations* as part of the bioeconomy (Ollinaho and Kröger 2023). The authors suggest a disintegration of the all-embracing concept of the bioeconomy to promote a radical and sustainable version of the bio-based economy against the extractive economy of plantations. The authors compare and contrast two types of land-use-based economies, the "plantation economy" and the "sociobiodiverse economy," highlighting the need to analyze how land-use practices impact biological and social diversity. They propose narrowing the concept of bioeconomy to practices that embrace diversity and support sociobiodiversity, such as agroecology or agroforestry. The paper also emphasizes the importance of avoiding onto-epistemic violence and colonialities in the conceptualization and implementation of bioeconomy, particularly in contexts where indigenous populations and other minorities face dire consequences (Ollinaho and Kröger 2023).

Along the lines of understanding the cohesion of policies and extending the rationale of an integrated bioeconomy concept, Muscat et al. (2021) employed qualitative content analysis, surveys, and focus groups to examine the alignment between bioeconomy and agri-food policy goals. The results indicated that these policy objectives are predominantly viewed as harmonious based solely on coherence scores. Moreover, the identified synergies between the two areas surpassed the trade-offs in both number and intensity. The findings underscore the necessity of recognizing the interplay between varied policy objectives across these sectors—moving away from "silo-thinking"—in the transition towards a bioeconomy. Achieving this requires clarity in policy formulation and a willingness for integrated policies to navigate and accommodate uncertainties and vagueness (Muscat et al. 2021).

When formulating and executing bioeconomy strategies, it is imperative for policymakers to adhere to the principles of strong sustainability; this involves a holistic approach that considers the social, economic, and environmental repercussions across all sectors associated with the evolution of the bioeconomy (Liobikiene et al. 2019). A critical aspect of this strategy lies in the careful evaluation of available land resources. To ensure sustainable development within the bioeconomy, it's essential that the land footprint remains within the boundaries of the region's biocapacity (Steffen et al. 2015; Liobikiene et al. 2019; Hickel 2020). Furthermore, for the judicious and sustainable utilization of resources within the bioeconomy, it's recommended to employ standardized measures of biomass potential. This will provide a consistent and reliable framework for resource allocation and use (Haberl et al. 2014; Liobikiene et al. 2019; Heide et al. 2023).

The regulatory dimension of the bioeconomy is as important as it is currently neglected. Innovations such as gene-edited animals, methane-reducing feed additives, and cultivated meats are at the forefront of various biotechnological and biomanufacturing advancements aimed at enhancing the sustainability of the livestock sector and related industries (Letti et al. 2021; Van Eenennaam 2023). For example, the authors claim that the regulatory landscape for these products in the United States is inconsistent. While intentional genomic alterations in animals and methane-reducing feed additives are subject to rigorous federal oversight through a costly and lengthy new animal drug-approval pathway, animal cells cultivated for human consumption undergo a more lenient "voluntary pre-market consultation" process (Letti et al. 2021). This drug-focused regulatory approach discourages investment in bioengineered animals and feed additives. As a result, the United States is trailing other nations in approving and integrating these innovative solutions, especially when there's a heightened emphasis on reducing the environmental impact of animal agriculture (Letti et al. 2021).

The inputs to the production processes of cultivated meat depend on several types of feedstock inputs, both plant-based (e.g., corn, soybean, cyanobacteria) and animal-based (e.g., collagen, serum) inputs (Santo et al. 2020). The overarching framework of the bioeconomy can play a critical role in rapidly developing and enhancing the quality and availability of this wide array of inputs. This brings us to the production costs and the larger economic impacts of cultivated meat production.

# 4 The Economic Dimension of Cultivated Meat Production

## *4.1 Production Costs and Scale-Up Implications*

Several studies have examined various aspects of the production costs of cultivated meat, including the cost of cell culture media, the scale of production facilities, and the potential for cost reduction through optimization and technological advancements. One study by Gomez Romero and Boyle (2023) focused on the cost of cell culture media in cultivated meat production. The study found that approximately \$20 out of every \$60 spent on cultivated meat production would be allocated to media costs. This highlights the importance of optimizing media formulations and exploring cost-effective alternatives to reduce production costs.

Another study by Negulescu et al. (2023) evaluated the impact of the production bioreactor scale on the cost of cultivated meat. The researchers conducted a techno-economic analysis and envisioned an industry with multiple facilities producing 100,000,000 kg of cultured beef per year, equivalent to roughly 0.14% of the global beef production in a given year. This study provides insights into the potential cost implications of scaling up cultivated meat production. Larger production bioreactors result in fewer facilities and bioreactors needed, leading to lower total facility-related costs. Materials, particularly media costs, make up a significant portion of the operating expenses (OPEX) in cultivated meat production. In the larger-scale scenarios, media costs account for 94–98% of the materials cost (Negulescu et al. 2023). Furthermore, Djisalov et al. (2021) discussed the challenges and costs associated with cultivated meat production, emphasizing the need for advanced monitoring options to optimize the production process and reduce costs. The authors suggest that further optimization of meat cultivation could help address the current high production costs.

In addition to these specific studies, there is a general consensus in the literature that cultivated meat production has unique constraints and challenges related to scale and cost. Bomkamp et al. (2022) highlighted the constraints of scale and cost in cultivated meat production, as well as the necessary attributes of the final product, such as texture and food safety. Similarly, Tuomisto and Teixeira De Mattos (2011) conducted a comparative study on the environmental impacts of cultivated meat production. They found that it requires significant energy, water, and land resources, although less than traditional meat production.

While the production costs of cultivated meat are currently high, there is optimism that technological advancements and optimization efforts can lead to cost reductions. Reiss et al. (2021) discussed the potential environmental benefits of cultivated meat production, including reduced water and land usage, as well as lower greenhouse gas emissions compared to conventional meat production. These environmental benefits could contribute to the overall cost-effectiveness of cultivated meat production.

A robust study developed by Humbird (2021) delves into the feasibility and scale-up economics of cultivated meat products. The research draws insights from technoeconomic analyses commonly found in industrial fermentation and biopharmaceuticals by focusing on the scalability inherent in animal cell production. Several economic and biological limitations are highlighted, challenging cultured meat production. These include factors such as the slow growth rate of the cells, metabolic inefficiencies, catabolite inhibition, damage to cells caused by shear forces, and the significant capital costs tied to the acquisition of equipment and establishment of facilities. These facilities must also ensure safeguards against microbial contamination, further escalating costs.

Further, the study underscores the elevated projected costs linked with the procurement of pure amino acids and protein growth factors, essential components in the cultured meat production process (Benny et al. 2022). From a financial perspective, detailed analyses of the capital and operational costs of conceptual facilities for cell-mass production suggest that producing cultured meat products at a price acceptable for regular consumption may not be feasible. The research concludes by emphasizing two pivotal points: the imperative need for enhancements in metabolic efficiency and the development of cost-effective media derived from plant hydrolysates. However, even the realization of these conditions might not be enough to position cultured meat in terms of cost as a true replacement for conventional meat (Humbird 2021; Lähteenmäki-Uutela et al. 2021; Benny et al. 2022).

## *4.2 Market Introduction, Price and Transition to Cultivated Meat Production*

Authors argue that in order to truly penetrate the market at a volume of 100 kTA (kilotonne per annum), which would mean around ten million people consuming the equivalent of 10 kg/year, cultivated meat would achieve a pricing status similar to an affordable luxury, something consumers might indulge in occasionally (Humbird 2021). Breaking down the figures further, the study sets a target: ~$25/kg of wet animal cell matter should translate, post the stages of growth, processing, packaging, and distribution (inclusive of profit margins), to at least ~$50/kg on the supermarket shelves. The author argues that exceeding this cost threshold might not entirely deter the adoption of cultured meat, but its potential to displace conventional meat diminishes with price increments. While enhancing metabolic efficiency and developing affordable hydrolysate media are undeniably crucial, they don't represent the complete solution to making cultured meat a household staple. Additionally, the reduction of capital costs, although secondary, still plays a considerable role in the affordability equation (Humbird 2021).

Extending across the broader economic implications of the emergent alternative protein sectors, Newton and Blaustein-Rejto (2021) examine the context of rural producers in the United States. Economic opportunities identified include the cultivation of crops for feedstock in cultivated meat, the use of animals for genetic material extraction, on-farm cultured meat production, transitioning to new sectors, and market potential for products that combine traditional and alternative meat sources, among others. On the flip side, Newton and Blaustein-Rejto (2021) claim that ranchers, livestock producers, and farmers who currently grow crops for animal feed might face a decline in income or even loss of livelihood. Additionally, transitioning to the cultivated meat sector can pose barriers, and some producers may find themselves excluded from these new economic opportunities. The study also points to the significant roles of universities, research institutions, government agencies, and non-profit organizations in shaping the trajectory of these emerging sectors, emphasizing their potential to both amplify economic benefits and mitigate risks. Based on the 37 interviews carried out by the authors, a prevailing view emerged: alternative meat is likely to exist in tandem with traditional animal agriculture, providing an expanded range of options for rural producers rather than replacing animal meat entirely. This view also aligns with the findings of Räty et al. (2023).

Now we turn to the environmental dimension.

# 5 The Environmental Dimension of Cultivated Meat Production

## 5.1 Environmental Opportunities and Impacts

The environmental impacts of cultivated meat have been a hot topic of interest and research in recent years. This innovative approach to meat production addresses several environmental concerns associated with conventional meat production. One of the key environmental benefits of cultivated meat is its potential to reduce greenhouse gas emissions. Conventional meat production significantly contributes to greenhouse gas emissions, accounting for an estimated 18% of all anthropogenic greenhouse gases (Tuomisto and Teixeira De Mattos 2011).

Cultivated meat has been shown to have substantially lower greenhouse gas emissions compared to conventionally produced meat. A study comparing the environmental impacts of cultured meat production to those of beef, pork, and poultry found that cultured meat had significantly lower greenhouse gas emissions (Mattick et al. 2015). Once on the market, it has been estimated that cultivated meat production could result in up to 45% decrease in energy use and 78–96% lower greenhouse gas emissions (Tuomisto and Teixeira De Mattos 2011; Shahin-Shamsabadi and Selvaganapathy 2022; Smetana et al. 2023). This reduction in emissions can be attributed to the fact that cultivated meat production does not require the same level of land use, feed production, and methane emissions associated with raising livestock.

In addition to greenhouse gas emissions, cultivated meat also reduces land use. Conventional meat production requires large amounts of land for grazing and feed production. In contrast, cultivated meat can be produced in a much smaller footprint, as it does not require the same amount of land for raising and feeding animals (Smetana et al. 2023). It is estimated that cultivated meat, at scale, will use 99% lower land area and 82–96% lower water consumption compared to conventional meat production (Tuomisto and Teixeira De Mattos 2011; Shahin-Shamsabadi and Selvaganapathy 2022; Smetana et al. 2023). A study comparing the land use of different meat substitutes found that lab-grown meat had the highest land use impact, but it was still significantly lower than conventional meat production (Smetana et al. 2015). This reduction in land use can help alleviate pressure on natural ecosystems and contribute to land conservation efforts.

Water consumption is another environmental impact associated with meat production. Conventional meat production requires significant amounts of water for animal hydration, feed production, and processing. Cultivated meat may significantly reduce water consumption compared to conventional meat production. This reduction in water consumption can help address water scarcity issues and promote more sustainable water management practices.

Furthermore, cultivated meat production reduces the use of antibiotics in comparison to the the animal meat industry. Conventional meat production often relies on antibiotics to promote animal growth and prevent disease outbreaks. This overuse of antibiotics contributes to the development of antibiotic-resistant bacteria, which poses a significant threat to public health. Cultivated meat production can be done in a sterile environment without antibiotics, reducing the risk of antibiotic resistance and associated public health issues (Post et al. 2020).

### *5.2 Environmental Challenges*

It is important to note that while cultivated meat has the potential to reduce certain environmental impacts, it is not without its own challenges and considerations. The production of cultivated meat requires energy-intensive processes, such as cell culture and tissue engineering, which can contribute to the overall environmental footprint (Tuomisto and Teixeira De Mattos 2011). However, advancements in technology and process optimization have the potential to mitigate these energy requirements and further reduce the environmental impacts of cultivated meat production (Negulescu et al. 2023).

Cultivated meat can significantly reduce the environmental impacts associated with traditional meat production. It has been shown to have lower greenhouse gas emissions, reduced land use, lower water consumption, and the potential to reduce the use of antibiotics. While there are still challenges to overcome and further research to be done, cultivated meat offers a promising alternative that can contribute to a more environmentally friendly food system. However, this is not disconnected from the social impacts around the cultivated meat supply chains

## 6 The Social Dimension of Cultivated Meat Production

### *6.1 Social Opportunities and Challenges*

The environmental benefits of cultivated meat outlined above can reflect significant social implications, as the environmental impact of meat production is a growing concern that drives consumption, choices and behavior for many individuals and communities across the globe (Bryant 2019; Bryant and Sanctorum 2021). Communities, particularly in vulnerable regions, can benefit from a reduced strain on local resources. With less land dedicated to feed crops and pastures, there's potential for land to be repurposed for other community needs or left in its natural state to support local biodiversity (Machovina et al. 2015).

Reduced water usage in meat production can lead to greater water availability for personal and community use, a pressing need in many areas facing water scarcity. Moreover, as communities worldwide become increasingly aware of the environmental footprint of their consumption choices, the demand for sustainable options grows. Cultivated meat can cater to this demand, strengthening community bonds and a unified movement towards environmentally conscious living. This shift not only aids the immediate environment but also supports a more globally sustainable ethos, encouraging further innovation and adaptation in other sectors.

Consumer acceptance and perception of cultivated meat are also important social factors. Studies have shown that consumer attitudes towards cultivated meat vary and are influenced by personal health benefits, taste, familiarity, and cultural and religious beliefs (Szejda et al. 2021a, b; Ho et al. 2023). While some consumers may be open to trying cultivated meat and recognize its potential benefits, others may have reservations or concerns about its safety, taste, and naturalness (Wilks et al. 2019; Bryant and Dillard 2019). Understanding and addressing these consumer perceptions and preferences is crucial for the successful adoption and integration of cultivated meat into the market (Stephens et al. 2018).

Furthermore, as the introduction of cultivated meat may disrupt traditional meat industries and supply chains, it will directly impact jobs and livelihoods. Studies have shown that the production of cultivated meat might create new and higher-skilled jobs (Morais-da-Silva et al. 2022b). The economic impact of cultivated meat will depend on various factors, including the scale of production, market demand, and the ability of traditional meat industries to adapt to new technologies and consumer preferences (Morais-da-Silva et al. 2022a). In particular, religious and cultural considerations are also important when discussing the social impacts of cultivated meat. Different religious and cultural groups may have specific dietary restrictions or beliefs regarding the consumption of meat (Bryant 2020). The acceptability and adoption of cultivated meat within these communities may depend on religious interpretations, cultural norms, and the perceived compatibility of cultivated meat with existing dietary practices (Ho et al. 2023).

Consider Brazil, for instance, one of the world's largest meat producers and exporters. Researchers have delved deep into the social ramifications of a shift from traditional meat sources to cultivated and plant-based alternatives. Through comprehensive interviews with 35 industry experts representing both animal production and the alternative protein sectors, the study has shed light on a series of potential opportunities and challenges that Brazil, and by extension many other nations, might face in this transformative journey (Luiz Morais-da-Silva et al. 2022). Brazil is poised to undergo transformative changes in its food production arena with the emergence of nine pivotal opportunities.

Firstly, agricultural producers in Brazil are presented with novel avenues to diversify and expand their operations. This not only ensures a more resilient agricultural sector but also paves the way for the second opportunity: the creation of new jobs within the emerging industry. These roles extend across the supply chain, potentially revitalizing regions and communities with new employment prospects. The third advantage lies in the potential stimulation for a more qualified workforce. As the industry evolves, there will be a growing demand for specialized skills, leading to more comprehensive training and education programs. This opportunity leads to the fourth and fifth opportunities: an anticipated rise in average salaries and the promise of vastly improved working conditions (Winders and Abrell 2021). These enhancements stand in stark contrast to the often precarious and high-risk conditions prevalent in traditional meat supply chains (Luiz Morais-da-Silva et al. 2022).

The sixth potential benefit is a fertile ground for entrepreneurial ventures. As the food production landscape shifts, innovative business ideas and models can find a conducive environment to take root and flourish. Alongside economic gains, the seventh opportunity focuses on public health and well-being, with the potential for greater access to protein sources, addressing the pressing issue of undernourishment prevalent in certain demographics. Furthermore, the eighth opportunity underscores the promise of products that are not just sustainably produced but also healthier for consumption. This aligns with global trends emphasizing health and well-being. Lastly, a move in this direction also signifies a more ethical stance towards non-human animals, heralding a future where production methods are more in line with contemporary ethical considerations (Luiz Morais-da-Silva et al. 2022).

While Brazil has several opportunities in its food production sector, it faces five pressing challenges that demand attention. First and foremost, there's a potential contraction in the demand for conventional meat products. As alternative and sustainable food production methods gain traction, traditional meat products could see a decline in market share. This leads to the second challenge: unemployment. The shift in consumer preferences might render certain roles in the traditional meat supply chain redundant, leading to job losses (Luiz Morais-da-Silva et al. 2022).

Thirdly, the existing low qualification levels of labor within these traditional meat supply chains present a significant hurdle. If the industry is to evolve and adapt, it will require a skilled and adaptable workforce, which is currently not the case. The fourth challenge is economic: the high prices associated with emerging food production methods could deter widespread adoption, at least initially. Lastly, and perhaps most crucially, is the challenge of consumer acceptance. New products and methods might face resistance from consumers, either due to unfamiliarity, perceived health concerns, or other factors (Luiz Morais-da-Silva et al. 2022).

The study particularly emphasizes the importance of stakeholder engagement. On the one hand, engaging with all relevant parties, from producers to consumers, could be the determining factor in leveraging the opportunities revealed by the study. On the other hand, a more reactive or passive stance might leave the industry ill-equipped to address and overcome these challenges. The implications of the study resonate beyond the immediate findings. For Brazil, and potentially other countries on a similar trajectory, there's a call for the formulation of robust and inclusive policies. These policies should support the transition and ensure it is just and equitable for all involved (Pretty et al. 2003; Harwatt 2019; Luiz Morais-da-Silva et al. 2022). The social challenges and opportunities of cultivated meat underscore the importance of a systemic approach to food system transitions, where environmental, market, and social factors are given their due consideration.

### 6.2 Ethical Considerations and Impacts on Animal Welfare

One of the key social impacts of cultivated meat is its potential to address ethical concerns surrounding animal agriculture and welfare. From issues of confinement in tight spaces to the use of growth hormones and antibiotics to the profound emotional and physical distress animals endure, numerous societal concerns arise when considering the ethics of meat consumption in the traditional sense. And this affects society as a whole. Cultivated meat offers a way to produce meat without traditional animal agriculture, thereby reducing animal suffering and the ethical dilemmas associated with raising and slaughtering animals for food (Moyano-Fernández 2023). This is particularly important for animal rights advocates who prioritize the reduction of animal suffering, even though some animals will probably still have to be reared to harvest cells for the production of cultivated meat (Chriki and Hocquette 2020).

The direct physical suffering that animals might experience in conventional farming setups is bypassed in a cultivated meat supply chain. Animals are no longer part of the equation in the same way. This addresses a major ethical concern that many people have regarding the palpable suffering experienced by animals. However, beyond just the physical, there are cognitive and emotional dimensions to consider. The animals we raise for food have demonstrated capabilities for emotional responses and social interactions (EFSA Panel on Animal Health and Welfare (AHAW) et al. 2020; Köllen and Schneeberger 2023). The conditions in many industrial farms often deprive them of natural behaviors, leading to stress, anxiety, and other negative emotional states. Cultivated meat production, by negating the need for these farms, might be seen as a way of preventing such emotional distress among animals.

For animal rights advocates, this might seem a revolutionary avenue. These individuals and groups recognize and deeply empathize with the sentience and suffering of animals, particularly in traditional meat supply chains. The prospect of cultivated meat allows them to reconcile the possible dietary preferences for meat with their stance against animal cruelty. In a world where ethical considerations weigh heavily on consumption choices, this harmonization of values and actions is immensely significant (Long and Murray 2013; Bodur et al. 2015; Goldstein et al. 2016).

## 7 Concluding Remarks and Future Perspectives

Cultivated meats offer a promising avenue to decrease reliance on traditional livestock for meat production. However, there are technological challenges that remain. In particular, there is also pressure to eliminate all animal-based inputs, notably fetal bovine serum. Despite the growing interest in cell-based meats, there's a scarcity of rigorous studies that evaluate their claimed benefits (Chriki and Hocquette

2020). As of now, many of the purported environmental and health advantages remain speculative as the sector develops and more information becomes available.

Many of the components of the environmental footprint of cultivated meat, for example, concerning pesticide use, are contingent on the components of the culture medium (Santo et al. 2020). If this medium relies on soy and corn-based inputs, the pesticide usage might align closely with traditional poultry production, which is large in the world. There's a clear call to action for researchers and policymakers: a comprehensive research effort is required to understand the myriad implications of cell-based meats. This encompasses public health, the environment, animal welfare, economic considerations, and policy ramifications (Santo et al. 2020). Furthermore, the wider socioeconomic and political consequences of transitioning from farmed meat to meat alternatives, including cell-based options, demand in-depth exploration.

Some potential guiding research questions for future research endeavors regarding the interplay we presented in this chapter (cultivated meat supply chains, the bioeconomy and the sustainability impacts) are outlined here:

- How will advancements in bioeconomy principles influence the scalability (Humbird 2021) and efficiency of cultivated meat supply chains in the next decade?
- How can we reconcile the sociobiodiverse view of the bioeconomy (Ollinaho and Kröger 2023) with the needs and requirements of forming cultivated meat supply chains?
- What specific sustainability metrics can be developed to evaluate the long-term impacts of integrating bioeconomy principles into cultivated meat production, which align with local, contextual specificity and broader global sustainability goals (Rockström et al. 2009; Liobikiene et al. 2019; Rosati et al. 2023)?
- How might consumer perceptions and behaviors evolve as the interplay between the bioeconomy and cultivated meat supply chains become more prominent (Stephens et al. 2018; Bryant and Barnett 2018)?
- What policy interventions and regulatory frameworks are needed to ensure that the fusion of bioeconomy principles and cultivated meat supply chains remains aligned with global sustainability goals (Dietz et al. 2018; Muscat et al. 2021; Palmer et al. 2022; Benny et al. 2022)?
- To what extent can the principles of the bioeconomy address potential challenges or bottlenecks in the cultivated meat supply chain (Mouat and Prince 2018; Santo et al. 2020; Chriki and Hocquette 2020; Benny et al. 2022; Negulescu et al. 2023)?

In this chapter, we set out to explore the intricate relationship between cultivated meat supply chains, bioeconomy principles, and their implications for sustainability. As the world grapples with the urgent need for sustainable food solutions, the potential of cultivated meat underpinned by the bioeconomy emerges as a promising avenue for research and action. However, it's clear that this field is full of uncharted territories and unanswered questions. As researchers, policymakers, and industry leaders move forward, it will be critical to continually reassess and refine our understanding. The proposed future research questions underscore the vast landscape awaiting investigation. It's a journey that promises not only innovative solutions but also a deeper understanding of the complex dance between technology, nature, and human aspirations.

## References


Abson DJ, Fischer J, Leventon J et al (2017) Leverage points for sustainability transformation. Ambio 46:30–39

Benny A, Pandi K, Upadhyay R (2022) Techniques, challenges and future prospects for cell-based meat. Food Sci Biotechnol 31:1225–1242

Bodur HO, Duval KM, Grohmann B (2015) Will you purchase environmentally friendly products? Using prediction requests to increase choice of sustainable products. J Bus Ethics 129:59–75

Bomkamp C, Skaalure SC, Fernando GF et al (2022) Scaffolding biomaterials for 3D cultivated meat: prospects and challenges. Adv Sci 9:e2102908

Bracco S, Calicioglu O, Juan MGS, Flammini A (2018) Assessing the contribution of bioeconomy to the total economy: a review of national frameworks. Sustain Sci Pract Policy 10:1698

Broad GM (2020) Making meat, better: the metaphors of plant-based and cell-based meat innovation. Environ Commun 14:919–932

Bryant CJ (2019) We can't keep meating like this: Attitudes towards vegetarian and vegan diets in the United Kingdom. Sustainability (Switzerland) 11. https://doi.org/10.3390/SU11236844

Bryant CJ (2020) Culture, meat, and cultured meat. J Anim Sci 98. https://doi.org/10.1093/jas/skaa172

Bryant C, Barnett J (2018) Consumer acceptance of cultured meat: A systematic review. Meat Sci 143:8–17

Bryant C, Dillard C (2019) The impact of framing on acceptance of cultured meat. Front Nutr 6:103

Bryant C, Sanctorum H (2021) Alternative proteins, evolving attitudes: Comparing consumer attitudes to plant-based and cultured meat in Belgium in two consecutive years. Appetite 161:105161

Carter CR, Meschnig G, Kaufmann L (2015) Moving to the next level: why our discipline needs more multilevel theorization. J Supply Chain Manag 51:94–102

Chriki S, Hocquette J-F (2020) The myth of cultured meat: a review. Front Nutr 7:7

Cockshaw R (2021) The end of factory farming. VIB 7. https://doi.org/10.52214/vib.v7i.8696

Coenen L, Benneworth P, Truffer B (2012) Toward a spatial perspective on sustainability transitions. Res Policy 41:968–979

Dietz T, Börner J, Förster JJ, von Braun J (2018) Governance of the bioeconomy: a global comparative study of national bioeconomy strategies. Sustain Sci Pract Policy 10:3190

Djisalov M, Knežić T, Podunavac I et al (2021) Cultivating multidisciplinarity: manufacturing and sensing challenges in cultured meat production. Biology 10. https://doi.org/10.3390/biology10030204

EFSA Panel on Animal Health and Welfare (AHAW), Nielsen SS, Alvarez J et al (2020) Welfare of pigs at slaughter. EFSA J 18:e06148

Goldstein B, Hansen SF, Gjerris M et al (2016) Ethical aspects of life cycle assessments of diets. Food Policy 59:139–151

Gomez Romero S, Boyle N (2023) Systems biology and metabolic modeling for cultivated meat: A promising approach for cell culture media optimization and cost reduction. Compr Rev Food Sci Food Saf 22:3422–3443

GOOD Meat (2022) GOOD Meat launches chicken satay with popular Singapore Hawker. https://www.goodmeat.co/all-news/good-meat-launches-chicken-satay-with-popular-singapore-hawker. Accessed 12 Oct 2023

Goven J, Pavone V (2015) The bioeconomy as political project: a Polanyian analysis. Sci Technol Human Values 40:302–337

Gunderson R (2013) From cattle to capital: exchange value, animal commodification, and barbarism. Crit Sociol 39:259–275
Haberl H, Erb K-H, Krausmann F (2014) Human appropriation of net primary production: patterns, trends, and planetary boundaries. Annu Rev Environ Resour 39:363–391
Harwatt H (2019) Including animal to plant protein shifts in climate change mitigation policy: a proposed three-step strategy. Clim Policy 19:533–541
Hearnshaw EJS, Wilson MMJ (2013) A complex network approach to supply chain network theory. Int J Oper Prod Manag 33:442–469
Heide M, Hauschild MZ, Ryberg M (2023) Reflecting the importance of human needs fulfilment in absolute sustainability assessments: Development of a sharing principle. J Ind Ecol. https://doi.org/10.1111/jiec.13405
Hickel J (2020) Quantifying national responsibility for climate breakdown: an equality-based attribution approach for carbon dioxide emissions in excess of the planetary boundary. Lancet Planet Health 4:e399–e404
Ho SS, Ou M, Vijayan AV (2023) Halal or not? Exploring Muslim perceptions of cultured meat in Singapore. Front Sustain Food Syst 7. https://doi.org/10.3389/fsufs.2023.1127164
Holling CS (2001) Understanding the complexity of economic, ecological, and social systems. Ecosystems 4:390–405
Humbird D (2021) Scale-up economics for cultured meat. Biotechnol Bioeng 118:3239–3250
Jönsson E, Linné T, McCrow-Young A (2019) Many meats and many milks? The ontological politics of a proposed post-animal revolution. Sci Cult 28:70–97
Köllen T, Schneeberger D (2023) *Avoiding unnecessary suffering*: Towards a moral minimum standard for humans' responsibility for animal welfare. Bus Ethics Environ Resp. https://doi.org/10.1111/beer.12565
Lähteenmäki-Uutela A, Rahikainen M, Lonkila A, Yang B (2021) Alternative proteins and EU food law. Food Control 130:108336
Letti LAJ, Karp SG, Molento CFM et al (2021) Cultivated meat: recent technological developments, current market and future challenges. Biotechnol Res Innov 5
Lewis E (2022) A food revolution—cultivated meat. Food Sci Technol 36:29–31
Liobikiene G, Balezentis T, Streimikiene D, Chen X (2019) Evaluation of bioeconomy in the context of strong sustainability. Sustain Dev 27:955–964
Long MA, Murray DL (2013) Ethical consumption, values convergence/divergence and community development. J Agric Environ Ethics 26:351–375
Luiz Morais-da-Silva R, Glufke Reis G, Sanctorum H, Forte Maiolino Molento C (2022) The social impacts of a transition from conventional to cultivated and plant-based meats: Evidence from Brazil. Food Policy 111. https://doi.org/10.1016/j.foodpol.2022.102337
Machovina B, Feeley KJ, Ripple WJ (2015) Biodiversity conservation: The key is reducing meat consumption. Sci Total Environ 536:419–431
Mattick CS, Landis AE, Allenby BR, Genovese NJ (2015) Anticipatory life cycle analysis of in vitro biomass cultivation for cultured meat production in the United States. Environ Sci Technol 49:11941–11949
Miranda BV, Picanco Rodrigues V (2023) Alternative proteins: organization of the agri-food system and sustainability outcomes. In: Tiwari BK, Healy LE (eds) Future proteins: sources, processing, applications and the bioeconomy. Elsevier, London, pp 425–445
Mollenkopf DA, Ozanne LK, Stolze HJ (2021) A transformative supply chain response to COVID-19. J Serv Manag 32:190–202
Morais-da-Silva RL, Reis GG, Sanctorum H, Molento CFM (2022a) The social impact of cultivated and plant-based meats as radical innovations in the food chain: Views from Brazil, the United States and Europe. Front Sustain Food Syst 6. https://doi.org/10.3389/fsufs.2022.1056615
Morais-da-Silva RL, Villar EG, Reis GG et al (2022b) The expected impact of cultivated and plant-based meats on jobs: the views of experts from Brazil, the United States and Europe. Humanit Soc Sci Commun 9:1–14

Mouat MJ, Prince R (2018) Cultured meat and cowless milk: on making markets for animal-free food. J Cult Econ 11:315–329

Moyano-Fernández C (2023) The moral pitfalls of cultivated meat: complementing utilitarian perspective with eco-republican justice approach. J Agric Environ Ethics 36:1

Muscat A, de Olde EM, Kovacic Z et al (2021) Food, energy or biomaterials? Policy coherence across agro-food and bioeconomy policy domains in the EU. Environ Sci Policy 123:21–30

Negulescu PG, Risner D, Spang ES et al (2023) Techno-economic modeling and assessment of cultivated meat: Impact of production bioreactor scale. Biotechnol Bioeng 120:1055–1067

Newton P, Blaustein-Rejto D (2021) Social and economic opportunities and challenges of plant-based and cultured meat for rural producers in the US. Front Sustain Food Syst 5. https://doi.org/10.3389/fsufs.2021.624270

OECD (2019) Development co-operation report 2019: a fairer, greener, safer tomorrow

Ollinaho OI, Kröger M (2023) Separating the two faces of "bioeconomy": Plantation economy and sociobiodiverse economy in Brazil. For Policy Econ 149:102932

O'Neill EN, Cosenza ZA, Baar K, Block DE (2021) Considerations for the development of cost-effective cell culture media for cultivated meat production. Compr Rev Food Sci Food Saf 20:686–709

Palmer E, Burton R, Gottschamer L (2022) Political influence structures in a future substitution bioeconomy transition: A focus on the food sector. J Clean Prod 358:131709

Post MJ, Levenberg S, Kaplan DL et al (2020) Scientific, sustainability and regulatory challenges of cultured meat. Nat Food 1:403–415

Pretty JN, Morison JIL, Hine RE (2003) Reducing food poverty by increasing agricultural sustainability in developing countries. Agric Ecosyst Environ 95:217–234

Rampasso IS, Quelhas OLG, Anholon R et al (2021) The Bioeconomy in emerging economies: a study of the critical success factors based on Life Cycle Assessment and Delphi and Fuzzy-Delphi methods. Int J Life Cycle Assess 1254–1266

Räty N, Tuomisto HL, Ryynänen T (2023) On what basis is it agriculture?: A qualitative study of farmers' perceptions of cellular agriculture. Technol Forecast Soc Change 196:122797

Reis GG, Heidemann MS, Borini FM, Molento CFM (2020) Livestock value chain in transition: Cultivated (cell-based) meat and the need for breakthrough capabilities. Technol Soc 62:101286

Reiss J, Robertson S, Suzuki M (2021) Cell sources for cultivated meat: applications and considerations throughout the production workflow. Int J Mol Sci 22. https://doi.org/10.3390/ijms22147513

Rockström J, Steffen W, Noone K et al (2009) Planetary boundaries: exploring the safe operating space for humanity. Ecol Soc 14:32

Rosati F, Rodrigues VP, Cosenz F, Li-Ying J (2023) Business model innovation for the sustainable development goals. Bus Strat Environ. https://doi.org/10.1002/bse.3334

Santo RE, Kim BF, Goldman SE et al (2020) Considering plant-based meat substitutes and cell-based meats: a public health and food systems perspective. Front Sustain Food Syst 4. https://doi.org/10.3389/fsufs.2020.00134

Scarlat N, Dallemand J-F, Monforti-Ferrario F, Nita V (2015) The role of biomass and bioenergy in a future bioeconomy: Policies and facts. Environ Dev 15:3–34

Shahin-Shamsabadi A, Selvaganapathy PR (2022) Engineering murine adipocytes and skeletal muscle cells in meat-like constructs using self-assembled layer-by-layer biofabrication: a platform for development of cultivated meat. Cells Tissues Organs 211:304–312

Shu C (2023) Good Meat approved to sell serum-free cultivated meat in Singapore. TechCrunch

Silvestre BS, Silva ME, Cormack A, Tavares TAM (2020) Supply chain sustainability trajectories: learning through sustainability initiatives. Int J Oper Prod Manag 40:1301–1337

Smetana S, Ristic D, Pleissner D et al (2023) Meat substitutes: resource demands and environmental footprints. Resour Conserv Recycl 190

Smetana S, Mathys A, Knoch A et al (2015) Meat alternatives: life cycle assessment of most known meat substitutes. Int J Life Cycle Assess 20:1254–1267. https://doi.org/10.1007/s11367-015-0931-6

Steffen W, Richardson K, Rockström J et al (2015) Planetary boundaries: guiding human development on a changing planet. Science 347. https://doi.org/10.1126/science.1259855

Stephens N, Di Silvio L, Dunsford I et al (2018) Bringing cultured meat to market: Technical, socio-political, and regulatory challenges in cellular agriculture. Trends Food Sci Technol 78:155–166

Szejda K, Bryant CJ, Urbanovich T (2021a) US and UK consumer adoption of cultivated meat: a segmentation study. Foods 10. https://doi.org/10.3390/foods10051050

Szejda K, Stumpe M, Raal L, Tapscott CE (2021b) South African consumer adoption of plant-based and cultivated meat: a segmentation study. Front Sustain Food Syst 5 https://doi.org/10.3389/fsufs.2021.744199

Tan ECD, Lamers P (2021) Circular bioeconomy concepts—a perspective. Front Sustain 2. https://doi.org/10.3389/frsus.2021.701509

The Good Food Institute (2023a) Aleph Farms make Europe's first cultivated meat approval submission to Swiss Regulators. https://gfieurope.org/blog/aleph-farms-make-europes-first-cultivated-meat-approval-submission-to-swiss-regulators/. Accessed 15 Oct 2023

The Good Food Institute (2023b) 2022 state of the industry report—cultivated meat and seafood

Tuomisto HL, Teixeira De Mattos MJ (2011) Environmental impacts of cultured meat production. Environ Sci Technol 45:6117–6123

UNDP (2023) Sustainable development goals. https://www.undp.org/sustainable-development-goals. Accessed 15 Oct 2023

UPSIDE Foods (2023) UPSIDE is approved for sale in the US! Here's what you need to know. https://upsidefoods.com/blog/upside-is-approved-for-sale-in-the-us-heres-what-you-need-to-know. Accessed 12 Oct 2023

US Food and Drug Administration (2023) FDA completes first pre-market consultation for human food made using animal cell culture technology. https://www.fda.gov/food/cfsan-constituent-updates/fda-completes-first-pre-market-consultation-human-food-made-using-animal-cell-culture-technology. Accessed 12 Oct 2023

Van Eenennaam AL (2023) What is a drug? Enabling the bioeconomy through updated Food and Drug Administration Policy. GEN Biotechnol 2:36–42

Visseren-Hamakers IJ (2020) The 18th sustainable development goal. Earth Syst Govern 3

Wieland A (2021) Dancing the supply chain: toward transformative supply chain management. J Supply Chain Manag 57:58–73

Wieland A, Stevenson M, Melnyk SA et al (2023) Thinking differently about supply chain resilience: what we can learn from social-ecological systems thinking. Int J Oper Prod Manag 43:1–21

Wilks M, Phillips CJC, Fielding K, Hornsey MJ (2019) Testing potential psychological predictors of attitudes towards cultured meat. Appetite 136:137–145

Winders DJ, Abrell E (2021) Slaughterhouse workers, animals, and the environment: the need for a rights-centered regulatory framework in the United States that recognizes interconnected interests. Health Hum Rights 23:21–33

Wood A, Queiroz C, Deutsch L et al (2023) Reframing the local-global food systems debate through a resilience lens. Nat Food 4:22–29

# Global and Regional Policies for Cultivated Meat

**Maria Marques, Rodrigo Luiz Morais-da-Silva, Jennifer C. Biscarra-Bellio, Mariana Hase Ueta, and Carla Forte Maiolino Molento**

**Abstract** The objective of this chapter is to study how public policies and governance actions may be addressed to support and accelerate the transition to cell-based foods as more sustainable and ethical means of protein production and consumption. In this regard, we assessed different impacts of public policies as maintainers of the current production system, the structure of policies for transformative innovations to be applied to cultivated meat, and the structural dimensions of planning, including supranational, national, and subnational plans. Besides, the selected concepts of governance actions such as dialogue among stakeholders, transparency, and inclusion were introduced, as they are regarded as essential for facilitating a successful transition. Finally, we address the challenges for the development of cell-based meat and conclude by emphasizing that public policies and governance actions are central to this transition, guiding investments, funding and decisions while considering social, ethical, and technological aspects.

**Keywords** Public policies · Transformative innovation policy · Planning · Governance actions · Cultivated meat

---

M. Marques (✉) · J. C. Biscarra-Bellio
Cellular Animal Science Laboratory, UFPR Federal University of Parana, Curitiba, Paraná, Brazil
e-mail: maria.marques@ufpr.br

R. L. Morais-da-Silva
School of Management, UFPR Federal University of Parana, Curitiba, Paraná, Brazil

M. H. Ueta
Department of Social Sciences, Wageningen University & Research, Wageningen, The Netherlands

C. F. M. Molento
Animal Welfare Laboratory & Cellular Animal Science Laboratory, Federal University of Paraná, Curitiba, Paraná, Brazil



C. R. Soccol et al. (eds.), *Cultivated Meat*,
https://doi.org/10.1007/978-3-031-55968-6_18

## 1 Introduction

A policy may be simply "anything a government chooses to do or not to do" (Dye 1972, p. 2) or, in a more specific definition, a "statement by the government of what it intends to do, such as a law, regulation, ruling, decision, order, or a combination of these. The lack of such statements may also be an implicit statement of a policy not to do something" (Birkland 2019, p. 6). While numerous alternative definitions exist, they typically refer to the actions or inactions of governments in addressing societal issues (Kraft and Furlong 2021). Public policies have a significant influence on production and consumption systems. Besides defining the government's active role as a highly relevant agent in society, public policies also affect various domains, encompassing aspects such as regulation and tax rates. Hence, public policies may be associated with changes in current systems, including the food systems.

Food production systems include at least the activities involved in food production, processing, packaging, distribution, marketing, and consumption (Ericksen 2008). Due to their vast scale and social and economic relevance, the food production systems may also have various undesirable impacts. Studies have shown that changing how food is produced and consumed is essential for a transition to a more sustainable landscape (Ammann et al. 2023; Cai et al. 2023; Eisen and Brown 2022; Moberg et al. 2021). In this context, the production of meat-based foods stands out for its many negative consequences for sustainability and common well-being. The effects are mainly related to the emission of greenhouse gases (Gerber et al. 2013), excessive use of water (Palhares et al. 2021) and soil (Pereira et al. 2020), inadequate animal welfare practices (Narayanan 2016), public health (Papier et al. 2021) and the risks of new pandemics (Wallace 2016).

Such severe impacts increasingly demand fundamental changes in the protein food production system, with cell-based foods and other alternative proteins such as plant-based and other meat analogues being promising proposals. The fact that the newly cultivated meats deliver real animal protein and not analogues puts this system in a special position. However, structural changes will be required as a radical innovation related to entering an entirely new territory (O'Connor and McDermott 2004) and altering previously known production and market patterns (Dean et al. 2022). These alterations may be related to existing institutional systems, such as laws and regulatory aspects, and social-economic factors, such as education, consumption patterns, food production, and distribution systems.

Considering the many aspects involved in establishing innovative food production systems, this chapter aims to study how public policies and governance action may be addressed to support the transition to cell-based foods as more sustainable and ethical means of protein production and consumption. To achieve this, we addressed public policies as maintainers of the status quo in food production, the perspective of transformative innovation policies, the emergence of cultivated meat, and the structural dimensions of planning and policy actions that may be employed to make the transition process more equitable.

## 2 Public Policies to Maintain the Status Quo in Food Production

Public policies have contributed to establishing the current pathways of food production and consumption. Historically, some of these policies, for instance, provided subsidies to ensure competitiveness in international trade while facilitating access to innovations in inputs and technologies to increase productivity.

According to Baffes et al. (2022), policy interventions in agricultural commodities began to increase after the First World War. As European imports declined, leading to a price drop, government measures were designed in the United States to protect American producers. In Western Europe, policy intervention started after the formation of the European Union in the early 1960s, with the creation of the Common Agricultural Policy. The goal was to keep farm incomes comparable to those of non-farm workers. Emerging markets and developing economies introduced high taxes on crop exports to keep food prices low and to promote food self-sufficiency. The same self-sufficiency strategy was observed in centrally planned economies through public ownership and control over production and trade. So, the main reasons to insert public policies into agricultural commodities were price control, food self-sufficiency and maintaining farmer incomes.

Recent data from the OECD (2022) reveals that the support the agricultural sector has received in recent years has grown. Over the period between 2019 and 2021, the total support reached US$ 817 billion annually for 54 countries, representing an increase of 13% over the US$ 720 billion support from 2018 to 2020. Of the total values, US$ 500 billion per year came from public budgets and US$ 317 billion through higher prices paid by consumers. The European Union and the United States, large agricultural producers, account for two-thirds of this total. Individual producers received 60% of this amount in the form of support through higher market prices paid by consumers or by taxpayers through budgetary transfers. In contrast, public investments in agricultural innovation systems, biosecurity services, and off-farm infrastructure accounted for only US$ 80 billion. These values are represented in Fig. 1. Despite their importance for climate change and food systems goals, they have been low compared to the sector's size.

These data indicate that public policies have played a fundamental role in supporting the current food production system despite its numerous problems. According to OECD (2021), the food system faces a triple challenge: to deliver food security and nutrition for a growing population, to provide incomes and livelihood for people involved in farming and other segments of the chain, and to do this in a sustainable manner. On balance, the agricultural policies do not address these three dimensions effectively.

Regardless of the benefits brought by alternative proteins and the need to transition from traditional production to solutions with less environmental impact and fewer issues concerning animal ethics, both in Europe and the United States, animal production remains widely subsidized by public policies and funding. Powerful vested interests exert their political influence to maintain the status quo. Animal

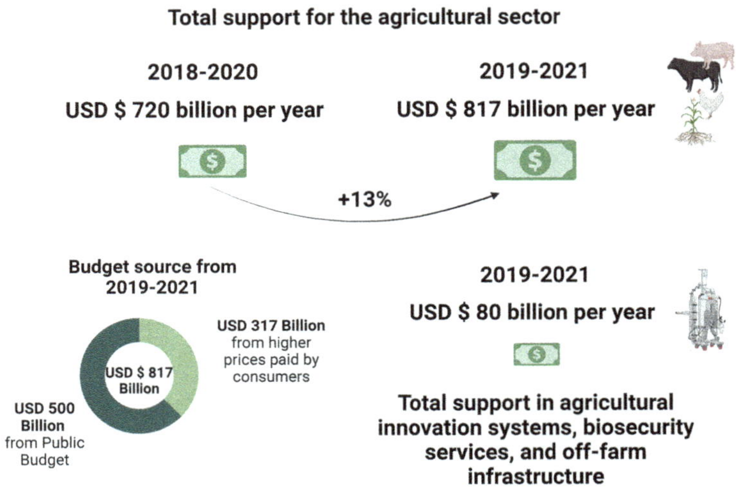


**Fig. 1** Total support for agricultural sector and agricultural innovation systems in 54 countries, adapted from OECD (2022). Created with BioRender.com

production receives the majority of public financial support for food producers and resists the transformation of the food system. Private companies lead the alternative protein products, while governments overlook their potential for climate mitigation. Public funding for new technologies is lower than traditional animal products by a factor of 1200 in the European Union (EU) and 800 in the United States (Vallone and Lambin 2023).

To achieve a higher degree in relation to the food system's sustainability and effectively mitigate climate change, it is imperative to systematically eliminate market price support and payments that have significant potential to harm the environment and disrupt markets and trade. Instead, redirecting budgetary support to encourage environmentally sustainable agricultural production, focusing income support on households facing the most acute needs, and implementing a robust pricing system for agricultural greenhouse gas emissions are crucial steps to incentivize the shift towards low-emission agriculture (OECD 2022). However, although governments have invested in innovations and started to alter regulations in recent years, they have preserved the status quo of animal production and consumption. Despite the need to improve the food system's sustainability, policies have failed to address the environmental impacts of animal production. Political influence has been used to keep the system unchanged and block the competition created by technological innovations (Vallone and Lambin 2023).

Therefore, despite the various challenges related to the continuity of meat food production in the conventional way, a profound change to a more sustainable standard will inevitably depend on changes in public policies. Public policies for solving societal problems are known in the literature as transformative innovation policies.

# 3 Transformative Innovation Policy

Transformative innovation policy may be considered a new paradigm within public policies to support innovation, science, and technology (Diercks et al. 2019). Its originality covers general and global challenges, such as environmental concerns, public health, sustainable food, and population aging (Haddad et al. 2022). Despite its importance, transformative innovation policy is a recent concept. The specific literature has highlighted at least three types of policies to boost science, technology, and innovation. The first is related to government support for research and development to fill private market failures that depend on new knowledge; the second is related to support for the national or regional innovation system in the formation of networks and the connection among agents; the third is the most distinctive of the three and is related to the transformative innovation policy, in which socio-technical systems of production and consumption related to the fulfillment of the Sustainable Development Goals (SDGs) are encouraged (Schot and Steinmueller 2018). The three types of policies are represented in Fig. 2.

The first two types of conventional innovation policies focus on the general increase in competitiveness and economic growth of nations (Haddad et al. 2022). Still, they are considered inefficient in facing global social and environmental challenges (Diercks et al. 2019). Although not replacing the previous types (Diercks et al. 2019), the transformative innovation policy presents the intentionality and directionality of incentives for innovative solutions that meet society's needs as relevant characteristics (Weber and Rohracher 2012). Thus, instead of focusing on a

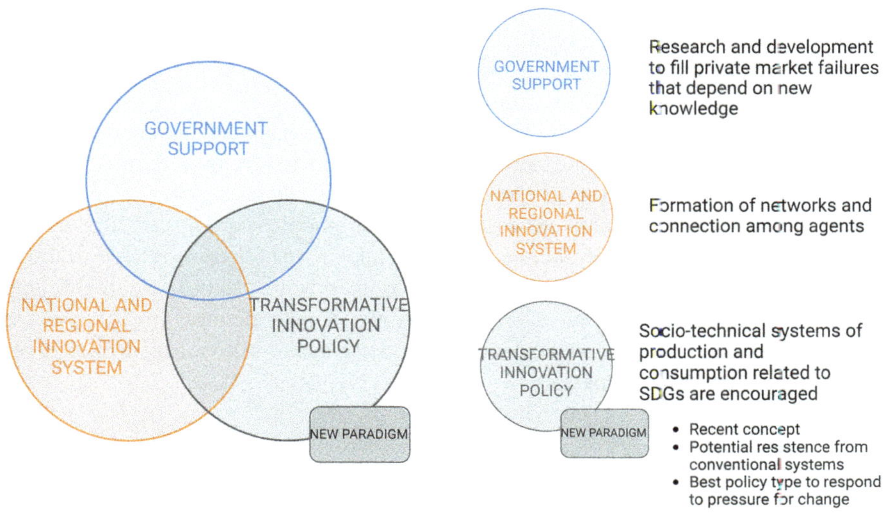


**Fig. 2** Types of policies to boost science, technology, and innovation according to Schot and Steinmueller (2018) and relevant information in the context of innovative meat production systems. Created with BioRender.com

specific domain of knowledge or on the development of an innovative technology that is being demanded by the market, the transformative innovation policy aims at a concrete, multidisciplinary challenge that generally affects the global context socially or environmentally.

By defending that innovation should not be sought only for economic development but mainly to solve major societal challenges (Haddad et al. 2022), transformative innovation policies usually encounter resistance from conventional systems. The literature has already emphasized that industries within dominant systems are likely to resist significant technological changes in their business models (Turnheim and Geels 2012). The dominant regimes begin to suffer when some actors abandon behaviors, beliefs, and values that sustain the current systems (Kainiemi et al. 2019). On the other hand, existing industries, although generally highly resistant to change, begin to abandon the existing system when institutional patterns, such as social behaviors, public policies, and economic issues, increase the pressure for change (Turnheim and Geels 2012).

Although the focus of transformative innovation policy is often evidenced mainly in supporting emerging niche innovations (Fagerberg 2018; Smith and Raven 2012), the implementation of measures to destabilize existing systems is also necessary if they represent obstacles to the socio-technical transition process towards sustainability to support more sustainable innovations (Kivimaa and Kern 2016). Turnheim and Geels (2012) agree that the transition to sustainable systems is multifaceted and includes destabilizing harmful systems or those that contribute little to sustainability. The destabilization of dominant systems is also essential to make room, or windows of opportunity, to advance niche technologies (Ghosh et al. 2021), especially if they are more sustainable and ethical than the previous ones. International organizations and recent studies (Brassesco et al. 2022; Bigliardi and Filippelli 2022; Herrero et al. 2020; Moberg et al. 2021) have warned that future technologies and systemic innovation are crucial for transforming the food system, one of the most relevant and complete socio-environmental impacts.

The literature on transformative innovation policies has been advancing in various directions, pointing to the need for direct action to strengthen and stimulate systems that contribute more to global sustainability, especially new clean energy generation systems and innovations that enhance people's quality of life. In this chapter, we argue that government agents may take a set of actions related to transformative innovation policy to facilitate the transition to a more sustainable and ethical food production system. These policies may accelerate the transition process and make it fairer and more inclusive, with supportive actions to mitigate challenges as changes disturb the status quo.

## 4 The Emergence of Alternative Proteins in a Complex Scenario

Alternative proteins comprise products generally made from plants and cell cultures that can replace conventional meat (Onwezen et al. 2021). However, substantial changes in public policies may be necessary for its insertion into the market. A study on cell-based meat policies reveals that countries and regions justified their support for cultivated meat to implement action plans for decarbonization and food security. However, the main policies observed are related to regulations and focus on safety and security (Miyake et al. 2022). Although these measures are already being taken in some regions around the world, they are still non-existent or incipient in many places.

Other studies also point towards implementing transformative innovation policies in the protein food production system. Herrero et al. (2020) propose accelerating a healthier food production system, including alternative proteins, to reduce the demand for animal farming products, which have a more significant environmental impact and are associated with animal suffering, slaughter, and excessive land use. These actions address specific incentives for alternative sectors, a friendly regulatory environment, and stakeholder dialogue.

Bryant and van der Weele (2021), Morais-da-Silva et al. (2022) and Newton and Blaustein-Rejto (2021) argued that countries may face challenges due to changes in the food production matrix, especially those with sizable conventional production. It is important to consider that challenges are likely inevitable because technological advancement is global, the reasons for changes are clear, and any specific disincentive actions will not be sufficient impediments to stop alternative protein advancement worldwide. However, advantages are not so straightforward if the challenges seem to be posed. To take most advantage of the potential benefits of the emerging sector, including economic, social, and other benefits, countries need to engage in joint efforts, starting with government public policies to strengthen the sector. In addition, superior policies tend to emerge as central actors are heard, such as universities, companies, and society (Morais-da-Silva et al. 2022).

Considering the imminent scenario of change (Gerhardt et al. 2020) and the reasons for the development of alternative proteins (Sinke and Odegard 2021; Heidemann et al. 2020; Nobre 2022), distinct actions are being carried out around the world regarding public policies to support the emerging sector. We highlight, in particular, structural dimensions of planning, including supranational, national, and local plans, as well as specific policy actions that have been applied or that seem important at this time, i.e., at the beginning of the transition to alternative food production systems.

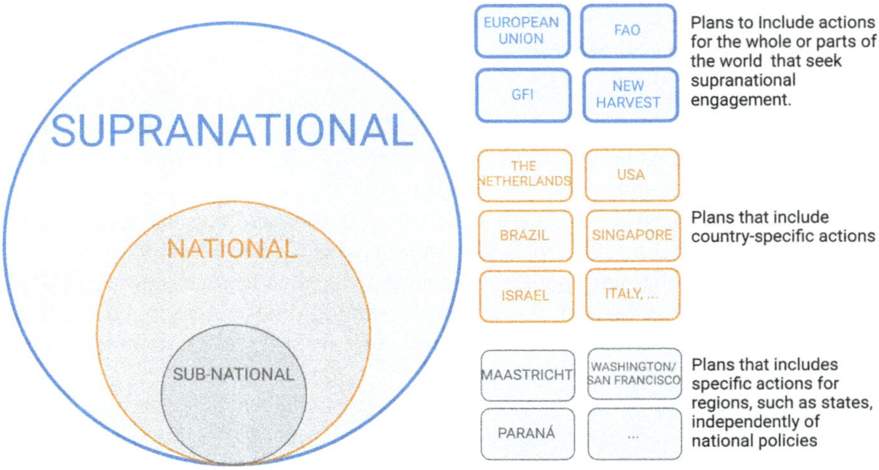


**Fig. 3** The three dimensions of supranational, national, and sub-national planning and examples of the diverse engagements of different government agencies and other institutions in the sector. Created with BioRender.com

# 5 Structural Dimensions of Planning

There are different dimensions of planning and policy development to advance alternative protein fields. In different parts of the world, the landscapes of policies around cellular agriculture seem to be taking different shapes. We present the three dimensions of supranational, national, and sub-national planning and show examples of the diverse engagements of different government agencies and other institutions in the sector. This is a very dynamic area, and this section presents a picture of the three dimensions in 2023, discussed to provide an inspiring and historical description (Fig. 3).

## *5.1 Supranational Plans*

Since transitioning to more sustainable food production systems is a global challenge, some institutions have been drawing up plans and engaging actively in different countries in global integration. For example, the Food and Agriculture Organization of the United Nations (FAO) focuses on supporting governments and development agencies by coordinating activities to improve and develop agriculture and food systems. It operates in 130 countries, and its goal is to achieve food security for all and guarantee that people have regular access to high-quality food. By providing expertise and facilitating international cooperation, the FAO plays an important role in the development of alternative proteins, such as publishing reports that aim to gather state-of-the-art knowledge in the field and making it accessible to different stakeholders (FAO 2023).

As a second case of supranational action, the European Community established the European Green Deal (EGD) in 2019 to be climate-neutral in 2050. The deal has 0.6 trillion € investment from the Next Generation EU recovery plan—part of the EU's 7-year budget (European Commission 2023a). The European community also presented The Farm to Fork in 2020 as a fundamental strategy for the previously mentioned agreement. It aims to make food systems fair, healthier, and environmentally friendly. The strategy seeks to accelerate the transition to a sustainable food system with neutral or positive environmental impact, reverse biodiversity loss, ensure food security, nutrition, and public health, and preserve food affordability while generating fairer economic returns. Although the initiative of plant protein is mentioned, there are more initiatives related to clean energy, as we can observe in the suggestion to the farmers to use renewable energy in their properties (European Commission 2023b). As the strategy has ambitious objectives, proportionally sized initiatives are required for key areas. So far, European countries have adopted different strategies regarding alternative proteins. France and Italy maintain a more conservative position, especially regarding cultivated meat, while The Netherlands, on the opposite side, has made significant investments to develop robust alternative protein chains. The country was one of the first to invest in cellular agriculture and, in 2022, recommitted 60 million € in funding for the development of the industry (GFI 2023a).

In the European context, Cellular Agriculture Europe (Cell Ag EU 2023) is a representative agency for the alternative protein sector. Cell Ag EU has affiliations with several startups in the industry, such as Mosa Meat and Super Meat and leading established firms, such as Nestle, which seek to actively contribute to the advancement of alternative proteins within the European territory. Besides the plans from traditional organizations, there is increasing action by non-governmental organizations, such as The Good Food Institute (GFI) and New Harvest. The GFI is a non-governmental organization founded in 2016 that promotes the transformation of the food system to a more sustainable, secure, and just protein supply (GFI 2023b). They have offices in the United States, Brazil, Israel, India, Asia-Pacific and Europe. The institute aims at changes in the food system with an impact on policies, industry, and science. With a broad range of industry analyses and knowledge-sharing through their reports, besides connecting players, they have made a relevant contribution to the alternative protein field. With an important participation in the definition of labelling, mainly with plant-based products and regulatory frameworks, GFI's contribution to alternative protein policies is present in all countries where they maintain offices.

New Harvest is also a non-governmental organization founded in 2004 to support open and public research on cultured meat. To maximize the impact of cellular agriculture, the organization provides funding for research through research fellowships and seed grants, among other initiatives (New Harvest 2023). New Harvest evaluates the cultured meat ecosystem and fosters collaboration between industry and academia through conferences and dialogues among various stakeholders. This collaborative effort helps build a core community for developing this emerging sector and its leaders. By engaging with diverse stakeholders and being a trusted organization, New Harvest is able to engage in productive discussions on crucial topics within this new sector. Initially operating in Canada and later in other countries, the organization played a significant role in facilitating the production of the first cultured hamburger in 2013 by Professor Mark Post, scientist and founder of Mosa Meat, a cultured meat company in the Netherlands. New Harvest coined the term "cellular agriculture"; interestingly, the term was cited in the United Nations' Intergovernmental Panel on Climate Change (IPCC) Sixth Assessment Report in 2022.

## 5.2 *National Plans*

Some countries have stood out internationally for their development plans for the alternative protein sector. The Netherlands has been in the cultured meat media spotlight since Mark Post's first cultured meat burger in 2013. In 2020, the Dutch Government released its National Protein Strategy (Ministerie van Landbouw, Natuur en Voedselkwaliteit 2020), which aims to consolidate the country's leadership in the alternative protein sector and implement a strategy focused on addressing climate change and biodiversity loss due to food production. According to the GFI (2023a) The Netherlands set the world record for funding cultivated meat and cellular agriculture capabilities by committing $66.2 million to "building a full cellular agriculture ecosystem". That investment will fund cultivated meat and fermentation research, commercialization, education, and workforce transition programs."

On the other side of the Atlantic, a powerful player in food production is also active. The United States attracted worldwide attention in 2023 as two companies (GOOD Meat and UPSIDE Foods) cleared the final regulatory processes and were authorized to be the first ones to commercialize cultivated chicken. These are positive developments since the Biden Administration's 2022 Executive Order on Advancing Biotechnology and Biomanufacturing Innovation for a Sustainable, Safe, and Secure American Bioeconomy (The White House 2023a) and the 2023 biomanufacturing workforce plan (The White House 2023b), which aims to expand training and education opportunities for American citizens in biotechnology and biomanufacturing. Among the specific objectives of these guiding plans, they aim to offer direction to different government organizations regarding the engagement of relevant federal education and training programs. They also recommend new efforts to promote multidisciplinary education programs. A few excerpts of their contents: (1) US education and training programs can meet the rapidly changing skill needs for good jobs and careers in the bioeconomy and increasing demand for workers; (2) a diverse pipeline of workers, including women, people of color, people living in rural communities, individuals with disabilities, justice-involved individuals, individuals returning to the workforce, and others underrepresented in emerging fields, can prepare for, secure, and advance in good bioeconomy jobs and careers;

(3) federal resources are directed at evidence-based education and training approaches that advance equity. In addition, in 2022, according to GFI (2023a) "US Congress appropriated nearly $6 million for alternative protein research for the USDA's Agricultural Research Service, representing a $1 million increase over the previous year."

In May 2022, China published the Five Year Plan for the Development of Bioeconomy, which encompasses synthetic protein as it promotes the development of the biological agriculture industry. The plan identifies four priority areas for the development of the bioeconomy, including biomedicine, bioagriculture, bioenergy, and biological materials, to be at the forefront of the bioeconomy by 2035. It also looks for improvements in market access policies, establishing a biological product and biotechnology system that considers environmental risk assessment (NDRC 2022). The nation's first 5-year plan for the bioeconomy includes exploring alternative proteins as novel foods (Morton et al. 2023).

Under the Government-wide New Zealand-Singapore Enhanced Partnership, the Ministry of Business, Innovation & Employment (MBIE) and the Agency for Science, Technology and Research (A*STAR) have jointly announced the launch of the New Zealand-Singapore Bilateral Research Programme on Future Foods in May 2019. This Programme aims to deepen bilateral cooperation with Singapore across five main pillars: trade and the economy; defense and security; people-to-people; science, technology and innovation; and climate change and sustainability. In 2022, the Enhanced Partnership was extended to include a new pillar focused on climate change cooperation called the "Green Economy" pillar. New Zealand's investments within the Program amount to NZ$11.8 million.

On December 1st, 2020, the Singapore Food Agency (SFA) approved the public sale of Eat Just's cell-grown chicken in Singapore. This was the world's first regulatory approval of a cultured meat product. Singapore's government is supporting innovation in the sector as part of a strategic vision of the city-state to produce 30% of its own food needs by 2030, known as the 30 by 30 goal (MFAT 2020).

In 2015, cultivated meat companies were created in Israel, consolidating their role as pioneers in alternative protein innovation worldwide. Currently, 58 companies and startups are active in the field, 15 of which are in the area of cultivated meat (GFI Israel 2023). The country's investment in alternative protein companies started with $14 million in 2018, growing to $114 million in 2020 and increased 450% in 2021, with Israeli startups raising some $623 million in investments, according to GFI Israel (2021).

In Brazil, the Ministry of Agriculture and Livestock—MAPA—the Ministry of Science, Technology, and Innovation—MCTI—along with the Financer of Studies and Projects—FINEP—a Brazilian public company promoting science, technology, and innovation, allocated up to BRL $50 million in projects for the productive chains of the bioeconomy (MCTI 2022) and to foster new ingredients, alternative proteins, and emerging technologies (MAPA 2022). The MCTI Bioeconomy Productive Chains Program, targeted at both scientific research and businesses, focused on addressing bottlenecks within its chains. Meanwhile, the MAPA FoodTech program aimed to develop innovative solutions aligned with the thematic areas of research networks, technology-based entrepreneurship, and technological challenges.

As with individual responses, countries also vary in their enthusiasm and acceptance towards cultivated meat. Some countries are experiencing an active movement of organizations that aim to slowdown or prevent the development of cellular agriculture. Perhaps the most emblematic case is that of Italy, which in 2023 witnessed the creation of an alliance that brought together 39 associations under the leadership of Coldiretti and Slow Food. They aim to oppose the introduction of cultured food in Italy, reclaiming the defense of conventional meat (Gaddi 2023). However, in October 2023, Italy officially withdrew the notification for the bill that prohibits the production and sale of cell-based meat and the use of names such as burgers, schnitzels, and the like for plant products. No ban can be approved or come into force because it has not been subject to scrutiny and approval by the EU. The withdrawal of the notification can be seen as a way to avoid an official rejection by the European Commission, and at the moment, it puts a stop to the measure (Pomo 2023).

In virtually all countries there are initiatives in opposition to the development of cultivated products and markets, but these are rarely dominant. For some countries, the discussion is delayed by silencing, which may be either an absence of understanding of the magnitude of the meat revolution that is on the horizon or a strategy to deal with novelty. It is logical to expect different attitudes to lead to different actions and results. Thus, the frequently asked question of whether or when cell-based products will reach price parity and be widely available for consumers has no pre-set answer, as this depends on decisions yet to be taken. Within this context, there is a major role for decisions on national public policies in terms of how policies that either favor the status quo or innovative systems will be maintained, abandoned or expanded, and implemented, especially considering timeframes.

A final note on the national dimension of governance is to mention the potential role of independent Cellular Agriculture Associations, nicknamed Cellags. They may result from efforts stemming from the industry, the scientific community, or both and are formed by those interested in contributing to a better understanding of the possibilities of cell-based foods and alternative proteins in general. Cellags have been constituted in many countries, such as the Netherlands, Germany, Portugal, France, Italy, Greece, Canada, Brazil, and Australia; the equivalent Cellular Agriculture Society (CAS) is in the United States. Their role is usually associated with harmonizing efforts, bringing actors together to catalyze results and avoid the hurdles of many people or institutions working individually, needlessly replicating similar initial steps. Depending on the degree of representativeness each Cellag achieves in its country of origin, they may influence public policies directly by providing high quality information and consultancy, and indirectly, through providing significant support at different levels and contexts relevant to developing cell-based food systems and chains.

### 5.3 *Subnational Plans*

It is common for countries to have regional specificities, which holds true regarding food production. For example, it was from Maastricht, in the Netherlands, that in 2013 came the first cell-based hamburger. Later, an important project on milk production by precision fermentation started in Wageningen. Are these initiatives pointing towards the establishment of regional cellular agriculture niches within the Netherlands? The government has put efforts into developing clusters of specialization in different regions of the country and, in this sense, making them poles to attract and concentrate professionals and investment, fostering the development of these regions. This is an interesting question also because it may show how key subnational plans may be, especially when considering key food production countries with vast geographical areas such as the United States, Australia and Brazil.

Some States are historically bolder in pushing for innovations in the United States, as epitomized by Silicon Valley in California. By 2023, two startups had the approval from the US Department of Agriculture (USDA) to market cell-based products. Good Meat, a subsidiary of Eat Just and Upside Foods got the approval to produce and commercialize cell-cultivated chicken, and their products are first going to be served in specific locations in Washington DC and San Francisco (CA).

Subnational planning is also emerging in Brazil, an important player in food-producing activities worldwide. Regarding funding, two States in Southern Brazil were pioneers in using public money to fund cellular agriculture projects: Rio Grande do Sul and Paraná, the latter being more recent and more substantial. State government support in Paraná allowed for the establishment of a network of research groups called the New Arrangement for Research and Innovation in Alternative Proteins (NAPI PA). This project, funded by the Fundação Araucária, the Paraná State agency that supports scientific research, technology and innovation, aims to strengthen the state of Paraná as a food producer and to promote cell-based products with added value by using local breeds of cattle and pigs.

Federal public policies may tend to the subnational level of planning if they foster regionally customized actions. In addition, state-level public policies may be more dynamic and agile, even if they are less robust in funding capacity, especially when state funding agencies are present. Regional policies may provide an interesting aid to the development of cellular agriculture, as they are inherently tailored to the local pre-existing strengths as well as specific bottlenecks. Their efficiency is likely highest when they are embedded in well-designed national policies.

## 6 Public Policy and Governance Actions

The public policy plans are, in general, guidelines for future actions, which in the case under consideration here concern alternative proteins. In addition, more specific governance actions may also be taken. Based on Biscarra-Bellio et al. (2023), fifteen distinct actions have been classified within three main domains: societal dialogue, changes in policies and regulations, and the design of market initiatives. We highlight four of such actions for their especially relevant role in the development of cellular agriculture. These are: the support of dialogue between university and industry, transparency, education and training, and support for conventional producers willing to transition.

## 6.1 Support for Dialogue Between University and Industry

The innovation development process is recognized as complex and non-linear, with the participation of several actors (Carayannis and Campbell 2009, Etzkowitz and Leydesdorff 2000). Despite this traditional understanding, several barriers still hinder collaboration between actors to develop innovations (Bruneel et al. 2010). We will especially highlight the difficulties and the potential for collaboration between industry and university, as they are at the frontier of developing products based on cultured meat. According to Wilkinson (2023), until recently the industry used to incorporate small changes to adapt to demands from society, frequently as alterations in the final stages of production such as labels with renewed terminology, a process known as delayed innovation. However, industries perceive benefits in embracing more radical innovation (Wilkinson 2023). This is followed by impressive private research funding, which tends to be more efficient when collaborating with academia; otherwise, the chances of publicizing results are much lower. If this is so, why is there a need to support the dialogue between university and industry?

On one hand, companies traditionally focus on capturing relevant knowledge to gain a competitive advantage in the market (Dasgupta and David 1994). For these organizations, the knowledge created is primarily hidden within the organization or disclosed superficially, mainly for patent registration and, consequently, market monopoly for some time (Dasgupta and David 1994). On the other hand, universities are more focused on general social welfare and knowledge sharing. Even though they are immersed in a system that is unfavorable to frontier information sharing, as the current scientific publication system adopts a posture that hampers the wide dissemination of new knowledge, including high costs to access scientific articles and the inability to access raw research data (Budzinski et al. 2020; Molloy 2011), academic institutions remain the stronghold of knowledge sharing. As more research data and articles are made freely available, the degree of transparency of the scientific process and the overall social benefit will increase (Molloy 2011).

Accordingly, attention has been drawn to the partnership between industry and university since this collaboration's potential contribution to innovation is recognized (Nsanzumuhire and Groot 2020). In the case of cultured meat as a radical innovation in food systems, there remain many challenges to be overcome through scientific research, mainly concerning technological development, costs, and ingredients. The literature has highlighted, for example, that the use of fetal bovine serum, something ethically controversial and costly, is one of the great challenges to overcome (Soccol et al. 2023). However, the company Eat Just recently announced that it had received approval from the Singapore Food Agency to use fetal bovine serum-free culture media (The Straits Time 2023). Therefore, the collaboration between the two groups of actors may evolve beneficially for all.

A major advantage is that current challenges in food production technology via cell cultivation may be more quickly resolved if university and industry collaboration were prioritized. Universities offer distinct scientific expertise, experience with cutting-edge research, already established laboratories and legitimacy in relation to research results, interpretation and conclusions. The industries offer real challenges that they have been facing in their development structures, in addition to support in the financing of collaborative research. Through specific public policies for this type of collaborative research, the government may provide its share of the necessary resources. In this way, all actors win and scientific progress in the field moves faster and more efficiently, as harmonized efforts build a solid pillar for this evolution.

## *6.2 Transparency*

Cellular agriculture technology has been mobilizing promises and expectations over the past decades. However, incipient and modest efforts to provide adequate information to the general public, capable of fostering science-based debate, make it difficult for an educated conversation on cultivated meat by society. Amcore (2020) pointed to the problem of the "black box" when constructing a public debate around the societal impacts and implications of technology. According to the same author, transparency and accountability are made difficult to access, reflect and discuss by the wider public due to the opacity of information given by technology developers. Also, according to Beumer and Edelenbosch (2019), when assessing potentially disruptive innovation, the research and the expectations are commonly articulated within the walls of individual organizations, making it difficult to have an informed societal debate.

An additional important factor to consider is that a significant portion of the entrepreneurs in the field of cellular agriculture either are or used to be scientists in the field. Thus, attention to the claims made by such actors benefits from considering both positionalities and the interests involved when the same individual is concomitantly a scientist and a capitalist entrepreneur. "When wearing a public entrepreneurial hat, they might make strident claims about the promise of their research, but when among research peers, they will be much more cautious and equivocal, though publicly still committed to the promises associated with their field" (Borup et al. 2006). One major challenge of the field regarding transparency is the unequally distributed access to knowledge, where information is contested and strategically used according to the different actors and interests that play unequal roles in the debate (Brown and Michael 2003). Finally, language is a bottleneck to be overcome as it can hinder transparency and fruitful debate. It is important to translate scientific language to comprehensible information for everyday conversational use. In addition, because English has become the official language in research and science, essential terms that are required to refer to aspects of new technologies are invariably in English. Thus, for countries where English is not the first language, there is a fundamental need to establish specific terminology in the local language, as this is a matter of local accessibility to the discussion (Garbin et al. 2024).

The importance of transparency and the pitfalls to avoid may be better understood by considering the lack of transparency in conventional animal farming. In part, the intensive animal farming sector has responded to the growing criticism it receives by hiding their practices and problems. The most symbolic example of this strategy is the passage of ag-gag laws in some American states, which contain provisions prohibiting taking or possessing photographs, video, or audio recordings from an animal farm without the farm owner's consent. Robbins et al. (2016) have studied how awareness of ag-gag laws eroded trust in farmers and demonstrated that even the intention to restrict access to information could undermine trust. A proactive stance fostering transparency is likely the best approach to a sound relationship between the food industry and society, allowing consumers full access to knowledge and understanding of all aspects of their food choices.

Therefore, we believe that efforts to provide full access to definitions, opportunities and challenges, as well as scientific advances related to alternative proteins, are warranted. As with any field, the essential knowledge to make informed decisions must be within the reach of public debate. This seems even more relevant because cellular agriculture involves human nutrition and deep-rooted cultural aspects.

## *6.3 Education and Training*

In addition to the relevant role of universities in knowledge sharing, they are the place where professionals get the needed qualifications and abilities for adequate performance in their jobs. In the special case of developing a completely new and worldwide food industry, a spate of new jobs requiring new education will likely involve all stages of the cell-based food production chain. Although novelty touches all stages, it is estimated that the biggest transformations occur upstream in the production chain, as cell-cultivation processes aim to substitute animal raising and slaughter processes. We take Brazil as a concrete case to discuss the challenges that seem to lie ahead in terms of education and training, for reasons related to the country's sizeable meat industry, the availability of scientific papers discussing the potential social impacts of cellular agriculture for Brazil, and our more direct knowledge regarding the context of this country.

The low qualification of the Brazilian workforce, especially in the initial stage of conventional animal food production, i.e., farms and slaughterhouses, may be a relevant bottleneck for the development of cellular agriculture. The arrival of alternative proteins may help raise worker qualifications since most job opportunities in the new production chains will be for workers with higher qualifications (Morais-da-Silva et al. 2022). It remains open to debate whether the necessary efforts, including public policies and governance, will be made to convert this bottleneck into an opportunity. This highlights the immeasurable importance of appropriate policies.

In alignment with research initiatives, Brazilian universities are gaining momentum in teaching and training activities. The Federal University of Paraná (UFPR), Brazil, in partnership with the GFI, offered three editions of the course "Introduction to cellular animal science" in the postgraduate Program in Veterinary Sciences, starting in 2020, and an optional course was approved in 2023 for undergraduate programs in veterinary medicine and animal sciences. The State University of Campinas (UNICAMP), in the State of São Paulo, offered the course "Alternative Proteins: Made of Plants, Fermentation and Cultured Meat," in the postgraduate program of the Faculty of Food Engineering in 2022. In other universities where cultivated-meat research is conducted, there is evidence of a ripple effect in education, as students get involved and the discussions ignite on campus, in a sort of teaching through the grapevine. For instance, the Federal University of Minas Gerais recently released its cell-based chicken proof of concept, as their students organized their participation in a network of students interested in cellular agriculture maintained by the GFI, regardless of the absence of an organized undergraduate course in this university so far.

Additional initiatives around education and training have been taken in Brazil by the GFI. For instance, the GFI offers an open-access online course "The Science of Alternative Proteins," in Portuguese. Thus, it is evident that there are increasing opportunities to learn about cellular agriculture in Brazil, including formal initiatives that can evolve to contribute to fulfilling the demand for new professionals in the nascent alternative protein industry.

Many countries have also initiated relevant initiatives. In the United States, The Cellular Agriculture & Bio Fabricated Foods, organized by Tufts University, is a permanent course in the Biomedical Engineering Department. In the Cellular Agriculture & Cultured Meat Lab, students designed and taught the first undergraduate lab course for cellular agriculture techniques through the Tufts Experimental College. The University of North Carolina (UNC) launched one of the first classes focused exclusively on cellular agriculture. The course was developed in partnership with students from the Chapel Hill Alt Protein Project, a student organization developed with support from the Good Food Institute. In Australia, the association "The Cellular Agriculture Australia (CAA)" launched the country's first online course on cellular farming in conjunction with industry, academia, and regulatory organizations.

In Singapore, Nanyang Technological University (NTU) launched the first university course on alternative protein. The undergraduate course "Future Foods—Introduction to Advanced Meat Alternatives" was developed in partnership with The Good Food Institute.

In the Netherlands, a proposal led by Wageningen University & Research, in collaboration with Maastricht University and Delft University of Technology, has been put forward to develop three courses about precision fermentation, cultured meat, and the societal and sustainability aspects of cellular agriculture.

In Israel, Technion (Israel Institute of Technology) announced the establishment of the “Sustainable Protein Research Center.” Addressing the global challenges of climate change, human health, and nutrition. Technion launched a multidisciplinary research center for sustainable proteins, which will function as a hub for fundamental and applied research in alternative proteins, commercialization, entrepreneurship, and research support for the industry and education. In Canada, the University of the Fraser Valley has also been presenting some initiatives around training.

Besides new courses on cellular agriculture, a more transversal approach seems also warranted. In Brazil, according to Biscarra-Bellio et al. (in preparation) there are some mandatory courses in veterinary medicine, animal science, bioengineering and food engineering programs that, despite their focus on conventional animal protein production or research, can be adapted and expanded to relate to different stages of the cell-based meat chain. Additionally, there are courses present in some programs that have more specific content directly related to the new technology of cell-based meat production, which can serve as a basis for creating new courses to be implemented in the curricula of all interested programs. The structuring of syllabi, courses and programs will be necessary for the preparation of qualified labor in order to mitigate the challenges and foster the opportunities of this new labor market.

## *6.4 Support for Producers*

The agribusiness sector is vital for certain countries and relevant to global food security. Therefore, any alteration in this sector can trigger numerous consequences for different stakeholders and perhaps higher risks for producers. Thus, it is common for public policies to support farmers. Created in 2003, the Brazilian “Plano Safra” is a successful example of a public policy that supports producers regarding financing. However, such policies must adapt to the increasingly prevalent new reality: producers who need to diversify or transition from animal production. For example, in 2022 the Dutch government established a policy of acquiring properties to reduce the emission of environmentally harmful gases banned by the European Union (Boztas 2022).

If cellular agriculture takes a prominent position in the total protein market, leading to progressive decreases in the market share of conventional producers, the most affected producers may be those on small farms due to the lack of scale and consequently less competitive advantage (Morais-da-Silva et al. 2022). Research conducted in 2022 and 2023 with cattle and broiler producers in Australia and Brazil, two of the main producers and exporters of animal protein, revealed that some producers are seeking alternatives to animal production. However, they reported a lack of knowledge on choosing a new business model and the absence of structured demand. In addition, there is no government support or policies to assist them in making this decision and economically surviving the risks of transitioning. As such, some farmers reported feeling at a loss and prone to failure, with no financially viable alternative business (Bogueva et al. 2023).

Public policies can be developed to assist producers in capitalizing on the opportunities of alternative proteins within the emerging market. As outlined in Bogueva et al. (2023), some key points were identified by the producers to mitigate risks in diversification. They are (1) identifying new demands for primary products considering liquidity, financial return, and suitability for the region's soil and climate, (2) sharing information and success stories in a manner accessible to producers, and (3) establishing the market and supply chain for cellular meat.

For identifying the primary products considering demand, liquidity, financial return, and suitability for the region's soil and climate, policies encouraging research and projects in collaboration with public institutions and a multidisciplinary team of professionals—economists, agronomists, and veterinarians—are necessary. Ideally, such a collaboration can identify market needs and assess the product's viability in the specific production region. As for sharing information and success stories in a manner accessible to producers, traditional outreach programs may be launched. Research findings can be disseminated to producers through pamphlets and local events, enabling them to access information on how best to address eventual challenges they may face. Likewise, promoting best practices is always welcome, fostering a virtuous cycle of success.

Finally, producers need some level of risk control to feel safe in diversifying their activities, and the establishment of the market and supply chain for cellular meat can aid risk reduction. The alternative protein sector is in its developmental stages, and policies that encourage this market, ensuring producers some demand, are essential. For instance, traditional agricultural products are promoted in the media and the stability of a strong alternative protein industry depends also on a similar level of promotion to consumers. This approach tends to increase information, exposition and awareness among consumers, thereby reducing potential neophobia. Additionally, the supply chain and other aspects of the alternative protein sector need further development. The collaboration among the government, market stakeholders, and producers to determine the optimal model for each stage of the chain is crucial for the scalability of this market.

## *6.5 A Broader View*

The establishment of new production chains for alternative proteins to feed humans and animals is based on biotechnological advances that have been considered the second domestication, which is the domestication of cells (Tubb and Seba 2019). This points to the enormity of the transformation that has just started, a major movement in the way humans produce food, only comparable with the domestication of animals 10,000 years ago. With this magnitude, the recruitment of a complex range of public policies and additional governance actions is required. Thus, a dynamic

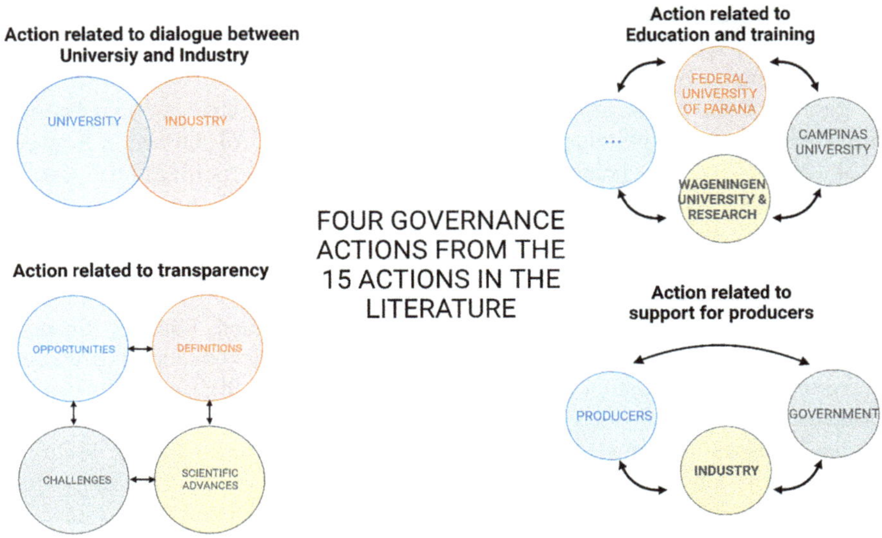


**Fig. 4** Summary of four governance actions selected from the 15 actions presented by Biscarra-Bellio et al. (2023). Created with BioRender.com

process of maturing the fifteen actions such as research and teaching, dialogue, transparency, awareness, inclusion, basic education, differentiate strategies, differentiate countries, inclusions as private goals, policies for transition, funding programs, marketing regulation, the inclusion of externalities, tax benefits and market incentives, proposed by Biscarra-Bellio et al. (2023) is to be expected. Investments in additional research on the fundamentals of governance seem strategically sound for this radical innovation to achieve its full disruptive potential in an equitable and just manner, which leaves no one behind. For the actions described, collaboration and developing a network that facilitates understanding gaps and needs are essential. It is not reasonable to expect that the incredible advances provided by biotechnology suffice to advance in relation to the many social needs involved; a proportional movement must be enacted in public policies and governance (Fig. 4).

## 7 Challenges and Future Developments

Forecasts indicate that a significant portion of the protein market will come from alternative proteins in 10–15 years. Some of these projections suggest that 22% of the market will be alternative proteins by 2035 (Witte et al. 2021), while others indicate 60% by 2040, with cultivated meat responsible for 35%, as stated by Gerhardt et al. (2020). Several variables can affect these predictions, including consumer acceptance, market supply and development, regulatory advancements, and levels of investment and public policies. Regarding public policies, they seem to require reforms to change the current food systems, as current public policies primarily focus on the existing production system, maintaining the status quo. Therefore, a natural first task is to identify the changes required to encourage alternative proteins through public policies and governance.

In Herrero et al. (2020), actions are proposed to accelerate the development of a more sustainable food system. The speed of innovation depends on appropriate incentives, regulations, and social license, which, in turn, are related to constructive dialogue with stakeholders and transparency regarding the paths to follow. Some actions are suggested to accelerate the transformation. To build trust among the stakeholders in the food system is a key point in this process. For systemic changes to occur, industries must identify the opportunity; governments must recognize the need for change to achieve prosperity and well-being; a dialogue with citizens must be conducted to change attitudes; and public policies and investments that encourage such changes must be implemented. Considering those points, governments can lead in facilitating and communicating why and how to innovate, and a high-level agreement on new directions seems fundamental (Herrero et al. 2020).

Despite the benefits, alternative proteins also pose challenges, frequently mentioning social issues. In Morais-da-Silva et al. (2022), both social opportunities and challenges were identified. Among the opportunities are the provision of inputs to the new sector, new jobs and better conditions and wages for workers, the possibility of reducing food insecurity, and a greater ethical consideration for both human and non-human animals. Among the challenges are the reduced demand for conventional animal protein and reduced employment in the first stages of the conventional chain (Morais-da-Silva et al. 2022), which may be felt in the long term. However, if planning is not carried out in time, the challenges may have a significant impact, as conventional animal production is currently of great importance to the income of numerous families and the economy of some countries, including Brazil. The study concludes that the threats will occur regardless of engagement with the new sector because the scenario depicts an imminent change in the global consumption pattern. In contrast, the opportunities will be seized only by the countries that invested and developed the alternative protein chain. Thus, public policy and governance for the development of cellular agriculture seem extremely important to reduce potential negative social impacts and create social opportunities.

## 8 Conclusion

Our study sought to clarify how public policies may be better addressed to support the transition to cell-based foods as more sustainable means of protein production and consumption. Our discussions led to a series of cases of supranational, national, and local agents developing medium and long-term planning actions and more specific ones.

The development of public policy and other aspects of governance are essential to guide public investments and highlight the alternative protein sector in social, investment, and technological development discussions. The public sector may also take specific actions in partnership with institutions of other dimensions, such as support for collaborative research and transparency about all aspects of cellular agriculture. We also argue that there is a need for significant advances in education and training so that people may seek jobs in the emerging sector. Finally, we emphasize that conventional farmers, primarily linked to family animal farming, need support to adapt and diversify their production to seize the possible opportunities. Thus, the role of public policies remains central in the transition to alternative proteins, even considering a partial transition. Governments worldwide have taken both general and more specific actions, but many countries have not yet presented decisive actions regarding the issue. As this is a complex path, it is likely that the sooner each country discusses and implements the best public policies and actions, the better prepared it will be for this emerging sector and all the accompanying opportunities.

As technology advances faster with increasing levels of financing or funding, it may become harder for inactive countries to jump in. This wave of technological development may represent numerous benefits in terms of sustainability, animal ethics, and food production capacity, but it also means new global economic arrangements. Countries that prefer to stay out of this new technology in food production may suffer the negative consequences of this potential transition without reaping part of the benefits due to the technological and implementation delays they will face. Therefore, it is highly recommended that a transformative innovation policy capable of directing technological development towards a level of collective well-being be designed to foster alternative proteins, along with proper governance actions.

**Acknowledgment** We would like to express our sincere gratitude to Professor John Wilkinson, Federal Rural University of Rio de Janeiro, for kindly reviewing this chapter.

## References

Ammann J, Arbenz A, Mack G, Nemecek T, El Benni N (2023) A review on policy instruments for sustainable food consumption. Sustain Prod Consumpt 36:338–353. https://doi.org/10.1016/j.spc.2023.01.012

Amoore L (2020) Cloud ethics: Algorithms and the attributes of ourselves and others. Contemp Polit Theory 21:118–121. https://doi.org/10.1057/s41296-021-00465-1

Baffes J, Kabundi A, Koh WC, Nagle P, Ohnsorge F, Vasishtha G, Yamazaki T, Zahid H (2022) Commodity markets: evolution, challenges, and policies. World Bank, Washington. https://openknowledge.worldbank.org/server/api/core/bitstreams/6b5eb213-7390-46a0-815a-bf6b9bceb03a/content

Beumer K, Edelenbosch R (2019) Hybrid potato breeding: A framework for mapping contested socio-technical futures. Futures 109:227–239. https://doi.org/10.1016/j.futures.2019.01.004

Bigliardi B, Filippelli S (2022) A review of the literature on innovation in the agrofood industry: Sustainability, smartness and health. Eur J Innov Manag 25(6):589–611. https://doi.org/10.1108/EJIM-05-2021-0258

Birkland TA (2019) An introduction to the policy process: Theories, concepts, and models of public policy making, 5th edn. Routledge, New York
Biscarra-Bellio JC, Oliveira GB, Marques MCP, Molento CFM (2023) Demand changes meat as changing meat reshapes demand: The great meat revolution. Meat Sci 196. https://doi.org/10.1016/j.meatsci.2022.109040
Bogueva D, Marques M, Molento CFM, Marinova D, Philips CJC (2023) Will the cows and chickens come home? Perspectives of Australian and Brazilian beef and poultry farmers towards diversification. Sustainability 15(16):12380. https://doi.org/10.3390/su151612380
Borup M, Brown N, Konrad K, Van Lente H (2006) The sociology of expectations in science and technology. Tech Anal Strat Manag 18(3-4):285–298. https://doi.org/10.1080/09537320600777002
Boztas S (2022) Up to 3,000 'peak polluters' given last chance to close by Dutch government. The Guardian. 30 Nov 2022. https://www.theguardian.com/environment/2022/nov/30/peak-polluters-last-chance-close-dutch-government. Accessed 26 July 2023
Brassesco ME, Pintado M, Coscueta ER (2022) Food system resilience thinking: from digital to integral. J Sci Food Agric 102(3):887–891. https://doi.org/10.1002/jsfa.11533
Brown N, Michael M (2003) A sociology of expectations: retrospecting prospects and prospecting retrospects. Tech Anal Strat Manag 15:3–18. https://doi.org/10.1080/0953732032000046024
Bruneel J, d'Este P, Salter A (2010) Investigating the factors that diminish the barriers to university–industry collaboration. Res Policy 39(7):858–868. https://doi.org/10.1016/j.respol.2010.03.006
Bryant CJ, van der Weele C (2021) The farmers' dilemma: Meat, means, and morality. Appetite 167:105605. https://doi.org/10.1016/j.appet.2021.105605
Budzinski O, Grebel T, Wolling J, Zhang X (2020) Drivers of article processing charges in open access. Scientometrics 124:2185–2206. https://doi.org/10.1007/s11192-020-03578-3
Cai W, Wang R, Zhang S (2023) Efficient food systems for greater sustainability. Nat Food 4(7):541–542. https://doi.org/10.1038/s43016-023-00780-3
Carayannis EG, Campbell DF (2009) 'Mode 3' and 'Quadruple Helix': toward a 21st century fractal innovation ecosystem. Int J Technol Manag 46(3–4):201–234. https://doi.org/10.1504/IJTM.2009.023374
Cellular Agriculture Europe—Cell Ag EU (2023) https://www.cellularagriculture.eu/about-us/. Accessed 22 Oct 2023
Dasgupta P, David PA (1994) Towards a new economics of science. Policy Res 23(5):487–521. https://doi.org/10.1016/0048-7333(94)01002-1
Dean T, Zhang H, Xiao Y (2022) The role of complexity in the Valley of Death and radical innovation performance. Technovation 109:102160. https://doi.org/10.1016/j.technovation.2020.102160
Diercks G, Larsen H, Steward F (2019) Transformative innovation policy: Addressing variety in an emerging policy paradigm. Res Policy 48(4):880–894. https://doi.org/10.1016/j.respol.2018.10.028
Dye TR (1972) Understanding public policy. Prentice-Hall, Englewood Cliffs, NJ
Eisen MB, Brown PO (2022) Rapid global phaseout of animal agriculture has the potential to stabilize greenhouse gas levels for 30 years and offset 68 percent of CO2 emissions this century. PLoS Climate 1(2):e0000010. https://doi.org/10.1371/journal.pclm.0000010
Ericksen PJ (2008) Conceptualizing food systems for global environmental change research. Glob Environ Chang 18(1):234–245. https://doi.org/10.1016/j.gloenvcha.2007.09.002
Etzkowitz H, Leydesdorff L (2000) The dynamics of innovation: from National Systems and "Mode 2" to a Triple Helix of university–industry–government relations. Res Policy 29(2):109–123. https://doi.org/10.1016/S0048-7333(99)00055-4
European Commission (2023a) A European green deal striving to be the first climate-neutral continent. https://commission.europa.eu/strategy-and-policy/priorities-2019-2024/european-green-deal_en. Accessed 12 Aug 2023
European Commission (2023b) Farm to fork strategy for a fair, healthy and environmentally-friendly food system. https://food.ec.europa.eu/horizontal-topics/farm-fork-strategy_en. Accessed 12 Aug

Fagerberg J (2018) Mobilizing innovation for sustainability transitions: A comment on transformative innovation policy. Res Policy 47(9):1568–1576. https://doi.org/10.1016/j.respol.2018.08.012

FAO (2023) Food and Agricultural Organization of the United Nation (2023). Food System. https://www.fao.org/food-systems/en/. Accessed 12 Aug

Gaddi M (2023) Coldiretti e Slow Food guidano la grande alleanza contro la carne coltivata. https://www.gamberorosso.it/notizie/coldiretti-e-slow-food-guidano-alleanza-contro-la-carne-coltivata/. Accessed 17 July 2023

Garbin V et al (Accepted in 2024) Ética animal e proteínas alternativas: questões atuais em terminologia da língua portuguesa. Archives of Veterinary Science

Gerber PJ, Steinfeld H, Henderson B, Mottet A, Opio C, Dijkman J, Falcucci A, Tempio G (2013) Tackling climate change through livestock: a global assessment of emissions and mitigation opportunities. Food and Agriculture Organization of the United Nations (FAO). http://www.fao.org/3/a-i3437e.pdf

Gerhardt C, Suhlmann G, Ziemssen F, Donnan D, Warshun M, Kühnle HJ (2020) How will cultured meat and meat alternatives disrupt the agricultural and food industry? Ind Biotechnol 16(5):262–270. https://doi.org/10.1089/ind.2020.29227.cge

GFI (2023a) 2020/22 State of global policy report: Public investment in alternative proteins to feed a growing world. https://gfi.org/wp-content/uploads/2023/01/State_of_Global_Policy_Report_2022.pdf. Accessed 13 Aug 2023

GFI (2023b) The Good Food Institute. https://gfi.org/. Accessed 12 Aug 2023

GFI Israel (2021) Israel State of Alternative Protein Innovation Report. https://gfi.org.il/wp-content/uploads/2021/04/InnovationReport_2021-5.pdf. Accessed 9 Oct 2023

GFI Israel (2023) Israel State of Alternative Protein Innovation. https://gfi.org.il/wp-content/uploads/2023/03/SOIR-2023-19.03.pdf. Accessed 6 Nov 2023

Ghosh B, Kivimaa P, Ramirez M, Schot J, Torrens J (2021) Transformative outcomes: assessing and reorienting experimentation with transformative innovation policy. Sci Public Policy 48(5):739–756. https://doi.org/10.1093/scipol/scab045

Haddad CR, Nakić V, Bergek A, Hellsmark H (2022) Transformative innovation policy: A systematic review. Environ Innov Societal Transit 43:14–40. https://doi.org/10.1016/j.eist.2022.03.002

Heidemann MS, Molento CFM, Reis GG, Phillips CJC (2020) Uncoupling meat from animal slaughter and its impacts on human-animal relationships. Front Psychol 11:1824. https://doi.org/10.3389/fpsyg.2020.01824

Herrero M, Thornton PK, Mason-D'Croz D et al (2020) Innovation can accelerate the transition towards a sustainable food system. Nat Food 1:266–272. https://doi.org/10.1038/s43016-020-0074-1

Kainiemi L, Eloneva S, Levänen J (2019) Transition towards a decentralised energy system: analysing prospects for innovation facilitation and regime destabilisation in Finland. Technol Anal Strat 31:1003–1015. https://doi.org/10.1080/09537325.2019.1582765

Kivimaa P, Kern F (2016) Creative destruction or mere niche support? Innovation policy mixes for sustainability transitions. Res Policy 45(1):205–217. https://doi.org/10.1016/j.respol.2015.09.008

Kraft ME, Furlong SR (2021) Public policy: politics, analysis, and alternatives, 7th edn. Sage CQ Press, Thousand Oaks, CA

MAPA (2022) Edital vai selecionar projetos de FoodTechs sobre novos ingredientes, proteínas alternativas e tecnologias inovadoras de alimentos. https://www.gov.br/agricultura/pt-br/assuntos/noticias-2022/edital-vai-selecionar-projetos-de-foodtechs. Accessed 12 Oct 2023

MCTI (2022) MCTI e FINEP anunciam editais no valor de R$40 milhões para cadeias produtivas da bioeconomia. https://www.gov.br/mcti/pt-br/acompanhe-o-mcti/noticias/2022/07/mcti-e-finep-anunciam-editais-no-valor-de-r-40-milhoes-para-cadeias-produtivas-da-bioeconomia. Accessed 12 Oct 2023

MFAT (2020) Singapore: alternative proteins market report. https://www.mfat.govt.nz/en/countries-and-regions/asia/singapore/. Accessed 09 Oct 2023

Ministerie van Landbouw, Natuur en Voedselkwaliteit (2020) Nationale Eiwitstrategie. Publicatie-nr. 0920-020. https://www.rijksoverheid.nl/documenten/kamerstukken/2020/12/22/nationale-eiwitstrategie. Accessed 23 Oct 2023
Miyake Y, Tachikawa M, Kohsaka R (2022) Policy frameworks and regulations for the research and development of cell-based meats: Systematic literature review. Food Res Int 167. https://doi.org/10.1016/j.foodres.2023.112599
Moberg E et al (2021) Combined innovations in public policy, the private sector and culture can drive sustainability transitions in food systems. Nat Food 2:282–290. https://doi.org/10.1038/s43016-021-00261-5
Molloy JC (2011) The open knowledge foundation: open data means better science. PLoS Biol 9(12):e1001195. https://doi.org/10.1371/journal.pbio.1001195
Morais-da-Silva RL, Reis GG, Sanctorum H, Molento CFM (2022) The social impact of the transition from conventional to cultivated and plant-based meats: evidence from Brazil. Food Policy 111:102337. https://doi.org/10.1016/j.foodpol.2022.102337
Morton J, Gandhi D, Pang V (2023) Alternative proteins in APAC: 2023 State of the Industry report. https://gfi-apac.org/the-state-of-apacs-alt-protein-industry-in-10-charts/?utm_source=reimagining_protein&utm_medium=email&utm_campaign=https%3A%2F%2Fgfi-apac.org%2Fthe-state-of-apacs-alt-protein-industry-in-10-charts%2F&utm_content=11.8.2023. Accessed 13 Nov 2023
Narayanan Y (2016) Where are the animals in sustainable development? Religion and the case for ethical stewardship in animal husbandry. Sustain Dev 24(3):172–180. https://doi.org/10.1002/sd.1619
NDRC (2022) The National Development and Reform Commission (NDRC) People's Republic of China. Development plan for the biological economy for the 14th five-year plan period. https://www.ndrc.gov.cn/xxgk/jd/jd/202205/t20220509_1324417_ext.html. Accessed 13 Nov 2023
New Harvest (2023). https://new-harvest.org/history/. Accessed 12 Aug 2023
Newton P, Blaustein-Rejto D (2021) Social and economic opportunities and challenges of plant-based and cultured meat for rural producers in the US. Front Sustain Food Syst 5:624270. https://doi.org/10.3389/fsufs.2021.624270
Nobre FS (2022) Cultured meat and the sustainable development goals. Trends Food Sci Technol 124:140–153. https://doi.org/10.1016/j.tifs.2022.04.011
Nsanzumuhire SU, Groot W (2020) Context perspective on University-Industry Collaboration processes: A systematic review of literature. J Clean Prod 258:120861. https://doi.org/10.1016/j.jclepro.2020.120861
O'Connor G, McDermott C (2004) The human side of radical innovation. J Eng Technol Manag 21:11–30. https://doi.org/10.1016/j.jengtecman.2003.12.002
OECD (2021) Agricultural policy monitoring and evaluation 2021: addressing the challenges facing food systems. OECD, Paris. https://doi.org/10.1787/ec6959aa-en
OECD (2022) Agricultural policy monitoring and evaluation 2022: reforming agricultural policies for climate change mitigation. OECD, Paris. https://doi.org/10.1787/7f4542bf-en
Onwezen MC, Bouwman EP, Reinders MJ, Dagevos H (2021) A systematic review on consumer acceptance of alternative proteins: Pulses, algae, insects, plant-based meat alternatives, and cultured meat. Appetite 159:105058. https://doi.org/10.1016/j.appet.2020.105058
Palhares JCP, Morelli M, Novelli TI (2021) Water footprint of a tropical beef cattle production system: The impact of individual-animal and feed management. Adv Water Resour 149:103853. https://doi.org/10.1016/j.advwatres.2021.103853
Biscarra-Bellio JC, Paolini G, Molento CFM (in preparation) Possibilities for higher education in alternative proteins in Brazil
Papier K, Fensom GK, Knuppel A et al (2021) Meat consumption and risk of 25 common conditions: outcome-wide analyses in 475,000 men and women in the UK Biobank study. BMC Med 19(1):1–14. https://doi.org/10.1186/s12916-021-01922-9

Pereira EJDAL, de Santana Ribeiro LC, da Silva Freitas LF, de Barros Pereira HB (2020) Brazilian policy and agribusiness damage the Amazon rainforest. Land Use Policy 92:104491. https://doi.org/10.1016/j.landusepol.2020.104491

Pomo C (2023) Clamoroso dietrofront su carne coltivata e censura ai prodoti veg! Carne coltivata e censura veg: l'Italia fa dietrofront! www.essereanimali.org. Accessed 20 Oct 2023

Robbins JA, Franks B, Weary DM, Von Keyserlingk MAG (2016) Awareness of Ag-Gag laws erodes trust in farmers and increases support for animal welfare regulations. Food Policy 61:121–125. https://doi.org/10.1016/j.foodpol.2016.02.008

Schot J, Steinmueller WE (2018) Three frames for innovation policy: R&D, systems of innovation and transformative change. Res Policy 47(9):1554–1567. https://doi.org/10.1016/j.respol.2018.08.011

Sinke P, Odegard I (2021) LCA of cultivated meat Future projections for different scenarios. https://canadacommons.ca/artifacts/2009833/lca-of-cultivated-meat-future-projections-for-different-scenarios/2762276/. Accessed 23 Oct 2023

Smith A, Raven R (2012) What is protective space? Reconsidering niches in transitions to sustainability. Res Policy 41(6):1025–1036. https://doi.org/10.1016/j.respol.2011.12.012

Soccol CR, Karp SG, Letti LAJ et al (2023) Is the development of low-cost media one of the greatest challenges to produce cultivated meat on an industrial scale? Biotechnol Res Innov J 6(2):e2022005. https://doi.org/10.4322/biori.20226201

The Straits Time (2023) Eat Just receives approval from SFA to produce serum-free cultivated meat. The Straits Time. https://www.straitstimes.com/singapore/eat-just-receives-approval-from-sfa-to-produce-serum-free-cultivated-meat. Accessed 20 Oct 2023

The White House (2023a) Executive order on advancing biotechnology and biomanufacturing innovation for a sustainable, safe, and secure American bioeconomy. https://www.whitehouse.gov/briefing-room/presidential-actions/2022/09/12/executive-order-on-advancing-biotechnology-and-biomanufacturing-innovation-for-a-sustainable-safe-and-secure-american-bioeconomy/. Accessed 22 Oct 2023

The White House (2023b) Fact sheet: Biden-Harris administration announces new bold goals and priorities to advance American biotechnology and biomanufacturing. https://www.whitehouse.gov/ostp/news-updates/2023/03/22/fact-sheet-biden-harris-administration-announces-new-bold-goals-and-priorities-to-advance-american-biotechnology-and-biomanufacturing/. Accessed 16 Oct 2023

Tubb C, Seba T (2019) Rethinking food and agriculture 2020–2030: the second domestication of the cow, and the collapse of industrial livestock farming. https://faunalytics.org/rethinking-food-and-agriculture-in-2019/. Accessed 23 Oct 2023

Turnheim B, Geels FW (2012) Regime destabilisation as the flipside of energy transitions: Lessons from the history of the British coal industry (1913–1997). Energy Policy 50:35–49. https://doi.org/10.1016/j.enpol.2012.04.060

Vallone S, Lambin EF (2023) Public policies and vested interests preserve the animal farming status quo at the expense of animal product analogs. One Earth 6:1–14. https://doi.org/10.1016/j.oneear.2023.07.013

Wallace R (2016) Big farms make big flu: dispatches on influenza, agribusiness, and the nature of science. NYU Press, New York

Weber KM, Rohracher H (2012) Legitimizing research, technology and innovation policies for transformative change: Combining insights from innovation systems and multi-level perspective in a comprehensive 'failures' framework. Res Policy 41(6):1037–1047. https://doi.org/10.1016/j.respol.2011.10.015

Wilkinson J (2023) O mundo dos alimentos em transformação, 1st edn. Appris, Curitiba. 235 p

Witte B et al (2021) Food for thought, the protein transformation. Report from BCG end Blue Horizon. https://web-assets.bcg.com/a0/28/4295860343c6a2a5b9f4e3436114/bcg-food-for-thought-the-protein-transformation-mar-2021.pdf. Accessed 25 July 2022

# Patents and Innovations in Cultivated Meat Production

**Susan Grace Karp, Leonardo Wedderhoff Herrmann, Giuliana Biagini, Ana Paula Boligon, and Carlos Ricardo Soccol**

**Abstract** This chapter presents the technological mapping of cultivated meat production. A patent search was carried out in the Derwent Innovations Index Database using a combination of synonyms of "cultivated meat" in the field Topic. From this search, 219 documents were selected and classified according to the publication year, the countries of publication, whether the document was submitted to an international application, the assignees, and the International Patent Classification (IPC) codes. The curve of patent publications per year started its exponential phase in 2019, and the trend line is still rising sharply, meaning that cultivated meat technologies are in the growth stage. China, South Korea and the United States contributed to most publications. The top assignees were Jiangnan University, Shanghai CellX Biotechnology, Nanjing Agricultural University, China Meat Research Center and Nanjing Joes Future Food Technology from China, Yonsei University Industry Foundation from South Korea, Aleph Farms from Israel, and Northwest University, Tufts College and Terasaki Biomedical Innovation Institute from the United States. The most frequent IPC codes were C12N-005/077 (58%) and A23L-013/00 (47%). The documents were grouped and analyzed within different technology areas to describe the recent advances and innovations focusing on culture media, bioreactors and cultivation methods, microcarriers and scaffolds, and product formulations.

**Keywords** Technological mapping · Patent search · Technological information · Innovation

---

S. G. Karp (✉) · G. Biagini · A. P. Boligon · C. R. Soccol
Bioprocess Engineering and Biotechnology Department, Federal University of Paraná, Curitiba, Paraná, Brazil
e-mail: susan.karp@ufpr.br

L. W. Herrmann
Bioprocess Engineering and Biotechnology Course, Polytechnic School, Positivo University, Curitiba, Paraná, Brazil


C. R. Soccol et al. (eds.), *Cultivated Meat*,
https://doi.org/10.1007/978-3-031-55968-6_19

# 1 Introduction

Patent documents are sources of technological information and contain data that can be used for technological mapping or technological forecasting. The maturity stage, the major market players, the market perspectives, and the future trends of technologies can be assessed or predicted from patent documents. Since the patent US6835390-B1 from 2001 mentioned the act of providing subjects with non-human meat products suitable for consumption (Vein 2001), offering strong indications of potential future developments in cultured meat, several technologies were developed. This chapter focuses on the technological mapping of cultivated meat production, with special attention to culture media, bioreactors and cultivation methods, microcarriers and scaffolds, and product formulations.

A patent search was carried out in the Derwent Innovations Index Database, which contains patent documents that have their titles and abstracts reviewed and rewritten according to the patent content. For this reason, it is a suitable database for patent searches based on keywords. The search was carried out on September 6th, 2023, using a combination of synonyms of "cultivated meat," namely [artificial or synthetic or in vitro or ex vivo or cell(ular)-based or (cell)-culture(d) or vat-grown or lab(oratory)-grown or animal-free or slaughter-free or animal-cell or test-tube or clean or engineered or cellular or in vitro cell(ular)-culture(d)] and [meat or beef or steak], and their orthographic variants, and the terms [cell(ular)-based food or cell(ular)-culture(d) food or cultured food or cultivated food or animal-cell food or in vitro cell(ular)-culture(d) food] in the field Topic (Title+Abstract), without limiting a time frame.

The patent search resulted in 834 patent documents, of which 219 were in fact related to cultivated meat. These 219 documents were selected by reading the titles, abstracts, and, when necessary, the claims of the patent documents. The technological mapping was performed by analyzing the publication year, the countries of publication and whether the document was submitted to an international application, the assignees (owners of intellectual property), and the International Patent Classification (IPC) codes, as defined by the World Intellectual Property Organization (WIPO 2023). Also, a set of patent documents were selected to describe the advances in culture media, bioreactors and cultivation methods, microcarriers and scaffolds, product formulations, and others.

# 2 Patent Search Results

## 2.1 *Number of New Patent Publications*

The timeline of patent publications, namely of the new patent documents published per year, is presented in Fig. 1. It is noticeable that, despite the first documents dating from 2001 (Vein 2001) and 2004 (Vein 2004), the increase in patent publications

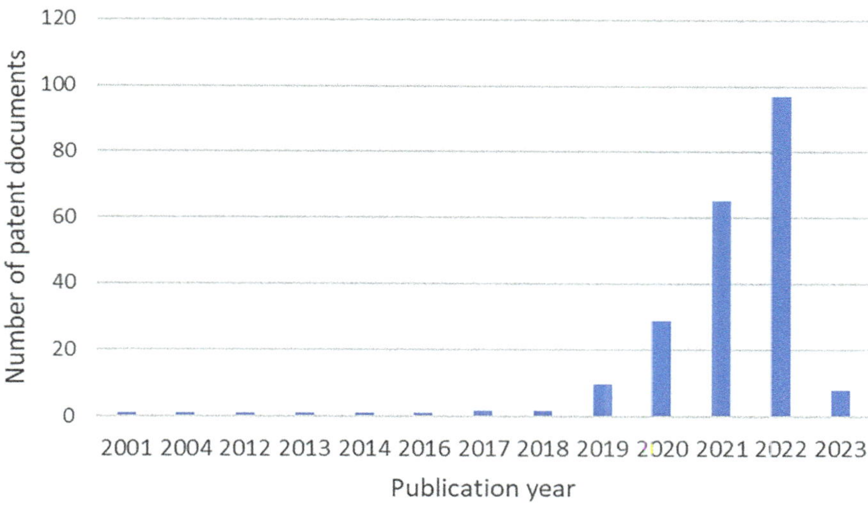


**Fig. 1** Number of new patent publications per year

related to cultivated meat started in 2019 and the trend line is still rising sharply. This means that cultivated meat technologies are in the growing stage and have not reached maturity. The decrease observed in 2023 is due to the secrecy period, i.e., the time interval between the filing date and the publication date of a patent document, which is usually 18 months.

## 2.2 *Classification of Patent Documents by Publication Country and International Applications*

The patent documents were classified according to the country of publication (Fig. 2). A major part of the documents (44%), however, was published as international applications (identified as WO) under the Patent Cooperation Treaty (PCT). This is usually an indication that intellectual property will be required in different countries once the international evaluation phase is accomplished. Among these international applications, 28% were originated in the United States, 19% in Israel, and 14% in South Korea. Concerning national publications, China accounts for 31% of the patent documents, followed by South Korea with 15% and the United States with 5%.

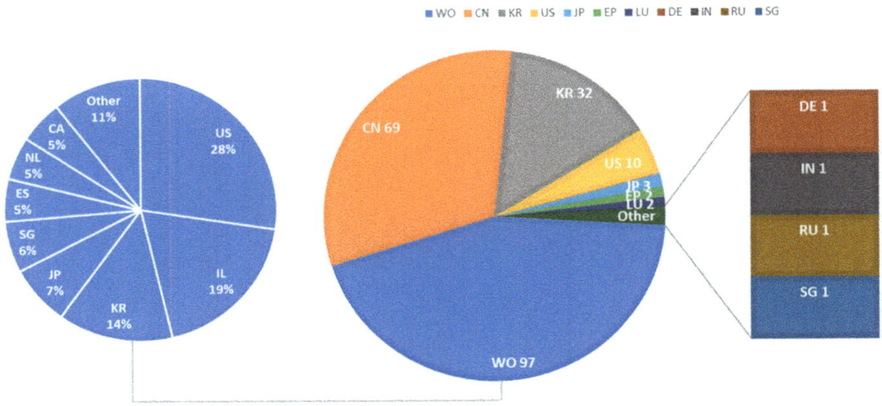


**Fig. 2** Distribution of patent documents by country of publication and international applications published under the Patent Cooperation Treaty (PCT). *Note*: International applications (WO) were classified by their countries of origin. WO, International PCT application; CN, China; KR, South Korea; US, United States; JP, Japan; EP, European Patent Office; LU, Luxembourg; DE, Germany; IN, India; RU, Russia; SG, Singapore; IL, Israel; ES, Spain; NL, Netherlands; CA, Canada

## 2.3 Top-10 Assignees

The top-10 assignees were Jiangnan University, Shanghai CellX Biotechnology, Nanjing Agricultural University, China Meat Research Center and Nanjing Joes Future Food Technology from China, Yonsei University Industry Foundation from South Korea, Aleph Farms from Israel, and Northwest University, Tufts College and Terasaki Biomedical Innovation Institute from the United States (Fig. 3).

## 2.4 Most Frequent IPC Codes

The IPC codes are used to classify the patent documents according to their technology areas. The most frequent IPC code, present in 58% of the selected documents, was C12N-005/077, which is classified within the area of "Undifferentiated human, animal or plant cells, e.g., cell lines; Tissues; Cultivation or maintenance thereof; Culture media therefor" (C12N-005/00), "Animal cells or tissues" (C12N-005/07), and more specifically "Mesenchymal cells, e.g., bone cells, cartilage cells, marrow stromal cells, fat cells or muscle cells" (C12N-005/077). In the second position, the IPC code A23L-013/00, representing "Meat products; Meat meal; Preparation or treatment thereof," was present in 47% of the documents. The distribution of the most frequent IPC codes is presented in Fig. 4.

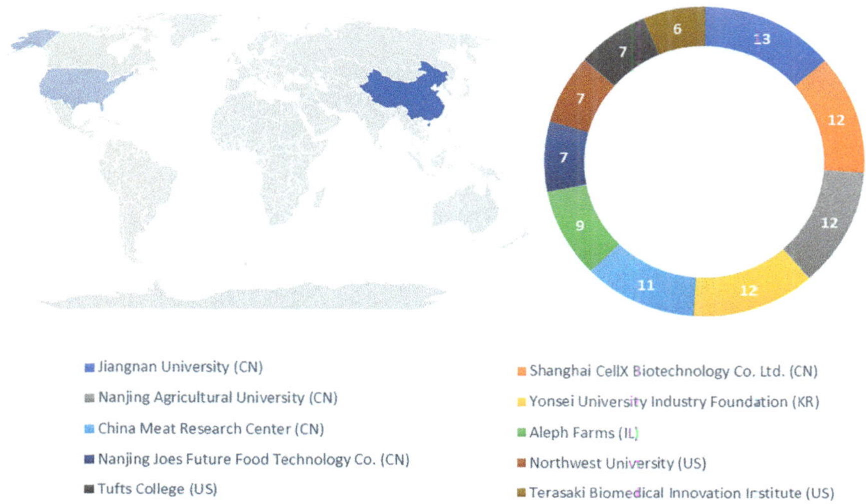


**Fig. 3** Top-10 patent assignees by institution/company and country, and the respective number of patent documents. *Note*: CN, China; US, United States; KR, South Korea; IL, Israel

## 3 Patents and Innovations in Culture Media

The culture media for proliferation and differentiation of animal cells used in the production of cultured meat is one of the main factors that contributes to the high cost of the process (Soccol et al. 2022). Due to the complexity of animal cells, the culture media must provide all the necessary nutrients for each stage of the production process, such as growth factors, salts, vitamins, amino acids, and many others, making the formulation costly. Furthermore, contrary to the original idea of avoiding the slaughter of animals, the most studied culture media for animal cells contain various components of animal origin such as fetal bovine serum (FBS), horse serum, enzymes, growth factors, and others (Good Food Institute 2022; Stout et al. 2022; Ye et al. 2022). Therefore, many studies focus on the elimination or reduction of the use of animal components in culture media while reducing process costs and maintaining favorable cell yields and multiplication rates.

Some patent documents describe culture media compositions for cultured meat production that aim to partially replace the use of animal serum, such as the document KR2021090560-A of the Sejong University, South Korea (Park et al. 2021), which presents a culture medium containing powdered milk serum as a substitute for FBS. This medium has been shown to be partially effective as a substitute for animal serum, with the potential to reduce costs in the production process of cultivated meat. However, it still contains ingredients of animal origin.

The document KR2022040417-A filed by Yonsei University, South Korea (Hong 2021a), describes a culture medium composition capable of improving the proliferation and differentiation rates of cells, containing egg white in combination with

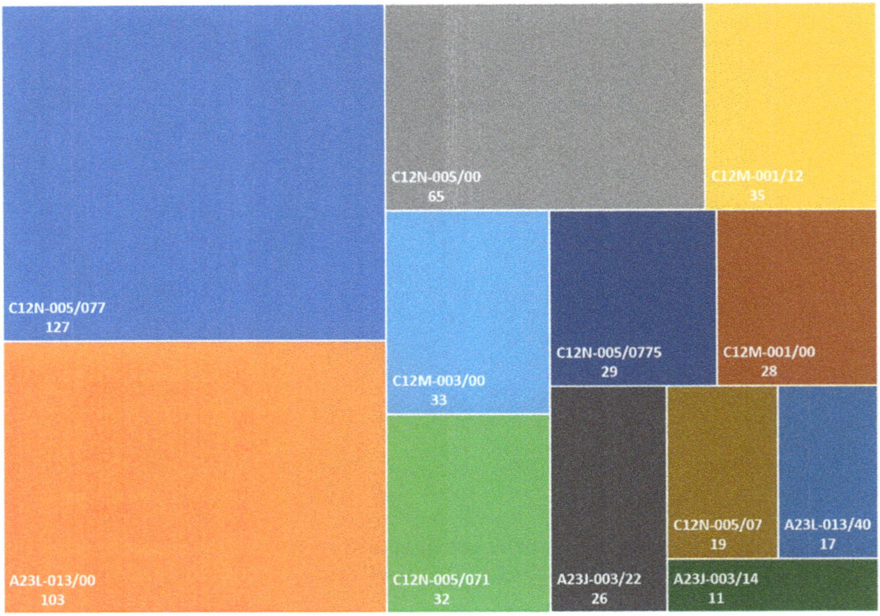


**Fig. 4** The most frequent IPC codes, present in >5% of the patent documents and the respective number of patent documents. *Note*: Definition of IPC codes; the middle dots represent the level of classification. A23J-003/14: (Working-up of proteins for foodstuffs) ·Vegetable proteins. A23J-003/22: (Working-up of proteins for foodstuffs) ·by texturizing. A23L-013/00: Meat products; Meat meal; Preparation or treatment thereof. A23L-013/40: ·containing additives. C12M-001/00: Apparatus for enzymology or microbiology. C12M-001/12: ·with sterilization, filtration, or dialysis means. C12M-003/00: Tissue, human, animal or plant cell, or virus culture apparatus. C12N-005/00: Undifferentiated human, animal or plant cells, e.g., cell lines; Tissues; Cultivation or maintenance thereof; Culture media thereof. C12N-005/07: ·Animal cells or tissues. C12N-005/071: ··Vertebrate cells or tissues, e.g., human cells or tissues. C12N-005/077: ···Mesenchymal cells, e.g., bone cells, cartilage cells, marrow stromal cells, fat cells or muscle cells. C12N-005/0775: ····Mesenchymal stem cells; Adipose-tissue derived stem cells

sericin and ingredients derived from microalgae, where the active ingredient obtained from microalgae is C-phycocyanin. Another patent that reports the use of animal-derived components in the culture medium for cultivated meat is the document WO2023127970-A1 filed by NH Foods, Japan (Segawa et al. 2022a), which presents the use of whey as a cell growth promoter, being economical and free of animal serum.

Compositions of culture media without the addition of animal components and derivatives were also identified in several patent documents. A serum-free and animal-component-free culture medium used in the expansion of muscle satellite cells was developed by Kaplan et al. (2021), with the use of recombinant albumin, which improves growth capability in at least 50% compared to the baseline B8 media. The medium, described in the patent document WO2022125809A1 filed by Tufts College, contains sugars (0.01–10 g/L), amino acids (0.001–5 g/L), vitamins

(0.001–1 g/L), minerals (0–15 g/L), trace elements (0.0001–10 mg/L), growth-stimulating and/or cell-signaling factors (0.01–100,000 ng/mL), and carrying proteins (0.01–1 g/L). The medium contains recombinant versions of interleukin-6, ethanolamine, curcumin, oleic acid and linoleic acid, and recombinant albumin in concentrations between 50 mg/L and 60 g/L. The medium may also contain L-ascorbic acid 2-phosphate, recombinant insulin, recombinant transferrin, sodium selenite, recombinant fibroblast growth factor 2, recombinant neregulin 1, and recombinant transforming growth factor beta 3.

The document WO2021148955-A1 filed by the Yissum Research and Development Company of the Hebrew University of Jerusalem and Future Meat Technologies, Israel (Nahmias et al. 2021b) describes a culture medium for cultivated meat containing plant proteins, such as vegetable albumin, vegetable catalase, vegetable insulin, among other proteins, as substitutes for the animal components found in sera. Another proposed alternative involves the use of recombinant animal growth factors associated with vegetable proteins, as described in the patent document WO2021245711-A1 filed by ORF Liftaekni Ehf, Iceland (Orvar et al. 2021).

The patent WO2023086502-A1 by Virginia Tech Intellectual Properties, United States (Ovissipour et al. 2022) describes a composition to be used in cultured meat production composed of algal protein hydrolysate, vegetable protein hydrolysate, fungus protein hydrolysate, insect protein hydrolysate, and multicellular marine organism protein hydrolysate. This composition can be used for reducing or eliminating the serum content of traditional basal media formulations while also increasing cell viability and prolonging cell lifespan. Another composition, obtained from algae and described in the document CN115287256-A filed by the Ocean University of China (Guo et al. 2022), contains an aqueous extract of blue algae that can completely replace animal serum in the medium for cultivated meat production (fish muscle). The inventors claim it is simple, safe, and suitable for industrial production.

The patent document WO2021148960-A1, filed by the Yissum Research and Development Company, Israel (Nahmias et al. 2021a) presents a culture medium composed of carbohydrates, amino acids, vitamins, minerals, and serum replenishment components such as albumin, growth factors, enzymes, attachment factors, and hormones. The document WO2022211461-A1 filed by the HyupSung University and Danagreen, South Korea (Han et al. 2022) reports a medium composed of turmeric, glycine, or insulin, which is useful for culturing muscle stem cells and preparing cultured meat. The advantages of this medium are the absence of basic fibroblast growth factor (bFGF) and the ability of the medium to maintain muscle stem cells in an undifferentiated state, making it cost-effective and allowing for a more appropriate selection of the time for muscle stem cell differentiation.

A method for cell differentiation with a defined chemical composition was developed by the Nanjing Joes Future Food Technology, China (CN115927172-A), to improve the efficiency of adipogenic differentiation of fibroblasts in vitro and induce the formation of larger and more numerous lipid droplets in cultured cell meat production (Li et al. 2022a). This culture medium is composed of a serum-free basal differentiation medium and supplemental factors for cell culture. As claimed by the inventors, the technology presents a more efficient, cost-effective, and safe method for cultivated meat to form more fat.

The method for cultivated meat preparation described in the document WO2023278317-A1, assigned to Air Protein, United States (Reed and Robertson 2022), involves adding a protein hydrolysate obtained from the biomass of a chemo-autotrophic microorganism to a culture medium for cultivated meat production without animal serum. This method can minimize the variability between batches of culture media and/or culture medium supplements, reduce the use of animal-derived products, and improve or enhance the flavor of a cultivated food product.

The patent documents CN114752554-A and CN114752554-B filed by Jiangnan University, China (Chen et al. 2022), describe the addition of growth factors to animal serum-free differentiation medium containing natural products derived from plants and other co-factors for cultivated meat production. This composition allows for protein synthesis comparable to that obtained in a differentiation medium containing 2% horse serum. Thus, it is possible to reduce the costs of cultivated meat production in the differentiation process and ensure the biosafety of the final product. The patent CN115141798-A filed by CellX Biotechnology, China (Huang and Wang 2022), presents a culture medium free of animal serum composed of basal medium, fatty acid-free albumin, ascorbic acid or its derivatives, insulin or its analogues, transferrin, selenite, growth factor, and neuregulin. This medium is useful for growing multiple generations of muscle stem cells for a long period, consequently reducing the cost of developing and producing cultivated meat.

## 4 Patents and Innovations in Bioreactors and Cultivation Methods

### *4.1 Bioreactors*

Among the patent documents retrieved from the patent search, 20 were dedicated to bioreactor models and cultivation systems. According to Nahmias and Wissotsky (2021), the two main processes for large-scale cultivation of cells are fed-batch and perfusion. In the first process, carried out in reactors of up to 25 $m^3$, nutrients are fed to the cells until the toxins' concentration reaches a limit (usually ammonia 5 mM) and cellular density reaches 30 million cells/mL. In perfusion processes, conducted in reactors of up to 5 $m^3$, the medium is continuously replaced by filtering the suspension, usually through hollow fibers, so the toxins are eliminated, and cell concentration reaches up to 270 million cells/mL.

The patent document WO2021152536-A1 filed by Future Meat Technologies and the Yissum Research Development Company of the Hebrew University of Jerusalem (Nahmias and Wissotsky 2021) described a receptacle comprising several hollow fibers, surrounded by an absorbent material and useful for filtering waste materials from fluids such as cell culture media and recycling cell culture media, reducing or eliminating waste molecules along the growth or differentiation of cultured cells. The waste materials are smaller than 60 kDa, including ammonia, lactate, toxins, and sodium salts. The receptacle can be placed inside a bioreactor to produce cultivated meat (also claimed). The patent document WO2018011805-A2, also filed by Yissum (Nahmias 2017), proposed a cultivation system based on perfusion for a small-scale (5 L) bioreactor. The system also contains a delivery system to deliver a solution to the bioreactor chamber for perfusion through the cells; a system for performing dialysis and a filter for reducing ammonia content in dialysate; a controller to circulate the perfusion solution out of the bioreactor through the dialyzer and back into the bioreactor, and to circulate the dialysate out of dialyzer through the filter and back into the dialyzer. The system provides that at least 90% of the perfusion solution that exits the bioreactor is recirculated into the bioreactor during one growth period.

A great concern in the large-scale production of cultivated meat is the high cost of nutrients and growth factors provided in the culture medium. In this sense, the patent document WO2023012523-A1 filed by Avant Meats (Chin et al. 2022) described a fed-batch bioreactor that automatically replenishes nutrients and removes growth inhibitors produced by cells while retaining unused growth factors. The system maintains a stable level of nutrients and a minimum level of growth inhibitors, reduces wastes discharged from the cell culturing system, and increases cell mass production. Cell growth can be carried out in a three-dimensional platform, in microcarriers, or in suspension. The bioreactor comprises a dialysis unit formed by a dialysis membrane of at least 500 Da configured to replenish nutrients to the cells and remove metabolic waste (ammonia and lactate) from the bioreactor to dialysate; a fresh medium unit connected to the dialysis unit, configured to supply the dialysate to the dialysis unit; and a waste removal unit connected to the fresh medium unit, configured to remove the metabolic waste from the dialysate. The waste removal unit is based on two biocatalysts: a nitrifying bacterium to remove ammonia and an enzyme to break down lactate. The resulting product is pyruvate that is fed to the culture medium to promote cell growth.

It is known that most cell lines used in the production of cultivated meat require a surface of contact, which is provided either by microcarriers or by structured scaffolds. The patent document WO2022097139-A1, filed by Aleph Farms (Lavon et al. 2021), described a cultivation system based on multiple scaffolds intended for the large-scale production of cultured cells. The scaffolds are distributed inside the bioreactor and a pump allows liquid circulation among the planar porous scaffolds (2–1.5 mm pores), with an inlet flow rate of about 0.5 volume exchanges per day. The thickness of the scaffolds ranges from 1 mm to 5 cm, and the area of each of the two faces of the scaffolds is at least 50 $cm^2$.

The document WO2020141236-A1, filed by Biotech Foods (Vila Juárez 2019), proposed a bioreactor model for the cultivation of adherent cells composed of a chamber, a cover, and a support formed by at least one tray covered with a three-dimensional polymer-based element. A generator of air bubbles is placed at the bottom of each tray, which should be submerged in the culture medium. The support comprises a series of electrical contacts capable of getting supplied with alternating or direct current to stimulate cell proliferation.

Two patent documents describing bioreactor models were filed by the China Meat Research Center. The first, WO2022068029-A1 (Wang et al. 2020a), described a 3D biological tissue culture tank and a liquid storage tank connected by a pipeline to enable the circulation of the culture medium. The opening of the 3D biological tissue culture tank is equipped with a sealing plug, which in turn is provided with multiple filling needles that penetrate the biological tissue during use. The second, WO2023016357-A1 (Liu et al. 2021), presented a device for meat cultivation comprising a culture medium adjusting tank, a cell proliferation tank, a muscle separating tank, a compression molding device, and a control system for each unit that allows automated control of the whole cultivation process.

## *4.2 Cultivation Methods*

The patent documents herein classified as "cultivation methods" are mostly related to cell culturing techniques to obtain high cell yield and productivity. As an example, the patent document WO2021092587-A1 filed by Memphis Meats (Genovese 2020) claimed a method to increase animal cell culture density and thickness in a structured substrate through the activation of a Transcriptional Enhanced Associate Domain (TEAD) protein or a homolog thereof in the cells, achieved by the introduction of a polynucleotide sequence encoding the TEAD protein into the cells. The patent document CN113862225-A assigned to the Nanjing Agricultural University (Tang et al. 2021) proposed a method to promote muscle stem cell proliferation and maintain stem gene expression under high-density conditions, useful to provide seed cells for large-scale cultured meat production. The method is based on the supplementation of activators and inhibitors of the Hippo pathway, specifically to activate the Hippo pathway effect factor—the yes-associated protein 1 (YAP) gene—and to inhibit the intermediate enzyme gene in muscle stem cell Hippo pathway—the Macrophage Stimulating 1 (MST1), the MST2, the Large Tumor Suppressor Kinase 1 (LATS1) or the LATS2.

Yonsei University in South Korea developed a series of methods to prepare cultured meat by preparing cell sheets coated with nanofilms, stacked as multi-layered cell sheets similar to a muscular structure. In the patent document WO2022019688-A1 (Hong et al. 2021), the nanofilm is formed by a positively charged material (chitosan, starch, collagen, gelatin, fibrinogen, silk fibroin, casein, elastin, laminin, or fibronectin) and a negatively charged material (hyaluronic acid, alginate, tannic acid, lignin, cellulose, heparin, carrageenan, agar, xanthan gum, arabic gum, glucomannan, carboxymethylcellulose or tara gum). In the patent document KR2022040424-A (Hong 2021b), the method involved introducing a porous polymer support to a substrate for cell culture, seeding cells on the porous polymer support, culturing the cells as a single-layered cell sheet, immobilizing the cell sheet to a substrate, and laminating the cell sheets. In this example, the porous polymer is alginate, gelatin, agar, starch, collagen, fibrinogen, chitosan and/or cellulose.

## 5 Patents and Innovations in Microcarriers and Scaffolds

Scaffolds and microcarriers (MCs) are tridimensional matrices used for cultivating adherent cells, which is the case of the cells used for cultivated meat production. The main difference between these matrices is that MCs are usually applied to expand cell culture and scaffolds are more useful for cell differentiation and tissue maturation (Singh et al. 2023). However, both are necessary to mimic the biological environment of the cells and each matrix composition and structure can induce specific cell responses (Biagini et al. 2021; Pisani et al. 2023). In the cultivated meat scenario, the scaffold could also influence the cultivated meat's characteristics, such as texture, flavor, appearance, and even nutritional value (Singh et al. 2023).

The significance of 3D matrices in cell culture also arises from the influence of scaffold rigidity on cellular responses. This phenomenon, known as mechanotransduction, is not entirely elucidated. Nonetheless, studies suggest that this process involves a synergy of mechanosensing molecules (such as integrins that act on cell-matrix adhesion), nuclear attributes, and the cell's genetic makeup (Song et al. 2020). Hence, the matrix's rigidity is important not only to recreate the extracellular matrix but also to control certain cell behaviors, such as differentiation and migration (Guimarães et al. 2020; Pisani et al. 2023). In cultivated meat production, this is crucial to ensure the product's quality and the efficiency of the production process.

There is no pattern to the scaffold and MC's composition and structure, and there should not be, as research has indicated that distinct cell types exhibit varying responses to the same scaffold configuration (Biagini et al. 2021). Thus, each 3D matrix should be adapted to the cell type used and the cell culture's specific purpose.

This phenomenon allows for a myriad of scaffold compositions to be formulated and explored. In the case of cultivated meat, it is also interesting that the scaffold exhibits edible properties. In fact, an analysis of 69 patents claiming scaffold compositions for cultivated meat production showed that only two of them proposed the use of synthetic non-edible polymers (Chintala et al. 2016; Johnson and Ohst 2020). Even so, they were biodegradable polymers that ideally would degrade in contact with saliva or before the meat is consumed. The pattern of scaffold composition found in the patent documents is illustrated in Fig. 5. Most patents described the use of more than one material and/or the combination of several materials from different sources to produce one scaffold. Therefore, the number of patents mentioned in the graph totals up to more than 69.

In general, the preferred composition is made up of natural proteins, polysaccharides, and polymers. Out of the 69 patents, 58 (84%) proposed the use of at least one of the aforementioned substrates, whether isolated or combined. The most commonly mentioned candidates were chitosan, alginate, and gelatin (Hu et al. 2020; Hong 2021c; Breemhaar et al. 2022; Feng et al. 2022).

Fig. 5 Number of patent documents relating to each proposed scaffold composition

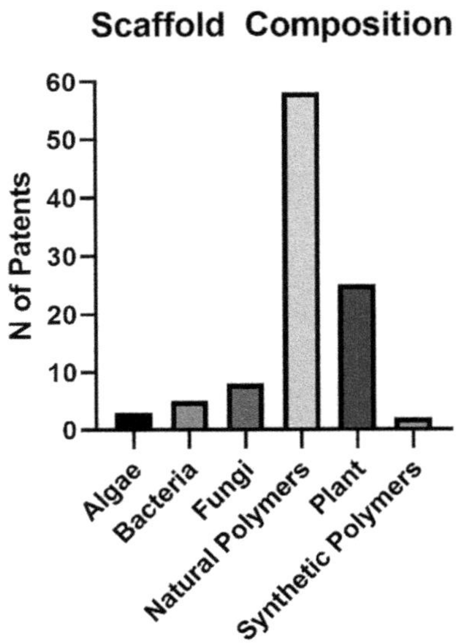


Another concern regarding the composition of these scaffolds is the origin of the materials, given that the main purpose of cultivated meat is to alleviate animal suffering. Hence, the materials used to produce cultivated meat should ideally be animal-free. Indeed, approximately 66% of the patents analyzed employed non-animal-derived materials. The most common alternative source in these cases were plant-derived substrates (24 patents), namely decellularized tissues (Do and Hong 2021; Pelling et al. 2022), soybean or peanut protein (Ben-Ayre and Levenberg 2018; Tang et al. 2022; Xiang and Li 2022a) and plant-derived silk protein (Liu et al. 2022). Another promising option among the patents were fungi ingredients (8 patents), such as mushroom mycelium (Beyer et al. 2019; Takle et al. 2022), mushroom chitosan (Xiang and Li 2022b) and yeast protein (Xiang 2022). Although more uncommon, other plausible alternatives included seaweed (Lee 2022), algal protein (Lavon and Kraus 2021), and bacterial cellulose (Park et al. 2022; Zhang et al. 2022). One patent differed from these patterns by suggesting the use of decellularized animal tissue (liver) in combination with nano-graphene (Kang et al. 2021).

It is noteworthy that mammalian cells require a positively charged and/or hydrophilic surface to attach adequately (Bodiou et al. 2020) and may not adhere to some MCs and scaffolds. Consequently, it is advised to either add a chemical component to the surface of the matrices to enhance their hydrophilicity or coat them with a layer of a biological component such as extracellular matrix proteins (Bodiou et al. 2020). A successful approach presented in several patents (Marga et al. 2014; Hong and Lee 2022; Levenberg et al. 2022) was to coat them with proteins containing the amino acid sequence RGD (Arginine-Glycine-Aspartate) because this domain is recognized by most of the known integrins that participate in cell adhesion (Bodiou et al. 2020).

In conclusion, scaffold composition plays a pivotal role in the development of cultivated meat not only because it is useful to recreate the cells' biological environment but also because it can offer different qualities to the meat. The field is currently undergoing extensive exploration, and it is evident that efforts are being directed toward achieving the highest standards of quality, sustainability, and scalability for cultivated meat.

## 6 Patents and Innovations in Product Formulations

Besides the methods for cultivating the cells and scaffolding, the final product also requires technology development and investment. The final characteristics and presentation of cultivated meat strongly determine public acceptance and market, especially regarding sensorial traits. The organoleptic properties of the product, such as the texture, flavor, and appearance, determine the final meat quality and how it must be adapted to reach the regular meat characteristics (Ng and Kurisawa 2021). Regarding the texture, a product is expected to be tender, juicy, and firm. Those three features depend on the type of cell, predominantly muscle tissue and adipocytes, the myofiber organization maintaining the structure, as well as an equilibrium of fat and water contents (Guinard and Mazzucchelli 1996; Lonergan et al. 2010). As for the appearance, besides the cell disposition, it is desirable to have a red-brownish "blood-related" color and opacity while raw (Hui et al. 2012). Finally, the fat content is the major component responsible for the flavor and depends on the composition, concentration, and degradation stages of the fatty acids (Calkins and Hodgen 2007; Khan et al. 2015).

Reproducing the regular meat characteristics is a huge obstacle in bioreactor cultivation, usually requiring a step for cell multiplication followed by a step for organization. The simple cell growth is not sufficient to reach a pleasant texture for consumption, requiring strategies for adjusting the cell disposition, such as the scaffolds. The use of 3D printing and the addition of nanoparticles and biopolymers are also resources for enhancing cell organization (Guan et al. 2021).

Besides fulfilling the organoleptic properties, the final product must also present the adequate nutritional value considering the contents of protein, fat, vitamins, and minerals. All those compounds can be either produced during cultivation or added after, modifying the final formulation. Other components, such as preservatives, are important to increase shelf-life and stability of the meat products during transport and sale (Ng and Kurisawa 2021).

In these aspects, there are patents that exclusively cover the cultivated meat product. The first one identified in the search dates to 2004, comprising a generic non-human tissue meat product (SG155930-A1) (Vein 2004). Other documents covered slight changes, such as the addition of flavor and coloring agents (KR2022005148-A) (Lee 2020), edible support particles (CN114711326-A) (Xiang and Fan 2021), filamentous fiber (CN115444071-A) (Xiang et al. 2022) or transdifferentiated adipocytes (WO2022232322-A1) (Arora and Valenzuela 2022).

Patent documents that presented significant changes described, for instance, the use of stacked layers of partially fused multicellular bodies containing non-human myocytes. These patents proposed a precise control of the cell's position during formation, achieving planar layers that highly improved texture (US2014093618-A1 and US2013029008-A1) (Forgacs et al. 2012, 2013). The patent document CN111165738-A defined a material carrier derived from a combination of collagen, elastin, hyaluronic acid and glycosaminoglycans incorporated into the product, resulting in edible bionic animal tissue (Zhao and Zhou 2020). Besides the conventional chicken, pork and cattle tissues, the patent document WO2020131661-A1 reported the use of insect muscle tissue to produce cultured meat. This approach aimed to offer a serum-free and food-safe substrate (Kaplan and Rubio 2019).

A few patent documents described the use of a combination of plant components to modify or enhance the characteristics of the meat. For instance, the document WO2018189738-A1 detailed a hybrid foodstuff that mixes a plant-originated substance with animal tissue, salt and fat, the plant material comprising up to 99.5% of the composition (Savir et al. 2018). The patent KR2023081976-A reported the use of grains as a surface to cultivate meat (Hong and Sohyun 2022). Both strategies, although not specifying the plant substance or grain, significantly altered the final product's composition and eliminated the need for fetal bovine serum. This opens various possibilities for formulating the resulting meat products.

## 7 Other Technologies

Not only are the patent applications associated with meat cultivation processes or products on the rise, but there is also an increase in patent filings related to indirect processes for cultivated meat production. Since 2019, these technologies have received attention, particularly following the initial release of cultured meat products in the market. Many of these patents employ current and recently developed biotechnology tools, ranging from molecular biology and genetic engineering to bioinformatics, to optimize and make the process more feasible.

During the cultivation of animal cells, myogenesis and adipogenesis are stimulated by several factors to provide different cells for adequate tissue formation. Few patent documents focused on the regulation or expression of those factors specifically for cultivated meat. For instance, the patent document WO2023139493-A1 (Biressi et al. 2023) reported a vector nucleic acid sequence for myogenesis regulatory factor, the patent KR2023079340-A (Rae et al. 2023) described a method to induce bovine muscle cells from somatic cells by expressing a myoblast determination protein (MyoD), and the patent CN111826451-A (Wang et al. 2020b) detailed a primer set for expressing both MyoD and MyoG.

In addition to the regulation factors, several components added to the culture to reach the desired product characteristics are under development. Hemoglobin is the most frequently mentioned compound. This metalloprotein is responsible for the blood-red color and the iron content in the meat, originally produced by animals to transport oxygen. One strategy for producing hemoglobin is to express the protein in a bacterium or yeast using vectors. The common hosts reported in patent documents were the yeast *Pichia pastoris* (CN116376733-A, CN114874929-A and CN113136349-A) (Zhao et al. 2021; Li et al. 2022b, 2023) and *Saccharomyces cerevisiae* (CN113549561-A) (Li et al. 2021). Besides hemoglobin, other components can be produced through different processes and then added to the bioreactor, such as enzymatic complexes with pronase, trypsin, and collagenase (CN114752590-A) (Zhou et al. 2022b), polysaccharide alginate hydrogel (WO2021158105-A1) (Arin et al. 2021), pectin and transglutaminase ink (CN115747197-A) (Yang et al. 2023), and phytoene, lycopene and beta-carotene (WO2020243695-A1) (Stout and Kaplan 2020).

Another set of technologies that support the meat cultivation process is related to cell differentiation. Since the cells are intended to be pluripotent and multiply rapidly at the onset of the culture, they must differentiate into the correct combination of tissues and cell types to achieve a product resembling meat. Typically, stem cells are stimulated to become muscle tissue or adipocytes, depending on various inducing agents or inhibitors. Most patents focused on differentiating cells into muscle, such as in the documents CN112708591-A (Zhou et al. 2020), WO2022097146-A1 (Brodie 2021) and CN114752558-A (Chen et al. 2021), or on controlling differentiation through an inhibitor (WO2023127960-A1) (Segawa et al. 2022b). Others were directed to adipogenesis, such as in patents WO2023003470-A1 (Mitic et al. 2022) and CN115044541-A (Zhu et al. 2022). All of them represent an effort to replace fetal bovine serum.

Other patents included devices, chemical compounds, or genetic engineering strategies to promote fast and cost-effective cultured meat production. For example, the patent CN217879010-U (Zhou et al. 2022a) refers to an electrode device developed to detect pH values in cultivated meat without compromising its structure. The patent CN114317394-A (Xiang and Liu 2021) comprised a biocompatible micro vector from gelling polysaccharide to stimulate cell growth and organization, while the document CN111513263-A (Liu et al. 2019) mentioned an exclusive method of mixing and stirring the animal cells in a way to improve cell morphology.

## 8 Conclusion

The search for patent documents based on keywords is a challenge, especially when several synonyms can be used to describe the topic of interest or when unspecific terms must be included to avoid losing information. The Derwent Innovations Index is an appropriate database for this kind of search since the titles and abstracts of the patent documents are rewritten according to the patent claims. In the patent search presented in this chapter, 26% of the retrieved documents were in fact related to cultivated meat production.

The technological mapping based on patent data indicated that the curve of patent publications started its exponential phase in 2019 and the trend line is still rising sharply, meaning that cultivated meat technologies are in the growth stage. China, South Korea and the United States contributed to most publications. The top assignees were Jiangnan University, Shanghai CellX Biotechnology, Nanjing Agricultural University, China Meat Research Center and Nanjing Joes Future Food Technology from China, Yonsei University Industry Foundation from South Korea, Aleph Farms from Israel, and Northwest University, Tufts College and Terasaki Biomedical Innovation Institute from the United States.

Among the retrieved patent documents, the most frequent IPC codes were C12N-005/077 (58%), i.e., "mesenchymal cells, e.g., bone cells, cartilage cells, marrow stromal cells, fat cells or muscle cells," and A23L-013/00 (47%), which is associated to "meat products; meat meal; preparation or treatment thereof." These IPC codes should be included in patent searches related to cultivated meat production, possibly associated with keywords.

The development of technologies focusing on culture media with reduced or absent use of animal-derived components, bioreactors and cultivation methods that allow high cell concentrations, separation of growth inhibitors, recycling of culture medium, microcarriers and scaffolds with novel materials and improved efficiency for cultivated meat production, and new product formulations and structures, as identified in the patent documents, demonstrates that several aspects of cultivated meat production are being explored. Another fact is that most of these patent documents were filed in association with companies, indicating that significant technological development and technology transfer can be expected for cultivated meat production in the following years.

**Acknowledgments** The authors are thankful to the Araucária Foundation (Fundação Araucária, NAPI Proteínas Alternativas), the National Council for Scientific and Technological Development (CNPq), and the Coordination for the Improvement of Higher Education Personnel (CAPES).

## References

Arin D, Post M, Spaans S, Dogan A (2021) Modified polysaccharide hydrogel as hydrogel, preferably as sacrificial biopolymer, comprises low molecular weight alginate, where alginate is conjugated with cell-adhesion peptides, for use in cultured meat applications. Patent document WO2021158105-A1

Arora N, Valenzuela R (2022) In-vitro cultured meat product, comprises population of cells comprising fibroblasts, and/or myoblasts, and population of cells transdifferentiated to express adipocyte phenotypes. Patent document WO2022232322-A1

Ben-Ayre T, Levenberg S (2018) Cultured meat compositions. Patent document US20200140810-A1

Beyer E et al (2019) Methods for generating mycelial scaffolds and applications thereof. Patent document CN20198009296A

Biagini G, Senegaglia AC, Pereira T et al (2021) 3D poly(lactic acid) scaffolds promote different behaviors on endothelial progenitors and adipose-derived stromal cells in comparison with standard 2D cultures. Front Bioeng Biotechnol 9. https://doi.org/10.3389/fbioe.2021.700862

Biressi M, Conti L, Fioravanti G, Lattanzi S (2023) New expression vector having temperature-responsive promoter operably linked to nucleic acid sequence encoding for regulatory factor of myogenesis and/or adipogenesis used to produce muscle and/or adipose tissue, preferably cultured meat. Patent document WO2023139493-A1

Bodiou V, Moutsatsou P, Post MJ (2020) Microcarriers for upscaling cultured meat production. Front Nutr 7. https://doi.org/10.3389/fnut.2020.00010

Breemhaar J et al (2022) Preparation of cell-containing hydrogel. Patent document WO2023009007

Brodie A (2021) Differentiating mesenchymal stromal cell (MSC) to non-MSC cell fate by contacting MSC with purified or isolated extracellular vesicles, purified or isolated matrix-bound vesicles or their combinations derived from another non-MSC source, or MSC cell differentiated to another non-MSC cell fate. Patent document WO2022097145-A1

Calkins CR, Hodgen JM (2007) A fresh look at meat flavor. Meat Sci 77:63–80. https://doi.org/10.1016/J.MEATSCI.2007.04.016

Chen J, Yan Q, Pan Z et al (2021) Inducing differentiation of myogenic stem cells in vitro involves differentiating and culturing myogenic stem cells in system containing flavonoids, where flavonoids include naringenin, epicatechin, genistein, luteolin and/or quercitrin and muscle-derived stem cells. Patent document CN114752558-A

Chen J et al (2022) Serum-free myogenic differentiation medium containing natural compounds and application thereof. Patent documents CN114752554-A and CN114752554-B

Chin P, Chan K, Li C, Spitters T (2022) System for producing cultivated meats, tissues and associated products from cells. Patent document WO2023012523-A1

Chintala R, Komarraju A, Rentala S (2016) 3-D scaffold matrix for meat production. Patent document IN201641015021-A1

Do J, Hong T (2021) Muscle cell culture using plant-derived 3D scaffold. Patent document KR2023059456-A

Feng X et al (2022) Preparation method of edible cell culture meat 3D scaffold using ionic cross-linking. Patent document CN115418023A

Forgacs G, Marga F, Jakab K (2012) Engineered meat comprises volume comprising stacked layers, where each layer has at least partially fused multicellular bodies comprising non-human myocytes, and adjacent to each other; and engineered meat is comestible and for ingestion. Patent document US2013029008-A1

Forgacs G, Marga F, Jakab K (2013) Engineered meat product comprises body having volume, where body contains stacked planar layers, where layers are partially fused and each planar layer comprises at least partially fused multicellular bodies containing non-human myocytes. Patent document US2014093618-A1

Genovese N (2020) Use of tead proteins for production of cellular biomass. Patent document WO2021092587-A1

Good Food Institute (2022) Deep dive: cultivated meat cell culture media. https://gfi.org/science/the-science-of-cultivated-meat/deep-dive-cultivated-meat-cell-culture-media/. Accessed 1 Oct 2023

Guan X, Lei Q, Yan Q et al (2021) Trends and ideas in technology, regulation and public acceptance of cultured meat. Future Foods 3:100032. https://doi.org/10.1016/j.fufo.2021.100032

Guimarães CF, Gasperini L, Marques AP, Reis RL (2020) The stiffness of living tissues and its implications for tissue engineering. Nat Rev Mater 5:351–370. https://doi.org/10.1038/s41578-019-0169-1

Guinard JX, Mazzucchelli R (1996) The sensory perception of texture and mouthfeel. Trends Food Sci Technol 7:213–219. https://doi.org/10.1016/0924-2244(96)10025-X

Guo H et al (2022) Blue algae water extract and application thereof as substitute serum. Patent document CN115287256-A

Han J et al (2022) Medium composition for culturing muscle stem cell comprising curcumin longa, glysin, or insulin for proliferation of muscle stem cell. Patent document WO2022211461-A1

Hong J (2021a) Culture solution composition useful in producing cultured meat for food additives. Patent document KR2022040417-A

Hong J (2021b) Producing cultured meat useful in food composition, comprises introducing porous polymer support to substrate, seeding cells on support, culturing seeded cells to prepare single-layered cell sheet, immobilizing cell sheet, and laminating. Patent document KR2022040424-A
Hong J (2021c) A porous cell culture support for the manufacture of cultured meat, a method for manufacturing cultured meat using the same, and cultured meat. Patent document KR2022040425A
Hong J, Lee M (2022) Scaffold assembly for cultured meat production. Patent document KR2023080366A
Hong J, Sohyun P (2022) Hybrid cultured meat comprises grains and cultured cell coating layer located on surface of grains. Patent document KR2023081976-A
Hong J, Park S, Hong J, Sohyun P (2021) Method for producing cultured meat on basis of cell sheet coating technique, and cultured meat produced thereby. Patent document WO2022019688-A1
Hu J, Yin J, Wang X (2020) A cross-linked hydrogel for muscle stem cell culture: preparation methods and applications. Patent document CN112778543A
Huang B, Wang F (2022) Serum-free culture medium and application and method thereof in culturing muscle stem cells. Patent document CN115141798-A
Hui YH, Aalhus JL, Cocolin L et al (2012) Handbook of meat and meat processing, 2nd edn. CRC, Boca Raton, FL
Johnson J, Ohst D (2020) Electrospun polymer fibers for cultured meat production. Patent document US20200245658-A1
Kang K et al (2021) Composition for preparing decellularized scaffold, comprising graphene nanostructure. Patent document WO2022114642-A1
Kaplan D, Rubio N (2019) Cultured meat product comprises confluent serum-free insect muscle cell culture seeded on food safe substrate. Patent document WO2020131661-A1
Kaplan D, Stout A, Yuen J, Mirliani A (2021) Supplemented serum-free media for cultured meat production. Patent document WO2022125809A1
Khan MI, Jo C, Tariq MR (2015) Meat flavor precursors and factors influencing flavor precursors—a systematic review. Meat Sci 110:278–284. https://doi.org/10.1016/J.MEATSCI.2015.08.002
Lavon N, Kraus D (2021) Edible protein products. Patent document WO2022097140-A1
Lavon N, Shalel Levanon S, Tal N (2021) A multi-scaffold system for large scale cultivation of cells. Patent document WO2022097139-A1
Lee J (2020) Animal cell cultured meat comprises fat raw material for imparting flavor and coloring agent for color improvement. Patent document KR2022005148-A
Lee H (2022) Scaffold for cell culture. Patent document WO2022250497-A1
Levenberg S et al (2022) 3D printable protein-enriched scaffolds. Patent document WO20222162662-A1
Li J, Chen J, Du G et al (2021) Construction and application of *Saccharomyces cerevisiae* strain for efficiently synthesizing hemoglobin or myoglobin from different sources. Patent document CN113549561-A
Li H et al (2022a) Improved culture medium with specific chemical components for inducing in vitro adipogenic differentiation and application thereof. Patent document CN115927172-A
Li J, Chen J, Du G et al (2022b) New recombinant bacteria, which uses *Pichia pastoris* to express heme synthesis key enzyme gene delta 70hem14 and/or delta 70hem15 with 2-71 amino acid truncated at carbon end useful for preparing heme or its derivative. Patent document CN114874929-A
Li J, Chen J, Du G et al (2023) New recombinant *Pichia pastoris* strain X33 used as expression vector for expressing soybean-derived hemoglobin and preparation of soybean hemoglobin-containing products, in which expression vector of the recombinant bacteria comprises pPICZ alpha A, pPICZ. Patent document CN116376733-A
Liu S, Wang X, Zhou J et al (2019) Method for improving animal cell source artificial meat morphology. Patent document CN111513263-A
Liu W, Li S, Li Y et al (2021) Apparatus for biologically culturing meat. Patent document WO2023016357-A1

Liu P et al (2022) A method for preparing plant-based fat cultured meat. Patent document CN114736851A
Lonergan EH, Zhang W, Lonergan SM (2010) Biochemistry of postmortem muscle—lessons on mechanisms of meat tenderization. Meat Sci 86:184–195. https://doi.org/10.1016/J.MEATSCI.2010.05.004
Marga F et al (2014) Edible and animal-product-free microcarriers for engineered meat. Patent document US20150079238-A1
Mitic R, Cantoni F, Hubalek S et al (2022) Differentiating adipogenic progenitor cell used for preparing cultured fat product and meat product, by culturing adipogenic progenitor cell in serum-free medium capable of differentiating adipogenic progenitor cell containing peroxisome proliferator-activated receptor gamma agonist, and hormone. Patent document WO2023003470-A1
Nahmias Y (2017) Systems and methods for growing cells in vitro. Patent document WO2018011805-A2
Nahmias Y, Wissotsky G (2021) Methods and devices for filtering cell culture media. Patent document WO2021152536-A1
Nahmias Y, Ayyash M, Pasitka L (2021a) Utilization of FGF activators in culture media. Patent document WO2021148960-A1
Nahmias Y, Cohen M, Caspi J (2021b) Utilization of plant protein homologues in culture media. Patent document WO2021148955-A1
Ng S, Kurisawa M (2021) Integrating biomaterials and food biopolymers for cultured meat production. Acta Biomater 124:108–129. https://doi.org/10.1016/J.ACTBIO.2021.01.017
Orvar B et al (2021) Growth factor composition for cell culture-produced meat. Patent document WO2021245711-A1
Ovissipour R, Duscher A, Batish I (2022) Hydrolyzed protein serum replacement compositions. Patent document WO2023086502-A1
Park S et al (2021) Producing powdered milk serum useful as an e.g. alternative to fetal bovine serum (FBS). Patent document KR2021090560-A
Park G et al (2022) Scaffold for producing cultured meat containing bio-cellulose powder and method for producing the same. Patent document KR246651-B1
Pelling A et al (2022) Methods for rapidly infiltrating 3D scaffolds with cells. Patent document WO2023004515-A1
Pisani S, Mauri V, Negrello E et al (2023) Assessment of different manufacturing techniques for the production of bioartificial scaffolds as soft organ transplant substitutes. Front Bioeng Biotechnol 11. https://doi.org/10.3389/fbioe.2023.1186351
Rae L, Lee S, Seungjin B, Ho P (2023) Inducing bovine muscle cells from bovine somatic cells for producing cultured meat by introducing exogenous bovine myoblast determination protein expression construct into bovine somatic cells. Patent document KR2023079340-A
Reed J, Robertson D (2022) Microorganism-derived protein hydrolysates and methods of preparation and use thereof. Patent document WO2023278317-A1
Savir I, Friedman S, Barak K (2018) Producing hybrid foodstuff, comprises combining plant-originated substance with cultured animal cells so as to enhance meat organoleptic and/or meat nutritional property in hybrid foodstuff and the animal cells do not form tissue. Patent document WO2018189738-A1
Segawa M, Furusawa Y, Nishiyama Y (2022a) Cell proliferation medium for producing cultured meat. Patent document WO2023127970-A1
Segawa M, Nishiyama Y, Furusawa Y (2022b) Inhibitor of differentiation of muscle tissue-derived cells used for producing cultured meat, comprises whey. Patent document WO2023127960-A1
Singh A, Kumar V, Singh SK et al (2023) Recent advances in bioengineered scaffold for in vitro meat production. Cell Tissue Res 391:235–247. https://doi.org/10.1007/s00441-022-03718-6
Soccol CR, Karp SG, Letti LAJ et al (2022) Is the development of low-cost media one of the greatest challenges to produce cultivated meat on an industrial scale? Biotechnol Res Innov 6:e2022005. https://doi.org/10.4322/biori.20226201

Song Y, Soto J, Chen B et al (2020) Cell engineering: Biophysical regulation of the nucleus. Biomaterials 234:119743. https://doi.org/10.1016/j.biomaterials.2019.119743

Stout A, Kaplan D (2020) New engineered cell that endogenously synthesizes phytoene, lycopene or beta-carotene and that comprises heterologous polynucleotide encoding phytoene synthase, phytoene desaturase, or lycopene cyclase, for preparing cultured meat product. Patent document WO2020243695-A1

Stout AJ, Mirliani AB, Rittenberg ML et al (2022) Simple and effective serum-free medium for sustained expansion of bovine satellite cells for cell cultured meat. Commun Biol 5. https://doi.org/10.1038/s42003-022-03423-8

Takle S, Sachdeva N, Srivastava A (2022) A scaffolding for cultivated meat and a process for making the same. Patent document WO2022208525

Tang C, Lin L, Ding S et al (2021) Culture medium for in vitro large-scale culture of meat seed cells and gene operation method. Patent document CN113862225-A

Tang C et al (2022) A plant-based culture based in plant protein scaffolds for the preparation of meat and its preparation method. Patent document CN114410578A

Vein J (2001) Providing nutrition to subject, by providing subject with non-human meat product suitable for consumption, produced by culturing non-human muscle cells ex vivo, seeding muscle cells onto support and growing non-human muscle cells. Patent document US6835390-B1

Vein J (2004) Non-human tissue engineered meat product, useful for food consumption, comprises muscle cells that are grown ex vivo, where the meat product is obtained by growing the muscle cells and attaching to a support structure. Patent document SG155930-A1

Vila Juárez M (2019) Bioreactor and method for the production of adherent cell cultures employing said bioreactor. Patent document WO2020141236-A1

Wang S, Li S, Li Y, Liu W (2020a) Special culture apparatus for 3D biological tissue, and method for preparing block-shaped cultured meat. Patent document WO2022068029-A1

Wang S, Li Y, Li S (2020b) New primer set comprising primer pair of two regulatory factors MyoD and MyoG and internal reference primer pair, used in kit for e.g. detecting expression of pork muscle regulatory factors. Patent document CN111826451-A

WIPO (2023) International Patent Classification (IPC). https://www.wipo.int/classifications/ipc/en/. Accessed 2 Oct 2023

Xiang N (2022) Method for preparing cell culture meat biological scaffolds through physical crosslinking. Patent document CN114438615

Xiang N, Fan X (2021) Artificial meat useful in preparation of single fat product and as a meat substitute and customizable protein food, formed by submerged printing using edible muscle printing ink in edible fat suspension medium. Patent document CN114711326-A

Xiang N, Li Y (2022a) Specifications for soy protein fiber scaffolds for cell cultured meat: modular production method. Patent document CN114438615

Xiang N, Li Y (2022b) Using mushroom chitosan fiber scaffolds to produce cell-cultured meat. Patent document CN114481363A

Xiang N, Liu Y (2021) Micro vector useful for three-dimensional cell culture, culturing cells, culturing cell-micro vector clumps and culturing meat, obtained by e.g. generating droplets from mixed solution containing ionomer-derived gelling polysaccharide. Patent document CN114317394-A

Xiang N, Wu S, Lin X (2022) Mixed type cell culture meat useful in biological and food field, comprises non-animal-derived structured protein and non-human animal cells physically combined with it, where the non-animal source structured protein is protein filamentous fiber, tissue. Patent document CN115444071-A

Yang F, Li S, Li Y et al (2023) Preparation raw material of the biological ink comprises pectin, transglutaminase, the first protein component and the second protein component. Patent document CN115747197-A

Ye Y, Zhou J, Guan X, Sun X (2022) Commercialization of cultured meat products: Current status, challenges, and strategic prospects. Future Foods 6. https://doi.org/10.1016/j.fufo.2022.100177

Zhang X, Tang Y, Xiang N (2022) Modified bacterial cellulose beneficial to cell culture and application thereof. Patent document CN114752067A

Zhao Y, Zhou Y (2020) Edible bionic animal tissue comprises natural material carrier and animal cells, which grows on natural material carrier through static or dynamic culture and secrete main cells including e.g. collagen, elastin, hyaluronic acid. Patent document CN111165738-A

Zhao X, Yu F, Wang Z et al (2021) New recombinant bacteria useful in preparing of myoglobin or hemoglobin or derivatives in the field of food processing including artificial meat comprising *Pichia pastoris*. Patent document CN113136349-A

Zhou G, Wu Z, Xu X et al (2020) Improved cell differentiation medium used for inducing and differentiating muscle stem cell in vitro, promoting differentiation of muscle stem cells in process of inducing differentiation in vitro, comprises definite chemical composition. Patent document CN112708591-A

Zhou G, Ding S, Jin S et al (2022a) Electrode device useful for detecting pH value of cell culture meat, comprises electrode supporting rod of hollow structure, electrode socket and plane electrode respectively set on upper and lower ends of supporting rod. Patent document CN217879010-U

Zhou J, Chen J, Li M et al (2022b) Complex enzyme useful in the field of cultured meat comprises pronase, trypsin, dispase and collagenase D. Patent document CN114752590-A

Zhu W, Hong P, Huang Z (2022) New high differentiation efficiency immortalized porcine adipose precursor cell line used in food development, virus culture or detection, vaccine development and production, and drug screening. Patent document CN115044541-A

# Future Feast: Mapping the Cultivated Meat Sector

**Cyntia Vilasboas Calixto Casnici and David Schulzmann**

**Abstract** The importance of the cultivated meat industry has been continuously growing, with the potential to substitute animal-based production systems and contribute to mitigating the effects caused by climate change. Despite the current uncertainty regarding scalability, reaching price parity with conventional meat and increasing consumer acceptance, we are witnessing an increasing number of cultivated meat companies joining the race and expanding and diversifying their product portfolio. Their rapid growth in size and technological advancement are supported by the developing competitive ecosystems.

This chapter analyses 18 globally leading cultivated meat companies, highlights differences and provides an outlook on the industry. Cultivated meat company horizontal diversification within conventional meats, rare or even extinct animals, and vertical expansion along the value chain supplying growth media and equipment. Cultivated meat companies have successfully demonstrated to attract large funds that can further push the novel industry even further. The increasing number of applications for regulatory approval provides a positive outlook to an upcoming race for more innovation and efficiency in the cultivated meat industry.

**Keywords** Cultivated meat · Climate change · Regulatory approval · Case studies · Investment grant

C. V. C. Casnici (✉)
University of Leeds, Leeds, United Kingdom
e-mail: c.casnici@leeds.ac.uk

D. Schulzmann
University of Leeds, Leeds, United Kingdom

Aalborg University, Aalborg, Denmark
e-mail: d.schulzmann@leeds.ac.uk


C. R. Soccol et al. (eds.), *Cultivated Meat*,
https://doi.org/10.1007/978-3-031-55968-6_20

# 1 Introduction

Animal-based production systems are one of the largest contributors to greenhouse gas emissions, responsible for about 14.5% of global emissions (Gerber et al. 2013). Cultivated meat production has the potential to dramatically reduce these emissions. A study from the University of Oxford suggested that cultivated meat could reduce greenhouse gas emissions by up to 96% compared to conventional meat production. Animal-based production systems are resource-intensive, requiring vast amounts of land and water. Cultivated meat production is significantly more efficient in this regard (Tuomisto and Mattos 2011). While the energy requirements for cultivated meat production are still a research subject, initial studies suggest that it could be more energy-efficient than conventional animal-based production systems, especially when renewable energy sources are utilized (Mattick et al. 2015).

The last few years have seen an exponential rise in the number of companies entering the cultivated meat space. In 2015, there were only a handful of such firms, but the landscape has changed dramatically, with dozens of startups now in operation globally (GFI 2021)—See Fig. 1 for the development of this industry. In these 156 companies, span a range of products, from red meat to poultry and even seafood. The cultivated meat industry is at a pivotal moment, seeing rapid technological advancements, regulatory milestones, and increasing consumer interest. Cultivated meat aims to provide a more ethical and sustainable alternative to conventional meat by growing animal cells in a controlled environment (GFI 2022).

Building upon the burgeoning momentum within this innovative industry, the primary aspiration of our investigative endeavour is to present a global outlook of

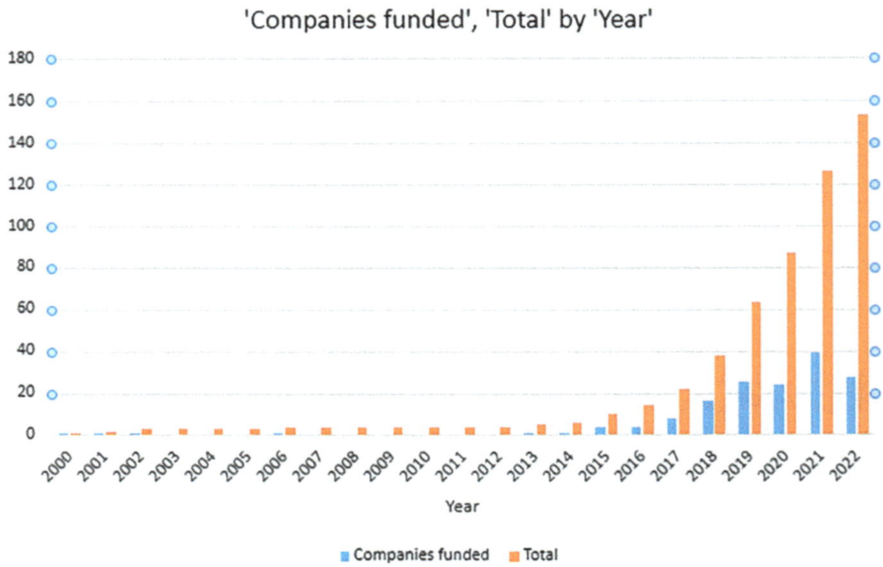


**Fig. 1** Companies funded per year. Source: adapted from GFI database, 2023

cultivated meat across varying geographical landscapes. Therefore, we examined different cases to illustrate the industry developments until October 2023, when this book chapter was written.

## 2 The Global Atlas of Cultivated Meat

As the industry matures, companies increasingly specialize within the areas along the value chain of cultivated meat. The cultivated meat value chain is complex and entails different stakeholders, processes and resources. Companies may focus on broader areas that include specific meats and fish such as poultry, beef or pork, and seafood. They may focus on a specific element such as fat or cell biosynthesis (e.g., Cellva). Other companies focus on value chain activities that include the supply of animal-free growth media (e.g., Clear Meat and Joes Future Food), certain types of meat as poultry and bacon (e.g., Good Meat and Uncommon) or seafood (e.g., BlueNalu).

The ecosystem has several companies in various stages of product development, pilot production, and commercial launch. This diversification allows for more targeted research and development, and commercialization efforts. To expedite growth and product development, many companies are entering strategic partnerships. For instance, Mosa Meat has collaborated with Nutreco, a global leader in animal nutrition and aqua feed, to develop growth media for cell cultivation.[1] Memphis Meats has received investments from big names in the animal-based food industry like Cargill and Tyson Foods, signaling an industry-wide interest in this new form of meat production.[2] It is important to note that the number of investors in cultivated meat and seafood grew by 19% to 679 investors total in 2022 (GFI 2022).

As shown in Fig. 2, the United States, Israel, and the United Kingdom are the leading countries on the number of startups founded in the last 5 years. However, in none of these countries, the commercialization of products was approved. The cultivated meat sector is at the intersection of food safety, biotechnology, and environmental policy, making its regulatory landscape incredibly complex. Each country has its own set of challenges and opportunities regarding commercialization approval. Singapore became the first country in the world to approve the commercial sale of cultivated meat in December 2020.[3] As the first in this space, Singapore had no existing frameworks to rely upon, but ensuring the safety of the new products was a prime concern.

Israel has been supportive but is still formalizing regulations. It works on coordinating different governmental bodies like health and agriculture departments. One example of Israel's engagement in this area is the most extensive state-supported

[1] "Nutreco and Mosa Meat renew partnership to develop cultivated meat", Nutreco Corporate Press Release, 2021

[2] "Memphis Meats Raises $161 Million in Funding, Aims to Bring Cell-Based Products to Consumers", Memphis Meats Press Release, 2020

[3] "Singapore becomes first country to approve lab-grown meat", BBC News, December 2020.

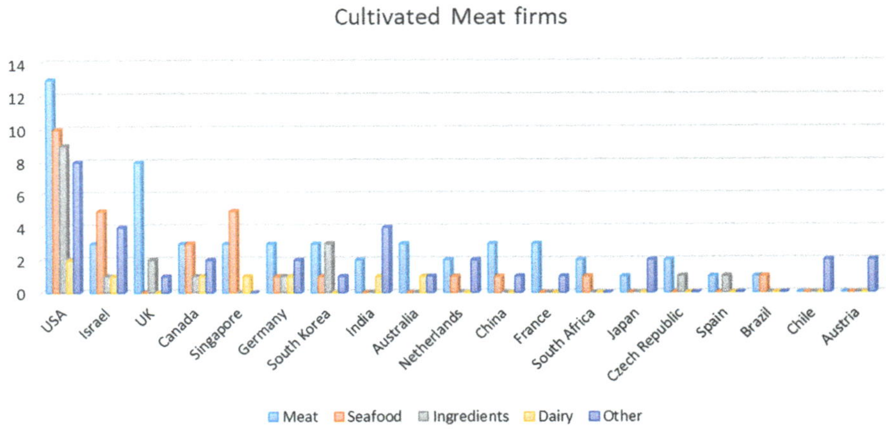


**Fig. 2** Cultivated meat firms around the world. Source: adapted from GFI database, 2023

consortium for cultivated meat thus far, which includes leading food manufacturers and academic research facilities (GFI 2022). In the case of the European Union, even though there are several startups interested in this segment, only The Cultivated B has applied to get access to the European Market. Many startups refer to stricter regulations and transparency requirements by the European Food Safety Authority (EFSA).

As in many other regions, cultivated meat falls under the "Novel Foods" category, requiring extensive safety assessments.[4] Finally, In June 2023, Upside Foods and Good Meat received final approval from the US Department of Agriculture to sell lab-grown meat, marking the advent of the USA's first-ever sales of such products.[5]

The growth of the cultivated meat industry has attracted cultivated meat companies from developed and emerging economies to adventure into a new agricultural industry. Based on our selection of the cultivated meat companies, we see a representation of both economies in our chapter. We selected 18 cases of cultivated meat companies that currently have pilot-scale (or larger) facilities to illustrate the development of the industry in different countries. In addition, we tried to present the most prominent cases in the media on cultivated meat, as well as outstanding regional startups. Thus, it will be possible to see an overview of the industry with a diverse pool of companies, different levels of maturity, product portfolio, and financial and regulatory conditions.

Table 1 summarises the case companies based on their product portfolio, uncommon characteristics, regulatory status and funding type. The selected case studies

[4] "EU Novel Food Regulations and Cultured Meat: What You Need to Know", New Food Magazine, 2021

[5] https://www.reuters.com/business/retail-consumer/upside-foods-good-meat-receive-final-usda-approval-sell-cultivated-meat-2023-06-21/

**Table 1** Overview of cultivated meat company product offering, regulatory and funding award from developed and emerging country

| Home country | Company | Product portfolio | Uncommon characteristics | Regulatory status | | | Latest funding type |
|---|---|---|---|---|---|---|---|
| | | | | Not applied | Applied | Success | |
| Australia | Vow |  | Mixed: elephant and mammoth | X | | | Series A (2022), Seed (2020), Pre-seed (2019) |
| Germany | The Cultivated B |  | Bioreactors, growth media and full service | | X | | |
| Israel | Aleph Farms |  | | | X | | Funding Round (2023), Series Unknown (2022), Series Unknown (2022), Series B (2021), Series A (2019) |
| Israel | Super Meat |  | | | X | | Grant (2022), Series A (2021), Seed (2018) |
| Israel | Stakeholder Foods |  | 3D bioprinters | X | | | post-IPO Equity (2023), post-IPO Equity (2022) |
| Netherlands | Mosa Meat |  | | | X | | Angel (2021), Series B (2021), Series B (2020), Series A (2020) |
| Singapore | Shiok Meats |  | | | X | | Seed (2021), Series A (2020) |
| South Korea | Cell Meat |  | Growth media | X | | | Series A (2023), Series A (2022), Seed (2021), Seed (2020) |
| UK | Ivy Farm |  | Pork fat | | X | | Series A (2022), Seed (2019) |
| UK | Uncommon |  | | X | | | Series A (2023), Seed (2021) |
| USA | BlueNalu |  | | X | | | Series unknown (2023), Series B (2020) |
| USA | GOOD Meat |  | Eggs | | X | X | Grant (2023), Series Unknown (2021) |

(continued)

**Table 1** (continued)

| Home country | Company | Product portfolio | Uncommon characteristics | Regulatory status: Not applied | Applied | Success | Latest funding type |
|---|---|---|---|---|---|---|---|
| USA | Upside Foods |  | | | X | X | Grant (2023), Series C (2022), Series B (2020), Series A (2018) |
| Brazil | Cellva | | Pork fat | X | | | Seed (2023) |
| China | Joes Future Food |  | Growth Media | X | | | Series A (2023), Series A (2021), Angel (2020) |
| India | Clear Meat |  | NASDAQ: STKH Growth Media | X | | | Seed (2021) |
| Russia | OKPI |  | | X | | | |
| South Africa | WildBio |  | | X | | | Pre-seed (2022) |

Source: Authors

demonstrate that cultivated meat companies focus on diversities of meats and fish, while only a few companies offer growth media as a specific product. Further it is interesting to observe that Wildbio and Vow offer unusual and even mixed meat from extinct animals, while Stakeholder Foods and The Cultivated B have extended their offerings vertically in the value chain to supply cultivated meat equipment.

# 3 Developed Economies

Developed economies tend to have strong institutional support with a focus on Sustainable Development Goals (SDGs), especially SDG 2—Zero Hunger, which addresses points to increase food security, improve nutrition, and promote more sustainable agriculture.

Businesses operating in developed economies have the support of educated and a large proportion of the wealthy population supporting sustainable business models. The acceptance of more sustainability-driven agriculture has nurtured a university and business research environment that has aided in building cross-disciplinary and international networks. The intertwined networks have been a major support in the early stages for startups and facilitated the better circulation of business, research and institutional collaboration. The formed ecosystem for cultivated meat has resulted in a larger number of diverse successful startups, especially in Israel and the United States, further progress on regulatory approval stages and more successful funding rounds.

These developments indicate a growing recognition and support for alternative proteins in developed countries, with significant investments and policy advocacy efforts to promote research, innovation, and sustainable practices in the food system.

## *3.1 Australia*

### 3.1.1 Vow

Established in 2019, the startup is focused not on mirroring inexpensive, common meats such as beef, pork, and chicken but rather on leveraging cultivated meat technology to devise new and enhanced proteins that industrial animal farming cannot produce. In 2022, Vow opened its first factory and is already planning a quick expansion to another one by 2024. Seeking to stretch the limits of culinary creativity, Forged/Vow embarked on crafting a meatball in 2023, blending woolly mammoth DNA with snippets of African elephant DNA, its nearest living kin, to achieve this endeavour (Vow website 2023). The firm has already submitted the dossier to Singapore and looks forward to obtaining regulatory approval.

Learn more at https://www.vowfood.com/

## 3.2 Germany

### 3.2.1 The Cultivated B

The company was founded in 2009 and is developing technologies related to cellular agriculture and precision fermentation with its research and development being located in Heidelberg, Germany and its manufacturing facilities in the USA and Canada (The Cultivated B 2023). It is a subsidiary emerging from the German multinational InFamily Foods, with additional subsidiaries focusing on meat sausages and plant-based meat alternatives (InFamily Foods 2023). Unlike many other companies, Cultivated B is not specialising in one area or specific meat. Still, it is positioning itself as a complete value chain player in cell lines and growth media development, production bioreactors and further scale and end-to-end solutions while working with leading research institutions and private multinationals (The Cultivated B 2023). Cultivated B is the first company to enter the pre-submission process towards EFSA approval for its cultivated sausages (Vegconomist 2023).

Learn more at https://www.thecultivatedb.com/

## 3.3 Israel

### 3.3.1 Aleph Farms

Founded in 2016 with experiments at Technion, Aleph Farms was able to produce the first cultivated steak in 2018. One year later, it received investment from big food companies—Migros and Cargill. Aleph Farms invested in promoting its beef into a space station and got recognition from the World Economic Forum and UNESCO for its technology. In 2021, it signed an MOU with Mitsubishi Corporation to sell cultivated meat in the future in Japan. In the same year, another agreement was made with BRF to sell the meat in Brazil, with the Thai Union to cover the Thai market, and Cheil Jedang for South Korea (Aleph Farms website 2023).

In 2023, Aleph Farms affirmed its cultivated stake is kosher, approved by a Chief Rabbi. The company is working towards securing regulatory approval for the commercial sale of its products, although it has not yet received such approvals.

Learn more at https://aleph-farms.com/

### 3.3.2 SuperMeat

SuperMeat, headquartered in Tel Aviv, was founded in 2015 and endeavours to provide the global market with premium poultry meat cultivated straight from chicken cells (SuperMeat website 2023). In 2022, SuperMeat revealed two very important Memorandum of Understanding (MOU). The first one is with PHW Group, one of Europe's largest poultry producers and the second is with Migros, Switzerland's leading retail supermarket chain and eminent meat manufacturer in Europe. Those MOUs are set to accelerate the production and commercial-scale distribution of cultivated meat.

SuperMeat stands as the initial B2B cultivated meat enterprise tackling the whole poultry meat category, offering a comprehensive solution for the production of cultivated meat. The arrangements outlined in the MOUs are furnishing the firms with chances to launch cultivated meat in the market under their respective brands.

Learn more at https://supermeat.com/

### 3.3.3 Stakeholder Foods

Founded in 2019 and focused on 3D bioprinting technology for beef, Stakeholder Foods, formerly MeaTech 3D Ltd., was the first cultivated meat company listed on the Nasdaq. In 2021, the firm has granted its first patent for systems and methods that enhance muscle fibre formation (Stakeholder Foods website 2023). It has facilities in Israel and Belgium (through the acquisition of Piece of Meat) and is currently expanding to the USA.

In 2023, Stakeholder Foods enlarged its portfolio of products to include 3D cultivated fish. Also, It has signed a Memorandum of Agreement for Strategic Cooperation (MOA) with a recognized governmental entity from the Gulf Cooperation Council (GCC) as its strategic ally. This partnership aims to enhance food security using Stakeholder Foods' innovative 3D printing technology. The MOA ultimately seeks to establish a pioneering large-scale production facility in the Persian Gulf region.

Learn more at https://steakholderfoods.com/cultured-meat-company-meatech-3d-becomes-steakholder-foods/

## *3.4 The Netherlands*

### 3.4.1 Mosa Meat

In 2006, researcher Mark Post and food expert Peter Verstrate first crossed paths via a government-sponsored initiative focused on the potential of lab-grown meat. This partnership was fruitful; years later, they produced the first lab-grown beef burger in 2013. To launch the concept and show the characteristics of this product, it was covered in a press conference in London (Mosa Meat 2023). This was a significant milestone in the field of cultivated meat.

In 2020, Mosa Meet opened its production facility in Maastricht and is still waiting for regulatory approval to sell its burgers (Green Queen 2023). However, a first step in this direction happened in July 2023 with a Dutch Government Agreement to Conduct Pre-Approval Tastings of Cultivated Meat in The Netherlands (Mosa Meat 2023)

Learn more at https://mosameat.com/

## *3.5 Singapore*

### 3.5.1 Shiok Meats

Shiok Meats was founded in 2018 and focused on developing products like shrimp, crab, and lobster, aiming to recreate the taste and texture of these items through lab-grown methods. The company aims to initially target the Asian market, where there is high demand for seafood but has ambitions for global distribution. While Shiok Meats had not received regulatory approval for commercial sales, it was actively working on fulfilling the requirements for such approvals, particularly in the Singapore market (Shiok Meats 2023).

Learn more at https://shiokmeats.com/

## *3.6 South Korea*

### 3.6.1 Cell Meat

Founded in 2019, the firm produces fetal bovine serum. This ingredient is crucial for growth media development. Cell Meat has developed mass cell culture technology for the economic viability of its current main products: shrimp and caviar.

Learn more at https://www.thecellmeat.com/index_en.php

## *3.7 United Kingdom*

### 3.7.1 Ivy Farm

Founded in 2019 at the University of Oxford, Ivy Farm aims to sell muscle and fat as raw ingredients to other food producers. 'Ivy' is an abbreviation for 'in vitro,' referring to a process occurring outside a living organism, also known as in vitro meat. The company already applied for patents in 12 countries (Ivy Farm 2023; Food Manufacture 2021).

In 2023, Ivy Farm inaugurated the largest operational cultivated meat pilot production facility in Europe in Oxford. However, as sales haven't yet been approved in the UK, the subsequent facility will be established overseas. This expansion will be undertaken in collaboration with Dennis Group, a US-based company specializing in the design and construction of food processing facilities due to easier approval compared to their home base in the UK (Food Navigator 2022).

Learn more at https://www.ivy.farm/news/

### 3.7.2 Uncommon

Founded in 2017 as Higher Stakes, it started producing alternative foie gras in 2018 and moved to pork belly and bacon in 2020. A year later, it raised £1M from Innovate UK and in 2023 opened its pilot plant and rebranded as Uncommon (Uncommon website 2023).

Learn more at https://uncommonbio.co/

## 3.8 *The United States*

### 3.8.1 BlueNalu

The company was founded in 2017 and focuses on developing cell-based seafood products. The company is working on various seafood products, such as finfish, crustaceans, and mollusks. Their first product offerings were expected to include species like mahi-mahi and bluefin tuna.

The company has been known to collaborate with research institutions (Nomad Foods, Nutreco), governments, and other industry players such as Mitsubishi Corporation to advance its technology and mission (BlueNalu website 2023).

Learn more at https://www.bluenalu.com/

### 3.8.2 GOOD Meat

Originally named Beyond Eggs, the company was founded in 2011 and focused on planted-based eggs. Later, research and development included cultivated chicken nuggets and in 2017 it was rebranded as Eat Just. In 2019, Eat Just opened its first manufacturing plant in the USA and 1 year later a subsidiary in Singapore. The expansion occurred through a joint venture with Proterra Investment Partners Asia (Shu 2020).

In December 2020, Singapore's government gave the green light to lab-grown meat produced by Eat Just, under the brand name GOOD Meat. The first establishment to offer this cultured meat was a Singaporean restaurant named 1880. Following this, Eat Just received further permission to market various poultry items like shredded and breast meat. Good Meat is available at Huber's Butchery in Singapore. More recently in June 2023, the FDA had also approved the company's lab-grown meat for sale in the United States (Good Meat website 2023).

Learn more at https://www.goodmeat.co/

### 3.8.3 Upside Foods

Founded in 2015 as Memphis Meat, Upsides Food created a diversified cultivate-meat portfolio, including beef meatballs, poultry, duck and seafood. The company attracted the attention of Bill Gates and Richard Branson and traditional meat companies such as Cargill and Tyson Foods (Upside website 2023).

In 2018, Upside Foods also developed a partnership with the North America Meat Institute to advocate for federal regulations on cultivated meat. A year later, co-funded Alliance for Meat, Poultry and Sea Food Innovation—a cultivated meat coalition (Upside website 2023).

After obtaining significant funding in 2020, Upside Food opened its first production facility in the USA. It reached the unicorn evaluation (over US$ 1B in 2022), the same year that obtained the green light from the FDA (Food and Drug Administration) for poultry products (Upside website 2023). Upside Food is still waiting for regulatory clearance to market 'unstructured' or ground items like nuggets, which are produced through a more scalable method that blends plant-based proteins with cell biomass cultivated in much larger bioreactors.

Learn more at https://upsidefoods.com/company

### 3.8.4 Emerging Economies

Emerging markets are earnestly attempting to align with Agenda 2030, particularly in terms of sustainable food systems, yet they confront a multifaceted set of challenges. Among these, the development of cultivated meat stands out as a significant opportunity to enhance its stature on the international stage. This innovative approach to protein production offers a pathway to meet sustainability goals, potentially reducing reliance on traditional, resource-intensive agriculture. However, firms in these markets are predominantly in the early stages of developing the cultivated meat industry. The path to regulatory approval remains uncharted for these firms, with most still engaged in the crucial phases of product validation and ongoing research and development. This lag in regulatory progress reflects not only the nascent state of the industry in these regions but also the broader developmental challenges facing emerging markets in high-tech, sustainable food production.

Moreover, the consumer dynamics in emerging markets present an additional hurdle. Unlike their counterparts in developed economies, customers in these regions exert comparatively less pressure on businesses for sustainable practices. There is a notable lack of a robust sustainability culture, and scepticism towards novel foods, like lab-grown meats, is more pronounced. This scepticism is compounded by a general lack of awareness and understanding of the benefits of such innovations. Consequently, firms in these markets face a dual challenge: not only do they need to advance in the technological and regulatory aspects of cultivated meat, but they also need to cultivate a market where consumers are more receptive to and supportive of sustainable, innovative food solutions. Bridging this gap is essential for these firms to not only contribute to global sustainability efforts but also to secure a competitive edge in the rapidly evolving global food industry.

## 3.9 *Brazil*

### 3.9.1 Cellva

Created in 2022 by very experienced founders in the food sector, Cellva aims to provide solutions for meat production in the food industry through fat production. In 2023, Cellva was selected to be part of the ProVeg incubator. This food tech startup aims to improve its business model during the acceleration programme. The plan is to be able to produce 11 tons of pork fat by 2030. Therefore, the firm is looking to have the production facility ready by 2025.

Learn more at https://cellva.com/

## 3.10 *China*

### 3.10.1 Joes Future Food

The company, founded in 2019, is China's first cultivated meat startup focusing on cultivated pork and serum-free growth media (Vegconomist 2022). It originated at Nanjing Agricultural University and is based in the national agricultural hi-tech industry demonstration area. A large success has been achieved through an awarded funding of $10.9 million in October 2021 to continue research and development efforts and further scale up production (Green Queen 2022). In addition to its investment backing, it is strongly embedded in university research networks in China and the Netherlands (Joes Future Food 2023). It has won several awards, among them the recognition of being one of the "50 Smartest Companies" by MIT Technology Review (EqualOcean 2023).

Learn more at http://joesfuturefood.com/en/

## 3.11 *India*

### 3.11.1 Clear Meat

The company is India's first cultivated meat technology startup, founded in 2018, focusing on poultry and growth media (Green Queen 2021). Its product portfolio includes chicken in minced and steak form as well as growth media (ClearMeat 2023). Growth media usually requires the extraction of necessary elements from foetal calves in pregnant cows that will be sent for slaughter (van der Valk et al. 2018). However, Clear Meat claims to base its growth media on being animal-free (The Vegan Indians 2022), which would be an important step to increase independence from the meat industry. It has been one of the successful startups that have gone through the ProVeg Incubator programme in 2019 and claims to have reached price parity on its chicken mince with conventional meat (ProVeg 2020; Vegconomist 2020).

Learn more at https://clearmeat.com

## 3.12 Russia

### 3.12.1 Ochakov Food Ingredients Plant (OKPI)

The company is focused on food additives and produced its first 40 mg cultivated meat beef sample in 2019 in Moskow. It has taken 2 years to produce the meat, given a cost of 900,000 rubles (ca. $9.800). They expect to launch a commercially viable product within the next few years (Forbes 2019).

Learn more at https://ochakovo-food.ru/

## 3.13 South Africa

### 3.13.1 Mogale Meat: Wild Bio

Mogale Meat, rebranding to WildBio, was founded in 2020 and it aims to make nutritious cell-based meat accessible in Africa. It focuses on driving innovation in cell cultivated meat production in Petroria to provide affordable, healthy products benefiting animals, people, and the planet (MogaleMeat website 2023). Its portfolio of products includes wild antelope, poultry and free-roaming livestock.

Learn more at https://mogalemeat.com/

Businesses operating in emerging economies are considered second-movers in the cultivated meat industry. With an increasing demand for meat in emerging economies compared….

# 4 Conclusions and Future Perspectives

The cultivated meat industry is still in its nascent stage but has shown enormous promise and received significant investments. In addition, it also shows great promise in mitigating the impact of conventional animal-based food production on climate change. However, it is crucial for the industry to continue to innovate and for regulatory frameworks to evolve to fully realize these benefits. All the companies analysed claimed that their main goals are to create a sustainable and ethical alternative to traditional meat production, which is often criticized for its environmental impact and ethical concerns related to animal welfare.

Cultivated meat companies have secured a large amount of funding in the last few years. This continuous trend displays confidence, a positive outlook and willingness to invest. It's important to note the increase in startups securing funds in Series A, which is used to optimize their product, conduct more aggressive market research, and structure the dossier for regulatory approval. In addition, we have observed expansion (even geographically) following substantial funding in selected cases.

Despite selected successes in regulatory approval in a handful of markets such as Singapore and the United States, the pipeline for approval is increasing, which will likely increase the competition on the consumer market in foreseeable 1 or 2 years in countries with existing food safety approvals and further expansion of additional countries approving cultivated meat in the near future. Regulatory approval in more countries and technological advancements will be crucial for its success. However, the industry must overcome challenges related to production cost, consumer acceptance, and supply chain complexities to realize its full potential. Hybrid solutions have proliferated in the alternative protein sector as food producers strive to integrate plant-based, fermentation, and cultivated meat technologies to develop products that achieve both taste and price parity with conventional meat (GFI 2022).

Like most academic studies, our study has some limitations; first, it only covers selected firms, which might represent a limitation concerning the generalizability of the industry. In addition, it is important to emphasize that this book chapter was written using only secondary sources. As such, the space dedicated to each selected company was determined not by the author's preferences but by the volume of information available online. We recognize the limitations of relying solely on desk research. However, the chapter aimed to provide key and most recent information about cultivated meat from around the globe until the publication of this book, which we believe the information provided was accurate.

Future studies may develop in-depth case studies with the selected companies through interviews and surveys. Another direction could be addressing the role of regulation on the internationalization of startups and foreign direct investment attraction policies. Since regulation is moving at a different pace around the world, we see the potential of global centres of excellence for cultivated meat.

## References

Aleph Farms (2023) Home. https://aleph-farms.com/. Retrieved 28 Sept 2023

BlueNalu (2023) Home. https://www.bluenalu.com/. Retrieved 28 Sept 2023

Cellva (n.d.) Home. https://cellva.com/. Retrieved 28 Sept 2023

ClearMeat (2023) Products. https://clearmeat.com/products/. Retrieved 28 Sept 2023

EqualOcean (2023) https://equalocean.com/news/2023042819696. Retrieved 28 Sept 2023

Food Manufacture (2021) Slaughter-free meat to hit plates by 2023, claims Ivy Farm, 17 May 2021. https://www.foodmanufacture.co.uk/Article/2021/05/17/Slaughter-free-meat-to-hit-plates-by-2023-claims-Ivy-Farm. Retrieved 28 Sept 2023

Food Navigator (2022) SuperMeat links with retail major Migros on cultivated meat distribution, 'We expect to be selling in Switzerland by 2025', 27 July 2022. https://www.foodnavigator.com/Article/2022/07/27/supermeat-links-with-retail-major-migros-on-cultivated-meat-distribution-we-expect-to-be-selling-in-switzerland-by-2025. Retrieved 28 Sept 2023
Forbes (2019) https://www.forbes.com/sites/davidebanis/2019/09/27/moscow-based-lab-produces-first-sample-of-russian-cultivated-meat/?sh=7490f88670ec
Gerber PJ, Steinfeld H, Henderson B, Mottet A, Opio C, Dijkman J, Falcucci A, Tempio G (2013) Tackling climate change through livestock—a global assessment of emissions and mitigation opportunities. Food and Agriculture Organization of the United Nations (FAO), Rome
GFI (2021) State of the industry report: plant-based meat, eggs, and dairy. The Good Food Institute
GFI (2022) State of the industry report: cultivated meat and seafood. The Good Food Institute
Good Meat (2023) Home. https://www.goodmeat.co/. Retrieved 28 Sept 2023
Green Queen (2021) ClearMeat, India's first cell-based chicken maker, files patent for cultivated meat technology. https://www.greenqueen.com.hk/clearmeat-indias-first-cell-based-chicken-maker-files-patent-for-cultivated-meat-technology/. Retrieved 28 Sept 2023
Green Queen (2022) Joes Future Food confirms first cultivated pork belly. https://www.greenqueen.com.hk/joes-future-food-confirms-first-cultivated-pork-belly/. Retrieved 28 Sept 2023
Green Queen (2023) Leonardo DiCaprio, Mosa Meat B Corp certified cultivated meat brand eyes regulatory approval. https://www.greenqueen.com.hk/leonardo-dicaprio-mosa-meat-b-corp-certified-cultivated-meat-brand-eyes-regulatory-approval/. Retrieved 28 Sept 2023
InFamily Foods (2023) FAQ. https://infamily-foods.com/en/faq/. Retrieved 28 Sept 2023
Ivy Farm (2023) News. https://www.ivy.farm/news/. Retrieved 28 Sept 2023
Joes Future Food (2023) About Us. http://joesfuturefood.com/en/about.aspx. Retrieved 28 Sept 2023
Mattick CS, Landis AE, Allenby BR, Genovese NJ (2015) Anticipatory life cycle analysis of in vitro biomass cultivation for cultured meat production in the United States. Environ Sci Technol 49(19):11941–11949
Mogale Meat Co (2023) Home. https://mogalemeat.com/. Retrieved 28 Sept 2023
Mosa Meat (2023) Home. https://mosameat.com/. Retrieved 28 Sept 2023
ProVeg International (2020) Cellular agriculture: 2020 in review. https://proveg.com/blog/cellular-agriculture-2020-in-review/. Retrieved 28 Sept 2023
Shiok Meats (2023) Home. https://shiokmeats.com/. Retrieved 28 Sept 2023
Shu C (2020) Eat Just partners with Proterra to launch a new subsidiary in Asia. TechCrunch. https://social.techcrunch.com/2020/10/20/eat-just-partners-with-proterra-to-launch-a-new-subsidiary-in-asia/. Retrieved 21 Oct 2020
Steakholder Foods (2023) Cultured meat company MeaTech 3D becomes Steakholder Foods. https://steakholderfoods.com/cultured-meat-company-meatech-3d-becomes-steakholder-foods/. Retrieved 28 Sept 2023
SuperMeat (2023) Home. https://supermeat.com/. Retrieved 28 Sept 2023
The Cell Meat (n.d.) R&D. https://www.thecellmeat.com/theme/sample02/html/r&d02_en.php. Retrieved 28 Sept 2023
The Cultivated B (2023) Home. https://www.thecultivatedb.com/. Retrieved 28 Sept 2023
The Vegan Indians (2022) India's Clear Meat develops animal-free growth medium that can cut culture cost by 80%. https://www.theveganindians.com/indias-clear-meat-develops-animal-free-growth-medium-that-can-cut-culture-cost-by-80/. Retrieved 28 Sept 2023
Tuomisto HL, Mattos MJTD (2011) Environmental impacts of cultured meat production. Environ Sci Technol 45(14):6117–6123
Uncommon Bio (2023) Home. https://uncommonbio.co/. Retrieved 28 Sept 2023
Upside Foods (2023) Company. https://upsidefoods.com/company. Retrieved 28 Sept 2023
van der Valk J, Bieback K, Buta C, Cochrane B, Dirks WG, Fu J, Hickman JJ, Hohensee C, Kolar R, Liebsch M, Pistollato F, Schulz M, Thieme D, Weber T, Wiest J, Winkler S, Gstraunthaler

G (2018) Fetal bovine serum (FBS): Past–present–future. ALTEX 35(1):99–118. https://doi.org/10.14573/altex.1705101
Vegconomist (2020) Indian cell-based meat company claims it has achieved price parity. https://vegconomist.com/company-news/indian-cell-based-meat-company-claims-it-has-achieved-price-parity/. Retrieved 28 Sept 2023
Vegconomist (2022) SuperMeat partners with Migros. https://vegconomist.com/cultivated-cell-cultured-biotechnology/supermeat-partners-with-migros/. Retrieved 28 Sept 2023
Vegconomist (2023) The Cultivated B becomes world's first company to pre-register cell-cultivated meat with European Food Safety Authority. https://vegconomist.com/company-news/the-cultivated-b-becomes-worlds-first-company-to-pre-register-cell-cultivated-meat-with-european-food-safety-authority/. Retrieved 28 Sept 2023
Vow (2023) Home. https://www.vowfood.com/. Retrieved 28 Sept 2023

# Challenges and Prospects for Cultivated Meat Production

**Germano Glufke Reis, Susan Grace Karp, Carla Forte Maiolino Molento, and Carlos Ricardo Soccol**

**Abstract** We envision the future landscape of cultivated meat and explore the key forces shaping this nascent industry. Delving into these pivotal forces, we ask a fundamental question: What are the key drivers likely to bring about the outcomes predicted by Gerhardt et al. (Ind Biotechnol 16(5):262–270, 2020) for 2040? We scrutinize these forces from diverse domains to understand how they may either propel or impede the progress of cultivated meat. These influential factors encompass consumer acceptance, price parity, production and distribution capacities, regulatory and legal considerations, environmental impact, animal ethics, social dynamics, and, crucially, support from institutional and entrepreneurial ecosystems. Additionally, we outline a set of priority capabilities essential for the industry's advancement—encompassing the intrinsic skills, knowledge, and experiences—individually and collectively—that stakeholders must cultivate to realize the anticipated outcomes. Strengthening these capabilities lays the groundwork for the nascent industry, paving the way for its projected future.

**Keywords** Alternative proteins · Capabilities · Food transition · Food production system · Future scenario · Radical innovation

---

G. G. Reis (✉)
School of Management, UFPR Federal University of Parana, Curitiba, Parana, Brazil

SFoodLab – Sustainable Food Value Chains Laboratory, UFPR Federal University of Parana, Curitiba, Parana, Brazil
e-mail: glufkereis@ufpr.br

S. G. Karp · C. R. Soccol
Bioprocess Engineering and Biotechnology Department, Federal University of Paraná, Curitiba, Paraná, Brazil

C. F. M. Molento
Animal Welfare Laboratory, UFPR Federal University of Parana, Curitiba, Parana, Brazil

Cellular Animal Science Laboratory, UFPR Federal University of Parana, Curitiba, Parana, Brazil


C. R. Soccol et al. (eds.), *Cultivated Meat*,
https://doi.org/10.1007/978-3-031-55968-6_21

In this chapter, we anticipate a future scenario (Ogilvy and Schwartz 2004; Spaniol and Rowland 2019) for cultivated meat and delve into key forces that will shape this novel industry. To do so, we begin by assuming that cultivated meat will account for 35% of the global meat market by the year 2040, as forecasted by Gerhardt et al. (2020) and the AT Kearney consulting team. According to their analysis, by 2030 the price of premium cultivated meat cuts will be around $40/kg for meat cuts, while processed cultivated products will be less costly. In 2035 cultivated meat cuts are expected to achieve price parity with animal meat, with costs being under $ 4/100 g (Gerhardt et al. 2020). In 2040, the cultivated meat industry is forecast to be consolidating.

Thus, we posit the following question: what are the key driving forces likely to lead to the outcomes forecasted by Gerhardt et al. (2020) for 2040? We identify and explore how these forces that emerge from several domains may stimulate or constrain the development of cultivated meat. These driving forces encompass consumer acceptance, price parity, production and distribution capacities, regulatory and legal aspects, environmental, animal ethics, and social aspects, and, lastly, support from institutional and entrepreneurial ecosystems (Fig. 1). As demonstrated in previous chapters, significant and positive advances have been achieved in those domains, paving the way forward and highlighting areas that require future development.

We also present a set of priority capabilities that should be strengthened in the industry as a whole for cultivated meat to garner further traction. Capabilities encompass the ingrained individual and collective skills, knowledge, and experiences (Andreoni and Lavopa 2022) that stakeholders must develop for the industry to deliver the expected outcomes. Thus, this nascent industry will be laying the building blocks for the projected future by strengthening these capabilities.

## 1 Driving Forces for the Growth of the Cultivated Meat Market

### *1.1 Consumer Acceptance*

Without increasing market demand and people interested in consuming cultivated meat, achieving those forecasts is unlikely. Therefore, the industry should ensure its products align with consumers' needs and expectations. As mentioned in chapter "Forecasting the Market Potential and Consumer Acceptance of Cultivated Meat," research on the acceptance, purchase intentions, and willingness to try cultivated meat has been ongoing for some time (Rombach et al. 2022; Siddiqui et al. 2022). Overall, these studies have shown optimism regarding consumer acceptance. Concerns related to the sustainability and ethical issues of animal-based meat can enhance consumer acceptance. Educating and informing consumers about these aspects may also stimulate purchase intentions. Aspects such as perceptions of

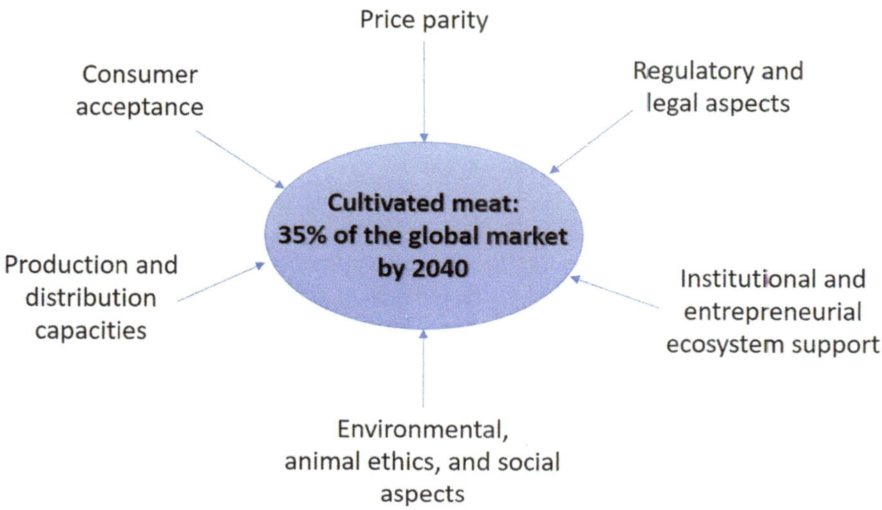


**Fig. 1** Driving forces for the projected growth of the cultivated meat market until 2040

unnaturalness and lack of familiarity should be managed to increase consumers' positive perceptions toward cultivated meat (Siddiqui et al. 2022).

However, consumer behavior studies that directly evaluate sensory responses to cultivated products are still lacking. Most of the research does not incorporate direct experiences with cultivated meat, with consumers trying the products. Only recently have some countries (e.g., the Netherlands and the United States) approved cultivated meat tastings (Deliza et al. 2023). Sensory tests and in-store research will enable a deeper understanding of consumers' perceptions regarding sensory attributes such as texture and taste. Complementary consumer acceptance studies should take measures to address social desirability bias (Larson 2019), particularly when consumers are informed about the environmental and animal ethics benefits of cultivated meat before assessing their perception and acceptance of the products.

Studies comparing cultivated and animal-based meats through an electronic tongue system show the need to enhance the bitterness and test growth media that may improve the similarity in taste to animal meat (Joo et al. 2022). Consequently, consumer acceptance concerns depend on technological advancements to improve products' taste, texture, appearance, quality, and similarity to conventional animal-based meat. Those aspects involve, for instance, the testing of scaffolds, new bioreactor types and sizes, growth medium, and the addition of ingredients and agents (Ben-Arye et al. 2020; Djisalov et al. 2021; Joo et al. 2022).

In addition, cultivated meat production allows for an increase in nutrients while decreasing the presence of fat components (Gerhardt et al. 2020), which illustrates the customization potential of the innovation and the possibility of addressing the enhancement of nutritional attributes as well. Customization can also be achieved through ingredients, additives, 3D printing, and the selection of cell lines; specific types of animal cells can be used to respond to local consumers' needs. Cultivated products may also be blended with plant-based components, fungi-based proteins, cultivated fat, and components produced through fermentation to improve taste, texture, and appearance. Thus, cultivated meat has a high potential for customization and addressing consumer expectations.

Other examples may also be mentioned regarding the technological side that should be considered for improving consumer acceptance. The proliferation and differentiation of cells, along with the subsequent processing strategies, play a pivotal role in determining the final characteristics of a product, which, in turn, can significantly impact consumer acceptance. The production of structured meat products involves the utilization of specific support constructs or 3D bioprinting, incorporating cultivated myocytes and adipocytes. Conversely, unstructured products like nuggets and hamburgers can be crafted using edible scaffolds, which also contribute to taste, texture, and nutritional quality. Regardless of the approach, downstream operations encompassing cell harvesting, dewatering, formulation, packaging, and storage must be executed meticulously to preserve the desired product structure, appearance, and organoleptic quality. The judicious selection of processes and materials, with a focus on consumer acceptance, is a pivotal factor in developing cultivated meat products.

However, to date, only a few companies worldwide have received approval for producing and selling their cultivated products. As a result, most firms have yet to transition from the laboratory to end consumers. Recently, some companies have announced that they are offering contract manufacturing facilities, which are likely to considerably facilitate startups in developing and scaling up their products. For instance, Extracellular, a UK company is developing a 10,000 L capacity contract facility to produce 50 tons of cultivated meat per year (Extracellular 2023). Moreover, this transition necessitates not only adapting cultivated meats to meet consumer expectations but also mastering the entry into the market, strategic product positioning and marketing, and effective communication with their consumers. Thus, we propose that the consumer acceptance driver force should be supported by the following capabilities (Table 1).

## *1.2 Price Parity*

Gerhardt et al. (2020) forecast that by 2030 the price of premium cultivated meat cuts will be around $40/kg for meat cuts. A techno-economic and life cycle assessment conducted by Sinke et al. (2023), in turn, estimates that cultivated meat will cost approximately $2.50 per pound in the same year. As Cai et al. (2024) point out, the price of cultivated beef meat has already dropped considerably. In 2013, its cost was approximately an impressive $300,000 per pound. By 2021, it was estimated to be around $10.00, a number that should be cautiously considered, given that products have not been significantly commercialized so far (Cai et al. 2024). In addition,

**Table 1** Capabilities for advancing cultivated meat's consumer acceptance

| Capabilities | What do they involve? |
|---|---|
| Consumer-oriented business modeling | Skills, processes, and infrastructure that allow for product development per consumers' needs and expectations (e.g., sensory, nutrition, price), including identifying consumer's needs. Product development and customization following global and local trends (e.g., cultural patterns, dietary habits, demographics, consumer segmentation). Being able to focus on cultivated meat's value proposition |
| Technological capabilities for end product development | Technological skills, knowledge, and infrastructure that allow for developing, testing, and scaling consumer-oriented products |
| Marketing capabilities | "Go-to-market" skills that enable the transition from conceptual design and product development to the delivery of end products that align with the value proposition of cultivated meats, as per consumer segmentation and product positioning strategies |

a techno-economic analysis (TEA) conducted by the Dutch consultancy CE Delft estimates that by 2030, the price of cultivated meat will be around $2.57 per pound. (Cai et al. 2024; Vergeer et al. 2021).

Several factors are expected to contribute to achieving price parity between cultivated and animal-based meat. Firstly, it's worth noting that the types of products will play a pivotal role in this context. Processed products like hamburgers and hot dogs tend to have lower production costs compared to complex cuts of meat, such as beef for barbecues. Additionally, as discussed in previous chapters, blended products that combine cultivated meat components with plant-based or fermentation-based ingredients (e.g., mycoptroteins) can lead to cost reductions. For instance, Future Meat Technologies claimed in 2021 that they could produce 4 oz. poultry meat blending cultivated and plant-based meat for US$ 4 (Brennan et al. 2021). As estimated by a major bioreactor producer (Quint 2023), cultivated meats blended with plant-based protein can already be price-competitive with animal-based products.

Secondly, there have been significant advancements in the development of more affordable and animal-free culture media, which have long been a cost bottleneck in the process (O'Neill et al. 2021). In addition, costs can be lowered by using food-grade media components instead of pharmaceutical-grade ones, reducing expenses; moreover, replacing animal-derived components in cultivated meat not only addresses animal ethics issues but also affordability and productivity optimization. Components such as growth factors, albumin, transferrin, and insulin are among the costliest, necessitating further research to find more cost-effective alternatives. As described by Hubalek et al. (2022), alternative manufacturing approaches for growth factors can lead to cost reduction, such as utilizing fermentation and vegetal-based methods instead of animal cell-based methods. Moreover, replacing the culture medium originally used in pharmaceutical applications with food-grade components (Hubalek et al. 2022; O'Neill et al. 2021) and recycling culture medium can further contribute to cost reduction. Furthermore, the scaling up of meat cultivation has necessitated the advancement of larger bioreactors capable of simultaneously supporting cell growth and efficiently producing larger quantities of cultivated meat (Hubalek et al. 2022; Negulescu et al. 2023).

Besides culture media ingredients, microcarriers and scaffolds used for cell adhesion in cultivated meat production are additional inputs that can significantly increase process costs. For example, the cost of Cytodex® 1 microcarrier beads can be higher than US$ 10,000 per kilogram. In this sense, the development of biomaterials based on plants, algae, fungi or bacteria, or even synthetic polymers, is essential to enable a feasible process. Another strategy, reported by Ianovici et al. (2023), is the association of scaffolds and growth factors, or recombinant proteins, that represent 99% of the cost of the culture medium. The scaffolds may act as slow-release platforms that reduce the use of growth factors or as constructs with morphogenic concentration gradients suitable for structured cultivated meat products.

The costs of bioreactors and analytical instruments to monitor cell viability and growth are other critical aspects of cultivated meat production. The cost of a bench-scale 10-L bioreactor, equipped with controllers and sensors for animal cell cultivation, can be higher than US$ 150,000. Of course, the increase in reactor cost is not a simple function of reactor volume. But it is known that the need to work with shear-sensitive cells in rich culture media in sterile conditions is a challenge for scale-up and requires specialized, well-adapted bioreactors and downstream equipment. Thus, we propose that the price parity driver force should be supported by the following capabilities (Table 2).

## 1.3 Production and Distribution Capacities

According to Gerhardt et al. (2020), cultivated meat is projected to comprise 35% of the global meat market by 2040. Considering that in recent years, world meat production has been approximately 340 million tonnes (Ritchie et al. 2019), and assuming an annual growth of 3% (Gerhardt et al. 2020) the total amount of produced cultivated meat is expected to be around 228 million tonnes by 2040.

Thus, a considerable production capacity should be achieved in the coming years. Given that a single facility is assumed to produce 10,000 metric tonnes/year

**Table 2** Capabilities to achieve price parity between cultivated meat and animal-based products

| Capabilities | What do they involve? |
|---|---|
| Capabilities for cost-effective input supplies development | Technological knowledge (e.g., engineering, biotechnological) oriented toward the development of cost-effective meat cultivation technologies (e.g., bioreactors, culture medium) customized for cultivated meat production. Includes the supply of raw materials, bioreactors, and downstream equipment for industrial-scale production |
| Efficient manufacturing capabilities | The design and implementation of production processes with decreased costs that allow the delivery of competitive products and economies of scale |

(Sinke et al. 2023) of cultivated meat, 22,800 manufacturing units worldwide will be needed to reach the projected amount for 2040. Each manufacturing facility will necessitate investments not only in cultivation equipment, processes, and inputs but also in access to water, energy, media recycling, and waste management (Chen et al. 2022). Bioreactors that allow scaling up cultivated meat production and achieving the abovementioned volumes are still in development. Furthermore, trained talent will be required to operate all the manufacturing processes proficiently.

To date, the large-scale cultivation of animal cells is majorly carried out in bioreactors of up to 25 $m^3$ that can reach $30 \times 10^6$ cells/mL. Smaller reactors of up to 5 $m^3$ operated in perfusion mode can reach around $270 \times 10^6$ cells/mL (Nahmias et al. 2022). According to Bodiou et al. (2020) *apud* Schlösser and Nold (2023), around $10^{12}$–$10^{13}$ cells are required to produce 10–100 kg of cultivated meat. So, the output of one bioreactor of 25 $m^3$, according to these considerations, would be around 7500 kg per batch. The duration of one process cycle (usually a fed batch) depends on the cell type. Myogenic and adipogenic progenitor cells have doubling times between 24 and 40 h, and after cell proliferation, which usually takes 4 days in the larger bioreactor, the necessary step of differentiation may take up to 7 days for myogenic differentiation and up to 28 days for adipogenic differentiation (Moutsatsou et al. 2023).

This data can be used to project scale-up procedures, equipment and facilities for cultivated meat production; however, attention should be given to the technical challenges associated with each stage. Cell multiplication, starting from a "test tube" to a large bioreactor, is carried out by successive scale-ups, usually at the volume ratio of 1:10. Thus, several bioreactors are necessary from the bench scale to the final production scale. Another important aspect is the proliferative capacity of the starting cells. The ex vivo expansion of adult animal cells is limited to approximately 30 population doublings (Moutsatsou et al. 2023). So, the development of more stable or immortalized cell lines is of utmost importance. Still in cell proliferation, the need for support for cell adhesion poses extra challenges of mass transfer, cell viability, and cell recovery and increases overall process costs. Finally, the scale-up stage for industrial bioreactors represents a critical step in the process. Even when scale-up criteria are strictly followed, the behavior of cell cultures in larger vessels may vary significantly as compared to bench scale vessels due to obstacles for mass transfer, shear stress, difficulty in accurately monitoring culture parameters, risk of microbial contamination.

The successful scale-up of cultivated meat production, with techno-economic feasibility, is strongly dependent on technological development in varied areas, including cell line development, culture media formulations, supports for cell adhesion, bioreactor design, instrumentation techniques, and of course requires highly skilled, trained personnel. This can only be reached with massive investments in infrastructure, research and development projects, and human resources.

As for distribution, to a certain extent pre-existing structures from animal meat processors or other intermediaries are likely to be used, especially when meat processors are supplied by meat cultivation companies (B2B model) or have their own cultivation facilities. For novel companies that will directly reach end consumers

(B2C), it will be necessary to develop the logistical and distribution infrastructure, which may be costly. The strategic role of retailers and supermarket chains is worth considering. Supermarkets can influence consumer purchases of sustainable products and may be cautious toward offering products with uncertain acceptance and sales. For example, in the case of plant-based meats (Reis et al. 2023), cultivated meat products will require entry strategies and strong partnerships in the retail sector to increase sales. Thus, we propose the following capabilities for developing production and distribution capacities (Table 3).

## 1.4 Regulatory and Legal Aspects

To realize the forecasted scenario, the relevant markets need to obtain clearance to produce and sell cultivated products. As of the publication of this book, to the best of our knowledge, only Singapore and the United States have approved cultivated meat products. In addition, the establishment of regulations will enhance safety conditions in cultivated meat production, reassuring consumers about the safety and benefits of these products. As mentioned in chapter "Regulatory Aspects of Cultivated Meat," each country has its unique regulatory framework with specific steps and timeframes. In some regions, regulatory schemes are still under development. Thus, despite global advancements in regulatory and food safety domains, local institutions must refine regulation guidelines, and companies need to enhance their production systems to become eligible to start the approval process. Furthermore, startups should develop expertise in complying with and responding to the requirements set by regulatory bodies.

When considering the legal aspects that may influence the growth of the cultivated meat market, it is essential to account for various factors that require effective management. These include using specific labels on packages, the terminologies used for products (such as whether to use the term "meat" or not), and the adherence to regulations that govern manufacturing, distribution, domestic trade, and

**Table 3** Capabilities for developing production and distribution capacities

| Capabilities | What do they involve? |
|---|---|
| Implementation of fully operating cultivated meat manufacturing facilities | Knowledge, talents, infrastructure, and financial resources to implement facilities with the projected production capacity; for instance, 10,000 metric tonnes/year (Sinke et al. 2023) |
| Distribution network architecture capabilities | The capacity to develop cultivated meat supply chains and establish relationships with retailers and other commercialization channels (e.g., food service). In-store marketing strategies for increasing purchases |
| Manufacturing capacity expansion | Expertise, financial resources, and technological capacity are required to implement multiple manufacturing systems, along with related supply and distribution chains, to achieve the projected global market participation of cultivated meat products |

international business. Moreover, international trade regulations might significantly impact exports, product pricing, investments in foreign facilities and distribution centers, as well as accessibility to suppliers and raw materials. In addition, all such factors suggest that broader questions of geopolitics, public policies, and governance are also involved, as discussed in chapters "Regulatory Aspects of Cultivated Meat" and "Global and Regional Policies for Cultivated Meat". Thus, we propose the following capabilities for advancing in the regulatory, food safety, and legal fields (Table 4).

## *1.5 Environmental, Animal Ethics, and Social Impacts*

As emphasized in previous chapters (e.g., chapter "Environmental Impacts of Cultivated Meat"), cultivated meat has several environmental benefits over conventional animal-based meat production. Furthermore, as technology development advances, fewer resources will be required in the production chain and key production-related environmental challenges are expected to be overcome (Rubio et al. 2020). For instance, improving the environmental benefits of cultivated meat will require the decarbonization of the energy involved in industrial production (Lynch and Pierrehumbert 2019), as well as reducing energy consumption (Mattick et al. 2015; Tuomisto and Teixeira de Mattos 2011). Cultivated meat production entails a relatively high industrial energy demand, with factors such as culture medium and bioprocess design significantly contributing to environmental impacts. The energy use of cultivated meat production is implicated in various processes (e.g., temperature management, cleaning and drying processes, production of culture medium, and other supplies). Implementing recycling practices, such as water and culture medium, and adopting renewable or low-emission energy sources can mitigate the environmental footprint of cultivated meat production. Gerhardt et al. (2020) point out that medium recycling can be used and energy savings are expected to reach 80% in cultivated meat production when customized bioreactors are

**Table 4** Capabilities for advancing in the regulatory, food safety, and legal fields

| Capabilities | What do they involve? |
|---|---|
| Food safety assessment capabilities | To establish a fully developed food safety and regulatory system in the market |
| Food safety assurance | To establish a comprehensive food safety program in the market that enables the attainment of rigorous food safety standards throughout the supply chain, manufacturing processes, and product distribution. Adherence to both local and international regulations facilitates the approval of production and sales in several markets |
| Marketing and trade legal capabilities | The knowledge and activities required to comply with domestic and international regulations concerning the manufacturing, marketing, and distribution of novel foods |

adopted. Concerning waste management, one potential approach is to either dispose of the generated effluents or enhance them by incorporating them into other production processes.

Animal ethics seems a powerful driving force for the acceleration of the development of cultivated meat production systems. Research on public opinions regarding alternative meats often points to animal welfare as a high-priority issue both for questioning conventional meat products and for supporting cultivated meat development (Valente et al. 2019; Specht et al. 2020; Bryant et al. 2020). However, it is common for animal needs not to be mentioned in many scenarios where a transition to alternative proteins is discussed. This phenomenon is yet to be studied, but it is probably related to some reasons that evolved in the relationship between conventional meat and animal ethics, such as cognitive dissonance (Festinger 1959), absent referent (Adams 2000), selective attention and strategic ignorance mechanisms (van der Weele 2021) and humane hushing (Hannan 2020). Thus, the recognition of animals as relevant stakeholders in the development of cultivated meat remains a challenge. Part of such a challenge is likely related to the fact that some key players in the cultivated meat development are professionals, industries, and regulators, which are also key players in the conventional meat market and are consequently also interested in silencing about the recognition of the main animal ethics issues at stake. Nevertheless, different from the conventional animal-based meat landscape in which the mentioned reasons evolved, bringing animal ethics to the fore may improve prospects for the development of the cultivated meat production chain.

According to Tubb and Seba (2019), based on data from the US Bureau of Labor Statistics, the American Veterinary Medical Foundation, and the American Feed Industry Association, jobs in animal feed and production activities, which will potentially be most affected, represent up to 23% of the jobs in the conventional animal-based meat chain (see Fig. 2). Most jobs in the conventional animal-based meat production chain are concentrated in processing and distribution activities and

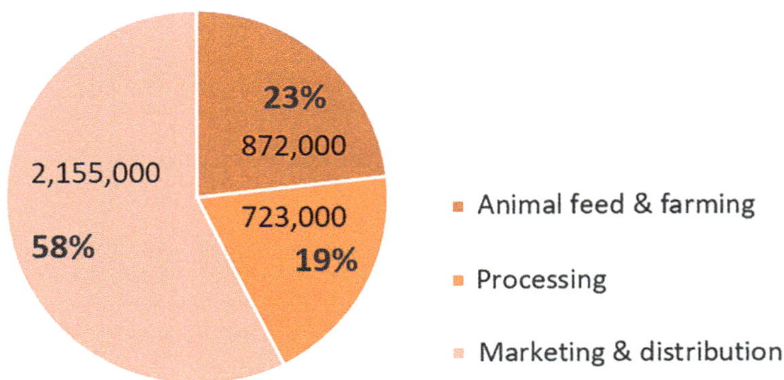


**Fig. 2** Estimated number of jobs per activity in the US animal-based meat production chain (based on Tubb and Seba 2019)

not in animal feed and farming, potentially the most impacted by the transition to cultivated meat.

Although those jobs may be affected, it is relevant to note that several other job opportunities are likely to be created, such as in the supply of equipment and infrastructure, as well as vegetable sources that can be used in culture media (Morais-da-Silva et al. 2022). These activities may provide opportunities for professionals who work in conventional animal production (e.g., animal scientists and veterinarians) to find new opportunities. To provide support to farmers during the transition process, a set of strategies can be implemented, backed by policies and public initiatives. These may include encouraging the diversification of activities among farmers and facilitating the development of skills and tools that assist them in transitioning to the new cultivated meat value chain. A notable recent example is the investment by the German government in this regard (GFI Europe 2023b). Thus, we propose the following capabilities for addressing environmental, animal ethics and social impacts (Table 5).

## 1.6 Institutional and Entrepreneurial Ecosystem Support

The support of institutions and entrepreneurial ecosystems has proven to be a sinequanon for the cultivated meat sector to flourish. An emblematic case is the experience of Singapore which was the first country to approve cultivated meat production, sales, and consumption. The case illustrates how governments can be successful in creating supportive structures for the growing alternative protein sector. These structures encompass enhancing access to funding, research

**Table 5** Capabilities for addressing environmental, animal ethics and social impacts

| Capabilities | What do they involve? |
|---|---|
| Capabilities for supporting the transition to alternative proteins | The commitment of knowledge and resources to ignite and sustain the transition to more sustainable sources of proteins, from cell cultivation, fermentation, and vegetal sources. It involves, for instance, funding research and diversification of the productive sector, enabling its entry into the alternative proteins chain (GFI Europe 2023b), and developing professionals for the cultivated meat and alternative proteins value chain |
| Environmental impacts management | To develop the knowledge and infrastructure to mitigate environmental issues related to cultivated meat production (e.g., energy consumption) and monitor all potential environmental impacts |
| Environmental and animal ethics benefits sensitization | To address consumers and other actors in the food sector (e.g., processors, retailers), educating and reinforcing the framing of cultivated meat and other alternative proteins as more sustainable and animal-ethics friendly. It also involves product development and positioning to emphasize environmental and animal ethics benefits |
| Transition-oriented regulatory frameworks | To develop legislation and policy incentives that stimulate the adoption of environmentally and animal ethics-friendly alternative protein products |

opportunities, startup incubators, infrastructures for product development, universities, networks, and regulatory frameworks that sustain the sector's development (Reis et al. 2022). Consequently, the country has witnessed an increase in inbound foreign investments toward the cultivated meat landscape. Thus, public support for product development (e.g., structures for proof-of-concept activities and scaling up manufacturing) is required for supply and manufacturing companies to grow. A coordinated network of ecosystem actors connecting the public, private, and non-profit sectors is also necessary. In addition to financial support, talent and knowledge development allow for the implementation of a cultivated meat ecosystem. High-quality ecosystems, with strong and consistent connections among their components, are indeed related to the creation of innovative and solid companies and their industries (Stam and Van de Ven 2022).

As for the political side, it is important to acknowledge that certain segments within the traditional animal-based meat production industry may resist the introduction of meat alternatives like cultivated meat and the implementation of supportive policies for innovation. According to Vallone and Lambin (2023) from the Stanford Doerr School of Sustainability, developed countries have shown modest support for innovations and regulatory progress in the emerging meat substitutes sector. Instead, they often prioritize investments in animal-based production and consumption, thereby strengthening the competitiveness of traditional products. Challenges in this regard can also lead to the imposition of regulatory barriers, as recently witnessed in Italy, where legislation has attempted to ban the development of cultivated meat products (GFI Europe 2023a). Showing a strategy that goes in the opposite direction, the German government has announced 38 million € in investments to boost the country's alternative protein sector as a whole (GFI Europe 2023b). Other countries are following a road similar to Germany. For instance, Denmark unveiled a strategy for the plant-based sector, committing a significant investment, while the United Kingdom has announced a research center for alternative proteins and is actively investing in the sector. In addition, France has allocated substantial funds for research and the expansion of the plant-based industry, and the Dutch government is significantly investing in cellular agriculture education, research, and upscaling facilities (TUDelft 2022).

Consequently, ongoing political efforts are essential to bolster investments in cultivated meat growth. Furthermore, it is worth noting that major meat-processing multinationals, such as JBS, have invested in cultivated meat, which will likely promote the adoption of political trends and policies that increasingly favor this innovation. The Brazilian company invests in a cultivated meat manufacturing plant in Spain and a research center in southern Brazil (Freitas 2023). Moreover, the recent approval of cultivated meat products by the FDA in the United States (Wiener-Bronner 2023) is likely to stimulate increasing investments and political support for the sector as a whole. Thus, we propose the following capabilities for achieving institutional and entrepreneurial ecosystem support (Table 6).

**Table 6** Capabilities for achieving institutional and entrepreneurial ecosystem support

| Capabilities | What do they involve? |
|---|---|
| Entrepreneurial ecosystem development | The capacity to develop and coordinate a supportive entrepreneurial ecosystem for the cultivated meat industry. It involves the articulation of domains such as formal institutions, stakeholder networks, physical resources, infrastructure, and financial resources. Leadership for collective action, human capital (i.e., individuals' skills and knowledge), and support for technological knowledge development are also essential components (Stam and Van de Ven 2021) |
| Political engagement capabilities | Political articulation that mobilizes key actors who can advocate for policies and regulations that support the innovation. Involves actors from the public, private, and nonprofit sectors, and from civil society as well, who may exert political influence in that regard |

## 2 Concluding Remarks

Cultivated meat represents a promising technology that may open new avenues for companies in the food sector. It will unlock highly attractive strategic possibilities as the technology becomes consolidated, with numerous products available in supermarkets at prices equivalent to animal-derived products. This includes minimizing the effects of efficiency loss in the conventional system related to continuous demand growth and gradual scarcity of resources at a low cost (e.g., water, land, crop supplies) and reducing other production costs associated with animal-based production (e.g., inputs, breeding, transportation, etc.). Entering the cultivated meat chain may also allow food processors to increase their bargaining power (e.g., in contracts and price negotiations) with their conventional suppliers (e.g., farmers) due to the diversification of processors' inputs; such diversification may lead to less dependency on conventional suppliers. There will be opportunities for vertical integration, greater control over the supply chain, and a reduction of transaction costs as well.

Furthermore, with a diversified portfolio that includes the option of cultivated meat, new opportunities for business risk management are likely to emerge. For instance, companies in the meat sector will be less exposed to international embargoes or fluctuations in prices for animal-based products. Moreover, in the long term, even regions that are not traditional meat producers will face fewer risks and costs in investing in (cultivated) meat facilities abroad compared to factories that involve animal supply, slaughter, and processing. On the other hand, possibilities for customizing products for various segments, regions, and cultural traditions are expanded. Ultimately, companies are increasingly entering scenarios of sustainable innovations aimed at addressing major global challenges, with the potential to tailor products to society's lifestyle preferences.

As presented, despite these and other opportunities, there are still challenges ahead, likely to be overcome in the coming years. Regarding startups entering this nascent industry, certain aspects may take priority. For instance, the need to have access to supportive infrastructure for product development (such as incubators and accelerators, contract manufacturing for proof of concept, or scaling up production). They may also face challenges related to abilities in developing their business value proposition, business model, and overall management systems, ensuring the nascent firm's sustainability. Additionally, there is a clear need to increasingly leverage firms' access to resources and investments. It is worth noting that the sector could benefit from increasing cooperation and knowledge-sharing strategies, which would contribute to accelerating the industry as a whole.

As a final note, it is worth mentioning that cultivated meat is one of the options in the alternative protein basket, which also includes plant-based alternatives and fermentation-based platforms. Thus, these approaches can maximize the opportunities and environmental and animal ethics benefits of reducing dependency on industrial large-scale animal-based products. All these approaches should be seen strategically as a whole and as collective initiatives involving private, public, nonprofit, and other civil sectors to pave the way for the expected scenario in 2040. In this regard, a study conducted by the Bavarian Competence Center for Nutrition (KErn 2022, in Germany, discusses the extent to which alternative proteins can disrupt the food industry, leading to necessary improvements. The study presents several scenarios ranging from no disruption at all and a scenario of disruption blocked by intricate and intimidating legislation, which are less likely to occur according to the study's outcomes, to more optimistic and inspiring ones. For instance, scenarios that lead to sustainable disruptions for global progress, disruptions in products that will be capable of driving advancements in the agribusiness sector as a whole, disruptions that could add value to conventional products, disruptions in biomass enabling agriculture to reinvent itself (e.g., through cellular agriculture), and finally, marketing disruptions offering new options for the lifestyle society. These scenarios represent possibilities for how each country or region can position alternative proteins in their agribusiness and food sectors. This positioning is contingent upon the strategic mobilization of local actors, as well as institutional support and resources.

In summary, a careful examination of the current horizon reveals significant upcoming changes that will impact our food production systems. This transformation will create a ripple effect, influencing environmental, animal ethics, and social aspects. However, the pace of these impending changes, the equitable global distribution of benefits, and the minimization of challenges all remain non-deterministic issues. In other words, crucial choices lie ahead, and these choices will determine whether the forecast for 2040 can be realized. As a final remark, we emphasize that our current decisions will shape future achievements and have the potential to accelerate the transition from conventional animal production to alternative protein food systems.

## References

Adams CJ (2000) The sexual politics of meat: a feminist-vegetarian critical theory, Tenth anniversary edition. Continuum, New York

Andreoni A, Lavopa A (2022) Building socio-economic resilience through industrial capabilities: What can we learn from the COVID-19 pandemic? Series: Impact of COVID-19: The future of industrialization in a post-pandemic world. UNIDO, United Nations Industrial Development Organization. https://iap.unido.org/articles/building-socio-economic-resilience-through-industrial-capabilities. Retrieved 10 Sept

Ben-Arye T, Shandalov Y, Ben-Shaul S, Landau S, Zagury Y, Ianovici I et al (2020) Textured soy protein scaffolds enable the generation of three-dimensional bovine skeletal muscle tissue for cell-based meat. Nat Food 1(4):210–220

Bodiou V, Moutsatsou P, Post MJ (2020) Microcarriers for upscaling cultured meat production. Front Nutr 7:10. https://doi.org/10.3389/fnut2020.00010

Brennan T, Katz J, Spencer B (2021) Cultivated meat: Out of the lab, into the frying pan. McKinsey & Company, 16. https://www.foodticker.co.nz/wp-content/uploads/2021/06/cultivated-meat-out-of-the-lab-into-the-frying-pan.pdf. Retrieved 10 Sept

Bryant C, van Nek L, Rolland NCM (2020) European markets for cultured meat: a comparison of Germany and France. Foods 9(9):1152. https://doi.org/10.3390/foods9091152

Cai J, Wang S, Li Y, Dong S, Liang J, Liu Y, Li S (2024) Industrialization progress and challenges of cultivated meat. J Future Foods 4(2):119–127

Chen L, Guttieres D, Koenigsberg A, Barone PW, Sinskey AJ, Springs SL (2022) Large-scale cultured meat production: Trends, challenges and promising biomanufacturing technologies. Biomaterials 280:121274

Deliza R, Rodríguez B, Reinoso-Carvalho F, Lucchese-Cheung T (2023) Cultured meat: a review on accepting challenges and upcoming possibilities. Curr Opin Food Sci 52:101050

Djisalov M, Knežić T, Podunavac I, Živojević K, Radonic V, Knežević NŽ et al (2021) Cultivating multidisciplinarity: Manufacturing and sensing challenges in cultured meat production. Biology 10(3):204

Extracellular (2023) Extracellular opens Europe's largest contract pilot facility for cultivated meat & seafood. Extracellular, 26 Sept 2023. https://www.extracellular.com/insights/extracellular-opens-europes-largest-contract-pilot-facility-for-cultivated-meat-seafood/#:~:text=A%20total%20facility%20capacity%20of%2010%2C000L%20is%20planned,100%20tons%20per%20year%20with%20predicted%20future%20capabilities. Retrieved 4 Oct

Festinger L (1959) A theory of cognitive dissonance. Stanford University Press, Stanford, CA

Freitas T (2023) The world's top beef supplier is building a lab-grown meat plant. Bloomberg, Markets, 7 June 2023. https://www.bloomberg.com/news/articles/2023-06-07/world-s-top-beef-supplier-jbs-is-building-a-lab-grown-meat-plant#xj4y7vzkg. Retrieved 10 Sept

Gerhardt C, Suhlmann G, Ziemßen F, Donnan D, Warschun M, Kühnle HJ (2020) How will cultured meat and meat alternatives disrupt the agricultural and food industry? Ind Biotechnol 16(5):262–270

GFI Europe (2023a) Italy bans cultivated meat, cutting itself off from innovation and blocking sustainable development, 16 Nov 2023. https://gfieurope.org/blog/italy-ban-on-cultivated-meat-cuts-itself-off-from-innovation-and-blocks-sustainable-development. Retrieved 16 Nov

GFI Europe (2023b) Federal budget 2024: Germany invests €38 million in the protein transition, setting sights on field leadership. The Good Food Institute Europe, 13 Nov 2023. https://gfieurope.org/blog/federal-budget-2024-germany-invests-38-million-euros-in-the-protein-transition-and-sets-out-to-become-a-leader-in-the-field/#:~:text=The%20Budget%20Committee%20of%20the%20German%20Bundestag%20have,path%20to%20international%20leadership%20in%20this%20growing%20field. Retrieved 18 Nov

Hannan J (ed) (2020) Meatsplaining: The animal agriculture industry and the rhetoric of denial. Sydney University Press

Hubalek S, Post MJ, Moutsatsou P (2022) Towards resource-efficient and cost-efficient cultured meat. Curr Opin Food Sci 47:100885

Ianovici I, Goldfracht I, Dvir YS, Shoyhat H, Levenberg S (2023) Tissue engineering using scaffolds in cultured meat product development. In: Post M, Connon C, Bryant C (eds) Advances in cultured meat technology, 1st edn. Burleigh Dodds Science, Cambridge, pp 209–260

Joo ST, Choi JS, Hur SJ, Kim GD, Kim CJ, Lee EY et al (2022) A comparative study on the taste characteristics of satellite cell cultured meat derived from chicken and cattle muscles. Food Sci Anim Resour 42(1):175
KErn (2022) Disruptionen: Wie sich die Rahmenbedingungen in der Ernährungswirtschaft ändern könnten. Cluster Ernährung am Kompetenzzentrum für Ernährung. https://www.kern.bayern.de/alternativeproteine. https://ourworldindata.org/meat-production. Retrieved 18 Nov 2023
Larson RB (2019) Controlling social desirability bias. Int J Market Res 61(5):534–547
Lynch J, Pierrehumbert R (2019) Climate impacts of cultured meat and beef cattle. Front Sustain Food Syst 5
Mattick CS, Landis AE, Allenby BR, Genovese NJ (2015) Anticipatory life cycle analysis of in vitro biomass cultivation for cultured meat production in the United States. Environ Sci Technol 49(19):11941–11949
Morais-da-Silva RL, Villar EG, Reis GG, Sanctorum H, Molento CFM (2022) The expected impact of cultivated and plant-based meats on jobs: the views of experts from Brazil, the United States and Europe. Humanit Soc Sci Commun 9(1):1–14
Moutsatsou P, Cameron P, Dages B, Nienow AW, Lye G, Theodosiou E, Hanga MP (2023) Scaling cell production sustainably in cultured meat product development. In: Post M, Connon C, Bryant C (eds) Advances in cultured meat technology, 1st edn. Burleigh Dodds Science, Cambridge, pp 261–295
Nahmias Y, Halaf G, Shidlovsky Z, Wissotsky G (2022) Systems and methods to recycle cell culture medium. Patent document WO2022238867A1
Negulescu PG, Risner D, Spang ES, Sumner D, Block D, Nandi S, McDonald KA (2023) Techno-economic modeling and assessment of cultivated meat: Impact of production bioreactor scale. Biotechnol Bioeng 120(4):1055–1067
Ogilvy J, Schwartz P (2004) Plotting your scenarios. Global Business Network, Emeryville, CA
O'Neill EN, Cosenza ZA, Baar K, Block DE (2021) Considerations for the development of cost-effective cell culture media for cultivated meat production. Compr Rev Food Sci Food Saf 20(1):686–709
Quint Y (2023) Scale: the final frontier. Cultivated meat companies have made incredible scientific progress; Now it's time to solve scale. 27 Sept 2023. https://www.ark-biotech.com/post/scale-the-final-frontier. Retrieved 4 Oct 2023
Reis GG, Villar EG, Gimenez FAP, Molento CFM, Ferri P (2022) The interplay of entrepreneurial ecosystems and global value chains: Insights from the cultivated meat entrepreneurial ecosystem of Singapore. Technol Soc 71:102116
Reis GG, Villar EG, Ryynänen T, Rodrigues VP (2023) David vs Goliath: The challenges for plant-based meat companies competing with animal-based meat producers. J Clean Prod. https://doi.org/10.1016/j.jclepro.2023.138705
Ritchie H, Rosado P, Roser M (2019) Meat and dairy production. Published online at OurWorldInData.org. https://ourworldindata.org/meat-production. Retrieved 24 Sept 2023
Rombach M, Dean D, Vriesekoop F, de Koning W, Aguiar LK, Anderson M et al (2022) Is cultured meat a promising consumer alternative? Exploring key factors determining consumer's willingness to try, buy and pay a premium for cultured meat. Appetite 179:106307
Rubio NR, Xiang N, Kaplan DL (2020) Plant-based and cell-based approaches to meat production. Nat Commun 11(1):6276
Schlösser S, Nold P (2023) Bioreactors for cell multiplication in cultured meat product development. In: Post M, Connon C, Bryant C (eds) Advances in cultured meat technology, 1st edn. Burleigh Dodds Science, Cambridge, pp 183–208
Siddiqui SA, Khan S, Farooqi MQU, Singh P, Fernando I, Nagdalian A (2022) Consumer behavior towards cultured meat: A review since 2014. Appetite. https://doi.org/10.1016/j.appet.2022.106314
Sinke P, Swartz E, Sanctorum H, van der Giesen C, Odegard I (2023) Ex-ante life cycle assessment of commercial-scale cultivated meat production in 2030. Int J Life Cycle Assess 28(3):234–254
Spaniol MJ, Rowland NJ (2019) Defining scenario. Futures Foresight Sci 1(1):e3

Specht AR, Rumble JN, Buck EB (2020) "You call that meat?" Investigating social media conversations and influencers surrounding cultured meat. J Appl Commun 104(1). https://doi.org/10.4148/1051-0834.2303

Stam E, Van de Ven A (2021) Entrepreneurial ecosystem elements. Small Bus Econ 56:809–832

Tubb C, Seba T (2019) Rethinking food and agriculture 2020–2030. https://www.rethinkx.com/food-and-agriculture. Retrieved 4 Oct 2023

TUDelft (2022) Dutch government confirms €60M investment into cellular agriculture. TUDelft – Delft University of Technology, News, 24 Oct 2022. https://www.tudelft.nl/en/2022/tnw/dutch-government-confirms-eur60m-investment-into-cellular-agriculture. Retrieved 18 Nov 2023

Tuomisto HL, Teixeira de Mattos MJ (2011) Environmental impacts of cultured meat production. Environ Sci Technol 45(14):6117–6123

Valente JPS, Fiedler RA, Sucha Heidemann M, Molento CFM (2019) First glimpse on attitudes of highly educated consumers towards cell-based meat and related issues in Brazil. PLoS One 4:e0221129

Vallone S, Lambin EF (2023) Public policies and vested interests preserve the animal farming status quo at the expense of animal product analogs. One Earth. https://doi.org/10.1016/j.oneear.2023.07.013

van der Weele C (2021) How to save cultured meat from ecomodernism? Selective attention and the art of dealing with ambivalence. In: Animals in our midst: the challenges of co-existing with animals in the anthropocene, The international library of environmental, agricultural and food ethics 33. Springer, Cham, pp 545–557. https://doi.org/10.1007/978-3-030-63523-7_30

Vergeer R, Sinke P, Odegard I (2021) TEA of cultivated meat—future projections of different scenarios. CE Delft, Nov 2021. https://cedelft.eu/publications/tea-of-cultivated-meat/. Retrieved 4 Oct 2023

Wiener-Bronner D (2023) Lab-grown meat is cleared for sale in the United States. CNN Business, 21 June 2023. https://edition.cnn.com/2023/06/21/business/cultivated-meat-us-approval/index.html. Retrieved 15 Sept

www.ingramcontent.com/pod-product-compliance
Ingram Content Group UK Ltd.
Pitfield, Milton Keynes, MK11 3LW, UK
UKHW022216051026
15624UKWH00001B/191

*9783031559679*